Architecture of the Central Brooks Range Fold and Thrust Belt, Arctic Alaska

Edited by

John S. Oldow
Department of Geology and Geological Engineering
University of Idaho
Moscow, Idaho 83844-3022

and

Hans G. Avé Lallemant
Department of Geology and Geophysics
Rice University
Houston, Texas 77005-1892

SPECIAL PAPER
324
1998

Published by The Geological Society of America, Inc.
3300 Penrose Place, P.O. Box 9140, Boulder, Colorado 80301

Printed in U.S.A.

GSA Books Science Editor Abhijit Basu

Library of Congress Cataloging-in-Publication Data
Architecture of the central Brooks Range fold and thrust belt, Arctic
    Alaska / edited by John S. Oldow and Hans G. Avé Lallemant.
        p.   cm. -- (Special paper ; 324)
    Includes bibliographical references and index.
    ISBN 0-8137-2324-8
    1. Geology, Structural -- Alaska -- Brooks Range.   2. Brooks Range
(Alaska)   I. Oldow, John S.   II. Avé Lallemant, Hans G.
III. Series: Special papers (Geological Society of America) ; 324.
QE627.5.A4A73   1998
551.8'09798'7—dc21   98-25089
                                                                CIP

Cover: Looking west at Mount Doonerak, the highest peak in the central Brooks Range and namesake for the Doonerak multiduplex. Dark rocks in the foreground and underlying Mount Doonerak constitute several thrust horses of the lowest part of the Doonerak multiduplex and are composed of lower Paleozoic clastic and volcanic rocks of the Apoon assemblage. Light gray to white rocks on the northern flank of Mount Doonerak are Carboniferous carbonate rocks of the Lisburne Group, which are internally imbricated and detached from the underlying Apoon horses and which constitute most of the northerly dipping Blarney Creek allochthon. In the distance to the northwest, gray Devonian clastic rocks of the Endicott Mountains allochthon rest on the Amawk thrust and structurally overlie the Blarney Creek allochthon.

10 9 8 7 6 5 4 3 2 1

Contents

Preface

INTRODUCTION

This volume represents the results of nearly a decade of geological research in the central Brooks Range of northern Alaska beginning in 1983. The contents of the book focus on various topics dealing with the evolution of rocks exposed in a north-to-south transect through the Brooks Range along the Dalton Highway. Like any undertaking of this scope, the work was built upon the cumulative work of earlier pioneering geologists.

Access to the Brooks Range is difficult in the best conditions, and the Arctic margin of Alaska lies in one of the most remote parts of the North America continent. Adventurer geologists such as E. de K. Leffingwell, A. G. Maddren, W. C. Mendenhall, J. B. Mertie, F. C. Schrader, I. M. Reed, and P. S. Smith were drawn to this mountain chain from the onset of the twentieth century to the Second World War and endured the harsh conditions to establish the geographic and geologic framework of the region. It was not until the practical utilization of fixed-wing aircraft and helicopters that the geology of the region was finally advanced to the state where a broad regional framework was established.

The next generation geologists are not singled out here because their work still constitutes the regional geologic framework for more detailed studies throughout the Brooks Range. Their prodigious contributions are apparent in the references cited throughout this book. This volume is a collaborative effort involving many individuals and strives to render a comprehensive view of the evolution of the central Brooks Range fold and thrust belt. Although of regional significance, we hope the exhaustive detail supplied in the constituent papers will serve as a broader purpose and illustrate structural processes involved in the formation of other con-

tractional orogenic belts around the world. Contributions to most of the chapters (14 of 17) were written by graduates, faculty, and former faculty of Rice University, but the results of important studies were solicited from researchers of other academic institutions and the U.S. Geological Survey. The work presented here makes no presumption of being the final word in the evolution of this intriguing thrust belt, but rather is viewed as part of a continuing process of assessing the intricacies of mountain building.

To aid the reader not familiar with the geology of Arctic Alaska, we will briefly summarize some of the salient points of the regional geology. More exhaustive and detailed renditions of various aspects of these topics are found in the chapters that follow.

REGIONAL FRAMEWORK

The east-west–trending Brooks Range of northern Alaska and the broad plain forming the North Slope stretching to the Arctic Ocean are underlain by sedimentary, metasedimentary, and igneous rocks of Precambrian to Cenozoic age. Owing to the Phanerozoic history of the region, these rocks are divided into three major unconformity sequences, which in ascending order are the Franklinian, Ellesmerian, and Brookian (e.g., Norris and Yorath, 1981).

The Franklinian sequence is best exposed in the northeastern Brooks Range and is composed of Precambrian to Upper Devonian volcanic and sedimentary rocks deposited in carbonate platform and siliciclastic basin environments. In the region underlying the North Slope and in exposures of the northeastern

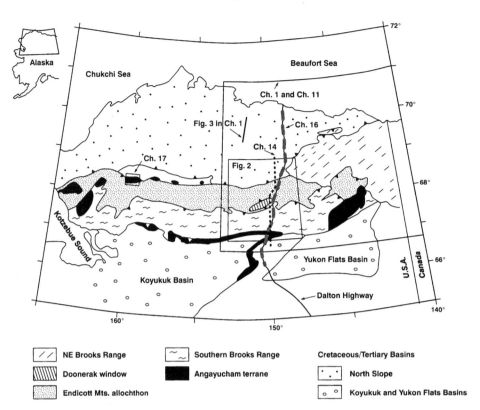

Figure 1. Simplified tectonic map of the Brooks Range fold and thrust belt (after Gottschalk, 1990). Medium-sized box is location of Figure 2. Also indicated are the locations of study areas discussed in Chapters 1, 11, 14, 16, and 17 (this volume).

Brooks Range, these rocks were intensely deformed and eroded prior to deposition of the clastic and carbonate rocks of the uppermost Devonian to Upper Jurassic Ellesmerian sequence. In the north, the Ellesmerian sequence overlies the Franklinian with pronounced angular unconformity, whereas farther south, in the interior of the Brooks Range, the contact no longer separates rocks with substantially different structural histories and apparently represents a regional disconformity. The upper bound of the Ellesmerian sequence is defined by the onset of deposition related to the active construction of the Brooks Range fold and thrust belt consisting of foredeep deposits of the Brookian sequence.

Structurally, the Brooks Range is underlain by a north-vergent fold and thrust belt of Middle Jurassic to Cenozoic age (e.g., Grantz et al., 1981; Oldow et al., 1987a; Crane, 1987; Hubbard et al., 1987). The orogenic belt can be divided into a southern "hinterland" of penetratively, polyphase deformed metamorphic and metasedimentary rocks and a nonmetamorphic, generally nonpenetratively deformed foreland belt in the north. The north-south shortening has been estimated to be in excess of 500 km (Oldow et al., 1987a).

LITHOTECTONIC ASSEMBLAGES

The Brooks Range fold and thrust belt (Figs. 1 and 2) has been divided into several, approximately east-west–trending lithotectonic assemblages or terranes by various authors on the basis of differences in lithology and/or structural position (e.g., Churkin and Trexler, 1981; Grantz et al., 1981; Jones et al., 1987; and Oldow et al., 1987a). The lithotectonic assemblages (Fig. 3) used here arise primarily from the structural setting in which the rocks now reside in the Brooks Range and are from north to south: (1) the North Slope foredeep, (2) North Slope (autochthonous to parautochthonous) sequence, (3) Endicott Mountains allochthon, (4) Doonerak duplex, (5) Skajit allochthon, (6) Schist belt, (7) Phyllite belt, (8) Rosie Creek allochthon, and (9) the Angayucham terrane. An overview of the stratigraphy of the central Brooks Range and the North Slope foredeep is presented by Handschy in Chapter 1 of this volume.

North Slope foredeep. The North Slope or Colville foredeep (Fig. 2) consists of Cretaceous and Tertiary conglomerate, sandstone, and shale (Patton, 1956; Molenaar, 1981a), which overlie Jurassic clastic rocks unconformably. The unconformity youngs progressively to the north (Detterman et al., 1975; Molenaar, 1981b). Whereas the source of the Jurassic sedimentary rocks is toward the north, the source of the Cretaceous and younger rocks is in the south indicating the emergence of the Brooks Range fold and thrust belt (Molenaar, 1981a, b).

North Slope assemblage. Post-Devonian rocks of the North Slope assemblage (Fig. 2) are mostly autochthonous, but toward the south, they are increasingly deformed during the Jurassic to Tertiary Brookian orogeny. The assemblage consists of highly deformed pre-Mississippian chert, phyllite, carbonate, and volcanic rock (Neruokpuk Formation), unconformably overlain by Mississippian conglomerate (Kekiktuk Conglomerate) and shale (Kayak Shale); Mississippian and Pennsylvanian carbonate, deposited in proximal environments (Lisburne Group); Permian to Jurassic clastic and carbonate rocks (Sadlerochit, Shublik, and Kingak Formations); and Cretaceous and Tertiary clastic rocks of the Colville foredeep (e.g., Dutro, 1981). The Neruokpuk Formation was deformed during the Late Devonian in a south-vergent fold and thrust belt (Oldow et al., 1987b) that probably is related to the Devonian Ellesmerian orogeny of the Canadian Arctic Islands (e.g., Norris and Yorath, 1981). Several granitic plutons were emplaced during the Devonian (e.g., Norris, 1981; Dillon et al., 1987).

Endicott Mountains allochthon. The Endicott Mountains allochthon (Figs. 2 and 3) consists of imbricate sheets of the calcareous clastic rocks of the Devonian Beaucoup Formation, the fine- to coarse-grained clastics of the Devonian to Mississippian Endicott Group, and the Mississippian and Pennsylvanian carbonate rocks of the Lisburne Group (Nilsen and Moore, 1984). The Lisburne carbonates were deposited in a distal environment (Armstrong et al., 1976). The Endicott Group consists of the Hunt Fork Shale, Kanayut Conglomerate, and Kayak Shale, which are discussed by Handschy (Chapters 2 and 3 in this volume).

Doonerak duplex. The Doonerak duplex (Figs. 2 and 3) consists of two imbricate stacks separated by the Blarney Creek thrust. The upper imbricated stack is composed of Kekiktuk, Kayak, Lisburne, and rare slices of lower Paleozoic rocks (see Seidensticker and Oldow, Chapter 6, this volume) lying above the Blarney Creek and beneath the Amawk thrust faults. The Lisburne apparently is of a more proximal facies than the Lisburne in the Endicott Mountains allochthon (Armstrong et al., 1976). The duplex is described in Chapter 4 (Phelps and Avé Lallemant) and Chapter 6 (Seidensticker and Oldow). The lower imbricate stack is composed of the (informal) Apoon assemblage, which consists of lower Paleozoic phyllite and volcanic and volcaniclastic rocks, and it is disrupted by several thrust faults. It is

described in Chapters 4, 5, and 6 by Phelps and Avé Lallemant, Julian and Oldow, and Seidensticker and Oldow, respectively.

Skajit allochthon. The Skajit allochthon (Figs. 2 and 3) consists of carbonate and clastic rocks of early Paleozoic age. It is generally metamorphosed and has greenschist-facies mineralogies. The stratigraphy and structural evolution are discussed in Chapters 7 and 8, respectively (Oldow et al.).

Schist belt. The Schist belt consists of phyllite, mica schist, and amphibolite. These rocks first underwent a high-pressure/low-temperature metamorphism (blueschist, eclogite), followed by a greenschist-facies retrograde metamorphism. The rocks are described in Chapters 9 (Gottschalk) and 12 (Gottschalk et al.).

Rosie Creek allochthon and Phyllite belt. The Rosie Creek allochthon (Figs. 2 and 3) consists of Devonian graywacke and phyllite. Oldow et al. (1987a) correlated this package with the Skajit allochthon. The Phyllite belt consists of undated, low-grade metamorphic phyllitic rocks. Both these assemblages are described in Chapter 12 (Gottschalk et al.).

Angayucham terrane. The Angayucham terrane (Figs. 2 and 3) consists generally of basalt, gabbro, peridotite, chert, shale, and limestone. Fossil ages are Devonian to Triassic (Jones et al., 1987); igneous ages are Middle Jurassic (Wirt et al., 1993). This assemblage is described in Chapters 12 (Gottschalk et al.) and 17 (Harris).

GEOCHRONOLOGY

Three chapters (10, 11, and 13) present new radiometric dates. Blythe et al. (Chapter 10) report on fission-track and $^{40}Ar/^{39}Ar$ analyses of samples collected along the entire Dalton Highway. O'Sullivan et al. (Chapter 11) present fission-track ages for samples from the Dalton Highway and from the Colville basin in the northeastern Brooks Range. Gottschalk and Snee (Chapter 13) present new $^{40}Ar/^{39}Ar$ ages for rocks collected in the Schist belt; the latter results are discussed in terms of the tectonics of the southern Brooks Range by Gottschalk et al. in Chapter 12.

STRUCTURAL GEOLOGY

The structural geology and its tectonic implication are discussed in most of the chapters. Particular emphasis has been given to the development of back folds (Chapter 14) by Avé Lallemant and Oldow and the evolution of the Kobuk fault zone, which is the southern boundary of the Brooks Range (Chapter 15) by Avé Lallemant et al.. A north-south cross section through the central Brooks Range, based on a seismic survey, is presented in Chapter 16 by Wissinger et al.

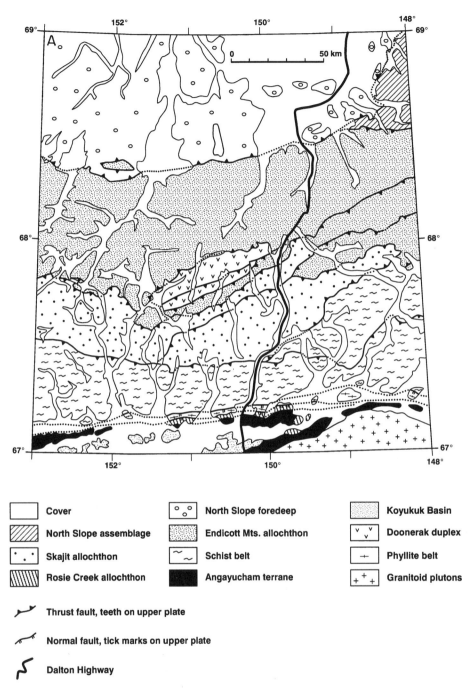

Cover

North Slope assemblage

Skajit allochthon

Rosie Creek allochthon

North Slope foredeep

Endicott Mts. allochthon

Schist belt

Angayucham terrane

Koyukuk Basin

Doonerak duplex

Phyllite belt

Granitoid plutons

Thrust fault, teeth on upper plate

Normal fault, tick marks on upper plate

Dalton Highway

Figure 2 (on this and facing page). A, Tectonic map of the central Brooks Range; location shown in Figure 1. Note that the Blarney Creek duplex is not indicated on the map because it is too narrow to be shown at the scale of the map; the Blarney Creek duplex occurs between the Doonerak duplex and the Endicott Mountains allochthon to the north. B, Location map (same area as in Figure 2A) for studies presented in Chapters 2 to 10, 12, 13, and 15. AT, Anguyucham terrane; CG, Cretaceous granite; DW, Doonerak window; EMA, Endicott Mountains allochthon; NSA, North Slope assemblage; NSF, North Slope foredeep; PB, Phyllite belt; RCA, Rosie Creek allochthon; SA, Skajit allochthon; SB, Schist belt.

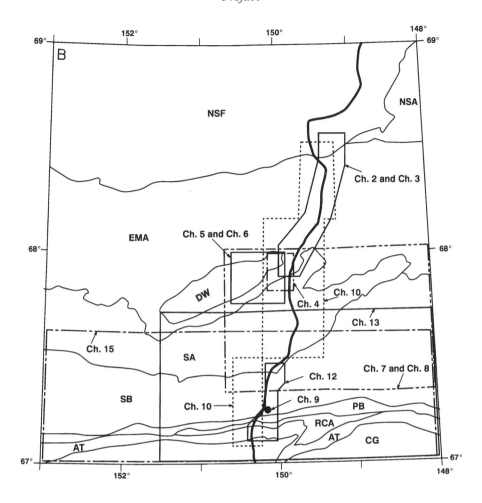

ACKNOWLEDGMENTS

This study was made possible by grants awarded to John S. Oldow, Hans G. Avé Lallemant, and Albert W. Bally from the Department of Energy (DE-AS05-83ER13124), the National Science Foundation (EAR-8517384), and the Rice University Alaska Industrial Associates Group. The Alaska Industrial Associates were Amoco Production Company, Arco Exploration Company, Chevron U.S.A., Inc., Gulf Oil Exploration and Production Company, Mobil Exploration and Producing Services, Inc., and The Standard Oil Company.

REFERENCES CITED

Armstrong, A. K., Mamet, B. L., Brosgé, W. P., and Reiser, H. N., 1976, Carboniferous section and unconformity at Mount Doonerak, Brooks Range, northern Alaska: American Association of Petroleum Geologists Bulletin, v. 60, p. 962–972.

Churkin, M., Jr., and Trexler, J. H., Jr., 1981, Continental plates and accreted oceanic terranes in the Arctic, *in* Nairn, A. E. M., Churkin, M., Jr., and Stehli, F. G., eds., The ocean basins and margins, Volume 5, The Arctic Ocean: New York, Plenum Press, p. 1–20.

Crane, R. C., 1987, Arctic reconstruction from an Alaskan point of view, *in* Tailleur, I., and Weimer, P., eds., Alaskan North Slope geology: Society of Economic Paleontologists and Mineralogists, Pacific Section, Publication 50, p. 769–783.

Detterman, R. L., Reiser, H. N., Brosgé, W. P., and Dutro, J. T., Jr., 1975, Post-Carboniferous stratigraphy, northeastern Alaska: U.S. Geological Survey Professional Paper 886, 46 p.

Dillon, J. T., Tilton, G. R., Decker, J., and Kelly, M. J., 1987, Resource implications on magmatic and metamorphic ages for Devonian igneous rocks in the Brooks Range, *in* Tailleur, I., and Weimer, P., eds., Alaskan North

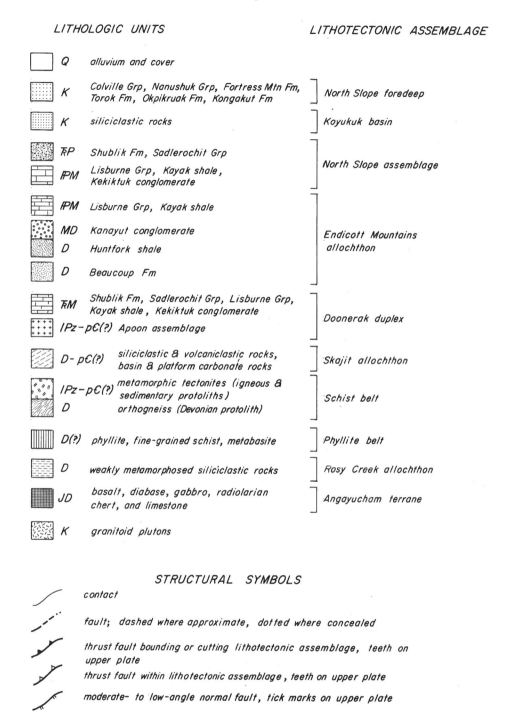

Figure 3 (on this and facing page). Geologic map of the central Brooks Range fold and thrust belt showing the distribution of major lithotectonic assemblages and stratigraphic units.

Slope geology: Society of Economic Paleontologists and Mineralogists, Pacific Section, Publication 50, p. 713–723.

Dutro, J. T., Jr, 1981, Geology of Alaska bordering the Arctic Ocean, *in* Nairn, A. E. M., Churkin, M., Jr., and Stehli, F. G., eds., The ocean basins and margins, Volume 5, The Arctic Ocean: New York, Plenum Press, p. 21–36.

Gottschalk, R. R., 1990, Structural evolution of the Schist belt, south-central

Brooks Range fold and thrust belt, Alaska: Journal of Structural Geology, v. 12, p. 453–470.

Grantz, A., Eittreim, S., and Whitney, O. T., 1981, Geology and physiography of the continental margin north of Alaska and implications for the origin of the Canada Basin, *in* Nairn, A. E. M., Churkin, M., Jr., and Stehli, F. G., eds., The ocean basins and margins, Volume 5, The Arctic Ocean: New York, Plenum Press, p. 439–492.

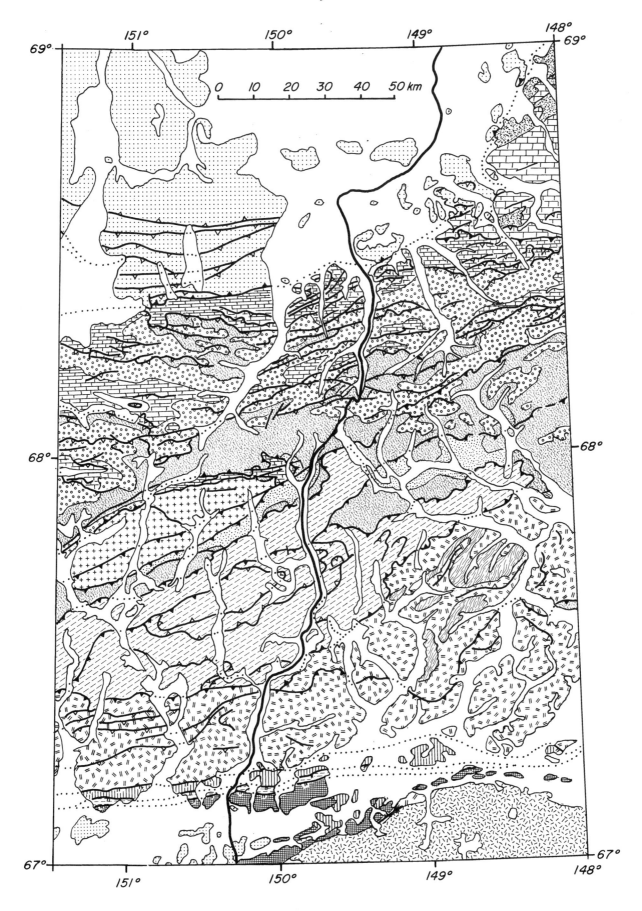

Hubbard, R. J., Edrich, S. P., and Rattey, R. P., 1987, Geologic evolution and hydrocarbon habitat of the "Arctic Alaska microplate": Marine and Petroleum Geology, v. 4, p. 1–92.

Jones, D. L., Silberling, N. J., Coney, P. J., and Plafker, G., 1987, Lithotectonic terrane map of Alaska (west of the 141st meridian): U.S. Geological Survey Miscellaneous Field Studies Map MF-1874-A, scale 1:2,500,000.

Molenaar, C. M., 1981a, Depositional history of the Nanushuk Group and related strata: U.S. Geological Survey Circular 823-B, p. B4–B6.

Molenaar, C. M., 1981b, Depositional history and seismic stratigraphy of Lower Cretaceous rocks, National Petroleum Reserve in Alaska and adjacent areas: U.S. Geological Survey Open-File Report 81-1084, 41 p.

Nilsen, T. H., and Moore, T. E., 1984, Stratigraphic nomenclature for the Upper Devonian and Lower Mississippian (?) Kanayut Conglomerate, Brooks Range, Alaska: U.S. Geological Survey Bulletin, v. 1529A, 64 p.

Norris, D. K., 1981, Geologic map of the Blow River and Davidson Mountains, Yukon Territory—District of Mackenzie: Geological Survey of Canada, Map 1516A, scale 1:250,000.

Norris, D. K., and Yorath, 1981, The North American plate from the Arctic Archipelago to the Romanzof Mountains, *in* Nairn, A. E. M., Churkin, M., Jr., and Stehli, F. G., eds., The ocean basins and margins: Volume 5, The Arctic Ocean: New York, Plenum Press, p. 37–103.

Oldow, J. S., Seidensticker, C. M., Phelps, J. C., Julian, F. E., Gottschalk, R. R., Boler, K. W., Handschy, J. W., and Avé Lallemant, H. G., 1987a, Balanced cross sections through the central Brooks Range and North Slope, Arctic Alaska: Tulsa, Oklahoma, American Association of Petroleum Geologists, Special Publication, 19 p., 8 pl.

Oldow, J. S., Avé Lallemant, H. G., Julian, F. E., and Seidensticker, C. M., 1987b, Ellesmerian(?) and Brookian deformation in the Franklin Mountains, northern Brooks Range, Alaska, and its bearing on the origin of the Canada Basin: Geology, v. 15, p. 37–41.

Patton, W. W., Jr., 1956, New and redefined deformations of Early Cretaceous age, in Gryc et al., eds., Mesozoic sequence in Colville River region, northern Alaska: American Association of Petroleum Geologists Bulletin, v. 40, p. 219–223.

Wirth, K. R., Bird, J. M., Blythe, A. E., and Harding, D. J., 1993, Age and evolution of western Brooks Range ophiolites, Alaska: Results from [40]Ar/[39]Ar thermochronometry: Tectonics, v. 12, p. 410–432.

John S. Oldow
Hans G. Avé Lallemant

Geological Society of America
Special Paper 324
1998

Regional stratigraphy of the Brooks Range
and North Slope, Arctic Alaska

James W. Handschy

Shell Exploration and Production Company, P.O. Box 481, Houston, Texas 77001-0481

ABSTRACT

The stratigraphy of northern Alaska records a complex paleogeographic and tectonic history. In general, the pre-Mississippian stratigraphic section is more complicated and controversial than the younger strata. It records a long period of mixed carbonate and clastic deposition, at least one pre-Mississippian collisional orogenic event, and a Late Devonian rifting event. Mississippian to Jurassic depositional patterns reflect the evolution of a south-facing passive margin, whereas Late Jurassic and younger sedimentation was controlled by plate convergence and formation of the Brooks Range fold-thrust belt. Understanding the paleogeographic distribution of stratigraphic facies in the Brooks Range is critical for deciphering the palinspastic history of the Brookian fold-thrust belt.

INTRODUCTION

The stratigraphy of the Brooks Range and North Slope provides valuable information about the paleogeography, deformational history, and lithologic factors that controlled thrust sheet emplacement in the Brooks Range fold-thrust belt. This chapter is a synopsis of the stratigraphy and how it varies between different structural provinces in the central Brooks Range and North Slope. To help simplify the lateral complications created by tectonic shortening, the stratigraphy is divided into six structural provinces (Fig. 1): (1) North Slope; (2) northeast Brooks Range; (3) Doonerak window of the central Brooks Range; (4) Endicott Mountains Allochthon; (5) southern Brooks Range; and (6) Angayucham terrane.

In general, these provinces are listed from autochthonous to most allochthonous and, thus, in paleogeographic order from north to south. For more details about specific stratigraphic intervals in the central Brooks Range see Chapters 2, 5, and 7 of this volume or Moore et al. (1994).

At the broadest scale the stratigraphy of northern Alaska is divided into three major sequences (Fig. 1): (1) *Franklinian Sequence:* Precambrian through Late Devonian; (2) *Ellesmerian*

Sequence: Late Devonian through Early Cretaceous; and (3) *Brookian Sequence:* Late Jurassic through Tertiary. As originally defined the three sequences were separated by major orogenic events and accompanied by dramatic changes in regional deposition patterns (Lerand, 1973). For the most part, this definition is unambiguous in the Canadian Arctic, the northeast Brooks Range, and in the subsurface of the North Slope where the Franklinian and Ellesmerian sequences are separated by a major angular unconformity. However, in the central and southern Brooks Range, where there is no evidence of a Late Devonian contractional orogeny, the original definition does not strictly apply. Specifically, Lerand (1973, p. 374) places the Late Devonian "very thick marine to non-marine clastic wedges shed southward from the tectonic highlands along and north of the present coast" in the Franklinian sequence. This is appropriate for the Imperial Formation in northwest Canada, but not for the Late Devonian through Early Mississippian(?) clastic wedge exposed in the Endicott Mountains allochthon that is part of the Ellesmerian sequence (Moore et al., 1994; Handschy, this volume, Chapter 2). In addition, even though the Angayucham terrane is age equivalent to the Ellesmerian sequence, it is an accreted oceanic assemblage with a distinctly different paleogeographic and depositional history.

Handschy, J. W., 1998, Regional stratigraphy of the Brooks Range and North Slope, Arctic Alaska, *in* Oldow, J. S., and Avé Lallemant, H. G., eds., Architecture of the Central Brooks Range Fold and Thrust Belt, Arctic Alaska: Boulder, Colorado, Geological Society of America Special Paper 324.

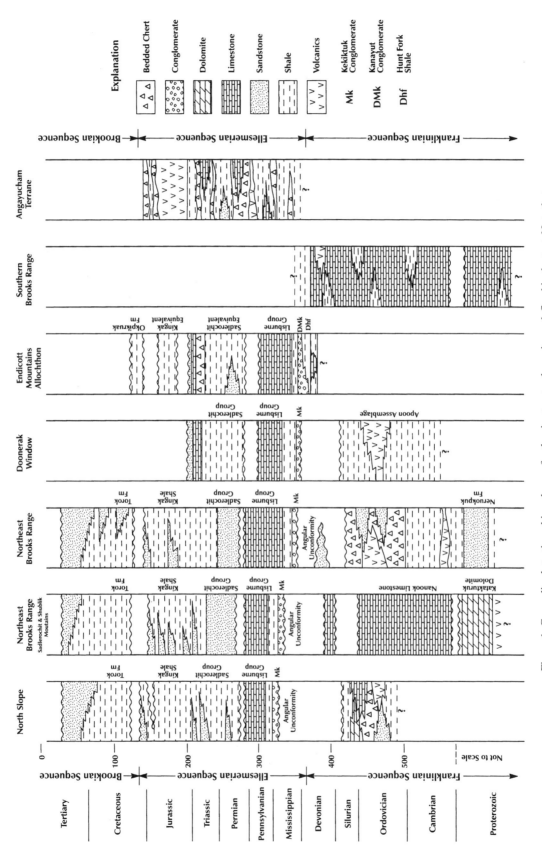

Figure 1. Generalized stratigraphic columns for the six structural provinces defined in the text. Note that there are two columns for the northeast Brooks Range. Facies distributions are schematic, especially for the Franklinian sequence rocks and the Angayucham terrane. Stratigraphic columns were compiled and modified from Moore et al. (1994), Bird and Molenaar (1992), Mull and Adams (1989), and Gyrc (1988).

FRANKLINIAN SEQUENCE

The Franklinian sequence is the most complex and least understood of the three depositional sequences in northern Alaska. It is composed predominantly of Proterozoic to Middle Devonian clastic and carbonate rocks that vary significantly from one structural province to another (Fig. 1). In part these variations are due to different amounts of Mississippian and pre-Mississippian erosion, but they also reflect differences in the original lithologic distribution plus variable amounts of pre-Mississippian and post-Mississippian deformation.

North Slope and northeast Brooks Range

Details about formation ages and initial facies distributions within the Franklinian sequence of the North Slope and northeast Brooks Range are still poorly understood. But, in general, the Franklinian rocks of both provinces share many similarities. The oldest known rocks in the area are Proterozoic-age metasedimentary rocks. In the Shublik and Sadlerochit Mountains of the northeast Brooks Range the Proterozoic Katakturuk Dolomite is a deep- to shallow-water carbonate sequence. Farther south, but still in the northeast Brooks Range, the equivalent-age rocks are quartz-rich turbidites of the Neruokpuk Formation (Leffingwell, 1919; Brosgé et al., 1962). In the subsurface of the North Slope the oldest rocks encountered by petroleum exploration wells are Ordovician to Silurian argillite, graywacke, limestone, dolomite, and chert (Carter and Laufeld, 1975). The equivalent age rocks in the Shublik and Sadlerochit Mountains are deep- to shallow-water carbonates of the Nanook Limestone (Clough et al., 1988). Farther south in the northeast Brooks Range the age-equivalent rocks are a complex assemblage of volcanics, graywacke, radiolarian chert, and argillite (Fig. 1; Moore and Churkin, 1984; Moore et al., 1994).

In both provinces pre-Mississippian contractional deformation is recorded by a major angular unconformity separating the Franklinian rocks from the overlying Ellesmerian sequence (Brosgé et al., 1962; Reiser et al., 1971, 1980). In most locations the Mississippian Kekiktuk Conglomerate directly overlies penetratively deformed and metamorphosed rocks (Brosgé et al., 1962; Oldow et al., 1987); however, in the Shublik and Sadlerochit Mountains the angularity of the unconformity is more subtle (Robinson et al., 1989). In a few locations farther south, tilted or gently folded Middle and Late Devonian terrigenous clastic rocks separate the strongly deformed lower Paleozoic rocks from the Kekiktuk Conglomerate suggesting that the most intense deformation ended during the Early or Middle Devonian (Dillon et al., 1987b; Anderson et al., 1993, 1994).

Doonerak window

In the Doonerak window of the central Brooks Range the Apoon assemblage is composed primarily of metavolcanics and argillaceous metasedimentary lithologies (Moore, 1987; Julian and Oldow, this volume, Chapter 5). Cambrian through Silurian fossils from the metasedimentary rocks (Dutro et al., 1984; Repetski et al., 1987) plus Ordovician K-Ar dates from the volcanics indicate the Apoon is age equivalent to the Nanook Limestone and other early Paleozoic formations in the northeast Brooks Range and North Slope. However, unlike its northern equivalents, the Apoon Assemblage was not penetratively deformed or strongly folded prior to the Mississippian (Oldow et al., 1987; Julian and Oldow, this volume, Chapter 5). In the Doonerak window the Mississippian unconformity is a disconformity with no angular discordance (Oldow et al., 1987; Seidensticker and Oldow, this volume, Chapter 6).

Southern Brooks Range

In the southern Brooks Range, the pre-Mississippian section is mixed carbonates, clastics and volcanics that were strongly deformed and metamorphosed during Brookian orogenesis (Oldow et al., this volume, Chapter 7; Gottschalk et al., this volume, Chapter 12; Moore et al., 1994). Metamorphic grade and deformation intensity generally increase southward reflecting both the more hinterland position in the fold-thrust belt and a deeper level of exposure by the modern erosion surface.

Age control is sparse in the southern Brooks Range. Fossils of Cambrian through Middle Devonian age have been extracted from various lithologies in the thrust sheets of the Skajit allochthon adjacent to the Dalton Highway (Dillon et al., 1987a; Dumoulin and Harris, 1987; Palmer et al., 1984; Dillon, 1989; Oldow et al., this volume, Chapter 7). A second, older assemblage of metamorphosed carbonates, clastics, and volcanics intruded by Proterozoic granitic plutons is exposed along strike to the west in the Baird Mountains Quadrangle (Karl et al., 1989; Turner et al., 1979; Mayfield et al., 1982). South of the Skajit allochthon, the Schist Belt and Rosie Creek allochthons are composed of thrust sheets containing metasedimentary and metaigneous rocks of Proterozoic through late Paleozoic age (Oldow et al., 1987; Gottschalk et al., this volume, Chapter 12; Moore et al., 1994; Dillon et al., 1986).

Pre-Mississippian igneous rocks

Franklinian rocks in most structural provinces were intruded by Late Proterozoic and Devonian igneous rocks. Age dates from the North Slope and northeast Brooks range are rare, but generally indicate intrusion of granitic rocks occurred during the Devonian (Moore et al., 1994). In the Doonerak window a 470 Ma metavolcanic sequence was intruded by mafic dikes at approximately 380 Ma (Dutro et al., 1976). In the southern Brooks Range granitic Proterozoic intrusives yielded 705 to 750 Ma ages (Karl et al., 1989) and younger intrusives with compositions ranging from granite to diorite crystallized between 366 and 402 Ma (Dillon, 1989; Dillon et al., 1980).

ELLESMERIAN SEQUENCE

The Ellesmerian Sequence is a thick succession of mixed carbonate and clastic rocks deposited on a south-facing continental margin that existed prior to the Brookian orogeny (Fig. 2). In the subsurface of the North Slope and in the northeast Brooks Range, the Ellesmerian Sequence unconformably overlies deformed and metamorphosed pre-Mississippian rocks of the Franklinian Sequence. In the central Brooks Range, the Early Mississippian unconformity beneath the Ellesmerian Sequence is (1) a disconformity in the Doonerak window (Julian and Oldow, this volume, Chapter 5), and (2) absent in the Endicott Mountains allochthon where the Upper Devonian and Lower Mississippian rocks are conformable (Handschy, this volume, Chapter 2; Moore et al., 1994).

The most significant difference between the Ellesmerian sequence in the different structural settings is in the Endicott Group at the base of the sequence. In the northeast Brooks Range, the Doonerak window, and most places beneath the North Slope, the Endicott Group is composed of the Early Mississippian Kekiktuk Conglomerate and Kayak Shale. The Kekiktuk is a transgressive package of fluvial, paralic, and shallow-marine conglomerate that fines upward into the marine Kayak Shale (LePain et al., 1994). In contrast, the Endicott Group in the Endicott Mountains allochthon, and possibly in some half grabens beneath the North Slope, is composed of thick Late Devonian to Early Mississippian(?) clastic rocks (Nilsen, 1981; Oldow et al., 1987; Grantz et al., 1990; Moore et al., 1994). In the Endicott Mountains allochthon the Hunt Fork Shale and Kanayut Conglomerate represent as much as 8 km of Late Devonian to Early Mississippian(?) basin fill (Handschy, this volume, Chapter 2). Palinspastic restoration of Brookian deformation requires that the northern edge of the Kanayut–Hunt Fork depositional basin was located south of the Apoon assemblage and Kekiktuk Conglomerate exposed in the Doonerak window (Oldow et al., 1987). Paleocurrents indicate that sediment entered the depositional basin from the northeast and facies distributions document that southward progradation in the lower Kanayut was followed by northward retrogradation in the upper Kanayut (Nilsen et al., 1980; Nilsen, 1981; Nilsen and Moore, 1982a, b; Moore and Nilsen, 1984; Moore et al., 1989). Like the Kekiktuk Conglomerate the uppermost Kanayut Conglomerate fines upward into the Kayak Shale.

The Kayak Shale interfingers with and is overlain by carbonates of the Mississippian through Early Permian Lisburne Group. Facies distribution within the Lisburne Group reflects a variety of carbonate environments ranging from near-shore deposition in the north to open-marine shelf and deeper marine anoxic deposition in the south (Armstrong, 1974; Armstrong and Bird, 1976; Armstrong and Mamet, 1977). Although the Lisburne Group contains many facies shifts associated with Carboniferous sea level fluctuations (Carlson and Watts, 1987; LePain et al., 1990), the overall facies distribution indicates northward transgression throughout the depositional history of the Lisburne Group (Armstrong and Bird, 1976). Timing of this transgression is documented by Early Pennsylvanian fossils at the base of the Lisburne Group in the northeast Brooks Range and North Slope, early Late Mississippian fossils at the base of the Lisburne Group in the Doonerak window, and Early Mississippian fossils at the base of the Lisburne Group in the Endicott Mountains allochthon, which restores to a position south of the Doonerak window (Fig. 1; Armstrong et al., 1976; Oldow et al., 1987).

Unconformably overlying the Lisburne Group is a series of Permian through Lower Cretaceous depositional cycles of the Sadlerochit Group, Shublik Formation, Kingak Shale, and their

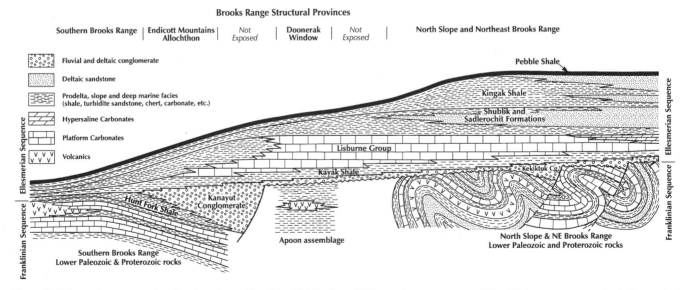

Figure 2. Schematic cross section showing the pre-Brookian distribution of Ellesmerian sequence and Franklinian sequence rocks in the central Brooks Range. Lisburne Group facies distribution was modified from Armstrong and Bird (1976) and Moore et al. (1994). Other facies distributions were compiled from Bird and Molenaar (1992), Mull and Adams (1989), Gyrc (1988), and Handschy (this volume, Chapter 2). Not to scale.

equivalents (Detterman et al., 1975). These cycles are composed predominantly of deltaic and slope facies that prograded southward. Individual progradational packages are separated by unconformities and transgressive shales. Facies variations in the Permian through Triassic sections of the North Slope, Doonerak window, and Endicott Mountains allochthon indicate that they represent shallow-water, slope, and deep-water facies, respectively (Fig. 1; Brosgé and Dutro, 1973). Southward progradation of the Jurassic through Lower Cretaceous Kingak Shale is not only documented by facies distribution, but is also visible on seismic profiles from the North Slope (Fig. 3; Molenaar, 1983, 1988; Kirschner et al., 1983). Overlying the Kingak Shale is the Pebble Shale condensed interval, which defines the top of the Ellesmerian Sequence.

Angayucham terrane

Based strictly on age the Angayucham terrane, which comprises both the structurally highest thrust sheets and the southern most thrust sheets in the Brooks Range, is part of the Ellesmerian sequence (Fig. 1). However, the paleogeographic history and accretionary tectonic history of the Angayucham terrane are significantly different from the Ellesmerian sequence. The Angayucharn terrane in the central Brooks Range is composed of fault-bounded packages that contain Late Devonian to Jurassic pillow basalt, diabase, radiolarian chert, argillite, and pelagic limestone (Jones et al., 1988; Pallister et al., 1989; Murchey and Harris, 1985). The geochemistry of the basaltic rocks plus the associated

siliceous sediments suggest that the Angayucham rocks formed in ocean-plateau settings (Barker et al., 1988; Pallister et al., 1989). Farther west the Angayucham terrane also contains serpentinite and, along the crest of the western Brooks Range, ophiolites (Jones et al., 1987; Harris, 1988, 1992, this volume, Chapter 17; Roeder and Mull, 1978; Wirth and Bird, 1992). Although there has been much debate in the literature about the presence, distribution, and completeness of ophiolites in the Brooks Range, most authors agree that the Angayucham terrane represents accretion and obduction of oceanic material along a plate boundary of Brookian age.

BROOKIAN SEQUENCE

The Brookian Sequence is a series of northward- and northeastward-prograding clastic depositional cycles that filled the foreland basin of the Brooks Range fold-thrust belt. Brookian sedimentary rocks onlap or downlap the Pebble Shale at the top of the Ellesmerian Sequence (Fig. 3). With the exception of the Okpikruak Formation, most Brookian formations have been defined on the basis of depositional facies. They often grade laterally into each other and are time transgressive.

Brookian thrusting started during the Jurassic and some Jurassic Brookian sedimentary rocks are exposed in the higher thrust sheets of the western Brooks Range (Mayfield et al., 1983; Moore et al., 1994). The oldest preserved Brookian rock unit in the central Brooks Range is the Early Cretaceous Okpikruak Formation, which crops out in thrust sheets of the Endicott Mountains allochthon

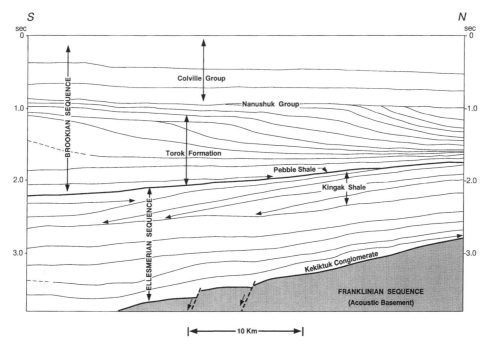

Figure 3. Line drawing of seismic line 29X-75 (seismic data published by Molenaar, 1988; Bird and Molenaar, 1992) showing northward progradation in the Brookian sequence and southward progradation in the Ellesmerian sequence. Note that sequence definitions are based on Lerand (1973) and do not follow the seismic stratigraphic sequence definitions of Mitchum et al. (1977).

(Gryc et al., 1951, 1956; Patton and Tailleur, 1964; Mull, 1982; Mayfield et al., 1988). The Okpikruak Formation is composed of turbidite beds that become more shaley toward the north (Brosgé et al., 1979; Siok, 1989). From the front of the Brooks Range northward to the Barrow arch the age of the base of the Brookian Sequence becomes progressively younger, reflecting northward migration of the foredeep. At the mountain front the Aptian(?) to Albian Fortress Mountain Formation overlies the Okpikruak Formation (Mull, 1985). Farther north, progressively younger strata of the Albian to Cenomanian Torok Formation onlap and downlap the Neocomian Pebble Shale (Fig. 3; Molenaar, 1985, 1988).

The Fortress Mountain Formation contains fan delta to coarse-grained submarine fan facies and represents the higher energy, more proximal deposits of a northward prograding flysch sequence (Crowder, 1987, 1989). The lower Torok Formation is the finer grained turbidite and basin plain equivalent of the Fortress Mountain Formation. Overlying and interfingering with the younger part of the Torok Formation are middle Albian to Cenomanian deltaics of the Nanushuk Group (Huffman et al., 1985; Molenaar, 1985, 1988). Seismic reflection profiles (Molenaar, 1985, 1988) illustrate the northward progradation of the Torok Formation and Nanushuk Group during the filling of the Brooks Range foredeep (Fig. 3). Subsequent deposition of the Late Cretaceous through Tertiary Colville Group and Sagavanirktok Formation follows a similar pattern of northeastward progradation of deltaic and slope facies (Molenaar, 1983).

Folding of the Colville Group in the southern and central North Slope indicates that Brookian deformation in the central Brooks Range continued into the Tertiary. In addition, fission track and $^{40}Ar/^{39}Ar$ ages confirm that much of the present physiography in the central Brooks Range is the result of Cenozoic uplift (Blythe et al., this volume, Chapter 10; O'Sullivan et al., this volume, Chapter 11; O'Sullivan et al., 1994).

PALEOGEOGRAPHIC SUMMARY

Although the Franklinian sequence is known from outcrops or subcrops in most of the structural provinces of the Brooks Range, the pre-Mississippian paleogeography of the Franklinian sequence is vague. In the most general sense it can be characterized as an intensely deformed terrane in the north and an undeformed terrane in the south. The undeformed terrane can be further subdivided into a northern mixed terrigenous clastic and volcanic region represented by the Apoon assemblage in the Doonerak window and a southern mixed carbonate, terrigenous clastic, and volcanic region exposed in the thrust sheets of the southern Brooks Range. Intense Brookian deformation in the southern Brooks Range precludes detailed pre-Mississippian paleogeography.

The base of the Ellesmerian sequence, like the pre-Mississippian history of the underlying Franklinian sequence, varies greatly with location. In the northeast Brooks Range and the subsurface of the North Slope the contact between the two sequences is an angular unconformity. The basal Ellesmerian unit is the Mississippian Kekiktuk Conglomerate. Farther to the southwest in the Doonerak

window the Kekiktuk Conglomerate is also the basal Ellesmerian formation, but the unconformity beneath the Kekiktuk is a disconformity and there is no evidence that the underlying Apoon assemblage (Franklinian sequence) was folded during a pre-Mississippian orogeny. In the Endicott Mountains allochthon, which restores to a position south of the Doonerak window, the basal Ellesmerian formations are the Hunt Fork Shale and Kanayut Conglomerate, which comprise a thick clastic wedge that prograded southward during the Late Devonian to Early Mississippian(?).

The southernmost, and structurally highest, thrust sheets in the Brooks Range are composed of rocks assigned to the Angayucham terrane. Although these rocks are Ellesmerian in age, they are an accreted oceanic assemblage that was depositionally unrelated to the Ellesmerian sequence.

In general, the thrust sheets that form the Brooks Range reflect north-directed shortening of an Ellesmerian south-facing passive continental margin and the underlying Franklinian basement, plus the accretion and obduction of an oceanic terrane. Brookian deformation started during the Jurassic. The Brookian sequence is composed of Jurassic through Tertiary terrigenous clastic sediments that were shed northward from the Brooks Range fold-thrust belt into the foreland basin. As the thrust belt migrated northward and Brookian sediments prograded northward, Brookian formations near the front of the Brooks Range were also involved in the deformation.

ACKNOWLEDGMENTS

Special thanks to J. S. Oldow, H. G. Avé Lallemant, A. W. Bally, C. G. Mull, J. Dillon and W. Wallace for many educational discussions about structural and stratigraphic relations in the Brooks Range. M. J. Baranovic, M. A. Chapin and K. W. Shanley provided valuable reviews of the manuscript. E. E. White and K. A. Wall drafted the figures.

REFERENCES CITED

Anderson, A. V., Mull, C. G., and Crowder, R. K., 1993, Mississippian terrigenous clastic and volcaniclastic rocks of the Ellesmerian sequence, upper Sheenjek River area, eastern Brooks Range, Alaska, *in* Solie, D. N., and Tannian, F., eds., Short notes on Alaskan geology 1993: Alaska Division of Geological and Geophysical Surveys, Geologic Report 113, p. 1–6.

Anderson, A. V., Wallace, W. K., and Mull, C. G., 1994, Depositional record of a major tectonic transition in northern Alaska: Middle Devonian to Mississippian rift-basin margin deposits, upper Kongakut River region, eastern Brooks Range, Alaska, *in* Thurston, D. K., and Fujita, K., eds., 1992 proceedings, International Conference on Arctic Margins: Anchorage, Alaska, U.S. Department of the Interior, Minerals Management Service, p. 71–76.

Armstrong, A. K., 1974, Carboniferous carbonate depositional models, preliminary lithofacies and paleotectonic maps, Arctic Alaska: American Association of Petroleum Geologists Bulletin, v. 58, p. 621–645.

Armstrong, A. K., and Bird, K. J., 1976, Carboniferous environments of deposition and facies, Arctic Alaska, *in* Miller, T. P., ed., Symposium on recent and ancient sedimentary environments in Alaska: Alaska Geological

Society Symposium Proceedings, p. A1–A16.

Armstrong, A. K., and Mamet, B. L., 1977, Carboniferous microfacies, microfossils and corals, Lisburne Group, Arctic Alaska: U.S. Geological Survey Professional Paper 849, 144 p.

Armstrong, A. K., Mamet, B. L., Brosgé, W. P., and Reiser, H. N., 1976, Carboniferous section and unconformity at Mount Doonerak, Brooks Range, northern Alaska: American Association of Petroleum Geologists Bulletin, v. 60, p. 962–972.

Barker, F., Jones, D. L., Budahn, J. R., and Coney, P. J., 1988, Ocean plateau-seamount origin of basaltic rocks, Angayucham terrane, central Alaska: Journal of Geology, v. 96, p. 368–374.

Bird, K. J., and Molenaar, C. M., 1992, The North Slope foreland basin, Alaska, *in* Macqueen, R. W., and Leckie, D. A., eds., Foreland basins and fold belts: American Association of Petroleum Geologists Memoir 55, p. 363–393.

Brosgé, W. P., and Dutro, J. T., Jr., 1973, Paleozoic rocks of north and central Alaska, *in* Pitcher, M. G., ed., Arctic geology: American Association of Petroleum Geologists Memoir 19, p. 361–375.

Brosgé, W. P., Dutro, J. T., Jr., Mangus, M. D., and Reiser, H. N., 1962, Paleozoic sequence in eastern Brooks Range, Alaska: American Association of Petroleum Geologists Bulletin, v. 46, p. 2174–2198.

Brosgé, W. P., Reiser, H. N., Dutro, J. T., Jr., and Detterman, R. L., 1979, Bedrock geologic map of the Phillip Smith Mountains Quadrangle, Alaska: U.S. Geological Survey Miscellaneous Field Studies Map MF-879B, 2 sheets, scale 1:250,000.

Carlson, R., and Watts, K. F., 1987, Shallowing upward cycles in the Wahoo Limestone, eastern Sadlerochit Mountains, ANWR, NE Brooks Range, Alaska: Geological Society of America Abstracts with Programs, v. 19, p. 364.

Carter, C., and Laufeld, S., 1975, Ordovician and Silurian fossils in well cores from the North Slope of Alaska: American Association Petroleum Geologists Bulletin, v. 59, p. 457–464.

Clough, J. G., Blodgett, R. B., Imm, T. A., and Pavia, E. A., 1988, Depositional environments of Katakturuk Dolomite and Nanook Limestone, Arctic National Wildlife Refuge, Alaska: American Association of Petroleum Geologists Bulletin, v. 72, p. 172.

Crowder, R. K., 1987, Cretaceous basin to shelf transition in northern Alaska: Deposition of the Fortress Mountain Formation, *in* Tailleur, I., and Weimer, P., eds., Alaskan North Slope geology: Society of Economic Paleontologists and Mineralogists, Pacific Section, p. 449–458.

Crowder, R. K., 1989, Deposition of the Fortress Mountain Formation, *in* Mull, C. G., and Adams, K. E., eds., Dalton Highway, Yukon River to Prudhoe Bay, Alaska: Alaska Division of Geological and Geophysical Survey Guidebook 7, v. 2, p. 293–301.

Detterman, R. L., Reiser, H. N., Brosgé, W. P., and Dutro, J. T., Jr., 1975, Post-Carboniferous stratigraphy, northeastern Alaska: U.S. Geological Survey Professional Paper 886, 46 p.

Dillon, J. T., 1989, Structure and stratigraphy of the southern Brooks Range and northern Koyukuk Basin near the Dalton Highway, *in* Mull, C. G., and Adams, K. E., eds., Dalton Highway, Yukon River to Prudhoe Bay, Alaska, Alaska Division of Geological and Geophysical Survey Guidebook 7, v. 2, p. 157–188.

Dillon, J. T., Pessel, G. H., Chen, J. A., and Veach, N. C., 1980, Middle Paleozoic magmatism and orogenesis in the Brooks Range, Alaska: Geology, v. 8, p. 338–343.

Dillon, J. T., Brosgé, W. P., and Dutro, J. T., Jr., 1986, Generalized geologic map of the Wiseman Quadrangle: U.S. Geological Survey Open-File Report 86-219, scale 1:250,000.

Dillon, J. T., Harris, A. G., and Dutro, J. T., Jr., 1987a, Preliminary description and correlation of lower Paleozoic fossil-bearing strata in the Snowden Mountain area of the south-central Brooks Range, Alaska, *in* Tailleur, I., and Weimer, P., eds., Alaskan North Slope geology: Society of Economic Paleontologists and Mineralogists, Pacific Section, p. 337–345.

Dillon, J. T., Tilton, G. R., Decker, J., and Kelly, M. J., 1987b, Resource implications of magmatic and metamorphic ages for Devonian igneous rocks in the Brooks Range, *in* Tailleur, I., and Weimer, P., eds., Alaskan North Slope geology: Society of Economic Paleontologists and Mineralogists,

Pacific Section, p. 713–723.

Dumoulin, J. A., and Harris, A. G., 1987, Lower Paleozoic carbonate rocks of the Baird Mountains Quadrangle, western Brooks Range, Alaska, *in* Tailleur, I., and Weimer, P., eds., Alaskan North Slope geology: Society of Economic Paleontologists and Mineralogists, Pacific Section, p. 311–336.

Dutro, J. T., Jr., Brosgé, W. P., Lanphere, M. A., and Reiser, H. N., 1976, Geologic significance of the Doonerak structural high, central Brooks Range, Alaska: American Association of Petroleum Geologists Bulletin, v. 60, p. 952–961.

Dutro, J. T., Jr., Palmer, A. R., Repetski, J. E., and Brosgé, W. P., 1984, Middle Cambrian fossils from the Doonerak anticlinorium, central Brooks Range, Alaska: Journal of Paleontology, v. 58, p. 1364–1371.

Grantz, A., May, S. D., and Hart, P. E., 1990, Geology of the Arctic continental margin of Alaska, *in* Grantz, A., Johnson, L., and Sweeney, J. F., eds., The Arctic Ocean region: Boulder, Colorado, Geological Society of America, The Geology of North America, v. L, p. 257–288.

Gryc, G., ed., 1988, Geology and exploration of the National Petroleum Reserve in Alaska 1974 to 1982: U.S. Geological Survey Professional Paper 1399, 940 p.

Gryc, G., Patton, W. W., Jr., and Payne, T. G., 1951, Present Cretaceous stratigraphic nomenclature of northern Alaska: Washington Academy of Science Journal, v. 41, p. 159–167.

Gryc, G., Bergquist, H. R., Detterman, R. L., Patton, W. W., Jr., Robinson, F. M., Rucker, F. P., and Wittington, C. L., 1956, Mesozoic sequence in Colville River region, northern Alaska: American Association of Petroleum Geologists Bulletin, v. 40, p. 209–254.

Harris, R. A., 1988, Origin, emplacement and attenuation of the Misheguk Mountain allochthon, western Brooks Range: Geological Society of America Abstracts with Programs, v. 20, p. A112.

Harris, R. A., 1992, Peri-collisional extension and the formation of Oman-type ophiolites in the Banda arc and Brooks Range, *in* Parson, L. M., Murton, B. J., and Browning, P., eds., Ophiolites and their modern oceanic analogs: Geological Society Special Publication 60, p. 301–325.

Huffman, A. C., Jr., Ahlbrandt, T. S., Pasternack, I., Stricker, G. D., and Fox, J. E., 1985, Depositional and sedimentologic factors affecting the reservoir potential of the Cretaceous Nanushuk Group, central North Slope, Alaska, *in* Huffman, A. C., Jr., ed., Geology of the Nanushuk Group and related rocks, North Slope, Alaska: U.S. Geological Survey Bulletin 1614, p. 61–95.

Jones, D. L., Coney, P. J., Harms, T. A., and Dillon, J. T., 1988, Interpretative geologic map and supporting radiolarian data from the Angayucham terrane, Coldfoot area, southern Brooks Range, Alaska: U.S. Geological Survey Miscellaneous Field Studies Map MF-1993, scale 1:63,360.

Jones, D. L., Silberling, N. J., Coney, P. J., and Plafker, G., 1987, Lithotectonic terrane map of Alaska: U.S. Geological Survey Miscellaneous Field Studies Map MF-1874A, scale 1:2,500,000.

Karl, S. M., Dumoulin, J. A., Ellersieck, I., Harris, A. G., and Schmidt, J. M., 1989, Preliminary geologic map of the Baird Mountains quadrangle, Alaska: U.S. Geological Survey Open-file Report 89-551, 65 p.

Kirschner, C. E., Gryc, G., and Molenaar, C. M., 1983, Regional seismic lines in the National Petroleum Reserve in Alaska, *in* Bally, A. W., ed., Seismic expression of structural styles, Volume 1: American Association of Petroleum Geologists Studies in Geology Series 15, p. 1.2.5-1–1.2.5-14.

Leffingwell, E. de K., 1919, The Canning River region, northern Alaska: U.S. Geological Survey Professional Paper 109, 251 p.

LePain, D. L., Crowder, R. K., and Watts, K. F., 1990, Mississippian clastic-to-carbonate transition in the northeastern Brooks Range, Alaska: Depositional cycles of the Endicott and Lisburne Groups: American Association of Petroleum Geologists Bulletin, v. 74, p. 703.

LePain, D. L., Crowder, R. K., and Wallace, W. K., 1994, Early Carboniferous transgression on a passive continental margin: Deposition of the Kekiktuk Conglomerate, northeastern Brooks Range, Alaska: American Association of Petroleum Geologists Bulletin, v. 78, p. 679–699.

Lerand, M., 1973, Beaufort Sea, *in* McCrossam, R. G., ed., The future petroleum provinces of Canada—their geology and potential: Canadian Society of Petroleum Geology Memoir 1, p. 315–386.

Mayfield, C. F., Silberman, M. L., and Tailleur, I. L., 1982, Precambrian metamorphic rocks of the Hub Mountain terrane, Baird Mountain Quadrangle, Alaska, *in* Conrad, W. L., ed., The U.S. Geological Survey in Alaska: Accomplishments during 1980: U.S. Geological Survey Circular 844, p. B11–B13.

Mayfield, C. F., Curtis, S. M., Ellersieck, I., and Tailleur, I. L., 1983, Reconnaissance geologic map of the De Long Mountains A3 and A4, and parts of B3 and B4 Quadrangles, Alaska: U.S. Geological Survey Open-File Report 83-183, 60 p.

Mayfield, C. F., Curtis, S. M., Ellersieck, I., and Tailleur, I. L., 1988, Stratigraphy, structure, and palinspastic synthesis of the western Brooks Range, northwestern Alaska, *in* Gryc, G., ed., Geology and exploration of the National Petroleum Reserve in Alaska 1974 to 1982: U.S. Geological Survey Professional Paper 1399, p. 143–186.

Mitchum, R. M., Jr., Vail, P. R., and Thompson, S., III, 1977, Seismic stratigraphy and global changes of sea level, Part 2: The depositional sequence as a basic unit for stratigraphic analysis, *in* Payton, C. E., ed., Seismic stratigraphy—applications to hydrocarbon exploration: American Association of Petroleum Geologists Memoir 26, p. 53–62.

Molenaar, C. M., 1983, Depositional relations of Cretaceous and lower Tertiary rocks, northeastern Alaska: American Association of Petroleum Geologists Bulletin, v. 67, p. 1066–1080.

Molenaar, C. M., 1985, Subsurface correlations and depositional history of the Nanushuk Group and related strata, North Slope, Alaska, *in* Huffman, A. C., Jr., ed., Geology of the Nanushuk Group and related rocks, North Slope, Alaska: U.S. Geological Survey Bulletin 1614, p. 37–59.

Molenaar, C. M., 1988, Depositional history and seismic stratigraphy of Lower Cretaceous rocks in the National Petroleum Reserve in Alaska and adjacent areas, *in* Gryc, G., ed., Geology and exploration of the National Petroleum Reserve in Alaska 1974 to 1982: U.S. Geological Survey Professional Paper 1399, p. 593–621.

Moore, T. E., 1987, Geochemistry and tectonic setting of volcanic rocks of the Franklinian assemblage, central and eastern Brooks Range, *in* Tailleur, I., and Weimer, P., eds., Alaskan North Slope geology: Society of Economic Paleontologists and Mineralogists, Pacific Section, Publication 50, p. 691–710.

Moore, T. E., and Churkin, M., Jr., 1984, Ordovician and Silurian graptolite discoveries from the Neruokpuk Formation (sensu lato), northeastern and central Brooks Range, Alaska: Paleozoic Geology of Alaska and Northwestern Canada Newsletter, no. 1, p. 21–23.

Moore, T. E., and Nilsen, T. H., 1984, Regional variations in the fluvial Upper Devonian and Lower Mississippian(?) Kanayut Conglomerate, Brooks Range, Alaska: Sedimentary Geology, v. 38, p. 465–497.

Moore, T. E., Nilsen, T. H., and Brosgé, W. P., 1989, Sedimentology of the Kanayut Conglomerate, *in* Mull, C. G., and Adams, K. E., eds., Dalton Highway, Yukon River to Prudhoe Bay, Alaska: Alaska Division of Geological and Geophysical Survey Guidebook 7, v. 2, p. 219–252.

Moore, T. E., Wallace, W. K., Bird, K. J., Karl, S. M., Mull, C. G., and Dillon, J. T., 1994, Geology of northern Alaska, *in* Plafker, G., and Berg, H. C., eds., The Geology of Alaska: Boulder, Colorado, Geological Society of America, The Geology of North America, v. G-1, p. 49–140.

Mull, C. G., 1982, Tectonic evolution and structural style of the Brooks Range, northern Alaska: An illustrated summary, *in* Blake, R. B., ed., Geologic studies of the Cordilleran thrust belt: Rocky Mountain Association Geologists, Denver, p. 1–45.

Mull, C. G., 1985, Cretaceous tectonics, depositional cycles, and the Nanushuk Group, Brooks Range and Arctic Slope, Alaska, *in* Huffman, C. A., Jr., ed., Geology of the Nanushuk Group and related rocks, North Slope, Alaska: U.S. Geological Survey Bulletin 1614, p. 7–36.

Mull, C. G., and Adams, K. E., eds., 1989, Dalton Highway, Yukon River to Prudhoe Bay, Alaska: Alaska Division of Geological and Geophysical Survey Guidebook 7, 309 p.

Murchey, B. L., and Harris, A. G., 1985, Devonian to Jurassic sedimentary rocks in the Angayucham Mountains of Alaska: Possible seamount or oceanic plateau deposits: Eos (Transactions, American Geophysical Union), v. 66, p. 1102.

Nilsen, T. H., 1981, Upper Devonian and Lower Mississippian redbeds, Brooks Range, Alaska, *in* Miall, A. D., ed., Sedimentation and tectonics in alluvial basins: Geological Association of Canada Special Paper 23, p. 187–219.

Nilsen, T. H., and Moore, T. E., 1982a, Sedimentology and stratigraphy of the Kanayut Conglomerate, central and western Brooks Range, Alaska—Report of the 1981 field season: U.S. Geological Survey Open-File Report 82-674, 64 p.

Nilsen, T. H., and Moore, T. E., 1982b, Fluvial facies model for the Upper Devonian and Lower Mississippian(?) Kanayut Conglomerate, Alaska, *in* Embry, A. F., and Balkwill, H. R., eds., Arctic geology and geophysics: Canadian Society of Petroleum Geologists, Memoir 8, p. 1–12.

Nilsen, T. H., Moore, T. E., and Brosgé, W. P., 1980, Paleocurrent maps for the Upper Devonian and Lower Mississippian Endicott Group, Brooks Range, Alaska: U.S. Geological Survey Open-File Report 80-1066, scale 1:1,000,000.

O'Sullivan, P. B., Murphy, J. M., Blythe, A. E., and Moore, T. E., 1994, Fission track evidence indicates that the present-day Brooks Range, Alaska, is a Cenozoic, not Early Cretaceous, physiographic feature: Eos (Transactions, American Geophysical Union), v. 75, p. 646.

Oldow, J. S., Seidensticker, M., Phelps, J. C., Julian, F. E., Gottschalk, R. R., Boler, K. W., Handschy, J. W., and Avé Lallemant, H. G., 1987, Balanced cross sections through the central Brooks Range, Alaska: American Association of Petroleum Geologists Special Paper, 19 p., 8 pl.

Pallister, J. S., Budahn, J. R., and Murchey, B. L., 1989, Pillow basalts of the Angayucham terrane: Oceanic plateau and island crust accreted to the Brooks Range: Journal of Geophysical Research, v. 94, p. 15901–15923.

Palmer, A. R., Dillon, J. T., and Dutro, J. T., Jr., 1984, Middle Cambrian trilobites with Siberian affinities from the central Brooks Range, northern Alaska: Geology of Alaska and Northwestern Canada Newsletter, no. 1, p. 29–30.

Patton, W. W., Jr., and Tailleur, I. L., 1964, Geology of the Killik-Itkillik region Alaska: U.S. Geological Survey Professional Paper 303-G, p. 409–500.

Reiser, H. N., Brosgé, W. P., Dutro, J. T., Jr., and Detterman, R. L., 1971, Preliminary geologic map of the Mt. Michelson Quadrangle, Alaska: U.S. Geological Survey Open-file report 71-237, scale 1:250,000.

Reiser, H. N., Brosgé, W. P., Dutro, J. T., Jr., and Detterman, R. L., 1980, Geologic map of the Demarcation Point Quadrangle, Alaska: U.S. Geological Survey Miscellaneous Investigations Series Map I-1133, scale 1:250,000.

Repetski, J. E., Carter, C., Harris, A. G., and Dutro, J. T., Jr., 1987, Ordovician and Silurian fossils from the Doonerak anticlinorium, central Brooks Range, Alaska, *in* Hamilton, T. D., and Galloway, J. P., eds., Geologic studies in Alaska by the U.S. Geological Survey during 1986: U.S. Geological Survey Circular 998, p. 40–42.

Robinson, M. S., Decker, J., Clough, J. G., Reifenstuhl, R. R., Dillon, J. T., Combellick, R. A., and Rawlinson, S. E., 1989, Geology of the Sadlerochit and Shublik mountains, Arctic National Wildlife Refuge, northeastern Alaska: Alaska Division of Geological and Geophysical Surveys Professional Report 100, scale 1:63,360.

Roeder, D., and Mull, C. G., 1978, Tectonics of Brooks Range ophiolites, Alaska: American Association of Petroleum Geologists Bulletin, v. 62, p. 1696–1713.

Siok, J. P., 1989, Stratigraphy and petrology of the Okpikruak Formation at Cobblestone Creek, north-central Brooks Range, *in* Mull, C. G., and Adams, K. E., eds., Dalton Highway, Yukon River to Prudhoe Bay, Alaska: Alaska Division of Geological and Geophysical Survey Guidebook 7, v. 2, p. 285–292.

Turner, D. L., Forbes, R. B., and Dillon, J. T., 1979, K-Ar geochronology of the southwestern Brooks Range, Alaska: Canadian Journal of Earth Sciences, v. 16, p. 1789–1804.

Wirth, K. R., and Bird, J. M., 1992, Chronology of ophiolite crystallization, detachment and emplacement: evidence from the Brooks Range, Alaska: Geology, v. 20, p. 75–78.

MANUSCRIPT ACCEPTED BY THE SOCIETY SEPTEMBER 23, 1997

Geological Society of America
Special Paper 324
1998

Sedimentology and paleogeographic significance of Upper Devonian and Lower Mississippian clastic rocks, Endicott Mountains allochthon, central Brooks Range, Alaska

James W. Handschy
Shell Exploration and Production Company, P.O. Box 481, Houston, Texas 77001-0481

ABSTRACT

Upper Devonian and Lower Mississippian rocks exposed in the Endicott Mountains allochthon of the central Brooks Range track the progressive evolution of a Late Devonian and Mississippian continental margin. Sedimentary facies in lower Upper Devonian rocks of the Beaucoup Formation delimit a volcanically active depositional basin, possibly related to a convergent arc system or an active rift. Volcaniclastic sedimentary rocks within the Beaucoup Formation were apparently derived from the south, whereas nonvolcanic sediments were derived from the north. By the late Late Devonian, the depositional basin had developed into a south-facing rifted continental margin. Southwestward progradation of the Upper Devonian Kanayut Conglomerate–Hunt Fork Shale deltaic system deposited thick conglomerate, sandstone, and shale on the margin and created a lithofacies pattern in which the Kanayut Conglomerate is thicker in the north and the Hunt Fork Shale is thicker in the south. Transgression of the Lower Mississippian Kayak Shale over the Kanayut Conglomerate occurred during the first major rise in Mississippian sea level. Subsequent transgressive-regressive cycles in the Lisburne Group indicate that the margin had evolved into a stable passive margin by the middle Mississippian.

INTRODUCTION

The Endicott Mountains allochthon is an imbricate stack of thrust sheets that constitutes much of the northern, nonmetamorphic, part of the late Mesozoic and Cenozoic Brooks Range fold-thrust belt in Arctic Alaska. Thrust sheets in the allochthon are composed predominantly of Upper Devonian and Mississippian clastic and carbonate rocks of the Endicott and Lisburne Groups (Brosgé et al., 1962, 1979; Tailleur et al., 1967; Mull, 1982). Typically, only the upper part of the Endicott Group and the lower part of the Lisburne Group are exposed at the surface. However, in the central Brooks Range, uplift of the Endicott Mountains allochthon by the Doonerak duplex (Oldow et al., 1987; Mull et al., 1987) has exposed the entire Endicott Group and a significant part of the underlying Beaucoup Formation.

Regional stratigraphy and previous work

The Endicott Group was named for exposures in the Endicott Mountains of the central Brooks Range and, in its type area, includes the Hunt Fork Shale, Kanayut Conglomerate, and Kayak Shale (Tailleur et al., 1967). Bowsher and Dutro (1957) described the type sections of the Kanayut Conglomerate and Kayak Shale near Shainin Lake. Chapman et al. (1964) defined the type section of the Hunt Fork Shale 100 km farther west near the Killik River. In the western Brooks Range, and locally in the central Brooks Range, the Endicott Group also includes the Noatak Sandstone, which lies between the Hunt Fork Shale and Kanayut Conglomerate (Dutro, 1952; Tailleur et al., 1967; Brosgé et al., 1988; Moore et al., 1989). In the northeastern Brooks Range and along the margins of the Doonerak window,

Handschy, J. W., 1998, Sedimentology and paleogeographic significance of Upper Devonian and Lower Mississippian clastic rocks, Endicott Mountains allochthon, central Brooks Range, Alaska, *in* Oldow, J. S., and Avé Lallemant, H. G., eds., Architecture of the Central Brooks Range Fold and Thrust Belt, Arctic Alaska: Boulder, Colorado, Geological Society of America Special Paper 324.

where the Kanayut Conglomerate and Hunt Fork Shale are absent due to nondeposition, the Endicott Group is composed of the Kekiktuk Conglomerate and Kayak Shale (Armstrong et al., 1976; Brosgé et al., 1962). The basal contact of the Endicott Group is a disconformity(?) that separates the Hunt Fork Shale from mixed clastic and carbonate rocks of the Beaucoup Formation (Dutro et al., 1979; Tailleur et al., 1967). The type section of the Beaucoup Formation is located on the northeast side of Beaucoup Creek approximately 100 km east of the study area (Dutro et al., 1979). The contact between the Kayak Shale, at the top of the Endicott Group, and the basal limestone of the Lisburne Group is gradational (Armstrong and Mamet, 1978).

Sedimentary facies, paleocurrents, clast size distribution, and provenance indicate a northern siliciclastic source for the Kanayut Conglomerate and Hunt Fork Shale. Clastic rocks in the Beaucoup Formation were apparently derived from both a northern siliciclastic source and a southern volcanic source. Age and compositional similarities between the Beaucoup Formation, at the bottom of the Endicott Mountains allochthon, and mixed carbonate and clastic rocks in the superjacent Skajit allochthon provide a tentative stratigraphic tie between the two allochthons (Oldow et al., this volume, Chapter 7; Dillon, 1989). Paleogeographic reconstructions based on these facies relations in the central Brooks Range and Devonian through Jurassic facies in the western and southern Brooks Range suggest that the tectonic setting of sedimentation changed from a back-arc or continental rift basin in the early Late Devonian, to a rifted continental margin in the latest Devonian and eventually to a passive margin by the middle Mississippian.

Aside from the preliminary work listed above, the only detailed study of Late Devonian stratigraphy in the Brooks Range is a regional analysis of the Kanayut Conglomerate (Nilsen, 1981; Nilsen and Moore, 1982a, b, 1984; Nilsen et al., 1980a, b, 1981a, b, 1982; Moore and Nilsen, 1984; Moore et al., 1989; Brosgé et al., 1988). This study is based on detailed examination of outcrops in the southern part of the Endicott Mountains allochthon between the Doonerak window and Atigun Pass plus reconnaissance work in the northern thrust imbricates of the allochthon between Atigun Pass and Galbraith Lakes (Fig. 1). It complements the work of Nilsen and his colleagues by focusing on marine and transitional marine facies distribution in the Kanayut Conglomerate–Hunt Fork Shale clastic wedge, depositional relationships between the Kanayut Conglomerate, Hunt Fork Shale, and Beaucoup Formation, and the significance of Upper Devonian and Lower Mississippian sequence boundaries in the central Brooks Range.

Structural setting

The structural geology of the central Brooks Range is dominated by a series of north-vergent thrust sheets. The Endicott Mountains allochthon structurally overlies younger clastic and carbonate rocks of the Ellesmerian sequence (Handschy, this volume, Chapter 1) along the front of the Brooks Range

and along the northern margin of the Doonerak window (Fig. 1; Mull, 1982; Mull et al., 1987, 1989; Seidensticker and Oldow, this volume, Chapter 6). At the front of the Brooks Range the Endicott Mountains allochthon was thrust over Mississippian through Jurassic rocks, whereas in the Doonerak window it structurally overlies Mississippian through Triassic rocks (Mull, 1982; Mull et al., 1987, 1989; Oldow et al., 1987). On the south side of the Doonerak window, the Endicott Mountains allochthon is in thrust contact with the Cambrian through Silurian Apoon assemblage (Fig. 1; Armstrong et al., 1976; Dutro et al., 1976; Oldow et al., 1987; Seidensticker and Oldow, this volume, Chapter 6), and is structurally overlain by internally imbricated, weakly metamorphosed carbonate and clastic rocks of the Precambrian(?) through Late Devonian Skajit allochthon (Oldow et al., 1987; Phelps, 1987). Farther to the southeast, the Endicott Mountains allochthon is structurally overlain by both the Skajit allochthon and rocks of the Schist belt (Oldow et al., 1987; Dillon, 1989). The ages of the protoliths of the Schist belt are poorly constrained, but deformed granitic plutons within the Schist belt have yielded Devonian ages (Dillon, 1989; Dillon et al., 1980, 1987b).

The present structural position of the Endicott Mountains allochthon is the product of northward translation of the allochthon and duplexing of the Apoon assemblage beneath the allochthon (Fig. 1; Oldow et al., 1987; Phelps and Avé Lallemant, this volume, Chapter 4). Balanced cross sections (Oldow et al., 1987) and facies reconstructions (Mull et al., 1987) indicate that the Endicott Mountains allochthon was originally located south of the Apoon assemblage. North-directed thrusting of the Endicott Mountains allochthon placed Upper Devonian rocks at the base of the allochthon over the Mississippian through Triassic sequence, which depositionally overlies the Apoon assemblage in the Doonerak window (Dutro et al., 1976; Mull et al., 1989). Duplexing of the Apoon assemblage was accompanied by partial detachment of the Mississippian Kekiktuk Conglomerate from the Apoon, and folding of the overlying stack of thrust sheets into an anticlinorium (Fig. 1; Oldow et al., 1987; Mull et al., 1987, 1989; Seidensticker and Oldow, this volume, Chapter 6).

STRATIGRAPHY AND DEPOSITIONAL ENVIRONMENTS

Five stratigraphic units are exposed in thrust sheets of the Endicott Mountains between the northeastern corner of the Doonerak window and Galbraith Lakes (Figs. 1 and 2). From bottom to top these five units represent three depositional sequences:

Figure 1. Geologic map and cross section of the central Brooks Range showing the location of the study area. The locations of the Doonerak window (DW), Atigun Pass (AP), and Galbraith Lakes (GL) are shown for reference. A, A′, and B are reference points used in the palinspastic reconstruction of this cross section (see Handschy, this volume, Chapter 3, Fig. 1). See the Preface of this volume for a regional location map.

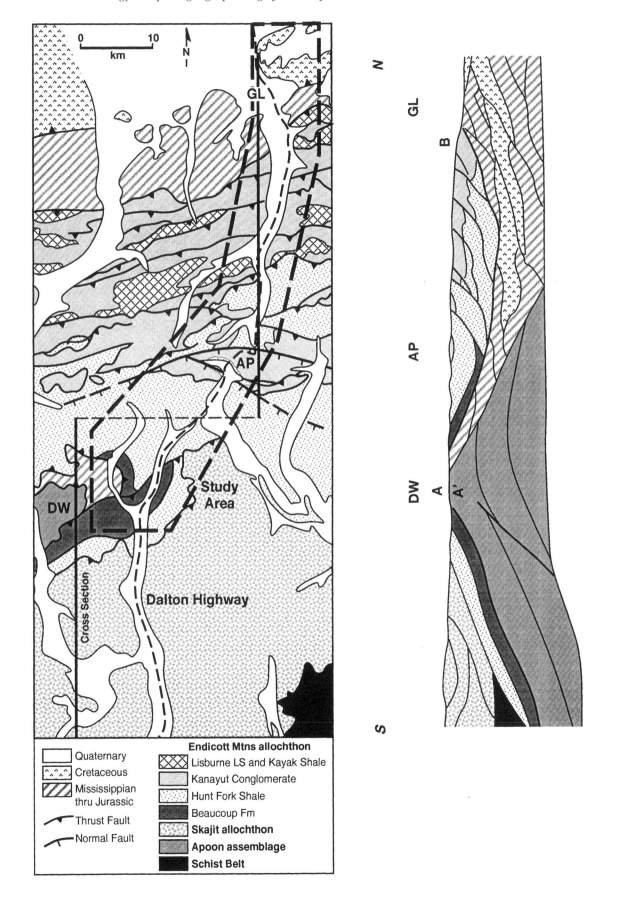

Quaternary

Cretaceous

Mississippian thru Jurassic

Thrust Fault

Normal Fault

Endicott Mtns allochthon

Lisburne LS and Kayak Shale

Kanayut Conglomerate

Hunt Fork Shale

Beaucoup Fm

Skajit allochthon

Apoon assemblage

Schist Belt

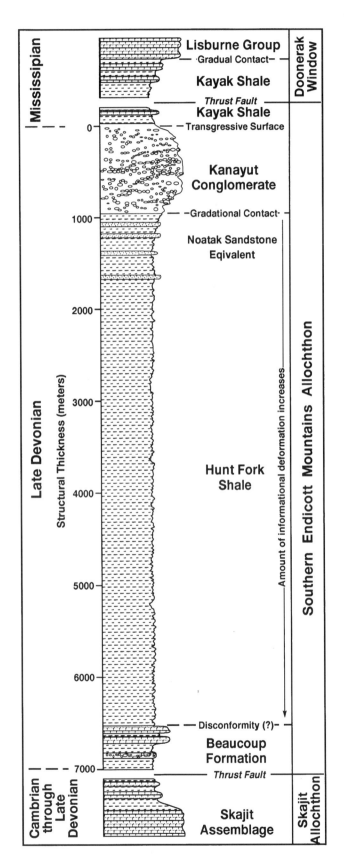

1. mixed carbonate and clastic rocks of the Upper Devonian Beaucoup Formation;

2. progradational shale, sandstone, and conglomerate of the Upper Devonian and lowest Mississippian(?) Kanayut Conglomerate and Hunt Fork Shale; and

3. transgressional sandstone and shale, and progradational carbonate rocks of the Lower Mississippian Kayak Shale and middle Mississippian through Lower Pennsylvanian Lisburne Group.

Beaucoup Formation

The top of the Beaucoup Formation is defined as the top of the uppermost carbonate buildup or, in the absence of massive carbonate beds, by the contact between the Hunt Fork Shale and brown to yellow weathering sandstone, shale, and calcareous shale "directly beneath the darker Hunt Fork Shale" (Dutro et al., 1979). Based on these criteria, Dutro et al. (1979) assigned the unnamed Upper Devonian brown calcareous clastic rocks in the Phillip Smith Mountains Quadrangle (Brosgé et al., 1979) to the Beaucoup Formation. In the vicinity of the Dalton Highway, however, the brown shale member of the unnamed unit (Brosgé et al., 1979) includes rocks belonging to both the Hunt Fork Shale and Beaucoup Formation and can be traced laterally into the Hunt Fork Shale as it is mapped by Brosgé et al. (1979). The actual Hunt Fork Shale–Beaucoup Formation contact is south of the contact between the Hunt Fork Shale and the unnamed brown shale mapped by Brosgé et al. (1979).

The Beaucoup Formation in the study area is a 0- to 1.5-km-thick sequence of interlayered shale, calcareous shale, sandstone, conglomerate, volcaniclastic rocks, carbonate buildups, and their low-grade metamorphic equivalents. Bedding thickness varies from less than 1 mm in shale and phyllite to greater than 5 m in massive limestone and amalgamated sandstone units. Most of the lithologies are thinly layered and strongly deformed. Carbonate buildups and sandstone units greater than 2 m thick are relatively undeformed.

Adjacent to the Dalton Highway in the southern part of the study area (Fig. 3), the Beaucoup Formation has a minimum tectonic thickness of 347 m (Fig. 4). The top of the Beaucoup Formation at this locality is a 5-m-thick, medium dark gray limestone unit containing colonial and solitary rugose corals, tabulate corals, and bryozoans (Table 1). The upper limestone unit is separated from a second, thinner (4 m thick) limestone unit containing corals, crinoids, stromatoporoids, and bryo-

---

Figure 2. General stratigraphy of the Endicott Mountains allochthon. Note that the vertical scale is tectonic thickness. The true stratigraphic thicknesses of the Hunt Fork Shale and Beaucoup Formation are unknown because of significant internal deformation. The stratigraphic thickness of the Kanayut Conglomerate varies from approximately 1 km in the southern Endicott Mountains allochthon (illustrated in this figure) to more than 2.5 km in the northern Endicott Mountains allochthon (see Fig. 5).

zoans by 8 m of intensely deformed yellowish gray to olive-black shale (Fig. 4). Below the second limestone unit the section consists of 330 m of poorly exposed, intensely deformed, olive-gray to olive-black shale, yellowish gray to light gray calcareous silt-shale, and mottled grayish red to greenish gray shale. A discontinuous grayish yellow to medium gray chert- and quartz-granule conglomerate interrupts this shaley sequence 20 m below the base of the lower limestone unit. A 2-m-thick interval of reddish brown volcaniclastic breccia, medium light gray chert pebble conglomerate, and grayish yellow fine-grained quartz arenite is located approximately 30 m above the lower fault contact (Fig. 4).

Intense deformation in the shale of the Beaucoup Formation makes it difficult to interpret all of the depositional environments represented. Where sedimentary structures are preserved, the shale is horizontal to wavy laminated and composed of silt or mud; graded beds are rare. Detrital components in both calcareous and noncalcareous shale are petrographically similar, with the exception of carbonate grains in the calcareous shale. Shale containing volcaniclastic detritus can be distinguished in the field by its mottled reddish brown and greenish gray color. Corals, crinoids, and other marine fossils (Table 1) collected from the carbonate buildups and some calcareous shale beds indicate deposition in a relatively shallow marine environment. Amalgamated, channelized sandstone and conglomerate occur at several stratigraphic levels in the Beaucoup Formation. Channel-fill pebble conglomerate is usually massive and composed of subrounded to subangular, light gray to dark gray chert and white quartz clasts. Medium gray carbonate clasts and reddish brown volcanic clasts are locally abundant. Beaucoup Formation sandstone beds are massive to trough cross-stratified, fine to very coarse grained, poorly to moderately sorted, quartz-rich to lithic-rich wackestones and arenites.

The age of the Beaucoup Formation is restricted to the early Frasnian by corals and brachiopods at the type section (Dutro et al., 1979). Age-equivalent corals, stromatoporoids, and conodonts have also been recovered from Beaucoup Formation limestone beds in the Chandalar and Wiseman Quadrangles (Dillon et al., 1987a; Dillon, 1989).

Dutro et al. (1979, p. A66) report that "the Beaucoup Formation at its type section is conformably overlain by the Hunt Fork Shale." Adjacent to the Dalton Highway the lower Hunt Fork Shale appears to be in depositional contact with different Beaucoup Formation lithologies, possibly indicating the presence of an unconformity at the base of the Hunt Fork Shale as originally interpreted by Tailleur et al. (1967). No bedding angularity can be measured across the Hunt Fork Shale–Beaucoup Formation contact, so the unconformity is probably a disconformity. Pervasive shear strain has caused top-to-the-north faulting along the Hunt Fork Shale–Beaucoup Formation contact, but there is no structural or stratigraphic evidence for a regionally significant fault between the Hunt Fork Shale and Beaucoup Formation.

Kanayut Conglomerate and Hunt Fork Shale

The Kanayut Conglomerate and Hunt Fork Shale are, respectively, the fluvial and marine parts of a large Late Devonian through Early Mississippian(?) fluvio-deltaic depositional system that is exposed in thrust sheets of the Endicott Mountains allochthon for approximately 950 km along the crest of the Brooks Range (Moore and Nilsen, 1984). Measured stratigraphic thicknesses of Kanayut Conglomerate in the study area vary from approximately 1 km in the southern nappe to more than 2.6 km in the northern thrust imbricates (Fig. 5), reflecting southwestward thinning of the Kanayut Conglomerate over the Hunt Fork Shale (Fig. 6; Moore et al., 1989). The age of the Kanayut Conglomerate–Hunt Fork Shale system is constrained by Famennian brachiopods and Early Mississippian(?) plant fragments in the upper Kanayut Conglomerate (Moore and Nilsen, 1984), Famennian brachiopods in the Noatak Sandstone (Brosgé et al., 1988), and Frasnian (early Late Devonian) brachiopods and corals in the lower Hunt Fork Shale (Dutro et al., 1979; Brosgé et al., 1979).

Nilsen and Moore (1984) divided the Kanayut Conglomerate into the Ear Peak, Shainin Lake, and Stuver Members, in ascending order (Fig. 5). The Stuver and Ear Peak Members generally consist of multiple fining-upward cycles and, at a regional scale, are interpreted as meandering river deposits (Nilsen and Moore, 1984; Moore and Nilsen, 1984). The intervening Shainin Lake Member is more conglomeratic and is interpreted as a braided river deposit (Moore and Nilsen, 1984). In the southern half of the study area all three members are dominated by conglomerate and sandstone, and appear to have been deposited primarily by braided rivers. Where all three members can be distinguished, maximum and average clast sizes are larger in the Shainin Lake Member, and sandstone/conglomerate ratios are higher in the Ear Peak and Stuver Members (Handschy, 1988).

Conglomerate beds in the Kanayut Conglomerate are light gray to grayish orange, organized, and clast supported. They generally have a bimodal clast distribution, with either pebble- to cobble-size framework clasts and a medium-grained sand to granule matrix, or granule- to pebble-size clasts and a fine- to coarse-grained sand matrix. Three distinct conglomerate lithofacies are recognized on the basis of sedimentary structures (Table 2). Large-scale trough cross-stratified conglomerate (Gt) is the most common lithofacies in the study area. Individual troughs range from 1 m to 5 m wide and 0.2 m to 1.5 m deep. Trough cross-stratified conglomerate beds are laterally continuous or form broadly lenticular bodies from 50 m to more than 500 m wide. Locally, individual troughs and trough cross-stratified beds are normally graded. Massive to crudely horizontally stratified conglomerate (Gm), the second most common conglomerate lithofacies in the Kanayut Conglomerate, occurs as single channel fill deposits or thick (>1 m) laterally continuous beds with erosional lower boundaries. Some Gm beds and channels are normally graded, but grain-size trends are usually absent. Minor occurrences of planar cross-stratified conglomer-

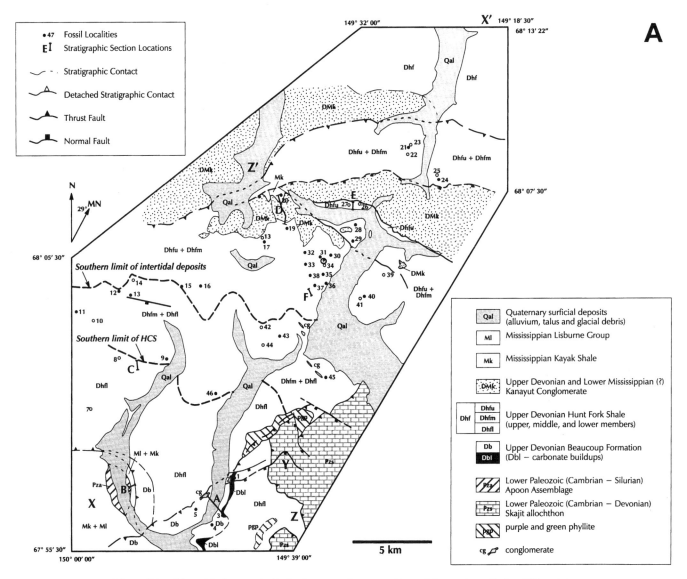

Figure 3 (on this and facing page). Geologic map and cross sections of the southern Endicott Mountains allochthon. a, Map showing fossil localities, facies boundaries within the Hunt Fork Shale, and locations of measured stratigraphic sections. Fossils are listed in Table 1. Measured stratigraphic section locations: A = Beaucoup Formation (Fig. 4); B = Kayak Shale in the northeast corner of the Doonerak window (Fig. 9); C = lower Hunt Fork Shale (Fig. 8); D = Kanayut Conglomerate (Fig. 5) and Kayak Shale (Fig. 9) west of Atigun Pass; E = upper Hunt Fork Shale (Fig. 8); F = middle Hunt Fork Shale (Fig. 8). b, Near-surface geologic cross sections. Cross section locations are shown on Figure 3a.

ate (Gp) are also found in the Kanayut Conglomerate. Where they occur, planar cross-stratified conglomerate beds are associated with trough cross-stratified conglomerate. Stratification in all three conglomerate lithofacies is defined by sandy layers and alignment of oblate pebbles. Imbricated clasts are present in Gm and Gt lithofacies.

Sandstone lithofacies comprise approximately 30% of the Kanayut Conglomerate in the northern Endicott Mountains section (Moore et al., 1989) and 40% of the section in the southern Endicott Mountains allochthon. Sandstone occurs in small lenses and channel-fill deposits within amalgamated conglom-

erate bodies, in the upper parts of fining upward cycles, and as laterally continuous, tabular to broadly lenticular bodies. Light olive gray to grayish yellow, trough cross-stratified, sometimes pebbly, very coarse to fine-grained sandstone (St) is the most common sandstone lithofacies in both sections (Fig. 5). Troughs are generally 0.5 m to 3 m wide and 0.1 m to 1.0 m deep. Pebbles are concentrated in the lower third of the troughs or, less commonly, scattered throughout the troughs. Horizontally laminated (Sh) and planar cross-stratified (Sp) sandstone is present, but not common.

Reddish brown mudstone lithofacies constitute approxi-

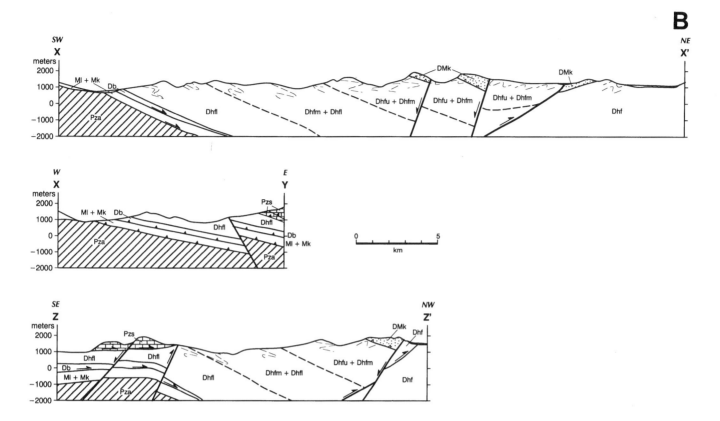

mately 30% of the northern Endicott Mountains section and less than 10% of the southern Endicott Mountains section. Massive (Fm) and horizontally laminated (Fl) siltstone, sandy mudstone, and mudstone are the only fine-grained lithofacies recognized. Some Fm beds are mottled, suggesting that they are burrowed or pedogenically altered. Fine-grained lithofacies occur as lenses within amalgamated sandstone and conglomerate beds, and as broadly lenticular sheets 100 m to more than 500 m wide and as much as 4 m thick. These laterally extensive sheets frequently contain channels filled with conglomerate or sandstone, and they are often encised by erosion surfaces along the bases of the overlying conglomerate beds.

The dominance of Gt, Gm, and St lithofacies in all three members is indicative of deposition by braided river systems. Lithofacies Gt is produced by migration of large three-dimensional gravel ripples in braided channels and by incision of channels into bars during flood stages (Rust, 1972; Miall, 1978). Lithofacies Gm is generally interpreted to represent longitudinal bars (Miall, 1977). Lithofacies St is the sandy braided river equivalent of lithofacies Gt, and is deposited by migration of sinuous dunes (Cant, 1978; Miall, 1981). Subordinate amounts of Gp and Sp lithofacies probably formed by accretion on the margins of longitudinal bars during waning flow and by the migration of gravelly and sandy transverse bars (Smith, 1974; Hein and Walker, 1977; Rust, 1978). Laterally extensive fine-grained lithofacies represent overbank mud deposited during waning flow (Williams and Rust, 1969),

whereas mudstone lenses in amalgamated conglomerate and sandstone beds are probably remnants of mud drapes deposited during waning flow within channels.

The combination of lithofacies in the Kanayut Conglomerate (Table 2) and the overall percentage of sandstone plus conglomerate (at least 70% in all three members) are characteristic of braided rivers with well-defined active channels, elevated partially active to inactive channels, and abundant sand and gravel bedload (Donjek-type braided rivers; Miall, 1978, 1981). Fining upward cycles, similar to the ones described by Moore et al. (1989) as meandering river cycles in the Ear Peak and Stuver Members, are also common in Donjek type braided rivers (Williams and Rust, 1969). All of the fining upward cycles in the Kanayut Conglomerate that were examined during this study are composed primarily of Gt and St lithofacies plus finer grained lithofacies indicative of waning flow. In addition, most of the fining upward cycles are defined by interbedded sequences of conglomerate, sandstone, and mudstone, not point-bar sequences. No complete point-bar sequences were observed, but if present were probably deposited in relatively low sinuosity channels. The absence of disorganized Gm and Gms lithofacies in the northernmost exposures also suggests that proximal alluvial fan environments are not represented in the study area.

Paleocurrents in the Kanayut Conglomerate of the southern Endicott Mountains allochthon are unimodal and indicate flow toward the south-southwest (Fig. 7A). This is consistent with paleocurrent data from the Atigun River valley in the

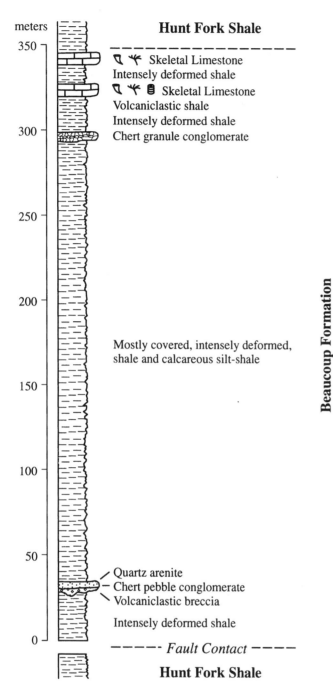

Figure 4. Measured section of the Beaucoup Formation adjacent to the Dalton Highway in the southern Endicott Mountains allochthon. Note that the section is intensely deformed and the measured thickness is a tectonic thickness, not a true stratigraphic thickness. Section location shown on Figure 3a (A).

northern Endicott Mountains allochthon and paleocurrents from the Kanayut Conglomerate in general (Moore et al., 1989). Maximum clast size decreases from approximately 30 cm in the northernmost exposures of Kanayut Conglomerate east of the Atigun River valley to 8 cm in the southern Endicott Mountains allochthon.

The contact between the Hunt Fork Shale and the Kanayut Conglomerate is gradational. Channel-fill conglomerate and sandstone, which cut into and are interbedded with intertidal deposits in the upper Hunt Fork Shale, can be traced laterally into amalgamated conglomerate and sandstone beds in the Kanayut Conglomerate at several locations along the crest of the Brooks Range in the study area.

Sedimentary structures and fossil assemblages allow the Hunt Fork Shale to be subdivided into upper, middle, and lower members (Fig. 8). The upper Hunt Fork Shale is composed of olive-gray to black, thinly laminated mudstone and siltstone, channelized grayish yellow quartz- and chert-rich sandstone and conglomerate, and olive-gray quartz-rich sheet sandstone. Thin (<10 cm) grayish orange calcareous shale beds are locally present. Mudcracks, wrinkle marks, ripples, algal mats, oncolites, brachiopods, Skolithos and Cruziana ichnofacies trace fossils, and rip-up breccias indicate deposition in an intertidal environment. Channels that cut into these intertidal shale beds are filled with trough cross-stratified fine- to coarse-grained sandstone and pebble conglomerate. Paleocurrent measurements from channels in the upper Hunt Fork Shale are more variable than paleocurrents in the Kanayut Conglomerate, possibly indicating greater channel sinuosity. Mean flow in both units, however, was approximately parallel (Fig. 7). Sheet sandstone in the upper Hunt Fork Shale is fine to medium grained and massive or trough cross-stratified.

The upper Hunt Fork Shale, as defined here, is stratigraphically equivalent to the Noatak Sandstone (Dutro, 1952) and partially equivalent to the "basal sandstone member" of the Kanayut Conglomerate mapped by Brosgé et al. (1979, 1988). Shale is the dominant lithology (sand/shale = 0.3) in this interval throughout the study area.

The middle Hunt Fork Shale is composed mostly of olive-gray to black, thinly laminated mud and silt shale, but also contains lenticular, trough cross-stratified, fine-grained sandstone, massive fine- to medium-grained sheet sandstone, fine- to coarse-grained hummocky cross-stratified sandstone, and thin graded beds (Fig. 8). All sandstone is composed mostly of quartz and chert grains and has a grayish yellow to light olive gray color. Light olive gray to medium gray channelized chert-pebble conglomerate occurs locally along discontinuous horizons (Fig. 3). Faunal assemblages in the middle Hunt Fork Shale are more diverse than in the upper Hunt Fork Shale and include rugose corals, tabulate corals, crinoids, bryozoans, gastropods, ammonoids, and several types of brachiopods (Table 1). Rip-up breccias and allogenic fossils are commonly present at the bases of hummocky cross-stratified units. Lenticular, east-west–trending, trough cross-stratified, fine- to medium-grained sandstone with east-west–trending paleocurrents (Fig. 7C) appear to be longitudinal offshore bar deposits. Allogenic rugose corals and crinoid fragments in lenticular carbonate beds suggest that small carbonate mounds or patch reefs were locally present in the middle Hunt Fork Shale. Skolithos and Cruziana ichnofacies trace fossils occur in some interbedded sandstone and shale intervals.

The absence of hummocky cross-stratified units, longitudinal bar deposits, and fossils distinguish the lower Hunt Fork Shale from the middle Hunt Fork Shale (Fig. 8). Most of the lower Hunt Fork Shale is composed of dark greenish gray to black, thin (0.5 to 4 cm), repetitious graded beds. Individual graded beds fine upward from siltstone or very fine grained sandstone to clay shale. Planolites and Chondrites are the only trace fossils recognized in the lower Hunt Fork Shale. Locally, sequences of graded beds are interrupted by laterally continuous, massive, light olive gray, fine-grained sandstone beds as much as 2 m thick. These sandstone beds have sharp upper and lower contacts, but do not appear to scour into the underlying shale. Less frequently, asymmetric channels as much as 3 m deep and 10 m wide cut into the underlying beds. Channels are filled with medium gray to greenish gray, massive or trough cross-stratified, fine- to medium-grained sandstone and rip-up clast conglomerate. Phelps (1987) also reports chert pebble conglomerate in channels within the lower Hunt Fork Shale farther to the west.

Deformation of the Hunt Fork Shale makes it impossible to accurately measure a complete stratigraphic section in the study area. The tectonic thickness of the Hunt Fork Shale in the southern Endicott Mountains allochthon, where both the top and bottom of the formation are exposed, is approximately 6 km (Fig. 2). Despite the intense intraformational deformation, the distribution of facies in the Hunt Fork Shale is in a normal stratigraphic order. Prodelta slope deposits of the lower Hunt Fork Shale are overlain by shelf facies of the middle Hunt Fork Shale; these shelfal deposits are in turn overlain by intertidal deposits of the upper Hunt Fork Shale, which grade laterally into the Noatak Sandstone and vertically into fluvial deposits of the Kanayut Conglomerate (Figs. 1 and 4).

Lithofacies, unimodal paleocurrents and the aerial extent of the Kanayut Conglomerate classify the Kanayut Conglomerate–Hunt Fork Shale delta as a braid delta (McPherson et al., 1987). The distribution of sedimentary facies in the Hunt Fork Shale, however, differs from previously described coarse-grained deltas in that turbidites are the only sediment-gravity flow structures recognized. Debris flows and slumps, which are common in the subaqueous parts of many coarse-grained deltas (Postma, 1984), are uncommon in the Hunt Fork Shale. Massive sheet sandstone in the lower Hunt Fork Shale may be grain-flow deposits or, alternatively, well-sorted, sand-rich turbidites deposited by unchannelized sheet flow down the delta front. The dearth of syndepositional deformation suggests that the Kanayut Conglomerate–Hunt Fork Shale delta front had a relatively shallow depositional dip.

Shoreline and subaqueous deposition in the Hunt Fork Shale was controlled by sediment influx rates, tidal range, wave energy, and wave direction. Intertidal deposits in the upper Hunt Fork Shale are cut by channelized sandstone and conglomerate suggesting that tidal flats developed between distrib-

utary channels. Sedimentary structures and paleocurrents in distributary channels, however, indicate unidirectional flow toward the south-southwest (Fig. 7B) discounting significant tidal influence within the channels. Variable flow rates are evidenced by fining-upward trends within single channels and by conglomerate in some channels and fine- to medium-grained sand in other channels. Laterally continuous siltstone and sandstone layers within the upper Hunt Fork Shale were deposited by sheet flood in interdistributary areas when flow rates exceeded channel capacity or when channels were clogged by high sediment loads.

Juxtaposition of intertidal/distributary channel deposits against delta-front shale without intervening beach and upper shoreface deposits suggests that fair-weather wave energy was low. East-west–trending offshore bars in the middle Hunt Fork Shale indicate that wave convergence was oblique to the shore line and created longshore currents. Paleocurrents in the middle Hunt Fork Shale of the study area (Fig. 7C) and thickening of upper shoreface and beach sandstone in the Noatak Sandstone farther to the west (Moore and Nilsen, 1984) suggest that longshore drift was from east to west. This combination of relatively high sediment influx, relatively low tidal influence, and low wave energy typifies modern fluvial-dominated deltas (Galloway, 1975). Hummocky cross-stratification and rip-up breccias, such as the ones preserved in the middle Hunt Fork Shale, are indicative of storm wave reworking of delta-front deposits below the fair-weather wave base (Chan and Dott, 1986). However, since fair-weather wave energy was apparently low, hummocky cross-stratified units could have been generated by small storms at relatively shallow depths. This is supported by intertidal units in close proximity to hummocky cross-stratified units and by interbedded hummocky cross-stratified units and trough cross-stratified longitudinal bar deposits in the middle Hunt Fork Shale.

Deposition of the lower Hunt Fork Shale occurred below the storm wave base, primarily by turbidity currents. The absence of Zoophycos and Nerites ichnofacies suggests that the lower Hunt Fork Shale was deposited in the sublittoral zone (<200 m; Crimes, 1975; Frey and Pemberton, 1984). However, since the *Planolites* and *Chondrites* trace fossils found in the lower Hunt Fork Shale are also common in bathyal deposits (Chamberlain, 1978), the maximum depth of deposition cannot be constrained.

Kayak Shale and Kekiktuk Conglomerate

The Mississippian Kayak Shale is exposed beneath the Endicott Mountains allochthon along the flanks of the Doonerak duplex (Fig. 3, Mk), at the top of the southern thrust nappe of the Endicott Mountains allochthon (Fig. 3), and in several thrust sheets in the northern part of the allochthon. Balanced cross sections through the central Brooks Range (Oldow et al., 1987; Handschy, this volume, Chapter 3) indicate that the original distribution of the Kayak Shale was such that the section in the northeastern Doonerak window was located approximately

J. W. Handschy

TABLE 1. FOSSIL ASSEMBLAGES IN THE SOUTHERN ENDICOTT MOUNTAINS ALLOCHTHON*

Location	Unit	Brachiopods	Corals	Bryozoans	Miscellaneous	Trace Fossils
1	Dbl		*Syringopora* sp. *Hexigoniaria* sp. (?) *Tabulophyllum* sp. (?) Unidentified zaphren-thids (colonial and solitary)	*Fenestrellina* sp. (?)		
2	Dbl		*Syringopora* sp. Unidentified zaphren-thids (colonial and solitary)			
3	Dbl		*Syringopora* sp.		Stromatoporoids (?) Crinoids	
4	Db	Unidentified spiriferid fragments				
5	Db	Unidentified spiriferid fragments				
6	Mk				Crinoid fragments	
7	Dhfl					*Planolites*
8	Dhfl					*Chondrites*
9	Dhfm		Unidentified rugose coral fragments			
10	Dhfm					*Skolithos* Ichnofacies
11	Dhfm	Unidentified spiriferid fragments	*Favosites* sp.			
12	Dhfm	*Cyrtospirifer* sp. *Eleutherokomma* sp. (?) *Atrypa* sp.				*Chondrites*
13	Dhfm				*Manticoceras* sp. (?) Ammonoids	
14	Dhfm					Plant impressions
15	Dhfu	*Centronella* sp. (?) and/or *Gypidula* sp. (?), terebratulids				
16	Dhfm	*Regelia* sp.	*Coenites* sp.			
17	Dhfu	Unidentified terebrat-ulid fragments			Algal rip-up clasts	
18	Dhfu					*Skolithos* Ichnofacies
19	Dhfm	Unidentified spiriferid				
20	Mk		Unidentified rugose corals		Crinold fragments	
21	Dhfm	Unidentified spiriferid fragments	*Favosites* sp.	*Fenestrellina* sp.	Algal rip-up clasts and oncolites	
22	Dhfu					*Skolithos* Ichnofacies
23	Dhfu					*Cruziana* Ichnofacies
24	Dhfm	Unidentified spiriferid fragments				
25	Dhfu					*Skolithos* Ichnofacies

TABLE 1. FOSSIL ASSEMBLAGES IN THE SOUTHERN ENDICOTT MOUNTAINS ALLOCHTHON* (continued - page 2)

Location	Unit	Brachiopods	Corals	Bryozoans	Miscellaneous	Trace Fossils
26	Dhfu					*Skolithos* Ichnofacies
27	Dhfu					*Skolithos* Ichnofacies
28	Dhfm	*Theodossia* sp. (?) *Productella* sp. (?) fragments, *Atrypa* sp.				*Cruziana* Ichnofacies
29	Dhfm					*Chondrites*
30	Dhfm	*Eleutherokomma* sp. *Cyrtospirifer* sp. (?)	*Coenites* sp. *Favosites* sp.	*Berenicea* sp. *Fenestrella* sp.	Algal rip-up clasts and oncolites Crinold fragments Planispiral gastropod fragments	
31	Dhfm	*Atrypa* sp. (?) fragments *Schizophoria* sp. fragments				
32	Dhfm	*Atrypa* sp. *Cyrtospirifer* sp.	*Favosites* sp. *Coenites* sp.	*Acanthoclema* sp.	Algal rip-up clasts and oncolites	
33	Dhfm	*Cyrtospirifer* sp.	*Favosites* sp.		Algal rip-up clasts and oncolites Crinold fragments	
34	Dhfm					Plant impressions
35	Dhfm	*Atrypa* sp. *Cyrtospirifer* sp.	*Favosites* sp. *Coenites* sp.	*Acanthoclema* sp.	Algal rip-up clasts and oncolites	
36	Dhfm	*Cyrtospirifer* sp.	*Favosites* sp.		Algal rip-up clasts Planispiral and turreted gastropods	
37	Dhfm	*Atrypa* sp. *Eleutherokomma* sp.	*Favosites* sp.	*Fenestrellina* sp.	Turreted gastropods Oncolites	
38	Dhfm	*Atrypa* sp. *Cyrtospirifer* sp.	*Favosites* sp.		Algal rip-up clasts and oncolites	
39	Dhfm					*Skolithos* Ichnofacies
40	Dhfm	Unidentified Spiriferid fragments	*Coenites* sp			
41	Dhfm					*Skolithos* Ichnofacies
42	Dhfm					*Chondrites*
43	Dhfm					
44	Dhfm					*Cruziana* Ichnofacies
45	Dhfm	Unidentified Spiriferid fragments				*Planolites*

*Location numbers correspond to Figure 3a. All fossils in this table were collected and identified by the author. Three locations (1, 9, and 40) are near fossil localities identified by other authors. Location 1 is near Middle to early Late Devonian Phacelloid rugose coral locality of Dillon et al., 1987a, location 24. Locations 9 and 40 are near Fransnian (early Late Devonian) brachiopod and coral localities of Brosgé et al., 1979, locations 147, 148, and 163. References used for fossil identification: Moore, 1953, 1956, 1965; Moore et al., 1952; Shimer and Shrock, 1944; Tasch, 1980; Warren and Stelck, 1956.

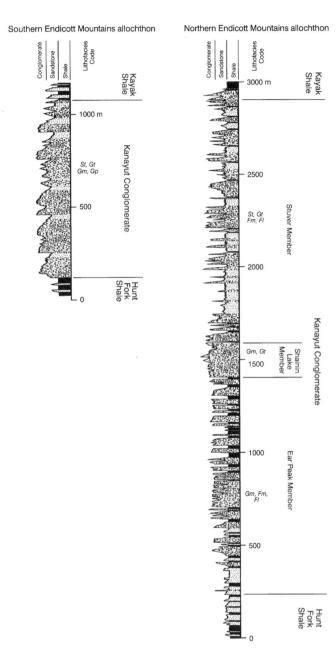

Figure 5. Stratigraphy of the Kanayut Conglomerate. Southern Endicott Mountains allochthon section after Handschy (1988). Northern Endicott Mountains allochthon section modified from Moore et al. (1989). Lithofacies are defined in Table 2. Southern Endicott Mountains allochthon section location shown on Figure 3a (D).

150 km north of the southern Endicott Mountains allochthon section. Outcrops in the northern imbricates of the allochthon restore to intermediate positions. Despite this initial paleogeographic separation, Kayak Shale lithofacies in the Doonerak window and the southern Endicott Mountains are very similar.

In the northeastern Doonerak window, the Kayak Shale is gradationally underlain by the Kekiktuk Conglomerate and is overlain by the Lisburne Group (Fig. 9). The age of the Kayak

Shale is constrained by early Visean (Early Mississippian) microfossils at the base of the Lisburne Group on Mount Doonerak, 30 km west of the study area, and by middle Tournaisian (Early Mississippian) microfossils 1 m above the base of the Kayak Shale exposed along Amawk Creek, 15 km west of the study area (Armstrong et al., 1976).

In the section measured along Kuyuktuvuk Creek (Figs. 2 and 8), disharmonic folds across the contact between the Kekiktuk Conglomerate and phyllites of the underlying Apoon indicate that the two units are locally separated by a detachment surface. The age of the Apoon assemblage is inferred to be Cambrian through Silurian on the basis of Middle Cambrian trilobites (Dutro et al., 1984; Palmer et al., 1984) and Ordovician through Silurian graptolites and conodonts (Repetski et al., 1987).

Massive to trough cross-stratified chert pebble conglomerate at the base of the Kekiktuk grades upward into massive, moderate- to well-sorted, coarse- to fine-grained quartz arenite immediately below the Kayak Shale (Fig. 9). The basal part of the Kayak Shale is composed of interlayered sandstone, siltstone, and mottled shale (Phelps, 1987). Above this basal unit, the Kayak Shale is dominated by black shale. The transition into the overlying Lisburne Group is marked by the appearance of crinoidal limestone beds approximately 30 m above the uppermost Kekiktuk sandstone bed.

The Kayak Shale in the southern Endicott Mountains allochthon differs from that of the northeastern Doonerak window because it overlies the Upper Devonian and Lower Mississippian(?) Kanayut Conglomerate instead of the Mississippian Kekiktuk Conglomerate and lower Paleozoic Apoon assemblage (Fig. 9). Famennian (late Late Devonian) brachiopods and Early Mississippian(?) plant fossils in the upper Kanayut Conglomerate (Moore and Nilsen, 1984; Moore et al., 1989), and Tournaisian foraminifera in the basal Lisburne Group (Armstrong et al., 1970) restrict the age of the Kayak Shale in the Endicott Mountains allochthon to Tournaisian.

At the base of the Kayak Shale in the southern Endicott Mountains allochthon a 2-m-thick interval of massive, horizontally laminated and ripple cross-laminated, well-sorted, fine-grained quartz arenite depositionally overlies a thin (<10 cm) pebble lag that appears to be reworked from the underlying Kanayut Conglomerate (Fig. 9). This quartz arenite interfingers with thinly laminated, bioturbated black shale. The highest sandstone bed (>1 cm) is located approximately 4.5 m above the top of the Kanayut Conglomerate. Black shale dominates the next 22.5 m, with thin (<1 cm) siltstone and sandstone laminations comprising less than 5% of the interval. Twenty-seven meters above the top of the Kanayut Conglomerate, black shale is interbedded with a 7.5-m-thick interval of medium to dark gray argillaceous limestone containing abundant crinoid fragments and rugose corals. These limestone beds are, in turn, overlain by 5.0 m of thinly laminated black shale (Fig. 9). The uppermost Kayak Shale is missing due to Holocene erosion.

The Kekiktuk Conglomerate in the northeastern Doonerak window, pebble conglomerate and sandstone beds in the lower

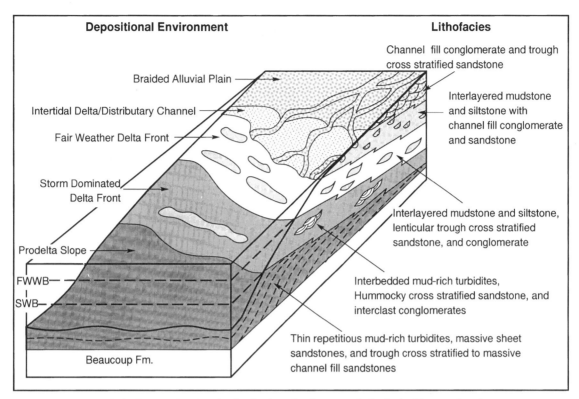

Figure 6. Block diagram showing the distribution of sedimentary facies in the Kanayut Conglomerate–Hunt Fork Shale delta system.

Kayak Shale near Galbraith Lakes (Moore et al., 1989), and the quartz arenite at the base of the Kayak Shale in the southern Endicott Mountains allochthon appear to be near shore transgressive deposits. All three contain poorly sorted, trough cross-stratified fluvial and paralic(?) conglomerate beds that are overlain by thin, horizontally laminated to ripple cross-laminated, well-sorted, quartz-rich sandstone. The sandstone, in turn, grades upward into shallow-marine shale. The disappearance of sandstone interbeds in the middle Kayak Shale indicates continued deepening and a transition to offshore deposition. Cyclical shale and limestone interbeds in the upper Kayak Shale and lower Lisburne Group indicate changes in the relative amounts of carbonate production and terrigenous sediment influx, and may reflect sea-level fluctuations or variations in clastic sediment influx.

SEQUENCE STRATIGRAPHY

In the southeastern part of the study area, east of the Dalton Highway, prodelta turbidites of the lower Hunt Fork Shale overlie fossiliferous carbonate buildups in the upper Beaucoup Formation (Fig. 4). Since the lower Hunt Fork Shale was deposited below the storm wave base and rugose coral–tabulate coral–stromatoporoid reefs grew within the influence of surface waves (Tasch, 1980, p. 181), the contact between the Beaucoup Formation and the Hunt Fork Shale records an increase in water depth. If the base of the Hunt Fork Shale is also an uncon-

formity, as postulated by Tailleur et al. (1967), the two formations were deposited in temporally separate depositional systems. However, if the Hunt Fork Shale–Beaucoup Formation contact is conformable (Dutro et al., 1979), it simply reflects a relative rise in sea level. Phelps (1987) and Handschy (1988) presented a scenario in which the Beaucoup Formation is a transitional facies between the Hunt Fork Shale and the upper Skajit Limestone/Whiteface Mountain volcanic assemblage. In their interpretation, quartz- and chert-rich conglomerate, sandstone, and shale in the Beaucoup Formation are inferred to be distal Hunt Fork Shale equivalents that interfinger with carbonates in the Skajit Limestone and volcaniclastic rocks of the Whiteface Mountain volcanics.

Neither field relations nor paleontological control are adequate to conclusively resolve these alternative depositional models. Dutro et al. (1979) report Frasnian ages for both the Beaucoup Formation and lower Hunt Fork Shale, and infer an early Frasnian age for the Beaucoup Formation on the basis of *Warrenella* species brachiopods. Thus, if an unconformity is present, the hiatus it represents is relatively short (<7 Ma). Similarly, the exact stratigraphic relationships between the Beaucoup Formation, Whiteface Mountain volcanics, and Skajit Limestone are unclear. The Beaucoup Formation and the Whiteface Mountain volcanics have yielded Frasnian fossils (Dillon, 1989), whereas both the Whiteface Mountain volcanics and the Skajit Limestone have yielded Middle Devonian ages (Dillon et al., 1987a; Dillon, 1989).

J. W. Handschy

TABLE 2. SUMMARY OF KANAYUT CONGLOMERATE LITHOFACIES

Facies Code	Lithofacies	Sedimentary Structures	Interpretation
Gm (organized)	Clast supported, poorly sorted, pebble-cobble conglomerate	Massive	Longitudinal gravel bars
Gt	Clast supported, poorly sorted, pebble-cobble conglomerate	Trough x-strata	Large 3D ripples and scour fills
Gp	Clast supported, poorly sorted, pebble-cobble conglomerate	Planar x-strata	Large 2D ripples and late-stage modification of gravel bars
St	Poorly to moderately sorted sandstone and pebbly sandstone	Trough x-strata	Large 3D ripples
Sp	Poorly to moderately sorted sandstone and pebbly sandstone	Planar x-strata	2D ripples and lateral accretion on bar margins during waning flow
Sh	Moderately sorted sandstone	Horizontal lamination	Planar bed flow
Fm	Mudstone	Massive	Overbank and waning flow; sometimes pedogenically altered
Fl	Mudstone	Horizontal lamination	Overbank and waning flow

*Modified from Miall, 1981, and Graham et al., 1986.

The influx of coarse- and fine-grained terrigenous clastic rocks during deposition of the Kanayut Conglomerate–Hunt Fork Shale delta system requires uplift of a northern source area and suggests that deepening across the Beaucoup Formation–Hunt Fork Shale contact may be due to subsidence. Alternatively, Devonian sea-level curves (Johnson et al., 1985) show a general transgression in the Frasnian that could also account for the increase in water depth across the Beaucoup Formation–Hunt Fork Shale contact (Fig. 10). As before, field relations and paleontology do not adequately resolve these options, but, if the Beaucoup Formation and Hunt Fork Shale are separated by an unconformity, the apparent absence of falls in eustatic sea level during the Frasnian (Fig. 10; Johnson et al., 1985) requires a tectonic origin for the unconformity.

Prograding and retrograding sequences in the Kanayut Conglomerate (Moore and Nilsen, 1984) follow the overall Famennian sea-level pattern of regression (progradation) in the lower part of the sequence and transgression (retrogradation) in the upper part of the sequence (Fig. 10). They do not, however, reflect the high-frequency fluctuations in the Famennian sea-level curve (Fig. 10). The biggest discrepancy between the sea-level curves and depositional patterns in the Kanayut Conglomerate–Hunt Fork Shale occurs in the latest Famennian and Early Mississippian. Nilsen and Moore (1984) reported a

possible Early Mississippian age for the uppermost Kanayut Conglomerate and show the retrograde cycle in the upper Kanayut Conglomerate to be continuous into the Kayak Shale, whereas the sea-level curves indicate a major regression in the latest Famennian (Fig. 10). This regression should have been accompanied by progradation of paralic and fluvial facies. Continuous retrogradation in the upper Kanayut Conglomerate may indicate that the rate of basin subsidence was equal to or greater than the rate of sea-level fall, that the sediment supply was decreasing faster than sea level was falling, or that the upper Kanayut Conglomerate is Famennian in age and separated from the Kayak Shale by a disconformity. Because the youngest fossils collected from the Hunt Fork Shale and Noatak Sandstone are Famennian in age (Nilsen and Moore, 1984; Brosgé et al., 1988), available fossil age control cannot resolve these options.

The implications of these alternative depositional models are important for interpreting the significance of tectonism during the latest Devonian. If the Kanayut Conglomerate–Hunt Fork Shale basin subsided at a rate equal to, or faster than, sea level was falling, then the basin must have been tectonically active during the latest Devonian. Conversely, decreasing sediment influx could indicate that topography, and possibly tectonic activity, in the source terrane was diminishing.

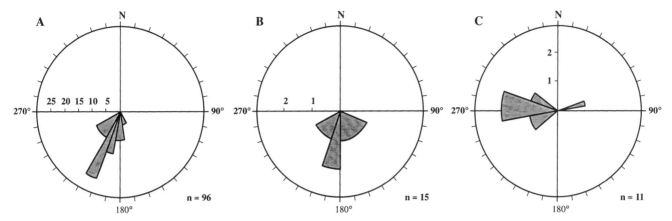

Figure 7. Paleocurrent measurements in the southern Endicott Mountains allochthon. A, Kanayut Conglomerate; B, upper Hunt Fork Shale channelized sandstone and conglomerate; C, middle Hunt Fork Shale lenticular sand bodies.

Younger-to-the-north onlap of the Kekiktuk Conglomerate and Kayak Shale over the Kanayut Conglomerate and Apoon assemblage reflects a marine transgression during the Early Mississippian (Fig. 10). Subsequent transgressive-regressive cycles in the upper Kayak Shale and in the Lisburne Group have been tentatively correlated to Mississippian and Pennsylvanian sea-level fluctuations (Carlson and Watts, 1987; Imm, 1989; LePain et al., 1990), which suggests that the Lisburne carbonate platform was deposited in a relatively quiet tectonic setting.

PROVENANCE

Chert is the principal conglomerate clast lithology in all of the formations in the Endicott Mountains allochthon. Pebbles in the Kekiktuk Conglomerate are composed of white and gray chert, mudstone, and milky quartz. Cobbles and pebbles in the Kanayut Conglomerate include gray, white, green, red, and black chert, milky and clear quartz, quartzite, mudstone, argillite, slate, and, rarely, granite. Channel fill conglomerate in the Hunt Fork Shale contains mudstone rip-up intraclasts, and black, gray, white, and green chert clasts. Conglomerate in the Beaucoup Formation is chert and quartz rich, carbonate rich, or volcanic rich. Chert pebble conglomerate and volcaniclastic conglomerate are occasionally interbedded, but chert and volcanic clasts are never mixed in the same channels.

The overwhelming abundance of chert, the most resistant lithology present, in most Kanayut Conglomerate, Hunt Fork Shale, and Beaucoup Formation conglomerates makes it difficult to evaluate the compositions of the respective source terranes on the basis of conglomerate clast composition alone. Provenance studies in other coarse-grained clastic wedges have shown that conglomerate clast compositions strongly reflect clast durability (Graham et al., 1986) and that accurate provenance analysis of conglomerate should include petrography of finer grained components that contain the more labile lithologies (Ingersoll et al., 1987).

One significant problem for interpreting detrital modes in sandstone is destruction of labile grains during reworking in high-energy, especially near-shore, environments (Mack, 1978). Because they were deposited in relatively high energy transgressive environments, quartz arenites and quartz-chert arenites in the Kekiktuk Conglomerate and lower Kayak Shale provide little information about the source of the detritus. Sandstone in the Kanayut Conglomerate, Hunt Fork Shale, and Beaucoup Formation are compositionally much less mature than sandstone in the Kekiktuk and Kayak Shale and do not appear to be significantly effected by reworking in high-energy environments.

Sandstone compositions in the Kanayut Conglomerate and Hunt Fork Shale reflect a mixed sedimentary, low-grade metamorphic and igneous source terrane. Quartz is the most common constituent in both units (Fig. 11). Chert and silicified argillite are the most common lithic fragments. Low-grade metamorphic rock fragments usually comprise less than 5% of the total lithics (Fig. 11), but quartz instability indices (Young, 1976; Basu et al., 1975) for undeformed medium-grained sandstone samples from the upper Kanayut Conglomerate suggest that 10 to 20% of the quartz grains also had a low-grade metamorphic source (Handschy, 1988). Metamorphic rock fragments include quartz-muscovite schist, ribbon quartz, and slate. Chert petrography indicates that 20 to 50% of the chert in the Kanayut Conglomerate is siliceous argillite or siliceous mudstone. The bulk of the chert is pure SiO_2 or contains scattered clay inclusions, suggesting deposition from a siliceous ooze or diagenetic precipitation from a nonargillaceous protolith. Many chert clasts also contain radiolarians (Anderson, 1987). Plagioclase is the most common feldspar, with trace amounts (<1%) of perthite, orthoclase, and microcline present in some samples (Handschy, 1988). Granitic rock fragments, mafic igneous rock fragments, and volcanic quartz are present, but rare (<1%), in samples from the northern Endicott Mountains allochthon.

The provenance of Beaucoup Formation sandstones is much more complex. Some Beaucoup Formation sandstone are

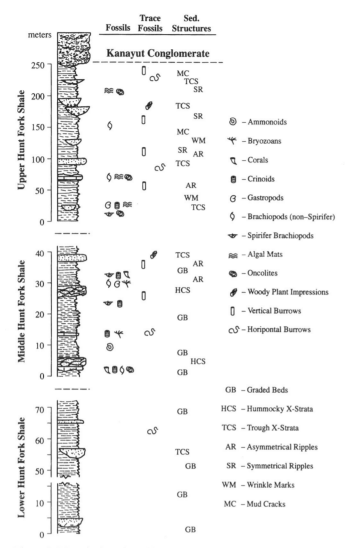

Figure 8. Measured stratigraphic sections in the upper, middle, and lower Hunt Fork Shale. Thicknesses are true stratigraphic thickness of undeformed sections. Due to internal deformation it is impossible to measure the true stratigraphic thickness of the Hunt Fork Shale. These sections illustrate sedimentary features of the different depositional facies in the Hunt Fork Shale. Section locations shown on Figure 3a (E, F, C).

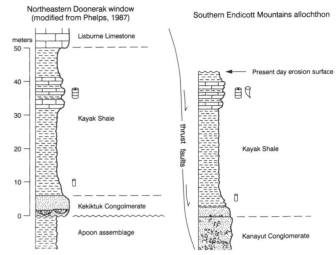

Figure 9. Measured stratigraphic sections in the Kayak Shale and adjacent formations. Northeastern Doonerak window section after Phelps (1987); southern Endicott Mountains allochthon section from Handschy (1988). Sections location shown on Figure 3a, where the stratigraphic section in the northeastern Doonerak window is labeled B and the section in the southern Endicott Mountains allochton is labeled D.

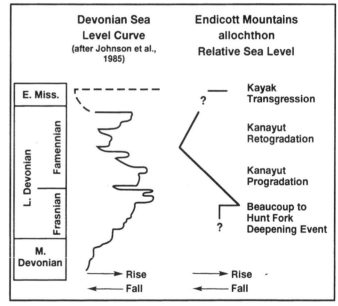

Figure 10. Eustatic sea level during the Late Devonian and earliest Mississippian after Johnson et al. (1985) and Krebs (1979) plus relative sea-level fluctuations in the Endicott Mountains allochthon interpreted from facies variations.

petrographically indistinguishable from Hunt Fork Shale and Kanayut Conglomerate sandstone, whereas others have an exclusively volcanic source (Fig. 11). Quartz and chert are the most common detrital grains in nonvolcanic arenites and wackestones. Small percentages of plagioclase, biotite, muscovite, mudrock fragments, metamorphic rock fragments, carbonate rock fragments, and volcanic rock fragments are also present in many Beaucoup Formation wackestones (Handschy, 1988). The mixture of compositionally distinct sandstone and conglomerate indicates that the clastic rocks in the Beaucoup Formation were derived from multiple sources.

Although no paleocurrent data are available, compositional similarities between some of the Beaucoup Formation, Hunt Fork Shale, and Kanayut Conglomerate sandstone suggest that

quartz- and chert-rich clastic rocks in the Beaucoup Formation were sourced in the north. Similarly, the increasing abundance of volcaniclastic rocks in the Beaucoup Formation and equivalent units southwest of the study area, and in the superjacent Skajit allochthon suggests that the volcanic source may have been to the south or southwest (Dillon, 1989). Mixing of volcaniclastic and siliciclastic grains in wackestones and mud-

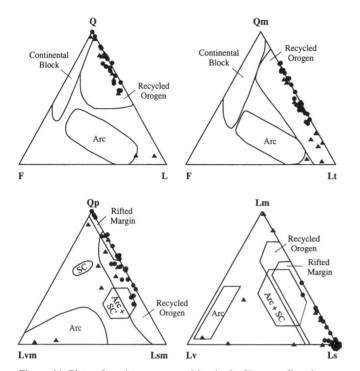

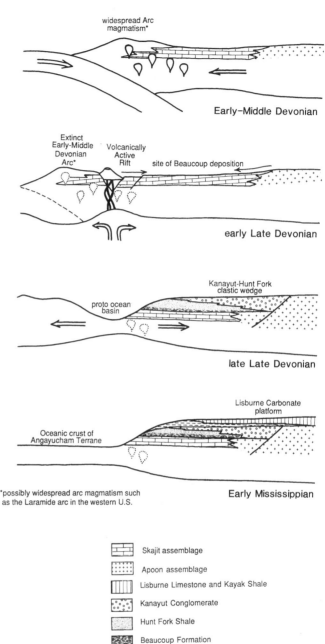

Figure 11. Plots of sandstone composition in the Kanayut Conglomerate (solid circles; 13 samples), Hunt Fork Shale (open circles; 23 samples), and Beaucoup Formation (solid triangles; 10 samples). Grain-type parameters are from Ingersoll et al. (1987). Tectonic fields are from Dickinson and Suczek (1979) and Ingersoll and Suczek (1979). Arc indicates Magmatic Arc Complex and SC denotes Subduction Complex tectonic fields. QFL plot: Q = total quartz grains, F = total feldspar grains, L = total lithic grains excluding polycrystalline quartz and chert. QmFLt plot: Qm = monocrystalline quartz grains, Lt = total lithic grains (including polycrystalline quartz and chert). LmLvLs plot: Lm = metamorphic rock fragments, Lv = volcanic rock fragments, Ls = sedimentary rock fragments. QpLvmLsm plot: Qp = polycrystalline quartz and chert, Lvm = Lv + metavolcanic rock fragments, Lsm = Ls + metasedimentary rock fragments.

*possibly widespread arc magmatism such as the Laramide arc in the western U.S.

Skajit assemblage
Apoon assemblage
Lisburne Limestone and Kayak Shale
Kanayut Conglomerate
Hunt Fork Shale
Beaucoup Formation

Figure 12. Conceptual paleogeographic cross sections for Late Devonian through Early Mississippian sedimentation in northern Alaska.

stone of the Beaucoup Formation indicate that material from both sources was entering the basin simultaneously.

Mixing of detritus from volcanic and siliciclastic sources could have occurred in several tectonic settings, such as a back-arc basin, a strike-slip system that juxtaposed different source terranes, a rift setting that exposed an older volcanic arc, a volcanically active rift basin, or a volcanically active strike-slip system. Volcanic flows and pillow basalts in the Beaucoup Formation in the southern Phillip Smith Mountains Quadrangle (Dutro et al., 1979), first-cycle volcaniclastic rocks in Upper Devonian mixed carbonate and clastic rocks in the Skajit allochthon (Hammond terrane; Anderson, 1987), and volcanic flows in the Skajit allochthon and the Schist belt (Dillon et al., 1986, 1987b) provide evidence for active volcanism within the depositional basin, eliminating tectonic settings that require reworking of material from older volcanic terranes. This restricts the probable tectonic setting to a back-arc basin, an active rift, or a volcanically active strike-slip fault system (Fig. 12). Uncertainties about the depositional setting of the

Skajit carbonate platform (passive margin versus intracratonic basin) and the tectonic affinities of the volcanics (arc versus intraplate) make it impossible to definitively distinguish between these options.

Middle and Upper Devonian volcaniclastic rocks (Whiteface Mountain volcanics) in the Skajit allochthon are partially age equivalent and compositionally similar to volcaniclastic rocks in the Beaucoup Formation (Dillon et al., 1987a; Dillon, 1989), suggesting that parts of the Skajit allochthon were laterally equivalent to the Beaucoup Formation prior to Brookian thrusting (Oldow et al., this volume, Chapter 7). Peraluminous two-mica granitic plutons in the Schist belt (Dillon et al., 1987b; Dillon, 1989), which restore to positions south of or below the Skajit allochthon (Oldow et al., 1987), have yielded late Early to early Late Devonian zircons (Dillon et al., 1980, 1987b). If these plutons are the intrusive equivalents of volcanics in the Beaucoup Formation and the Whiteface Mountain volcanics in the Skajit allochthon, as postulated by Dillon (1989; Dillon et al., 1987b), then they may be remnants of a marginal volcanic arc. If, however, the volcanics in the Beaucoup Formation are not the extrusive equivalents of the plutons, then the two igneous systems may represent different tectonic settings and the volcanics could be rift related. An intracratonic rift setting seems unlikely for the peraluminous Devonian plutons in the southern Brooks Range because igneous activity in continental rifts is dominantly alkaline or peralkaline (Barberi et al., 1982). Mafic igneous rocks reported by Dillon et al. (1986) could occur in either a back-arc or intracratonic rift setting.

PALEOGEOGRAPHIC IMPLICATIONS

Since the northern margin of the Kanayut Conglomerate–Hunt Fork Shale basin is not exposed, it is difficult to assess the width or orientation (in present coordinates) of the basin. The present-day east-west distribution of the Kanayut Conglomerate and Hunt Fork Shale along the crest of the Brooks Range suggests that the Late Devonian through earliest Mississippian shoreline may have had a general east-west strike (Moore and Nilsen, 1984), but these units have been displaced a minimum of 150 km to the north during late Mesozoic and Cenozoic (Brookian) thrusting (Handschy, this volume, Chapter 3; Oldow et al., 1987; Mull et al., 1987) and may not accurately reflect the prethrust orientation of the basin. Because of the lateral continuity and constant strike of the Kanayut Conglomerate outcrops along the crest of the Brooks Range, it is assumed that thrust sheets were not rotated significantly during emplacement. However, regionally consistent south-southwest–directed paleocurrents and a southwestward decrease in clast size in the Kanayut Conglomerate (Moore and Nilsen, 1984) suggest that the margin of Kanayut Conglomerate–Hunt Fork Shale basin had a west-northwest strike instead of an east-west strike (Fig. 13). If the Late Devonian and earliest Mississippian shore line did trend west-northwest, then the paucity of Kanayut Conglomerate in the western Brooks Range may be the product of oblique truncation of facies boundaries in the Kanayut Conglomerate–Hunt Fork Shale system by east-west–striking Brookian thrusts (Fig. 13), not the product of westward-decreasing relief in the source area as implied by Moore and Nilsen (1984).

The absence of the Beaucoup Formation, Hunt Fork Shale, and Kanayut Conglomerate between the Apoon assemblage and the Kekiktuk–Kayak Shale transgressional sequence along the margins of the Doonerak window indicates that the Kanayut Conglomerate–Hunt Fork Shale clastic wedge has been thrust past the northern margin of its own depositional basin (Handschy,

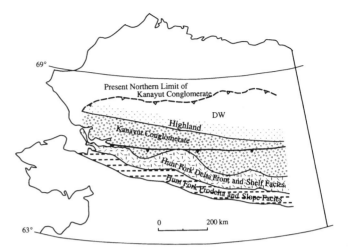

Figure 13. Conceptual paleogeographic map of the distribution of facies in the Kanayut Conglomerate and Hunt Fork Shale (modified from Moore and Nilsen, 1984). The present (allochthonous) northern edge of the Kanayut Conglomerate is shown as a dashed thrust fault. The postulated pre-thrust location of the same Kanayut Conglomerate beds are shown as a solid thrust fault. The location of the Doonerak window (DW) is shown for reference. Note that the restored positions of facies boundaries do not account for an unknown amount of shortening in the Apoon assemblage.

this volume, Chapter 3; Oldow et al., 1987; Mull et al., 1987). In addition, the dearth of proximal alluvial fan environments in the Kanayut Conglomerate of the central Brooks Range suggests that the most proximal part of the Kanayut Conglomerate–Hunt Fork Shale system was left on the lower plate during thrusting. Compositional plots for the Kanayut Conglomerate and Hunt Fork Shale indicate that these units were derived from a recycled orogenic belt (Fig. 11). The exact tectonic setting of deposition is less certain, however, since the age of the orogenic belt is unknown. The classical interpretation is that it that the Kanayut Conglomerate and Hunt Fork Shale were deposited in a foredeep along the southern edge of a fold-thrust belt (Nilsen, 1981). The primary evidence cited for a foredeep setting includes:

1. the lateral continuity of the Kanayut Conglomerate;

2. the distribution of sedimentary facies (alluvial in the north and marine in the south);

3. the mixed igneous, metamorphic, and sedimentary provenance of the Kanayut Conglomerate (Nilsen et al., 1981a);

4. presence of deep marine chert in the Kanayut Conglomerate source terrane; and

5. the presence of pre-Mississippian folds in the lower Paleozoic rocks exposed in the Doonerak window (Dutro et al., 1976).

Of these criteria, only the existence of pre-Mississippian folds in the Doonerak window would provide direct evidence for a Late(?) Devonian contractional orogenic belt in the central Brooks Range. Structural analysis in the Doonerak window, however, has shown that all of the fold phases in the lower Paleozoic Apoon assemblage are related to late Mesozoic and Cenozoic (Brookian) orogenesis (Julian and Oldow, this volume, Chapter 5). This presents a significant problem for the foredeep model because it requires transport of the coarse clastics in the Kanayut Conglomerate from the fold-thrust belt, across the undeformed Apoon assemblage, to a distant foreland basin.

Foreland basins associated with both modern (Andes, Himalayas, Taiwan, etc.) and ancient (Alps, Appalachian-Ouachita, Canadian Rocky Mountains, etc.) fold-thrust belts developed immediately adjacent to the fold-thrust belt. In addition, syntectonic sediments deposited in a foreland basin are often involved in deformation along the leading edge of the fold-thrust belt. Thus, separation of the foreland basin from the fold-thrust belt by an undeformed region, as required for the Kanayut Conglomerate–Hunt Fork Shale system, is atypical of foreland basins. From an analytical perspective, separation of a fold-thrust belt from its associated foreland is impossible because foreland basins are the product of thrust sheet and syntectonic sediment loading and must be located immediately adjacent to the fold-thrust belt (Beaumont et al., 1982).

A second depositional setting compatible with the distribution of facies and along-strike continuity of the Kanayut Conglomerate and Hunt Fork Shale is a rift setting. Rift systems (e.g., East Africa, Rio Grande, Rhine Graben) are often continuous for hundreds to thousands of kilometers and filled with thick clastic wedges (Ramberg and Morgan, 1984). Clastics that fill rift basins are derived from rocks exposed during extensional faulting and, as such, their compositions often plot in tectonic fields (as defined by Dickinson and Suczek, 1979; and Ingersoll and Suczek, 1979) that do not accurately represent the tectonic setting of the depositional basin (Mack, 1984). Because the age of the orogenic belt recycled into the Kanayut Conglomerate–Hunt Fork Shale delta is unknown, it is possible that the clastic rocks in these units were derived from an exhumed pre-Late Devonian orogenic belt instead of a syndepositional orogenic belt. The topography required to generate cobbles in the Kanayut Conglomerate could have been produced by down-to-the-south normal faults along the northern margin of the rift basin. Similar down-to-the-south normal faults, which bound pre-Mississippian or Early Mississippian extensional basins beneath the North Slope, are visible on seismic lines from the National Petroleum Reserve (Mauch, 1985; Oldow et al., 1987; Bird and Molenaar, 1992).

As rift-fill sediments are derived primarily from erosion of uplifted areas along the margins of the rift basin, an extensional setting for Kanayut Conglomerate–Hunt Fork Shale deposition does not require coarse clastics to be transported from a northern fold-thrust belt, past the undeformed Apoon assemblage, to a southern foreland basin. Extensional disaggregation of an older fold-thrust belt could easily explain why Kanayut Conglomerate and Hunt Fork Shale sandstones plot in the recycled orogen field (Fig. 11), and it does not require pre-Mississippian deformation of the Apoon assemblage.

The distribution of facies in the Kanayut Conglomerate and

Hunt Fork Shale (Fig. 6), the across-strike width of the delta system (>100 km), and regionally consistent paleocurrents in the Kanayut Conglomerate (Moore and Nilsen, 1984) are fairly typical of both foredeep clastic wedges and rift-phase/early drift-phase continental margin clastic wedges. They are atypical of narrower continental rifts in which flow is perpendicular to the rift axis near the margins of the basin, but parallel to the rift axis in the center of the basin (Potter, 1978). Deposition in a rifted continental margin setting is consistent with all of the data currently available (Handschy, 1988; Anderson and Wallace, 1991; Anderson et al., 1992, 1993, 1994).

Stratigraphic relations in the Lisburne Group and obducted Mississippian through Jurassic oceanic crust in the southern Brooks Range (Patton, 1973; Patton and Box, 1989; Roeder and Mull, 1978; Silberling and Jones, 1984; Harris, 1992) suggest that the rift-phase margin, on which the Kanayut Conglomerate–Hunt Fork Shale delta system was deposited, had evolved into a drift-phase margin by the middle Mississippian. The spatial extent (300+ km wide and 1,000+ km long) and distribution of sedimentary facies (restricted environment in the north and open marine in the south; Armstrong, 1974; Imm, 1989) in the Early Mississippian through Pennsylvanian Lisburne Group are typical of passive margin carbonate platforms. Progressive, younger-to-the-north onlap of the Kayak Shale and Lisburne Group (Armstrong et al., 1976), and transgressive-regressive cycles in the Lisburne (Carlson and Watts, 1987; Watts et al., 1989; Imm, 1989), also suggest that tectonic influences on sedimentation had diminished by the middle Mississippian. Obducted Mississippian through Jurassic oceanic crust in the Angayucham terrane (Patton and Box, 1989; Barker et al., 1988; Harris, 1992), which restore to positions south of the Skajit allochthon (Oldow et al., 1987), indicate the presence of an ocean basin south of the Lisburne shelf edge from Mississippian to Jurassic (Handschy, this volume, Chapter 1; Churkin et al., 1979; Patton and Box, 1989; Pallister et al., 1989).

At a broader scale, the Paleozoic evolution of depositional systems in the central Brooks Range is similar to other parts of the Cordilleran miogeocline. Cambrian through Devonian carbonate/shale platform deposits, analogous to some of the rocks in the Skajit allochthon, are common along the Cordillera from Mexico to northern Canada (Cook and Bally, 1975). In western Canada, the miogeocline was disrupted by extensional or transtensional faulting and igneous activity during the Late Devonian and Early Mississippian (Abbott, 1986; Gordey et al., 1987). At several localities in the Yukon and British Columbia, chert-rich conglomerate, sandstone, and shale of the Devonian Earn sequence overlie or are intercalated with mafic to felsic igneous rocks and are inferred to represent deposition on a rifted continental margin or in smaller transtensional basins (Gordey et al., 1987; Templeman-Kluit, 1979).

Farther to the south, the miogeocline is disrupted by the Late Devonian and Early Mississippian Antler orogeny (Roberts et al., 1958; Speed and Sleep, 1982), and by early Late Mississippian to Permian extension or transtension (Miller et al., 1984; Little, 1987). Within the Late Devonian through Permian Schoonover sequence of Nevada, volcaniclastic rocks derived from a western source are intercalated with siliciclastic sedimentary rocks derived from the Antler orogenic belt to the east (Miller et al., 1984). Late Mississippian basalt flows in the Schoonover sequences are coeval with basaltic volcanism and subsidence in the "overlap sequence," which unconformably overlies the Roberts Mountain allochthon (Little, 1987), and are interpreted to signal extensional disaggregation of the Antler orogenic belt during the Late Mississippian (Little, 1987; Miller et al., 1984).

Intermediate between the Earn sequence of northwestern Canada and the Schoonover sequence of Nevada, pre-Mississippian deformation in rocks of the Kootenay arc is evidenced by foliated clasts in the Mississippian through Lower Pennsylvanian Milford Group, which unconformably overlies the internally imbricated Lardeau Group (Kelpacki and Wheeler, 1985; Gehrels and Smith, 1987). Gehrels and Smith (1987) have interpreted this angular unconformity and Ordovician through Devonian plutons within the Kootenay arc (Okulitch, 1985) to indicate that the Antler orogenic belt extended at least as far north as southern British Columbia. Stratigraphic similarities between the Milford and the Late Mississippian "overlap sequence" may also indicate that post-Antler extension extended into British Columbia.

Clearly, there are correlation problems between the Beaucoup Formation–Hunt Fork Shale–Kanayut Conglomerate sequence in the Brooks Range, the Earn sequence and Milford Group of western Canada, and the Schoonover sequences of Nevada. First, coarse clastic rocks in the Earn sequence were derived from the west (Gordey et al., 1987), whereas the Kanayut Conglomerate and Hunt Fork Shale were derived from the north and northeast. Second, extension apparently began in the Late Devonian in the Brooks Range, in the latest Devonian or Early Mississippian in northwestern Canada, and in the middle Mississippian in Nevada. Despite these problems, Late Devonian through Early Mississippian sedimentation in the central Brooks Range is similar to Late Devonian through Late Mississippian sedimentation in British Columbia and Nevada in that convergent tectonics (an arc system in the Brooks Range and an arc system plus collisional orogenic belt in British Columbia and Nevada) is apparently followed by extension (Kanayut Conglomerate–Hunt Fork Shale system in the Brooks Range, Schoonover and "overlap" sequences in Nevada, and possibly the Milford Group in British Columbia).

CONCLUSIONS

Upper Devonian and Lower Mississippian clastic rocks in the Endicott Mountains allochthon form three depositional sequences. Each sequence is characterized by specific depositional environments, facies distributions, provenance, and tectonic settings. Mixed carbonate and clastic rocks of the Upper Devonian Beaucoup Formation were deposited in a volcanically active, probably back-arc or rift, basin. Facies distribution and detrital modes suggest that volcaniclastic rocks were derived from the south, and terrigenous clastic rocks were derived from the north (Fig. 12).

The laterally extensive Late Devonian to Early Mississippian(?) Kanayut Conglomerate–Hunt Fork Shale delta system was apparently deposited on a south-facing rifted continental margin (Fig. 12). Southward progradation of the Kanayut Conglomerate over the Hunt Fork Shale produced a lithofacies pattern in which the Kanayut Conglomerate is thickest in the north and thins southward (Fig. 6). Coarse-grained lithofacies in the Kanayut Conglomerate define the Kanayut Conglomerate–Hunt Fork Shale delta as a braid delta. The distribution of sedimentary structures and fossil assemblages in the Hunt Fork Shale document intertidal and storm-influenced portions of the delta front and a turbidite dominated prodelta. Sedimentation rates in the Kanayut Conglomerate–Hunt Fork Shale system apparently overwhelmed high-frequency sea-level cycles in the Late Devonian, but larger eustatic fluctuations are represented by progradational and retrogradational cycles in the Kanayut Conglomerate (Fig. 10).

Transgression of the Kekiktuk Conglomerate and Kayak Shale over the Kanayut Conglomerate–Hunt Fork Shale delta and the lower Paleozoic Apoon assemblage reflects subsidence and/or a sea level rise in the Early Mississippian (Fig. 10). Carbonate facies in the Lisburne Group (Armstrong, 1974) and oceanic crust in the Angayucham terrane suggest that the latest Devonian rifted continental margin had evolved into a passive margin by the middle Mississippian (Fig. 12).

ACKNOWLEDGMENTS

This study was funded by the U.S. Department of Energy (grant DE-AS05-83ER13124), the National Science Foundation (grant EAR-8517384), and the Rice University Alaska Industrial Associates Program (Amoco, Arco, Chevron, Gulf, Mobil, and Sohio). Discussions with J. S. Oldow, J. T. Dillon, W. K. Wallace, H. G. Avé Lallemant, C. L. Hanks, T. E. Moore, R. R. Gottschalk, F. E. Julian, C. M. Seidensticker, J. C. Phelps, and K. W. Boler greatly improved my understanding of Brooks Range geology. R. Guthrie and K. Hewlett assisted with field work. K. Wall, K. Kimbro, and L. A. Beckmann drafted the figures. J. S. Oldow, A. W. Bally, and H. G. Avé Lallemant reviewed an early version of the manuscript. Special thanks to T. E. Moore and R. B. Blodgett for providing thorough, informative reviews that greatly improved this paper.

REFERENCES CITED

Abbott, J. G., 1986, Devonian extension and wrench tectonics near MacMillan Pass, Yukon Territory, Canada, *in* Turner, R. J. W., and Einaudi, M. T., eds., The genesis of stratiform sediment hosted lead and zinc deposits, Conference Proceedings: Stanford, California, Stanford University Press, p. 85–89.

Anderson, A. V., 1987, Provenance and petrofacies of the Endicott and Hammond terranes, Phillip Smith Mountains and Arctic Quadrangles, Brooks Range, Alaska: Geological Society of America Abstracts with Programs, v. 19, p. 354.

Anderson, A. V., and Wallace, W. K., 1991, Middle Devonian to Early Mississippian stratigraphic record of the formation of a passive continental margin in northeastern Alaska: Geological Society of America Abstracts with Programs, v. 23, p. A436.

Anderson, A. V., Wallace, W. K., and Mull, C. G., 1992, Stratigraphic variation across a mid-Paleozoic rift-basin margin, Endicott Group, eastern Brooks Range, Alaska: Geological Society of America Abstracts with Programs, v. 24, p. 3.

Anderson, A. V., Mull, C. G., and Crowder, R. K., 1993, Mississippian terrigenous clastic and volcaniclastic rocks of the Ellesmerian sequence, upper Sheenjek River area, eastern Brooks Range, Alaska, *in* Solie, D. N., and Tannian, F., eds., Short notes on Alaskan geology 1993: Alaska Division of Geological and Geophysical Surveys, Geologic Report 113, p. 1–6.

Anderson, A. V., Wallace, W. K., and Mull, C. G., 1994, Depositional record of a major tectonic transition in northern Alaska: Middle Devonian to Mississippian rift-basin margin deposits, upper Kongakut River region, eastern Brooks Range, Alaska, *in* Thurston, D. K., and Fujita, K., eds., 1992 Proceedings, International Conference on Arctic Margins: Anchorage, Alaska, U.S. Department of the Interior, Minerals Management Service, p. 71–76.

Armstrong, A. K., 1974, Carboniferous carbonate depositional models, preliminary lithofacies and paleotectonic maps, Arctic Alaska: American Association of Petroleum Geologists Bulletin, v. 58, p. 621–645.

Armstrong, A. K., and Mamet, B. L., 1978, Microfacies of the Carboniferous Lisburne Group, Endicott Mountains, Arctic Alaska, *in* Stelck, C. R., and Chatterton, B. D. E., eds., Western and Arctic Canadian biostratigraphy: Geological Association of Canada Special Paper 18, p. 333–362.

Armstrong, A. K., Mamet, B. L., and Dutro, J. T., Jr., 1970, Foraminiferal zonation and carbonate facies of the Mississippian and Pennsylvanian Lisburne Group, central and eastern Brooks Range, Alaska: American Association of Petroleum Geologists Bulletin, v. 54, p. 687–698.

Armstrong, A. K., Mamet, B. L., Brosgé, W. P., and Reiser, H. N., 1976, Carboniferous section and unconformity at Mount Doonerak, Brooks Range, northern Alaska: American Association of Petroleum Geologists Bulletin, v. 60, p. 962–972.

Barberi, F., Santacroce, R., and Varet, J., 1982, Chemical aspects of rift magmatism, *in* Palmason, G., ed., Continental and oceanic rifts: American Geophysical Union Geodynamics Series, v. 8, p. 223–258.

Barker, F., Jones, D. L., Budahn, J. R., and Coney, P. J., 1988, Ocean plateauseamount origin of basaltic rocks, Angayucham terrane, central Alaska: Journal of Geology, v. 96, p. 368–374.

Basu, A., Young, S. W., Suttner, L. J., James, W. C., and Mack, G. H., 1975, Reevaluation of the use of undulatory extinction and polycrystallinity in detrital quartz for provenance studies of sedimentary rocks: Journal of Sedimentary Petrology, v. 45, p. 873–882.

Beaumont, C., Keen, C. E., and Boutilier, R., 1982, A comparison of foreland and rift margin sedimentary basins: Philosophical Transactions of the Royal Society of London, Series A, v. 305, p. 295–317.

Bird, K. J., and Molenaar, C. M., 1992, The North Slope foreland basin, Alaska, *in* Macqueen, R. W., and Leckie, D. A., eds., Foreland basins and fold belts: American Association of Petroleum Geologists Memoir 55, p. 363–393.

Bowsher, A. L., and Dutro, J. T., Jr., 1957, The Paleozoic section in the Shainin Lake area, central Brooks Range, Alaska: U.S. Geological Survey Professional Paper 303-A, p. 1–39.

Brosgé, W. P., Dutro, J. T., Jr., Mangus, M. D., and Reiser, H. N., 1962, Paleozoic sequence in eastern Brooks Range, Alaska: American Association of Petroleum Geologists Bulletin, v. 46, p. 2174–2198.

Brosgé, W. P., Reiser, H. N., Dutro, J. T., Jr., and Detterman, R. L., 1979, Bedrock geologic map of the Phillip Smith Mountains Quadrangle, Alaska: U.S. Geological Survey Miscellaneous Field Studies Map MF-879B, 2 sheets, scale 1:250,000.

Brosgé, W. P., Nilsen, T. H., Moore, T. E., and Dutro, J. T., Jr., 1988, Geology of the Upper Devonian and Lower Mississippian (?) Kanayut Conglomerate in the central and eastern Brooks Range: U.S. Geological Survey Profes-

sional Paper 1399, p. 299–316.

Cant, D. J., 1978, Development of a facies model for sandy braided river sedimentation: Comparison of the south Saskatchewan River and the Battery Point Formation, *in* Miall, A. D., ed., Fluvial sedimentology: Canadian Society of Petroleum Geologists Memoir 5, p. 627–640.

Carlson, R., and Watts, K. F., 1987, Shallowing upward cycles in the Wahoo Limestone, eastern Sadlerochit Mountains, ANWR, NE Brooks Range, Alaska: Geological Society of America Abstracts with Programs, v. 19, p. 364.

Chamberlain, C. K., 1978, Trace fossil ichnofacies of an American flysch, *in* Chamberlain, C. K., ed., A guidebook to the trace fossils and paleoecology of the Ouachita geosyncline: Tulsa, Oklahoma, Society of Economic Paleontologists and Mineralogists, p. 23–37.

Chan, M. A., and Dott, R. H., Jr., 1986, Depositional facies and progradational sequences in Eocene wave-dominated deltaic complexes, southwestern Oregon: American Association of Petroleum Geologists Bulletin, v. 70, p. 415–429.

Chapman, R. M., Detterman, R. L., and Mangus, M. D., 1964, Geology of the Killik-Etivluk Rivers region, Alaska: U.S. Geological Survey Professional Paper 303-F, p. 325–407.

Churkin, M., Jr., Nokleberg, W. J., and Huie, C., 1979, Collision deformed Paleozoic continental margin, western Brooks Range, Alaska: Geology, v. 7, p. 379–383.

Cook, T. D., and Bally, A. W., 1975, Stratigraphic atlas of North and Central America: Princeton, New Jersey, Princeton University Press, 272 p.

Crimes, T. P., 1975, The stratigraphic significance of trace fossils, *in* Frey, R. W., ed., The study of trace fossils: New York, Springer-Verlag, p. 109–130.

Dickinson, W. R., and Suczek, C. A., 1979, Plate tectonics and sandstone composition: American Association of Petroleum Geologists Bulletin, v. 63, p. 2164–2182.

Dillon, J. T., 1989, Structure and stratigraphy of the southern Brooks Range and northern Koyukuk Basin near the Dalton Highway, *in* Mull, C. G., and Adams, K. E., eds., Dalton Highway, Yukon River to Prudhoe Bay, Alaska, Bedrock geology of the eastern Koyukuk Basin, central Brooks Range and east-central Arctic Slope: Alaska Geological and Geophysical Surveys Guidebook 7, p. 157–187.

Dillon, J. T., Pessel, G. H., Chen, J. A., and Veach, N. C., 1980, Middle Paleozoic magmatism and orogenesis in the Brooks Range, Alaska: Geology, v. 8, p. 338–343.

Dillon, J. T., Brosgé, W. P., and Dutro, J. T., Jr., 1986, Generalized geologic map of the Wiseman Quadrangle: U.S. Geological Survey Open-File Report 86-219, 1 sheet, scale 1:250,000.

Dillon, J. T., Harris, A. G., and Dutro, J. T., Jr., 1987a, Preliminary description and correlation of lower Paleozoic fossil-bearing strata in the Snowden Mountain area of the south-central Brooks Range, Alaska, *in* Tailleur, I., and Weimer, P., eds., Alaskan North Slope geology: Society of Economic Paleontologists and Mineralogists, Pacific Section, Publication 50, p. 337–345.

Dillon, J. T., Tilton, G. R., Decker, J., and Kelly, M. J., 1987b, Resource implications of magmatic and metamorphic ages for Devonian igneous rocks in the Brooks Range, *in* Tailleur, I., and Weimer, P., eds., Alaskan North Slope geology: Society of Economic Paleontologists and Mineralogists, Pacific Section, Publication 50, p. 713–723.

Dutro, J. T., Jr., 1952, Stratigraphy and paleontology of the Noatak and associated formations: U.S. Geological Survey Open-File Report 33, 154 p.

Dutro, J. T., Jr., Brosgé, W. P., Lanphere, M. A., and Reiser, H. N., 1976, Geologic significance of the Doonerak structural high, central Brooks Range, Alaska: American Association of Petroleum Geologists Bulletin, v. 60, p. 952–961.

Dutro, J. T., Jr., Brosgé, W. P., Reiser, H. N., and Detterman, R. L., 1979, Beaucoup Formation, a new Upper Devonian stratigraphic unit in the central Brooks Range, northern Alaska: U.S. Geological Survey Bulletin, 1482-A, p. A63–A69.

Dutro, J. T., Jr., Palmer, A. R., Repetski, J. E., and Brosgé, W. P., 1984, Middle Cambrian fossils from the Doonerak anticlinorium, central Brooks Range,

Alaska: Journal of Paleontology, v. 58, p. 1364–1371.

Frey, R. W., and Pemberton, S. G., 1984, Trace fossil facies models, in Walker, R. G., ed., Facies models: Geological Association of Canada, Geoscience Canada Reprint Series 1, p. 189–208.

Galloway, W. E., 1975, Process framework for describing the morphologic and stratigraphic evolution of deltaic systems, in Broussard, M. L., ed., Deltas: Houston, Texas, Houston Geological Society, p. 87–98.

Gehrels, G. E., and Smith, M. T., 1987, "Antler" allochthon in the Kootenay arc?: Geology, v. 15, p. 769–770.

Gordey, S. P., Abbott, J. G., Tempelman-Kluit, D. J., and Gabrielse, H., 1987, "Antler" clastics in the Canadian Cordillera: Geology, v. 15, p. 103–107.

Graham, S. A., Tolson, R. B., DeCelles, P. G., Ingersoll, R. V., Bargar, E., Caldwell, M., Cavazza, W., Edwards, D. P., Follo, M. F., Handschy, J. W., Lemke, L., Moxon, I., Rice, R., Smith, G. A., and White, J., 1986, Lithology of source terranes as a determinant in styles of foreland sedimentation, in Allen, D. A., and Homewood, P., eds., Foreland basins: International Association of Sedimentologists Special Publication 8, p. 425–436.

Handschy, J. W., 1988, Sedimentology and structural geology of the Endicott Mountains allochthon, central Brooks Range, Alaska [Ph.D. thesis]: Houston, Texas, Rice University, 172 p.

Harris, R. A., 1992, Peri-collisional extension and the formation of Oman type ophiolites in the Banda Arc and Brooks Range, in Parson, L. M., Murton, B. J., and Browning, P., eds., Ophiolites and their modern oceanic analogs: Geological Society of London Special Publication 60, p. 301–325.

Hein, F. J., and Walker, R. G., 1977, Bar evolution and development of stratification in the gravelly, braided Kicking Horse River, British Columbia: Canadian Journal of Earth Sciences, v. 14, p. 562–570.

Imm, T. A., 1989, Stratigraphic variations in both north-south and east-west directions of the Lisburne Group carbonate platform, northeastern Brooks Range, Alaska: Geological Society of America Abstracts with Programs, v. 21, p. 96.

Ingersoll, R. V., and Suczek, C. A., 1979, Petrology and provenance of Neogene sand from Nicobar and Bengal fans, DSDP sites 211 and 218: Journal of Sedimentary Petrology, v. 49, p. 1217–1228.

Ingersoll, R. V., Cavazza, W., Graham, S. A., and Indiana University Graduate Field Seminar Participants, 1987, Provenance of impure calclithites in the Laramide foreland of southwestern Montana: Journal of Sedimentary Petrology, v. 57, p. 995–1003.

Johnson, J. G., Klapper, G., and Sandberg, C. A., 1985, Devonian eustatic fluctuations in Euramerica: Geological Society of America Bulletin, v. 96, p. 567–587.

Klepacki, D. W., and Wheeler, J. O., 1985, Stratigraphic and structural relations of the Milford, Kaslo, and Slocan Groups, Goat Range, Lardeau and Nelson map areas, British Columbia: Geological Survey of Canada Paper 85-1, p. 277–286.

Krebs, W., 1979, Devonian basinal facies: Paleontological Association of London Special Papers, no. 23, p. 125–139.

LePain, D. L., Crowder, R. K., and Watts, K. F., 1990, Mississippian clastic to carbonate transition in the northeastern Brooks Range: American Association of Petroleum Geologists Bulletin, v. 74, p. 703.

Little, T. A., 1987, Stratigraphy and structure of metamorphosed upper Paleozoic rocks near Mountain City, Nevada: Geological Society of America Bulletin, v. 98, p. 1–17.

Mack, G. H., 1978, Survivability of labile light-mineral fluvial, eolian, and littoral marine environments: the Permian Cutler and Cedar Mesa Formations: Sedimentology, v. 25, p. 587–604.

Mack, G. H., 1984, Exceptions to the relationship between plate tectonics and sandstone composition: Journal of Sedimentary Petrology, v. 54, p. 212–220.

Mauch, E. A., 1985, A seismic stratigraphic and structural interpretation of the middle Paleozoic Ikpikpuk-Umiat Basin, National Petroleum Reserve, Alaska [M.A. thesis]: Houston, Texas, Rice University, 220 p.

McPherson, J. G., Shanmugam, G., and Moiola, R. J., 1987, Fan-deltas and braid deltas: Varieties of coarse-grained deltas: Geological Society of America Bulletin, v. 99, p. 331–340.

Miall, A. D., 1977, A review of the braided river depositional environment: Earth Science Reviews, v. 13, p. 1–62.

Miall, A. D., 1978, Lithofacies types and vertical profile models in braided rivers: a summary, in Miall, A. D., ed., Fluvial sedimentology: Canadian Society of Petroleum Geologists Memoir 5, p. 597–604.

Miall, A. D., 1981, Analysis of fluvial depositional systems: American Association of Petroleum Geologists, Educational Course Notes No. 20, 75 p.

Miller, E. L., Holdsworth, B. K., Whiteford, W. B., and Rodgers, D., 1984, Stratigraphy and structure of the Schoonover sequence, northeastern Nevada: Implications for Paleozoic plate margin tectonics: Geological Society of America Bulletin, v. 95, p. 1063–1076.

Moore, R. C., 1953, Treatise on Invertebrate Paleontology, Part G, Bryozoa: Geological Society of America, University of Kansas Press, 253 p.

Moore, R. C., 1956, Treatise on invertebrate paleontology, Part F, Coelenterata: Geological Society of America, University of Kansas Press, 498 p.

Moore, R. C., 1965, Treatise on invertebrate paleontology, Part H, Brachiopoda 2: Geological Society of America, University of Kansas Press, 404 p.

Moore, R. C., Lalicker, C. G., and Fisher, A. G., 1952, Invertebrate fossils: New York, McGraw-Hill, 766 p.

Moore, T. E., and Nilsen, T. H., 1984, Regional variations in the fluvial Upper Devonian and Lower Mississippian(?) Kanayut Conglomerate, Brooks Range, Alaska: Sedimentary Geology, v. 38, p. 465–497.

Moore, T. E., Nilsen, T. H., and Brosgé, W. P., 1989, Sedimentology of the Kanayut Conglomerate, in Mull, C. G., and Adams, K. E., eds., Dalton Highway, Yukon River to Prudhoe Bay, Alaska, Bedrock geology of the eastern Koyukuk Basin, central Brooks Range and east-central Arctic Slope: Alaska Geological and Geophysical Surveys Guidebook 7, p. 219–252.

Mull, C. G., 1982, Tectonic evolution and structural style of the Brooks Range, northern Alaska: An illustrated summary, in Powers, R. B., ed., Geologic studies of the Cordilleran thrust belt: Denver, Colorado, Rocky Mountain Association of Geologists, v. 1, p. 1–45.

Mull, C. G., Adams, K. E., and Dillon, J. T., 1987, Stratigraphy and structure of the Doonerak fenster and Endicott Mountains allochthon, central Brooks Range, Alaska, in Tailleur, I., and Weimer, P., eds., Alaskan North Slope geology: Society of Economic Paleontologists and Mineralogists, Pacific Section, Publication 50, p. 663–679.

Mull, C. G., Adams, K. E., and Dillon, J. T., 1989, Stratigraphy and structure of the Doonerak fenster and Endicott Mountains allochthon, central Brooks Range, in Mull, C. G., and Adams, K. E., eds., Dalton Highway, Yukon River to Prudhoe Bay, Alaska, Bedrock geology of the eastern Koyukuk Basin, central Brooks Range and east-central Arctic Slope: Alaska Geological and Geophysical Surveys Guidebook 7, p. 204–217.

Nilsen, T. H., 1981, Upper Devonian and Lower Mississippian redbeds, Brooks Range, Alaska, in Miall, A. D., ed., Sedimentation and tectonics in alluvial basins: Geological Association of Canada Special Paper 23, p. 187–219.

Nilsen, T. H., and Moore, T. E., 1982a, Sedimentology and stratigraphy of the Kanayut Conglomerate, central and western Brooks Range, Alaska—Report of the 1981 field season: U.S. Geological Survey Open-File Report 82-674, 64 p.

Nilsen, T. H., and Moore, T. E., 1982b, Fluvial facies model for the Upper Devonian and Lower Mississippian(?) Kanayut Conglomerate, Alaska, in Embry, A. F., and Balkwill, H. R., eds., Arctic geology and geophysics: Canadian Society of Petroleum Geologists, Memoir 8, p. 1–12.

Nilsen, T. H., and Moore, T. E., 1984, Stratigraphic nomenclature for the Upper Devonian and Lower Mississippian(?) Kanayut Conglomerate, Brooks Range, Alaska: U.S. Geological Survey Bulletin, 1529-A, p. A1–A64.

Nilsen, T. H., Moore, T. E., and Brosgé, W. P., 1980a, Paleocurrent maps for the Upper Devonian and Lower Mississippian Endicott Group, Brooks Range, Alaska: U.S. Geological Survey Open-File Report 80-1066, scale 1:1,000,000.

Nilsen, T. H., Moore, T. E., Dutro, J. T., Jr., Brosgé, W. P., and Orchard, D. M., 1980b, Sedimentology and stratigraphy of the Kanayut Conglomerate, central and eastern Brooks Range, Alaska—Report of the 1978 field season: U.S. Geological Survey Open-File Report 80-888, 40 p.

Nilsen, T. H., Brosgé, W. P., Dutro, J. T., Jr., and Moore, T. E., 1981a, Depositional model for the fluvial Upper Devonian Kanayut Conglomerate, Brooks Range, Alaska: U.S. Geological Survey Circular 823-B, p. B20–B21.

Nilsen, T. H., Moore, T. E., Brosgé, W. P., and Dutro, J. T., Jr., 1981b, Sedimentology and stratigraphy of the Kanayut Conglomerate and associated units Brooks Range, Alaska—Report of the 1979 field season: U.S. Geological Survey Open-File Report 81-506, 37 p.

Nilsen, T. H., Moore, T. E., Balin, D. F., and Johnson, S. Y., 1982, Sedimentology and stratigraphy of the Kanayut Conglomerate, central Brooks Range, Alaska—Report of the 1980 field season: U.S. Geological Survey Open-File Report 82-199, 81 p.

Okulitch, A. V., 1985, Paleozoic plutonism in southeastern British Columbia: Canadian Journal of Earth Sciences, v. 22, p. 2813–2857.

Oldow, J. S., Seidensticker, C. M., Phelps, J. C., Julian, F. E., Gottschalk, R. R., Boler, K. W., Handschy, J. W., and Avé Lallemant, H. G., 1987, Balanced cross sections through the central Brooks Range, Alaska: American Association of Petroleum Geologists Special Publication, 19 p., 8 pl.

Pallister, J. S., Budahn, J. R., and Murchey, B. L., 1989, Pillow basalts of the Angayucham terrane: Oceanic plateau and island crust accreted to the Brooks Range: Journal of Geophysical Research, v. 94, p. 15901–15923.

Palmer, A. R., Dillon, J. T., and Dutro, J. T., Jr., 1984, Middle Cambrian trilobites with Siberian affinities from the central Brooks Range, northern Alaska: Alaska Geological Society, Geology of Alaska and Northwestern Canada Newsletter, no. 1, p. 29–30.

Patton, W. W., Jr., 1973, Reconnaissance geology of the northern Yukon-Koyukuk province, Alaska: U.S. Geological Survey Professional Paper 774-A, 17 p.

Patton, W. W., Jr., and Box, S. E., 1989, Tectonic setting of the Yukon-Koyukuk Basin and its borderlands, western Alaska: Journal of Geophysical Research, v. 94, p. 15807–15820.

Phelps, J. C., 1987, Stratigraphy and structure of the northeastern Doonerak window area, central Brooks Range, northern Alaska [Ph.D. thesis]: Houston, Texas, Rice University, 171 p.

Postma, G., 1984, Slumps and their deposits in fan-delta front and slope: Geology, v. 12, p. 27–30.

Potter, P. E., 1978, Significance and origin of big rivers: Journal of Geology, v. 86, p. 13–33.

Ramberg, I. B., and Morgan, P., 1984, Physical characteristics and evolutionary trends of continental rifts: Tectonics, v. 7, p. 165–216.

Repetski, J. E., Carter, C., Harris, A. G., and Dutro, J. T., Jr., 1987, Ordovician and Silurian fossils from the Doonerak anticlinorium, central Brooks Range, Alaska: U.S. Geological Survey Circular 998, p. 40–42.

Roberts, R. J., Hotz, P. E., Gilluly, J., and Ferguson, H. G., 1958, Paleozoic rocks of north-central Nevada: American Association of Petroleum Geologists Bulletin, v. 42, p. 2813–2857.

Roeder, D., and Mull, C. G., 1978, Tectonics of Brooks Range ophiolites, Alaska: American Association of Petroleum Geologists Bulletin, v. 62, p. 1696–1713.

Rust, B. R., 1972, Structure and process in braided river: Sedimentology, v. 18, p. 221–245.

Rust, B. R., 1978, Depositional models for braided alluvium, *in* Miall, A. D., ed., Fluvial sedimentology: Canadian Society of Petroleum Geologists Memoir 5, p. 605–625.

Shimer, H. W., and Shrock, R. R., 1944, Index fossils of North America: New York, John Wiley and Sons, 837 p.

Silberling, N. J., and Jones, D. L., 1984, Lithotectonic terrane map of the North American Cordillera: U.S. Geological Survey Open-File Report 84-523, p. A1–A12.

Smith, N. D., 1974, Sedimentology and bar formation in the upper Kicking Horse River, a braided outwash stream: Journal of Geology, v. 82, p. 205–224.

Speed, R. C., and Sleep, N. H., 1982, Antler orogeny and foreland basin: A model: Geological Society of America Bulletin, v. 93, p. 815–828.

Tailleur, I. L., Brosgé, W. P., and Reiser, H. N., 1967, Palinspastic analysis of Devonian rocks in northwestern Alaska, *in* Oswald, D. H., ed., International symposium on the Devonian system: Calgary, Canada, Alberta Society of Petroleum Geologists, p. 1345–1361.

Tasch, P., 1980, Paleobiology of the invertebrates: New York, John Wiley and Sons, 975 p.

Templeman-Kluit, D. J., 1979, Transported cataclasite, ophiolite, and granodiorite in Yukon: Evidence of an arc-continent collision: Geological Survey of Canada Paper 79-14, 27 p.

Warren, P. S., and Stelck, C. R., 1956, Devonian faunas of western Canada: Geological Association of Canada Special Paper, No. 1, 73 p.

Watts, K. F., Imm, T. A., Harris, A. G., 1989, Stratigraphy and paleogeographic significance of the Carboniferous Wahoo Limestone—reexamination of the type section, northeastern Brooks Range: Geological Society of America Abstracts with Programs, v. 21, p. 157.

Williams, P. F., and Rust, B. R., 1969, The sedimentology of a braided river: Journal of Sedimentary Petrology, v. 39, p. 649–679.

Young, S. W., 1976, Petrographic textures of detrital polycrystalline quartz as an aid to determining crystalline source rocks: Journal of Sedimentary Petrology, v. 46, p. 595–603.

MANUSCRIPT ACCEPTED BY THE SOCIETY SEPTEMBER 23, 1997

Geological Society of America
Special Paper 324
1998

Spatial variation in structural style, Endicott Mountains allochthon, central Brooks Range, Alaska

James W. Handschy

Shell Exploration and Production Company, P.O. Box 481, Houston, Texas 77001-0481

ABSTRACT

The Endicott Mountains allochthon in the central Brooks Range of northern Alaska is an east-west–striking stack of north-northwest–vergent thrust sheets. It is composed of mid-Paleozoic clastic and carbonate rocks that were deformed during late Mesozoic and Cenozoic (Brookian) orogenesis. Fundamental differences in structural style within the allochthon are related to the original distribution of lithostratigraphic units, the rheologies of the units, and the structural depth at which deformation occurred. In the north, the allochthon is characterized by imbricate thrust sheets and large single-phase folds. In the south, it is composed of an 8-km-thick thrust nappe that exhibits a vertical strain gradient. Strain variation in the southern nappe is recorded by a systematic change from single-phase folds at the top of the nappe to polyphase folds at the bottom. First-phase fold axes change from strike-parallel at the top of the southern nappe to dip-parallel at the bottom, and the angle between first-phase axial planes and the basal thrust decreases from approximately 70° at the top of the nappe to less than 10° near the base. In other fold-thrust belts around the world this type of fold rotation has been attributed to simple shear. Geometric analysis of folds and strain markers in the southern Endicott Mountains allochthon suggests that deformation was more complex and included components of layer parallel shortening, simple shear, and layer normal flattening of the shear zone.

INTRODUCTION

The Endicott Mountains allochthon (Mull, 1982; Oldow et al., 1987) is part of a family of structural sheets that constitute much of the northern, nonmetamorphic part of the late Mesozoic and Cenozoic Brooks Range fold-thrust belt in northern Alaska. Internally, the Endicott Mountains allochthon is composed of several east-west–striking, north-northwest–vergent thrust sheets containing Upper Devonian and Lower Carboniferous clastic and carbonate rocks. Individual thrust sheets attain thicknesses in excess of 8 km and the allochthon is continuous for approximately 50 km across-strike from the southern, metamorphic part of the orogen to the northern limit of the mountain range. Along the front of the Brooks Range, the Endicott Mountains allochthon structurally overlies an imbricate stack of Carboniferous carbonates of the Lisburne Group (Brosgé et al., 1979; Oldow et al.,

1987) and, near Galbraith Lakes, is in close proximity to mid-Cretaceous foredeep deposits (Fig. 1). Paleocurrent indicators plus clast composition data from the foredeep deposits suggest that the middle Cretaceous Fortress Mountain Formation and ostensibly the finer grained Torok Formation were derived from the Endicott Mountains allochthon (Crowder, 1987, 1989; Handschy, 1988). To the south and east, the Endicott Mountains allochthon is structurally overlain by lower Paleozoic and Precambrian(?) metasedimentary rocks of the Skajit allochthon, which is approximately 10 km thick (Oldow et al., 1987; Oldow et al., this volume, Chapter 8).

Thrust faults and folds are well exposed throughout the study area (Fig. 2). The basal detachment of the Endicott Mountains allochthon, the Amawk thrust (Mull, 1982), is exposed both in the north and, more importantly, in the south where the base of the allochthon is uplifted and exposed by the late-stage Doonerak

Handschy, J. W., 1998, Spatial variation in structural style, Endicott Mountains allochthon, central Brooks Range, Alaska, *in* Oldow, J. S., and Avé Lallemant, H. G., eds., Architecture of the Central Brooks Range Fold and Thrust Belt, Arctic Alaska: Boulder, Colorado, Geological Society of America Special Paper 324.

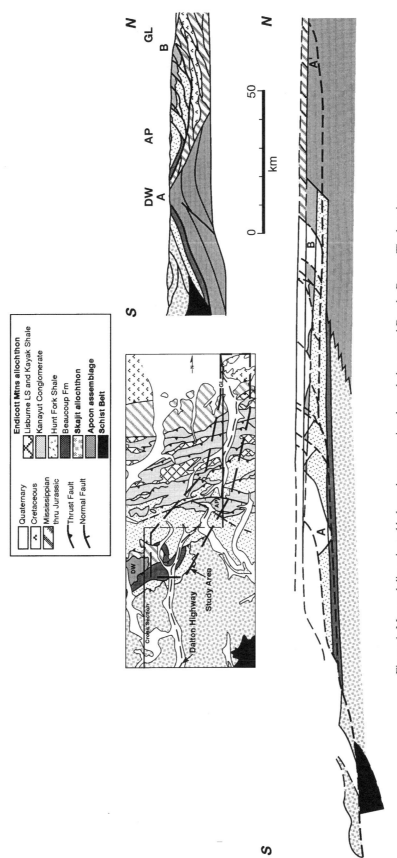

Figure 1. Map and dip-oriented balanced cross section through the central Brooks Range. The location map shows the study area and the surrounding tectonostratigraphic units in the central Brooks Range. The locations of the Doonerak window (DW), Atigun Pass (AP), and Galbraith Lakes (GL) are shown for reference. A and B indicate correlative positions in the deformed and restored cross sections Note that the steeply dipping normal faults shown on the map and in Figure 2 are relatively insignificant for the reconstruction and have been omitted on this cross section.

duplex (Julian and Oldow, this volume, Chapter 5; Seidensticker and Oldow, this volume, Chapter 6). The allochthon extends around the closure of the east-northeast–plunging duplex and the Amawk thrust is exposed both on the northern and southern flanks of the uplift (Mull, 1982; Phelps, 1987; Phelps and Avé Lallemant, this volume, Chapter 4). Uplift and erosion of the Endicott Mountains allochthon around the Doonerak window offers a unique opportunity to investigate differences in structural style over an essentially continuous structural section approximately 8 km thick and to compare vertical strain variations with changes in structural style across strike for a distance of over 50 km.

STRATIGRAPHY

Fold-thrust nappes of the Endicott Mountains allochthon are composed of five stratigraphic units ranging in age from Late Devonian to Early Carboniferous (Handschy, this volume, Chapter 2). These five stratigraphic units are:

1. mixed carbonates and siliciclastics of the Middle to Late Devonian Beaucoup Formation;

2. marine shale and sandstone of the Late Devonian Hunt Fork Shale;

3. nonmarine conglomerate sandstone and shale of the Late Devonian and Early Mississippian(?) Kanayut Conglomerate;

4. marine shale, sandstone, and limestone of the Early Carboniferous Kayak Shale; and

5. marine limestone and shale of the Carboniferous Lisburne Group.

The Beaucoup Formation, Hunt Fork Shale, and Kanayut Conglomerate make up the bulk of the Endicott Mountains allochthon.

Beaucoup Formation

The Beaucoup Formation (Dutro et al., 1979) exposed in the southern Endicott Mountains allochthon is a sequence of interlayered shale, calcareous shale, sandstone, conglomerate, volcaniclastics, carbonate build-ups, and their low-grade metamorphic equivalents (Phelps, 1987; Handschy, this volume, Chapter 2). Bedding varies from less than 1 mm thick in shale and phyllite to greater than 5 m thick in massive reefal limestone units and amalgamated sandstone beds. Most of the lithologies are thinly layered and strongly deformed with the notable exception of limestone and sandstone units greater than 2 m thick, which are relatively undeformed internally. The top of the Beaucoup Formation is defined as the top of the uppermost reefal limestone unit (Dutro et al., 1979). The bottom of the Beaucoup Formation is not exposed in the Endicott Mountains allochthon.

Kanayut Conglomerate and Hunt Fork Shale

Facies relations in the Kanayut Conglomerate and Hunt Fork Shale reflect deposition in a fluvio-deltaic system (Moore and Nilsen, 1984; Nilsen and Moore, 1984; Handschy, this volume,

Chapter 2). Southwestward progradation of the fluvial and non-marine deltaic facies of the Kanayut Conglomerate over the marine Hunt Fork Shale produced a lithofacies pattern in which the Kanayut Conglomerate is thicker in the north and the Hunt Fork Shale is thicker in the south (Handschy, this volume, Chapter 2). The contact between the Kanayut Conglomerate and the Hunt Fork Shale is gradational. Channel-fill conglomerates and sandstone, which cut into and are interbedded with intertidal shale of the upper Hunt Fork Shale, can be traced laterally into amalgamated conglomerate and sandstone beds in the Kanayut Conglomerate.

Sedimentary structures and fossil assemblages allow the Hunt Fork Shale to be divided into upper, middle, and lower depositional facies (Handschy, this volume, Chapter 2). The upper Hunt Fork Shale is composed of intertidal mudstone, plus channelized sandstone and conglomerate. The middle Hunt Fork Shale is composed mostly of thinly laminated shale, but also contains lenticular trough cross-stratified sandstone, massive sheet sandstone, hummocky cross-stratified sandstone, and thin graded beds. The lower Hunt Fork Shale is similar to the middle Hunt Fork Shale except that it does not contain hummocky cross-stratified units, graded beds are common, and thick intervals of repetitious graded beds are more common than in the middle Hunt Fork Shale. Deformation in the Hunt Fork Shale makes it impossible to measure stratigraphic thickness in the study area. The tectonic thickness of the Hunt Fork Shale in the southern nappe, where both the top and bottom of the formation are exposed, is approximately 6 km.

STRUCTURAL RELATIONS

The style and intensity of deformation in the Endicott Mountains allochthon changes with lithology, across-strike position, and depth. Variations in fold style, fold orientation, and the number of superposed fold phases allow the allochthon to be divided into five structural domains (Fig. 3a). Domain 1 includes all of the northern thrust imbricates between the front of the Brooks Range and the vicinity of Atigun Pass (Fig. 3a). Domains 2, 3, 4, and 5 are within the southern nappe and their boundaries approximately coincide with lithofacies boundaries within and between the Kanayut Conglomerate, Hunt Fork Shale, and Beaucoup Formation.

Structural domains

Crosscutting cleavages and fold superposition allow discrimination of four phases of folds in the Endicott Mountains allochthon. In Domains 1 through 4, where fold vergence can be determined, first-phase folds (D1) are north vergent. In the northern imbricates (Domain 1) and the upper part of the southern nappe (Domain 2), D1 is characterized by a single phase of north-vergent inclined folds (Fig. 4). In the lower part of the southern nappe (Domains 3, 4, and 5), D1 is divided into sequentially superposed structures, D1a and D1b. Superposition

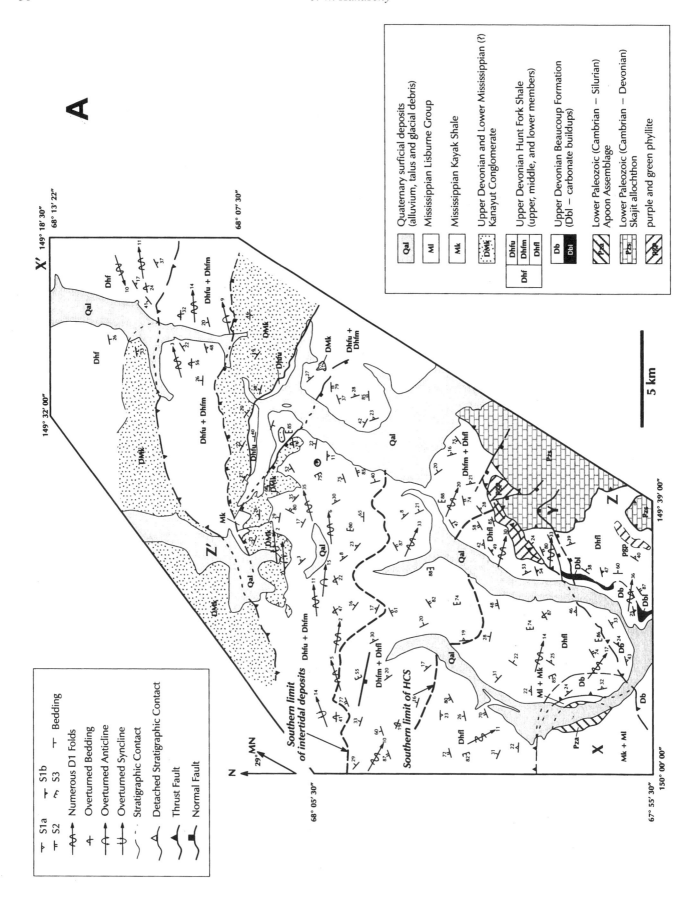

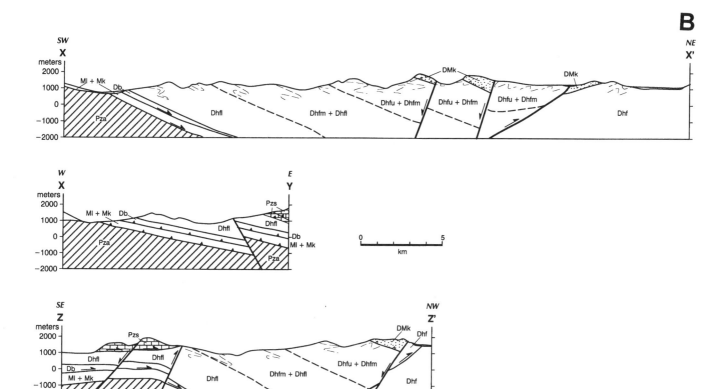

B

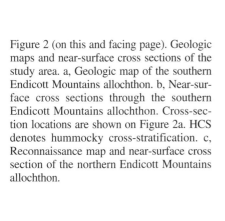

Figure 2 (on this and facing page). Geologic maps and near-surface cross sections of the study area. a, Geologic map of the southern Endicott Mountains allochthon. b, Near-surface cross sections through the southern Endicott Mountains allochthon. Cross-section locations are shown on Figure 2a. HCS denotes hummocky cross-stratification. c, Reconnaissance map and near-surface cross section of the northern Endicott Mountains allochthon.

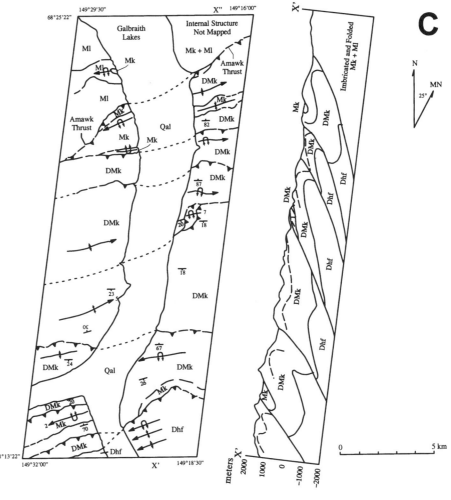

C

of D1b folds on D1a folds occurs where D1a interlimb angles approach 0° and the angle between D1a axial planes and the basal thrust of the allochthon is approximately 10°. Upright northeast-trending D2 folds and cleavage (S2) are superposed on D1 folds in the Hunt Fork Shale and Beaucoup Formation of Domains 3, 4, and 5 in the southern nappe. Small D3 folds and a subvertical north-northwest– to north-northeast–striking crenulation cleavage (S3) postdate D2. High-angle, south-dipping normal faults (Fig. 2) also postdate D2 folds, but the temporal relations between D3 and these normal faults is not clear. Dynamic analysis of tension cracks and fault plane lineations suggests that D3 and the high-angle faults may be genetically related (Phelps et al., 1987; Phelps and Avé Lallemant, this volume, Chapter 4).

Thrust imbricates in Domain 1 (Figs. 2c and 3a) expose the upper and middle Hunt Fork Shale, Kanayut Conglomerate, Kayak Shale, and part of the Lisburne Group. Exposures of the Hunt Fork Shale are restricted to the southern part of Domain 1, whereas the Kayak Shale and Lisburne Group are more common in the northern part of the domain. A single phase of open to close (Fleuty, 1964), north-vergent folds characterizes the structure in Domain 1. Fold amplitudes range from 0.2 km to 1 km. Fold axes are subhorizontal and trend east-northeast or west-southwest (Fig. 3b). Axial planes dip to the south (Fig. 3b) and are inclined 30 to 70° with respect to the nearest imbricate thrust. Usually all of the stratigraphic units in Domain 1 are folded together, but in the southern part of the domain, the Kanayut Conglomerate is locally detached from the Hunt Fork Shale. Where the Kanayut Conglomerate–Hunt Fork Shale contact is detached the folds in both units have similar amplitudes, wavelengths, and orientations, but the units are separated by a relatively flat structural contact.

Domain 2 includes the Kanayut Conglomerate and Kayak Shale exposed at the top of the southern thrust nappe of the Endicott Mountains allochthon (Fig. 3a). It is separated from Domain 1 by a thrust fault that places Hunt Fork Shale over Kanayut Conglomerate (Figs. 2a and 3a) and from Domain 3 by a detached stratigraphic contact that lies within the facies transition from Kanayut Conglomerate to Hunt Fork Shale (Fig. 5a). As in Domain 1, only one phase of north-vergent folds is present in Domain 2 (Fig. 3b). Fold amplitudes are between 0.5 km and 1 km. Fold axes in Domain 2 preferentially plunge to the east-northeast because of their position on the northeast-plunging flank of the Doonerak duplex. Axial planes dip 20 to 50° toward the south (Fig. 3b) and are inclined 40 to 70° relative to the Kanayut Conglomerate–Hunt Fork Shale contact, which dips 20° toward the north-northeast (Fig. 5a). Axial-planar cleavage (S1) is well developed only in the hinges of folds in Domain 2.

The detachment surface along the Kanayut Conglomerate–Hunt Fork Shale contact separates open to close, moderately south-dipping asymmetric D1 folds in Domain 2 from tight, inclined to recumbent, north-vergent D1 folds with amplitudes of as much as 1 km in Domain 3 (Fig. 5a). Most D1 axial planes have dips between 10° toward the north and 10° toward the south

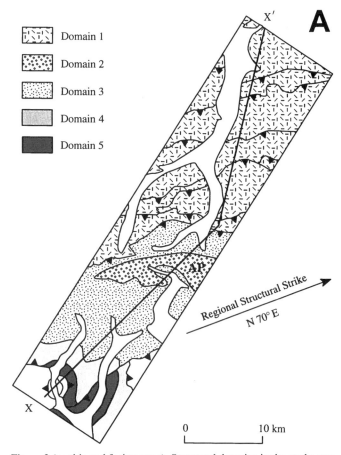

Figure 3 (on this and facing page). Structural domains in the study area. a, Structural domain map. Northeast to southwest line shows the location of Figure 4 sections. Note that the steeply dipping normal faults shown in Figure 2 have been omitted on this map for simplicity. b, Lower hemisphere, equal-area stereographic projections of structural data from the Endicott Mountains allochthon. S0 is bedding, Fn are folds (solid symbols are axial planes, open symbols are axes), Dn are cleavages (solid symbols are planes, open symbols are intersection lineations and crenulation cleavage lineations).

and are inclined 10 to 30° relative to the Kanayut Conglomerate–Hunt Fork Shale contact, which dips 20° toward the north-northeast (Fig. 3b). D1 fold axes in Domain 3 plunge moderately in various directions, but most commonly plunge toward the northeast (Fig. 3b). Superimposed on the D1 folds in Domain 3 are small, open, upright northeast-trending D2 folds with amplitudes less than 1 m and a crenulation cleavage (S2). D2 fold axes plunge toward the east-northeast.

Domain 4 corresponds to the lower Hunt Fork Shale and differs from Domain 3 in two ways. First, the maximum fold amplitudes observed in Domain 4 are less than 500 m. Second, D1 structures are divided into two phases, D1a and D1b (Fig. 3b). Tight D1a folds have a well-developed, closely spaced, axial-planar cleavage (S1a). D1a structures are folded by open, inclined, north-vergent, asymmetric D1b folds. D1b folds also have an associated cleavage (S1b), which cuts and locally crenulates S1a. Mesoscopic D1a folds measured in Domain 4 have

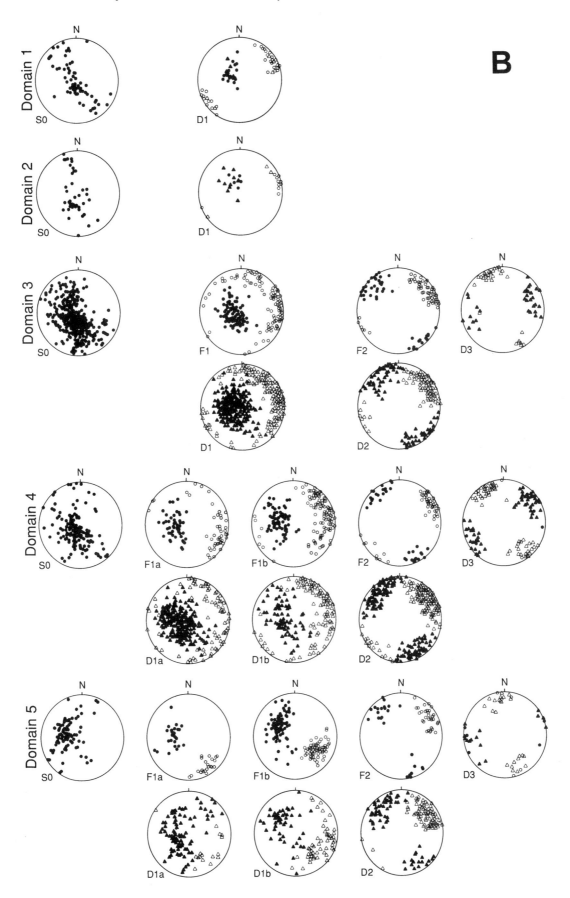

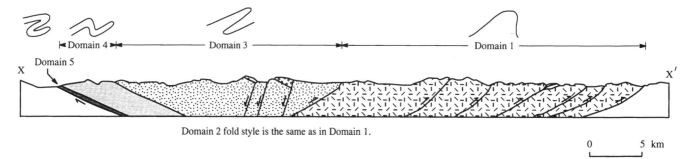

Domain 2 fold style is the same as in Domain 1.

0   5 km

Figure 4. Cross section through the Endicott Mountains allochthon showing spatial distribution of D1 fold style. See Figure 3a for cross section location and domain patterns.

amplitudes between 0.5 m and 10 m, whereas D1b folds have amplitudes between 0.5 m and 3 m. Superimposed on D1a and D1b folds are small, open, upright, northeast-trending D2 folds and an associated (S2) crenulation cleavage. Because Domain 4 is folded around the northeast-plunging axis of the Doonerak duplex, the orientation of fold axes, axial planes, and cleavage planes varies with location. North of the duplex axis, the boundary between Domains 3 and 4 is approximately parallel to both the Amawk thrust and the Kanayut Conglomerate–Hunt Fork Shale contact (Fig. 3a). Northeast and east of the duplex, the strike of D1 structures mimics the strike of the Amawk thrust where it plunges into the subsurface around the northeastern end of the Doonerak window. Throughout Domain 4, D1b axial planes and S1b are inclined 20 to 50° relative to the Amawk thrust and, after the effects of D1b and D2 folding have been removed, D1a axial planes and S1a are inclined approximately 10° relative to the Amawk thrust.

The contact between the lower Hunt Fork Shale and the underlying Beaucoup Formation is the boundary between Domains 4 and 5. In Domain 5, D1a and D1b folds are tight to isoclinal (Fig. 5c). Bedding, S1a, and S1b are all subparallel except in the hinges of folds. The largest D1 folds observed in Domain 5 have amplitudes of less than 100 m. Superimposed on the D1a and D1b folds are open, upright D2 folds with amplitudes less than 1 m and a crenulation cleavage (S2). Domain 5, like Domain 4, is folded around the northeastern end of the Doonerak duplex. D1a and D1b axial planes are inclined between 0 and 10° relative to the Amawk thrust. D1a and D1b fold axes preferentially plunge southeast toward the approximately parallel to stretching lineations (Fig. 3b). As in all of the domains, D2 fold axes trend east-northeast.

Deformation in Domain 5 is strongly lithology dependent. Shaley intervals, which comprise most of the domain, are intensely deformed. Limestone units and sandstone/conglomerate beds thicker than 2 m are relatively undeformed. Top indicators, where found, indicate that the more competent limestone and coarse-grained siliciclastic units are consistently upright. Unit boundaries are either faults or shear zones that exhibit increasing amounts of shear strain into adjacent shaley units.

Fold style and cleavage

Fold style varies with lithology, structural position, and bulk strain within the Endicott Mountains allochthon. In Domains 1 and 2, D1 structures are all large, open to close, parallel folds (class 1b; Ramsay, 1967, p. 367). Parasitic mesoscopic D1 folds are locally abundant in the Hunt Fork Shale and Kayak Shale, but are rare in the Kanayut Conglomerate and Lisburne Group. In the Kanayut Conglomerate, S1 cleavage is developed only in the hinges of folds. In all lithologies, S1 cleavage is zoned (Type C; Gray, 1978). D2 and D3 structures were not observed in Domains 1 and 2.

In Domain 3 macroscopic D1 folds have approximately the same amplitudes as in Domains 1 and 2, but the amplitude/wavelength ratio is higher (Table 1). Several orders of parasitic mesoscopic D1 folds are common on the limbs of the macroscopic folds. Most macroscopic D1 folds are class 1c folds. The style of mesoscopic D1 folds varies from class 1c to class 3 depending on the lithologies involved and bedding thickness. Beds greater than 1 cm thick and amalgamated units of siltstone and sandstone generally form asymmetric class 1c folds. Siltstone or sandstone beds less than 1 cm thick and flanked by shale often form class 2 or rootless class 3 folds. S1 cleavage is always closely spaced (Type B) and accompanied by syntectonic phyllosilicates.

Macroscopic D1a folds in Domain 4 have amplitude/wavelength ratios similar to D1 folds in Domain 3, but the maximum amplitude is smaller by a factor of two. D1a folds in Domain 4 are more difficult to identify than D1 structures in Domain 3. Most D1a folds in Domain 4 are mesoscopic class 1c and class 2 folds with amplitudes less than 100 m. Small rootless class 3 folds are infrequent because discrete, thin sandstone and siltstone beds are rare. D1b folds in Domain 4 generally belong to class 1c. S1a cleavage is always closely spaced (Type B) and accompanied by recrystallized phyllosilicates along the cleavage planes. S1b cleavage is variable between closely spaced (Type B) and zoned (Type C). Where zoned, S1b crenulates S1a. Recrystallized phyllosilicates are not as well developed along S1b cleavage planes.

D1a and D1b folds in Domain 5 are mesoscopic class 1c and 2 folds with amplitudes less than 100 m. Amplitude/wavelength

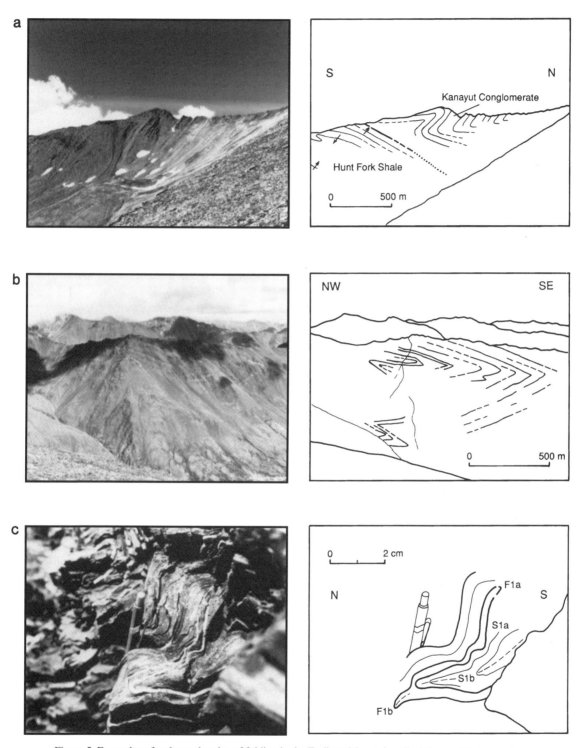

Figure 5. Examples of styles and scales of folding in the Endicott Mountains allochthon. a, Disharmonic
folding across the Domain 2–Domain 3 boundary. Note the transposed bedding (D1) in the Hunt Fork
Shale and the open asymmetrical D1 fold in the overlying Kanayut Conglomerate. b, Recumbent D1
folds in Hunt Fork Shale of Domain 3. c, Small D1 folds in the Beaucoup Formation (Domain 5).

TABLE 1. FOLD CHARACTERISTICS IN
THE ENDICOTT MOUNTAINS ALLOCHTHON*

	Fold Class	A	A/λ	Interlimb Angle (°)
Domain 1				
D1	1b	≤ 1 km	0.5 – 1.0	120 – 45
Domain 2				
D1	1b	≤ 1 km	0.5 – 1.0	120 – 45
Domain 3				
D1	1c	≤ 1 km	1.5 – 3.0	45 – 10
D2	1b – 1c	≤ 1 m	0.2 – 0.5	≥ 120
D3	c.c. only			
Domain 4				
D1a	2	≤ 500 m	> 5.0	≤ 10
D1b	1c	≤ 10 m	0.5 – 1.5	90 – 30
D2	1b – 1c	≤ 1 m	0.2 – 0.5	≥ 120
D3	c.c. only			
Domain 5				
D1a	2 – 3	≤ 500 m	≥ 100	≤ 2
D1b	2	≤ 100 m	≥ 5.0	≤ 10
D2	1b – 1c	≤ 1 m	0.2 – 0.5	≥ 120
D3	c.c. only			

*Fold classification is from Ramsay, 1967. A = fold amplitude; λ = fold wavelength; c.c. = crenulation cleavage only.

ratios are very large. S1a and S1b cleavages are equally well developed and only distinguishable in the hinges of D1b folds.

D2 structures are bivergent upright to inclined, parallel (class 1b) folds with amplitudes less than 1 m and amplitude/wavelength ratios of approximately 2/3. S2 cleavage is variable from discontinuous (Type A), to closely spaced (Type B), to zoned (Type C) and locally crenulates S1. D2 structures are equally well developed in Domains 3 through 5 south of Atigun Pass, they decrease northward in Domain 3 north of Atigun Pass, and they are apparently absent in Domains 1 and 2. The lack of D2 structures in the Kanayut Conglomerate of Domain 2 appears to be a function of lithology. Cleavage is rare in the well-cemented conglomerate and sandstone beds that comprise most of the Kanayut Conglomerate, and, possibly because amalgamated packages of beds in the Kanayut Conglomerate are several meters thick, small-scale folds with amplitude/wavelength ratios greater than 1/10 are not developed. The northward decrease of D2 in Domain 3 and the absence of D2 structures in the Hunt Fork Shale of Domain 1 is more difficult to explain. Based on crosscutting relationships between faults and D2 structures adjacent to the roof thrust of the Doonerak duplex, Seidensticker et al. (1987) presented a model in which the northward decrease in the intensity of D2 is a function of location relative to the front of the duplex. In their model (Seidensticker et al., 1987), the frontal culmination of the Doonerak duplex acted as a buttress causing D2 shortening south of the culmination to be accommodated by shortening along conjugate shears, thus explaining the bivergent nature of D2 structures in Domains 3 through 5, the decrease in

the intensity of D2 structures in Domain 3, and the absence of D2 structures in Domain 1.

D3 is usually a steeply dipping, zoned (Type C) crenulation cleavage. Inclined D3 folds are also locally present in Domains 4 and 5. D3 folds are open to close parallel (class 1b) folds with amplitudes less than 2 m and amplitude/wavelength ratios between 2/3 and 1/1. D3 axial planes strike north-northwest to north-northeast and dip toward both the east and the west (Figs. 3b).

Thrust faults

In addition to the basal thrust fault of the Endicott Mountains allochthon (Amawk thrust; Mull, 1982), there are five major imbricate thrust faults in Domain 1 (Fig. 2c). All of the imbricate thrusts place Hunt Fork Shale over Kanayut Conglomerate, Kanayut Conglomerate over Kanayut Conglomerate, or Kanayut Conglomerate over Kayak Shale and the Lisburne Group (Fig. 2). Northward displacements along these faults range from greater than 100 km along the Amawk thrust (Oldow et al., 1987; Mull et al., 1987) to less than 5 km along some of the imbricate thrusts (Fig. 2c).

Within the southern thrust nappe, intraformational thrusts and bedding parallel detachment surfaces are common. Although most of the identifiable intraformational thrust faults in Domains 3, 4, and 5 are approximately parallel to the Kanayut Conglomerate–Hunt Fork Shale contact and the Amawk thrust, some faults ramp up-section from south to north. Fault ramps often truncate previously folded intervals and cause fault-bend folding of superjacent faults. Intraformational thrust sheet thickness in the southern nappe varies from less than 10 m to more than 1 km. Along intraformational thrust faults where both the hanging-wall and footwall cut-offs of the same beds can be identified, the displacement is between 1 m and 500 m. Unfortunately, the monotonous interbedded nature of shale and sandstone in Domains 3, 4, and 5 makes it impossible to reliably measure the amount of displacement along many intraformational thrusts. Likewise, it is often difficult to trace intraformational faults laterally for more than 1 km.

Intraformational thrust faults in Domains 3, 4, and 5 are commonly concentrated in zones separated by internally folded, but relatively unfaulted intraformational thrust sheets. In Domain 3 intraformational thrust sheets are as much as 1 km thick. Deeper in the allochthon maximum intraformational thrust thickness decreases to less than 500 m in Domain 5. This depth-related decrease in intraformational thrust sheet thickness is accompanied by a decrease in maximum fold size from Domain 3 to Domain 5 (Table 1).

The youngest thrust fault in the study area is a relatively steep south-dipping thrust that places Domain 5 over Domain 4 (Fig. 3a). This fault cuts the Amawk thrust near the eastern end of the Doonerak duplex, postdates D1 folding, and may be related to D2 deformation. Oldow et al. (1987) interpret this fault to be an out-of-sequence thrust that cuts up-section through the Doonerak duplex, off-setting the Amawk thrust.

Boudinage

Two distinct morphologies of boudinage are exposed in Domains 3 and 4. The most conspicuous boudins occur in sandstone beds between 10 cm and 1 m thick. Intermediate boudin axes are often several meters long and separated by straight to sigmoidal tension cracks that indicate less than 10% extension. These large boudins occur on both limbs of macroscopic folds and have long axes that are approximately parallel to bedding-S1 intersection lineations and D1 fold axes. In Domain 3 most boudin axes trend northeast, whereas in Domain 4 they trend east or southeast. On the limbs of large overturned folds in Domain 3 the orientation of sigmoidal tension cracks indicates top-to-the-north shear in upright limbs and top-to-the-south shear in overturned limbs.

The second boudin morphology is found in thin sandstone and siltstone layers contained within a shaley matrix. These boudins occur on both upper and lower limbs of mesoscopic folds. Characteristically they are bound on two sides by cleavage planes and have long axes consistently parallel to bedding-S1 intersections. In Domain 3, most boudin axes trend northeast, whereas in Domain 4 boudin axes trend southeast. The distance between boudins is commonly many times greater than the length of the intermediate boudin axes, indicating several hundred percent extension. In some localities, the intermediate axes of the boudins within an individual bed are parallel to each other. In other locations, the boudins are systematically rotated. Boudin rotation varies with location, but in Domain 3 the rotation usually indicates top-to-the-north shear in the upper limbs and top-to-the-south shear in the overturned limbs of macroscopic D1 folds. In Domains 4 and 5, and adjacent to intraformational thrusts, boudin rotation almost always indicates top-to-the-north shear.

Strain indicators

Sheared Skolithos burrows, stretched conglomerate clasts, stretched breccia clasts, and deformed fossils provide information about the strain intensity in the Endicott Mountains allochthon (Fig. 6). When evaluated with fold orientation, cleavage orientation, and fault data, these strain indicators also give a more complete picture of deformation within the allochthon.

Skolithos burrows in the upper Hunt Fork Shale of Domains 2 and 3 provide shear strain information in two structural settings. Along the contact between the Kanayut Conglomerate and Hunt Fork Shale immediately west of Atigun Pass Skolithos burrows in upright, thinly bedded sandstone are sheared between 45 and 70° toward the north-northwest demonstrating that shear strain within the sandstone beds is as high as γ = 2.75. Slickenside lineations on bedding planes in the thin shale partings between the sandstone beds are parallel to the direction of shear, indicating that top-to-the-north shear in the overall interval is partitioned into both brittle and ductile components. Below the Kanayut Conglomerate–Hunt Fork Shale contact farther to the southwest, sheared Skolithos burrows are found on both limbs of

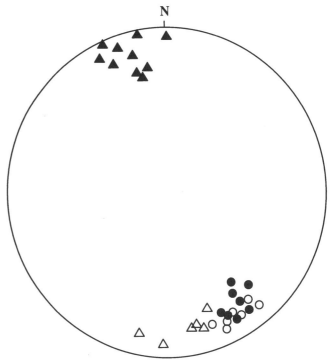

Figure 6. Stretching lineations in the Endicott Mountains allochthon. Lower hemisphere equal-area stereographic projection. Solid triangles are from sheared Skolithos burrows in Domain 3, open triangles are sheared corals in Domain 5, open circles are stretched breccia clasts in Domain 4, and solid circles are stretched pebbles in Domain 5.

macroscopic north-vergent folds. Burrows in the upper limbs are sheared as much as 60° (with respect to the bedding surface) toward the north-northwest, whereas burrows in the overturned limbs are sheared between 42° toward the north-northwest and 55° toward the south-southeast. Axial planes and cleavage in all of the folds dip to the south-southeast. Folds that have opposite senses of shear on different limbs are consistent with flexural slip induced shear strain in the sandstone beds. Folds with the same sense of shear on both limbs suggest that the shear strain was induced by pervasive top-to-the-north shear or the overturned limbs experienced a complex shear strain history of flexural slip followed by a greater amount of top-to-the-north shear.

Breccia clasts in Domain 4 near the base of the Skajit allochthon, stretched conglomerates in Domain 5, and consistently aligned, deformed corals along the bases of carbonate beds in Domain 5 all indicate north-northwest or south-southeast stretching (Fig. 6). Rf/θ′ analysis indicates that undeformed conglomerate clasts had axial ratios between 1.0 and 1.6 and that there has been as much as 400% extension parallel to stretching lineations (Table 2).

Cleavage in Hunt Fork Shale and Beaucoup Formation mudrocks is apparently the product of dissolution of relatively soluble minerals (e.g., quartz and chert) and the subsequent concentration and alignment of less soluble phyllosilicates along dissolution planes. Microlithons between cleavage planes con-

**TABLE 2. MEASURED AND ESTIMATED STRAIN
IN THE ENDICOTT MOUNTAINS ALLOCHTHON***

	Ψ (°)	Θ (°)	Rs Meso	Rs Micro	α
Domain 1		50 ± 20			0.6 to 0.7
Domain 2		60 ± 10			0.6 to 0.7
Domain 3	45 to 70	20 ± 5	2.9	< 1.2	
Domain 4		10 ± 5	3.2	1.6	
Domain 5		8 ± 5	5.1	2.1	

*Ψ = shear strain measured directly from sheared Skolithos burrows; Θ = the angle between the Kanayut Conglomerate–Hunt Fork Shale contact and D1a axial planes in Domains 4 and 5. Note that the Kanayut Conglomerate–Hunt Fork Shale contact and the Amawk thrust are approximately parallel on the north flank of the Doonerak duplex (Figs. 1, 2, and 4). Rs meso is the x:z axial ratio of stretched conglomerate and breccia clasts. Rs micro is the x:z axial ratio of determined by thin section analysis of deformed mudrocks using the Fry method. α = the shortening by folds in the Kanayut Conglomerate of Domains 1 and 2 where bed thickness is relatively constant and cleavage is rare.

tain abundant quartz and chert grains in a phyllosilicate-rich matrix. Cleavage planes are composed almost exclusively of phyllosilicates. Contacts between sand- and silt-sized grains within deformed mudrocks are rare, but where present, are irregular or sutured, indicating removal of material by pressure solution (Durney, 1972). Grains that are separated from other grains by phyllosilicate-rich cleavage planes are often elongate and their long axes are parallel to the cleavage planes. In some samples, grain boundaries adjacent to cleavage planes are pitted, suggesting dissolution. In other samples, there is no evidence for dissolution along grain boundaries and the elongate character of the grains probably has a detrital origin. Alignment of grains parallel to cleavage planes in these samples may be the product of grain rotation.

It was initially hoped that Fry analysis (Fry, 1979; Ramsay and Huber, 1983) of sand- and silt-sized grains in cleaved mudrocks would provide useful strain data in areas where other types of strain indicators are absent. Although Fry analysis on mudrocks from Domains 3, 4, and 5 indicate that the principal extension direction is parallel to stretching lineations and that the principal flattening direction is perpendicular to S1 cleavage planes, axial ratios consistently indicate lower strains than measured in stretched conglomerates and breccia (Table 2). Pressure shadows composed of quartz and, less frequently, quartz plus phyllosilicates, are commonly developed on detrital grains within microlithons. Pressure shadow lengths vary from grain to grain and do not always indicate consistent extension within an individual sample. They do, however, usually indicate more extension than measured using the Fry method (Handschy, 1988). Because the Fry technique commonly yields lower strain values than stretched clasts and pressure shadows, it is believed that entire grains have been dissolved along cleavage planes and the

strain measured using the Fry technique is only a fraction of the actual strain (Handschy, 1988; Onasch, 1986; Dunne et al., 1990). As noted above, dissolution of grains is evidenced by pitted, irregular, and sutured grain boundaries, but it is impossible to evaluate whether entire grains are missing.

STRUCTURAL ANALYSIS

Seven fundamental observations provide the basis for interpreting deformation in the Endicott Mountains allochthon:

1. Domains 1 and 2 are composed mostly of Kanayut Conglomerate, whereas Domains 3, 4, and 5 are composed mostly of Hunt Fork Shale and Beaucoup Formation mudrocks;

2. provenance of mid-Cretaceous syntectonic sediments near Galbraith Lakes indicates that the Kanayut Conglomerate was exposed during emplacement of the Endicott Mountains allochthon;

3. the preserved northern limit of the Skajit allochthon does not overlap the frontal imbricates of the Endicott Mountains allochthon;

4. lateral and vertical facies relations in the Kanayut Conglomerate, Hunt Fork Shale, and Beaucoup Formation in the southern Endicott Mountains allochthon are in a normal stratigraphic succession;

5. the number of fold phases and the amplitude/wavelength ratio of D1 folds increase with depth in Domains 2 through 5;

6. bedding parallel shear zones are preferentially developed along contacts between mechanically different lithologies and increase in frequency with depth; and

7. quantitative strain measurements indicate that strain increases with depth in Domains 3 through 5.

Controls on structural style

The change in structural style from imbricate thrust sheets and large single-phase folds in Domains 1 and 2 to heterogeneous intranappe deformation in Domains 3 through 5 appears to be a function of the mechanical differences between the Kanayut Conglomerate and the underlying shale, the relative thicknesses of the units, the depth of deformation, and the amount of strain. Mechanical differences between the units are indicated by disharmonic folds across the Kanayut Conglomerate–Hunt Fork Shale contact and, locally, by a decoupling surface between the two units (Fig. 5a). Detachment of the Kanayut Conglomerate from the Hunt Fork Shale is less common and less conspicuous in the northern imbricates than in the southern nappe because folds in both units have similar geometries and orientations. Locally, however, the contact between folded Kanayut Conglomerate and folded Hunt Fork Shale is flat, indicating slip between the two units. Fold kinematics and small-scale shear indicators (e.g., Skolithos burrows, slickensides, fault gouge, etc.) along the contact consistently indicate top-to-the-north shear.

The northward thickening wedge geometry of the Kanayut Conglomerate (Handschy, this volume, Chapter 2) causes the thrust imbricates in Domain 1 to be composed mostly of relatively rigid, well-cemented conglomerate and the southern nappe to be composed mostly of shale. Therefore, it is inferred that imbricate thrusting is the dominant structural style in Domain 1 because of the greater thickness of the relatively rigid Kanayut Conglomerate. However, if deformation was controlled exclusively by these mechanical differences, changes in fold style and orientation characteristic of the Hunt Fork Shale in Domain 3 would also be developed in the Hunt Fork Shale of Domain 1. The consistency of fold style and orientation in the Hunt Fork Shale of Domain 1 suggests that other factors in addition to the relative proportions of Kanayut Conglomerate and Hunt Fork Shale contributed to the change in structural style from the northern imbricates to the southern nappe. One possible difference between the northern imbricates and the southern nappe is the thickness of overburden during deformation. Mid-Cretaceous syntectonic clastic sedimentary rocks exposed along the front of the central Brooks Range were derived almost exclusively from the Kanayut Conglomerate (Crowder, 1987, 1989; Handschy, 1988), indicating that the northern imbricates were exposed at the surface during thrusting. In contrast, the modern erosional limit of the Skajit limestone to the south, east, and northeast of the Doonerak window (Figs. 1 and 2), plus a southward increase in the amount of chlorite and other micas in mudrocks of the southern Endicott Mountains allochthon, suggest that at least part of the southern nappe was overlain by the Skajit allochthon during deformation. Since the Skajit allochthon is as much as 10 km thick (Oldow et al., this volume, Chapter 8), loading by the Skajit allochthon would have caused formation temperatures during deformation to be higher in the southern Endicott Mountains allochthon, thus favoring more ductile deformation.

Throughout the southern nappe, formation contacts, sedimentary facies boundaries, domain boundaries, intraformational thrusts, and the Amawk thrust (Fig. 2) are approximately parallel. Sedimentary facies are, for the most part, in normal stratigraphic order despite intranappe deformation. This suggests that the southern nappe deformed in a layered fashion, that the layers were bounded by detachment surfaces (thrust faults) subparallel to the initial bedding planes, and that the depth related increase in deformation intensity reflects a strain gradient.

Changes in D1 fold style and orientation are the most obvious indicators of strain-depth relationships within the southern Endicott Mountains allochthon (Fig. 7). As the depth below the Kanayut Conglomerate–Hunt Fork Shale contact increases, the angles between axial planes, cleavage, and the Amawk thrust decrease (Fig. 7a). D1 fold interlimb angles become tighter and D1 fold axes change from a strike-parallel point concentration in Domain 2, to girdle distributions in Domains 3 and 4, and then to a dip-parallel point concentration in Domain 5 (Fig. 8). This systematic change in fold orientation from Domain 3 to Domain 5 suggests that the shear strain intensity increases downward from the base of the Kanayut Conglomerate to the Amawk thrust. The appearance of D1b folds in Domain 4 coincides with very small D1a interlimb angles (Table 1; Fig. 7b) and increasing alignment

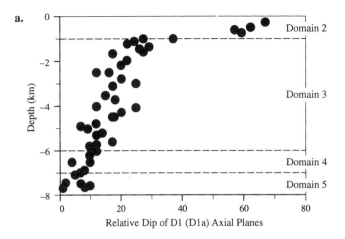

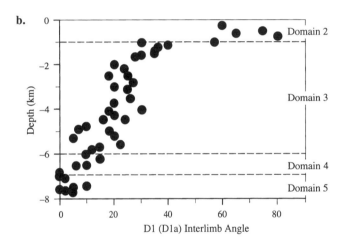

Figure 7. Graphs of depth-related changes in D1 fold orientation and tightness in the southern Endicott Mountains allochthon. The approximate depths of Domain boundaries are plotted for reference. a, Angles between D1 axial planes and the Kanayut Conglomerate–Hunt Fork Shale in Domains 2 and 3, and between D1a axial planes and the Amawk thrust in Domains 4 and 5. Note that the Kanayut Conglomerate–Hunt Fork Shale contact and the Amawk thrust are approximately parallel on the north flank of the Doonerak duplex (Figs. 2b and 3). b, D1 (D1a) interlimb angles.

of D1a fold axes with stretching lineations, apparently reflecting refolding of D1a folds as they became approximately parallel to the shear zone boundaries.

Deformation timing

Unlike D1 structures, which are folded by the Doonerak duplex, the attitudes of D2 axial planes and cleavage are similar on both the northern and southern flanks of the duplex. D2 axial planes and cleavage strike northeast to southwest and S2 × S1 intersection lineations trend northeast in all domains (Fig. 3b). Based on consistently oriented and equally well developed D2

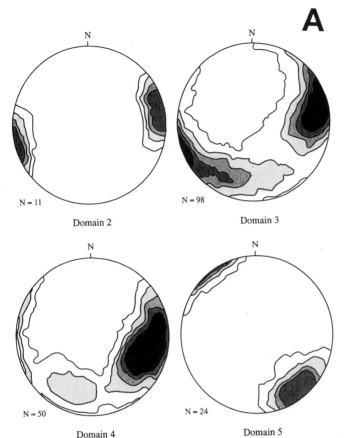

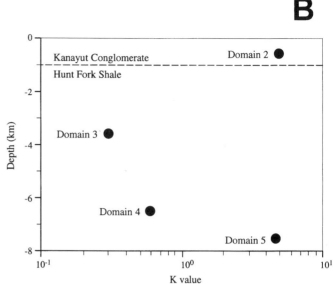

Figure 8. Analysis of D1 folds. a, Structurally corrected lower hemisphere equal-area plots of D1 (D1a) fold axes. The effects of D2 folding have been removed from all domains and D1b folding has been removed from Domains 4 and 5. Contours are 5% of data per 5% of area. Dark areas represent greatest concentrations, light areas are smaller concentrations and white areas contain no fold axes. b, K values (Woodcock, 1977) versus depth for D1 and D1a fold axes. High K values indicate point concentrations; low K values indicate girdle distributions. Depth values are for domain midpoints.

structures in all of the structural sheets above and below the Amawk thrust. Seidensticker et al. (1987) postulated that D2 was genetically linked to the last stages of duplex formation. This idea—combined with the observation that consistently oriented, bivergent D2 structures are superimposed on D1 folds and cleavage in Domain 3, and on both D1a and D1b folds and cleavage in Domains 4 and 5—suggests that D1 in the southern Endicott Mountains allochthon ended before the latest stages of duplex formation. In addition, exposure and erosion of the Kanayut Conglomerate during deposition of mid-Cretaceous syntectonic sediments along the front of the Brooks Range, Albian to Paleocene uplift ages for the northern imbricates of the Endicott Mountains allochthon (Murphy et al., 1994), and late Oligocene uplift of the Doonerak duplex (Blythe et al., this volume, Chapter 10) indicate that emplacement of the Endicott Mountains allochthon occurred significantly earlier than formation of the Doonerak duplex and, thus, D1 and D2 were probably formed by temporally distinct deformation events.

Discussion

D1 fold styles in Domains 3, 4, and 5 are transitional, they are not separated by sharp boundaries or major thrust faults that juxtapose distinctly different depositional facies. Rotation of D1 folds increases with depth in the southern nappe and super-

position of D1b folds on D1a folds occurs in Domains 4 and 5 where D1a interlimb angles approach zero and the angle between the Amawk thrust and D1a axial planes is approximately 10°. Similarly, the interlimb angles of D1b folds and the dip angle between D1b axial planes and the Amawk thrust progressively decrease from the top of Domain 4 to the base of Domain 5. The transitional nature of the Domains 3, 4, and 5 implies that D1, D1a, and D1b folds probably formed during a semicontinuous, progressive deformation.

The change from upright to recumbent folds and from strike-parallel to dip-parallel fold axes is well documented in simple shear zones at many scales (Ramsay and Graham, 1970; Escher and Watterson, 1974; Sanderson, 1979; Mitra and Elliott, 1980). Numerous studies of strain in thrust sheets (e.g., Carmignani et al., 1978; Coward and Kim, 1981; Ramsay et al., 1983) point out that variations in fold orientation are consistent with simple shear models of progressive intranappe deformation. If the primary layering is inclined such that simple shear will cause shortening of the layers, folding of the layers can result from simple shear alone (Ramsay et al., 1983). However, if the initial layering in the thrust sheet is essentially parallel to the basal thrust during deformation, then intranappe folding also requires a component of layer-parallel pure shear (Sanderson, 1982).

The combined effects of pure and simple shear have been recognized in many thrust sheets and, if the sequence of deformation is known, the total strain can be factored into pure shear (α) and simple shear (γ), or volume change (Δv) and simple shear components by plotting strain ratio (Rs) against the angle (θ') between the x-strain axis and the shear direction (Coward and Kim, 1981; Kligfield et al., 1981; Sanderson, 1982). Although the exact deformational history of the Endicott Mountains allochthon is not known, the sequence of deformation is partially constrained by the relative timing of deformational phases, regional crosscutting relationships, and balanced cross sections (Oldow et al., 1987). Four distinct, but partly synchronous, deformational events are recognized (Fig. 9): (1) emplacement of the Skajit allochthon over the southern portion of the Endicott Mountains allochthon, (2) internal imbrication of the Endicott Mountains allochthon, (3) northward displacement of the Endicott Mountains allochthon, and (4) formation of the Doonerak duplex by imbrication of the Apoon assemblage beneath the Endicott Mountains allochthon (Oldow et al., 1987; Julian and Oldow, this volume, Chapter 5). As noted above, D2 structures postdate all D1 structure and are related to formation of the Doonerak duplex.

If the problem is simplified further by assuming plane strain and no significant bulk volume loss during deformation[1], the relative roles of pure shear and simple shear can be evaluated by comparing Rs and θ' (Fig. 10). Sanderson (1982) postulated that layer-parallel pure shear in a thrust sheet is constant with a value $\alpha < 1$ (i.e., layer-parallel shortening) and that the commonly observed depth related decreases in θ' are a function of increasing shear strain. In the Endicott Mountains allochthon, however, measured values for Rs and θ' from mesoscopic strain indicators (Table 2) deviate substantially from a predicted $\alpha =$ constant curve and suggest that the amount of pure shear (α) may also vary with depth in the nappe (Fig. 10).

Proper interpretation of Fig. 10 requires detailed understanding of field relationships and the limitations of the strain measurements. The southern nappe of the Endicott Mountains allochthon is not a continuous media; intraformational thrust faults (discontinuities) are common throughout the nappe. The frequency of intraformational thrusts and bedding-parallel detachments increases with depth in the nappe, possibly indicating that the amount of fault-distributed shear strain also increases with depth. Adjacent to intraformational thrusts, shear strain is heterogeneous. Between faults, different strain indicators often yield different strain values. For example, in Domain 3, Skolithos burrows are sheared from 45 to 70°, which correspond to strain ratios from Rs ≅ 2.6 to Rs ≅ 9.5 if plane strain deformation is assumed. In the same domain, Rf/ϕ analysis of stretched pebbles yield strain ratios less than Rs = 1.2 and boudinage indicates less

than 10% to greater than 600% stretching. These inconsistencies are partly related to lithology, but also reflect heterogeneous strain distribution within the same lithologies.

Since the preserved finite strain represents superposition of multiple, incremental finite strains, the principal uncertainty in this analysis is whether the measured finite strains accurately represent the strains required to produce the observed folds. Geometric modeling of fold axis reorientation (Handschy, 1988) suggests that mesoscopic strain in Domain 3 is adequate to produce the observed fold arcuation. However, measured strains in Domains 4 and 5 are approximately four times smaller than model strains required to produce the observed amounts of fold arcuation. This is not surprising since all of the strain markers are in relatively well cemented, relatively strong lithologies surrounded by highly deformed shale. In this case, where the strain markers have different mechanical properties than the surrounding lithologies, fold geometry and orientation probably provide a more accurate estimate of strain than the finite strain markers do.

Clearly, the use of geometric modeling is limited. If the deformation history and deformation mechanisms are poorly constrained, then more assumptions must be made. For the Endicott Mountains allochthon Handschy (1988) assumed plane strain, no bulk volume loss, and an initial fold axis arcuation of 10 to 15°. Therefore, the models are poorly constrained and should be considered approximations, not accurate estimates. In other areas, where the deformation mechanisms and initial fold geometries are better constrained (e.g., Yang and Nielsen, 1995), geometric modeling may provide accurate strain estimates.

In the Endicott Mountains allochthon both mesoscopic strain indicators and strain ratios required to produce the observed fold axis arcuation suggest that α increases with depth and that the total pure shear component changes from layer-parallel shortening ($\alpha < 1$) in Domain 2 to layer-normal shortening ($\alpha > 1$) in Domains 3 through 5. The change from $\alpha < 1$ to $\alpha > 1$ occurs abruptly across a bedding-parallel detachment that separates the Kanayut Conglomerate from the Hunt Fork Shale (Fig. 5a) and reflects a major change in deformational style across what could easily be mistaken for a normal stratigraphic contact. Within the Hunt Fork Shale and Beaucoup Formation strain is locally heterogeneous, but it appears to increase systematically with depth.

CONCLUSIONS

The change in structural style from single-phase folds and imbricated thrust sheets in Domain 1 to heterogeneous intranappe strain and polyphase folding in Domains 2 through 5 is the product of: (1) rheologic differences between the Kanayut Conglomerate, Hunt Fork Shale, and Beaucoup Formation; (2) the original distribution of sedimentary facies; (3) the depth of deformation; and (4) the strain intensity. The greater thickness of Kanayut Conglomerate, its relative rigidity, and the lack of an overriding thrust sheet favored thrust imbrication in the north, whereas more shale and deformation beneath the Skajit allochthon favored formation of a strain gradient in the south.

---

[1]As a first approximation layer-parallel shortening by folding (pure shear) and volume loss across axial planar cleavage have similar effects on bed length. The principal difference is that pure shear also thickens the layer, whereas dissolution removes mass. Based on field and thin section observations the actual deformation included components of both processes, but it is impossible to quantify the relative proportions. It is also unclear whether material dissolved along cleavage planes was totally removed from the system, or reprecipitated elsewhere in the allochthon (Handschy, 1988).

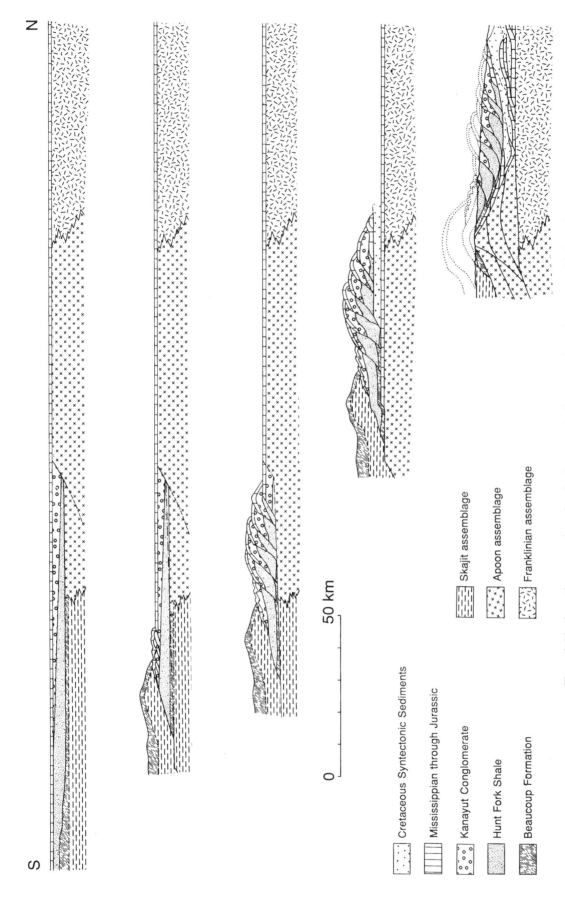

Figure 9. North-south cross sections showing interpreted paleogeography and sequence of deformation for the central Brooks Range. The Endicott Mountains allochthon is composed of the Kanayut Conglomerate, Hunt Fork Shale, and Beaucoup Formation. The top cross section shows the distribution of pre-Jurassic rocks prior to Brookian thrusting. The bottom cross section shows the present-day structure of the central Brooks Range. The four deformational events described in the text are represented in the bottom four cross sections.

Cretaceous Syntectonic Sediments

Mississippian through Jurassic

Kanayut Conglomerate

Hunt Fork Shale

Beaucoup Formation

Skajit assemblage

Apoon assemblage

Franklinian assemblage

0   50 km

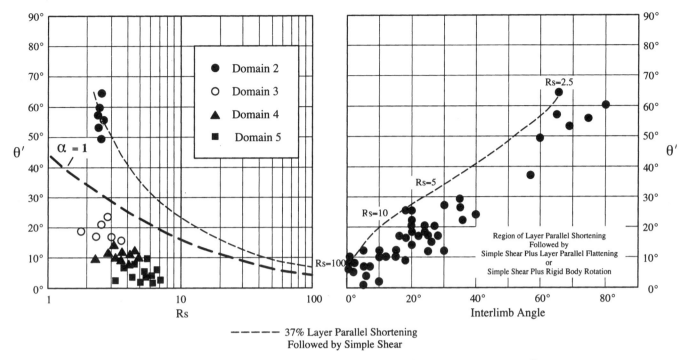

Figure 10. Strain measurements and theoretical strain paths. a, Strain ratio (Rs) versus angle (θ') plot. Rs and θ' in Domain 2 are from geometric analysis of large folds. In Domains 3, 4, and 5 Rs and θ' are derived from deformed clast measurements. Objects deformed by simple shear alone should follow the $\alpha = 1$ line. Deformation above the $\alpha = 1$ line is a combination of simple shear and shortening parallel to the shear direction. Deformation below the $\alpha = 1$ line is a combination of simple shear and extension parallel to the shear direction. The modeled layer-parallel shortening followed by simple shear deformation path assumes that Domain 2 represents the starting point and that folding was the product of a simple two-increment, constant volume deformation. b, θ' versus fold interlimb angle plot. The layer-parallel shortening followed by simple shear curve and Rs values are the same as in Figure 10a. Since most of the measured folds have lower θ' angles and tighter interlimb angles than the simple theoretical model, it is inferred that folding also included a component of rigid body rotation and flattening perpendicular to the shear zone boundaries.

Unlike strain gradients documented in the Helvetic Alps (Ramsay et al., 1983), the Apennines (Carmignani et al., 1978), and the Moine Thrust Zone (Coward and Kim, 1981) where the deformation was apparently dominated by simple shear, deformation in the southern Endicott Mountains allochthon was more complex and included components of layer-parallel shortening, simple shear, and layer-normal flattening of the shear zone.

ACKNOWLEDGMENTS

This study was funded by the U.S. Department of Energy (grant DE-AS05-83ER13124), the National Science Foundation (grant EAR-8517384), and the Rice University Alaska Industrial Associates Program (Amoco, Arco, Chevron, Gulf, Mobil, and Sohio). Discussions with J. S. Oldow, J. T. Dillon, W. K. Wallace, H. G. Avé Lallemant, C. L. Hanks, T. E. Moore, R. R. Gottschalk, F. E. Julian, C. M. Seidensticker, J. C. Phelps, and K. W. Boler greatly improved my understanding of Brooks Range geology. R. Guthrie and K. Hewlett assisted with field work. K. Wall, K. Kimbro, and L. A. Beckmann drafted the figures. J. S. Oldow, A. W. Bally, and H. G. Avé Lallemant reviewed an early version of the manuscript. Special thanks to K. C. Nielsen and F. A. Diegel for providing thorough, informative reviews that greatly improved this paper.

REFERENCES CITED

Brosgé, W. P., Reiser, H. N., Dutro, J. T., Jr., and Dettermnan, R. L., 1979, Bedrock geologic map of the Phillip Smith Mountains quadrangle, Alaska: U.S. Geological Survey Miscellaneous Field Studies Map MF-879B, 2 sheets, scale 1:250,000.

Carmignani, L., Giglia, G., and Kligfield, R., 1978, Structural evolution of the Apuane Alps: An example of continental margin deformation in the northern Apennines, Italy: Journal of Geology, v. 86, p. 487–504.

Coward, M. P., and Kim, J. H., 1981, Strain within thrust sheets, *in* McClay, K., and Price, N. J., eds., Thrust and nappe tectonics: Geological Society London, Special Publication, p. 275–292.

Crowder, R. K., 1987, Cretaceous basin to shelf transition in northern Alaska: Deposition of the Fortress Mountain Formation, *in* Tailleur, I., and Weimer, P., eds., Alaskan North Slope geology: Society of Economic Mineralogists, Pacific Section, Publication 50, p. 449–458.

Crowder, R. K., 1989, Deposition of the Fortress Mountain Formation, *in* Mull,

C. G., and Adams, K. E., eds., Dalton Highway, Yukon River to Prudhoe Bay, Alaska: Alaska Division of Geological and Geophysical Survey Guidebook 7, v. 2, p. 293–301.

Dunne, W. M., Onash, C. M., and Williams, R. T., 1990, The problem of strain marker centers and the Fry method: Journal of Structural Geology, v. 12, p. 933–938.

Durney, D. W., 1972, Solution-transfer, an important geological deformation mechanism: Nature, v. 235, p. 315–316.

Dutro, J. T., Jr., Brosgé, W. P., Reiser, H. N., and Detterman, R. L., 1979, Beaucoup Formation, a new Upper Devonian stratigraphic unit in the central Brooks Range, northern Alaska: U.S. Geological Survey Bulletin, 1482-A, p. A63–A69.

Escher, A., and Watterson, J., 1974, Stretching fabrics, folds and crustal shortening: Tectonophysics, v. 22, p. 223–231.

Fleuty, M. J., 1964, The description of folds: Geological Association Proceedings, v. 75, p. 461–492.

Fry, N., 1979, Random point distributions and strain measurements in rocks: Tectonophysics, v. 60, p. 89–105.

Gray, D. R., 1978, Cleavages in deformed psammitic rocks from southeastern Australia: Their nature and origin: Geological Society of America Bulletin, v. 89, p. 577–590.

Handschy, J. W., 1988, Sedimentology and structural geology of the Endicott Mountains allochthon, central Brooks Range, Alaska [Ph.D. thesis]: Houston, Texas, Rice University, 172 p.

Kligfield, R., Carmignani, L., and Owens, W. H., 1981, Strain analysis of a Northern Appenine shear zone using deformed marble breccias: Journal of Structural Geology, v. 3, p. 421–436.

Mitra, G., and Elliott, D., 1980, Deformation of basement in the Blue Ridge and the development of the South Mountain cleavage, *in* Wones, D. R., ed., The Caledonides in the USA: Virginia Polytechnical Institute/State University, Memoir 2, p. 307–311.

Moore, T. E., and Nilsen, T. H., 1984, Regional variations in the fluvial Upper Devonian and Lower Mississippian(?) Kanayut Conglomerate, Brooks Range, Alaska: Sedimentary Geology, v. 38, p. 465–497.

Mull, G. C., 1982, Tectonic evolution and structural style of the Brooks Range, northern Alaska: An illustrated summary, *in* Powers, R. B., ed., Geologic studies of the Cordilleran thrust belt: Denver, Colorado, Rocky Mountain Association of Geologists, v. 1, p. 1–45.

Mull, C. G., Adams, K. E., and Dillon, J. T., 1987, Stratigraphy and structure of the Doonerak fenster and Endicott Mountains allochthon, central Brooks Range, Alaska, *in* Tailleur, I., and Weimer, P., eds., Alaskan North Slope geology: Society of Economic Paleontologists and Mineralogists, Pacific Section, Publication 50, p. 663–679.

Murphy, J. M., O'Sullivan, P. B., and Gleadow, A. J. W., 1994, Apatite fission-track evidence for episodic Early Cretaceous to late Tertiary cooling and uplift events, central Brooks Range, Alaska, *in* Thurston, D. K., and Fujita, K., eds., 1992 Proceedings, International Conference on Arctic Margins: Anchorage, Alaska, U.S. Department of the Interior, Minerals Management Service, p. 257–262.

Nilsen, T. H., and Moore, T. E., 1984, Stratigraphic nomenclature for the Upper Devonian and Lower Mississippian(?) Kanayut Conglomerate, Brooks Range, Alaska: U.S. Geological Survey Bulletin, 1529-A, p. A1–A64.

Oldow, J. S., Seidensticker, C. M., Phelps, J. C., Julian, F. E., Gottschalk, R. R., Boler, K. W., Handschy, J. W., and Avé Lallemant, H. G., 1987, Balanced cross sections through the central Brooks Range, Alaska: American Association of Petroleum Geologists Special Publication, 19 p., 8 pl.

Onasch, C. M., 1986, Ability of the Fry method to characterize pressure-solution deformation: Tectonophysics, v. 122, p. 187–193.

Phelps, J. C., 1987, Stratigraphy and structure of the northeastern Doonerak window area, central Brooks Range, northern Alaska [Ph.D. thesis]: Houston, Texas, Rice University, 171 p.

Phelps, J. C., Avé Lallemant, H. G., Seidensticker, C. M., Julian, F. E., and Oldow, J. S., 1987, Late-stage high-angle faulting, eastern Doonerak window, central Brooks Range, Alaska, *in* Tailleur, I., and Weimer, P., eds., Alaskan North Slope geology: Society Economic Paleontologists Mineralogists, Pacific Section, Publication 50, p. 685–690.

Ramsay, J. G., 1967, Folding and fracturing of rocks: New York, McGraw-Hill, 568 p.

Ramsay, J. G., and Graham, R. H., 1970, Strain variation in shear belts: Canadian Journal of Earth Sciences, v. 7, p. 786–813.

Ramsay, J. G., and Huber, M. I., 1983, The techniques of modern structural geology, Volume 1: Strain analysis: London, Academic Press, p. 1–307.

Ramsay, J. G., Casey, M., and Kligfield, R., 1983, Role of shear in development of the Helvetic fold-thrust belt of Switzerland: Geology, v. 11, p. 439–442.

Sanderson, D. J., 1979, The transition from upright to recumbent folding in the Variscan fold belt of southwest England: a model based on the kinematics of simple shear: Journal of Structural Geology, v. 1, p. 171–180.

Sanderson, D. J., 1982, Models of strain variation in nappes and thrust sheets—a review: Tectonophysics, v. 88, p. 201–233.

Seidensticker, C. M., Julian, F. E., Oldow, J. S., and Avé Lallemant, H. G., 1987, Kinematic significance of antithetic structures in the central Brooks Range, Alaska: Geological Society of America Abstracts with Programs, v. 19, p. 449.

Woodcock, N. H., 1977, Specification of fabric shapes using an eigenvalue method: Geological Society of America Bulletin, v. 88, p. 1231–1236.

Yang, Q., and Nielsen, K. C., 1995, Rotation of fold hinge lines associated with simple shear during southerly directed thrusting, Ouachita Mountains, southeastern Oklahoma: Journal of Structural Geology, v. 17, p. 803–817.

MANUSCRIPT ACCEPTED BY THE SOCIETY SEPTEMBER 23, 1997

Geological Society of America
Special Paper 324
1998

Out-of-sequence thrusting and structural continuity of the Endicott Mountains allochthon around the eastern end of the Doonerak window, central Brooks Range, Alaska

James C. Phelps
Chevron Petroleum Technology Company, P.O. Box 446, La Habra, California 90633-0446
Hans G. Avé Lallemant
Department of Geology and Geophysics, Rice University, Houston, Texas 77005-1892

ABSTRACT

The Doonerak window in the central Brooks Range fold and thrust belt is a northeast-southwest–trending, doubly plunging antiform that is underlain by two duplexes: (1) the lower Doonerak duplex, consisting of slices of lower Paleozoic rocks; and (2) the higher Blarney Creek duplex, consisting primarily of slices of upper Paleozoic rocks. Overlying the latter is the Endicott Mountains allochthon, consisting of Devonian calcareous and clastic sedimentary rocks. The Endicott Mountains allochthon is continuous around the eastern end of the window, but it is breached by an out-of-sequence thrust fault (Eekayruk fault). Overlying the Endicott Mountains allochthon is the Skajit allochthon, which consists of lower to mid-Paleozoic calcareous and clastic rocks. All allochthons were deformed by three generations of folds and at least four generations of faults; these structures formed progressively as the result of northwest-directed thrusting in Late Jurassic to Tertiary time.

INTRODUCTION

The Brooks Range of northern Alaska is a north-vergent fold and thrust belt of mid-Jurassic to Tertiary age (e.g., Grantz et al., 1981; Crane, 1987; Oldow et al., 1987). It is a classical fold and thrust belt with a hinterland of penetratively, polyphase deformed, metamorphic rocks in the south and a nonmetamorphic, generally nonpenetratively deformed foreland belt with a mildly deformed foreland basin in the north. Contraction in the belt started in the south during the Middle Jurassic with the northward emplacement of ophiolitic thrust sheets (e.g., Wirth et al., 1993). Thrusting and folding continued as the deformation front migrated northward through Cretaceous and Tertiary time (e.g., Oldow et al., 1987). One of the outstanding problems in the Brooks Range (Figs. 1 and 2 in Preface, this volume) relates to the significance of the Doonerak window, which occupies the core of an northeast-southwest–trending, doubly plunging anti-form. Originally (e.g., Mull et al., 1976), the window was thought to consist of lower and upper Paleozoic rocks, overthrust along the Amawk thrust fault by another package of upper Paleozoic rocks. The lower Paleozoic rocks (here informally called the Apoon assemblage; Oldow et al., 1987; Julian and Oldow, Chapter 5, this volume) were thought to have been deformed during Middle to Late Devonian contraction (the Ellesmerian orogeny of the Canadian Arctic Archipelago; e.g., Norris and Yorath, 1981) and the contact with the overlying upper Paleozoic rocks (Endicott and Lisburne Groups) was generally considered to be an unconformity (e.g., Brosgé and Reiser, 1971). Oldow et al. (1984), however, suggested that the contact between the lower and upper Paleozoic rocks of the Doonerak window is a thrust fault (Blarney Creek thrust fault).

The Amawk thrust fault between the window and the overlying upper Paleozoic rocks has also been controversial. Dutro et al. (1976) suggested that the southern thrust contact was a

Phelps, J. C., and Avé Lallemant, H. G., 1998, Out-of-sequence thrusting and structural continuity of the Endicott Mountains allochthon around the eastern end of the Doonerak window, central Brooks Range, Alaska, *in* Oldow, J. S., and Avé Lallemant, H. G., eds., Architecture of the Central Brooks Range Fold and Thrust Belt, Arctic Alaska: Boulder, Colorado, Geological Society of America Special Paper 324.

52   *J. C. Phelps and H. G. Avé Lallemant*

north-vergent (south-dipping) thrust fault and the northern one a south-vergent (north-dipping) thrust fault, whereas Mull et al. (1976) believed that the entire contact was a north-vergent thrust fault that was deformed into an antiformal structure as a result of a basement uplift.

The main goals of the study presented in this chapter were to investigate (1) the contact between the Apoon assemblage and the overlying rocks, to find out whether this contact is indeed an unconformity or rather a fault; (2) the Amawk thrust fault separating the window from its cover sequence (Is it one fault, as suggested by Mull et al., 1976; or are there two faults as suggested by Dutro et al., 1976?); and (3) the deformational history of the rocks within the window and those in the cover rocks. (Did the Apoon rocks undergo a Late Devonian deformation?) Answers to these questions are important for reconstruction of the Brooks Range fold and thrust belt, because alternative models require drastically different displacements or displacement histories.

To these ends, the two contacts were followed around the entire eastern end of the Doonerak window (Figs. 1 and 2). Geologic reconnaissance maps of the area by Brosgé and Reiser (1964, 1971) and Brosgé and Patton (1982) were used, as well as a then-unpublished geologic map by Dillon et al. (1988). A detailed map at a scale of 1:31,680 was made and structural analysis was performed (Phelps, 1987). The results of this study are (1) that, in general, the upper contact of the Apoon assemblage is a fault (Blarney Creek fault; see also Seidensticker and Oldow, Chapter 6, this volume); (2) that the entire upper contact of the window is a north-vergent thrust fault (Amawk fault); and (3) that the rocks of the Apoon assemblage were not penetratively deformed during the Ellesmerian. Furthermore, a thrust fault was identified that breaches the window (Eekayruk thrust fault, Figs. 1 and 2).

STRATIGRAPHY

The Brooks Range fold and thrust belt has been divided into numerous, approximately east-west–trending terranes, subterranes, or tectono-stratigraphic assemblages (e.g., Churkin and Trexler, 1981; Grantz et al., 1981; Jones et al., 1987; Oldow et al., 1987). The nine tectono-stratigraphic provinces used here are from north to south: (1) North Slope foredeep, (2) North Slope assemblage, (3) Endicott Mountains allochthon, (4) Blarney Creek and Doonerak duplexes, (5) Skajit allochthon, (6) Schist belt, (7) Phyllite belt, (8) Rosie Creek allochthon, and (9) Angayucham terrane. These belts are described in the Preface to this volume (see Figs. 1, 2A, and 2B therein). The tectono-stratigraphic belts of interest here are the Doonerak and Blarney Creek duplexes and the Endicott Mountains and Skajit allochthons, detailed descriptions of which follow hereafter.

Doonerak duplex

The Doonerak duplex is underlain by lower Paleozoic phyllite and volcanic and volcaniclastic rocks and locally by Upper Devonian to Lower Mississippian clastic rocks. It is disrupted by several thrust faults (Julian and Oldow, Chapter 5, this volume).

The pre-Upper Devonian rocks of the Doonerak duplex (Figs. 1, 2, and 3) have been named informally the Apoon assemblage (Oldow et al., 1987). In the central part, the Doonerak duplex consists of four imbricate sheets (Julian and Oldow, Chapter 5, this volume). At the eastern end of the window, however, only two of these sheets are exposed: a tectonically lower unit of volcanic and volcaniclastic rocks and a higher unit of phyllite and argillite.

The volcanic unit is at least 150 m thick (bottom not exposed) and consists mostly of fine-grained volcanic and volcaniclastic rocks interbedded with gray and black argillites and slates. A few dikes and sills also occur. The volcanic rocks are altered to greenstone and consist of feldspar and clinopyroxene phenocrysts in a fine-grained matrix of the same minerals; alteration to chlorite, white mica, and clay minerals is common. The composition of these rocks is basaltic to andesitic (Julian and Oldow, Chapter 5, this volume). The sedimentary rocks are rich in quartz and chlorite with minor graphite and white mica.

The upper unit has a tectonic thickness of 2.4 to 2.7 km and consists of argillite, phyllite, and siltstone with occasional horizons (10 to 50 cm thick) of coarser-grained sandstone and conglomerate. The fine-grained rocks are mineralogically similar to the argillite and slate of the volcanic unit; the coarse-grained rocks consist of fragments of shale, metamorphic rock, chert, and sandstone and grains of quartz, feldspar, and chlorite.

The contact relations with the overlying Kekiktuk Formation are variable. Generally, the contact is the Blarney Creek thrust fault. However, occasionally, the Apoon rocks coarsen upward and grade into typical Kekiktuk conglomerate. In such cases, the contact seems to be conformable and the Blarney Creek thrust fault lies above the Kekiktuk Formation.

Rarely, Kekiktuk Conglomerate is infolded into the Apoon assemblage rocks (Fig. 4). These folds are decapitated by the Blarney Creek thrust. The axial planar cleavage in the Apoon assemblage rocks is not parallel to the Blarney Creek thrust fault or the bedding planes and cleavages in the overlying Kekiktuk Conglomerate, Kayak Shale, and Lisburne Group limestone (see also Fig. 2). These features are easily misinterpreted as indicating an angular unconformity between the Apoon assemblage and the upper Paleozoic rocks. Below, the significance of these structures will be discussed (see also Seidensticker and Oldow, Chapter 6, this volume).

No fossils were found in the area during the present study and radiometric dates were not acquired. Fossils of Middle Cambrian (Brosgé et al., 1962; Dutro et al., 1984), and Ordovician and Silurian (Moore and Churkin, 1984) ages have been found elsewhere in the Apoon assemblage. Dutro et al. (1976) reported two Ordovician and Devonian K/Ar ages of mafic dikes, intruding into the volcanic rocks in the Doonerak window.

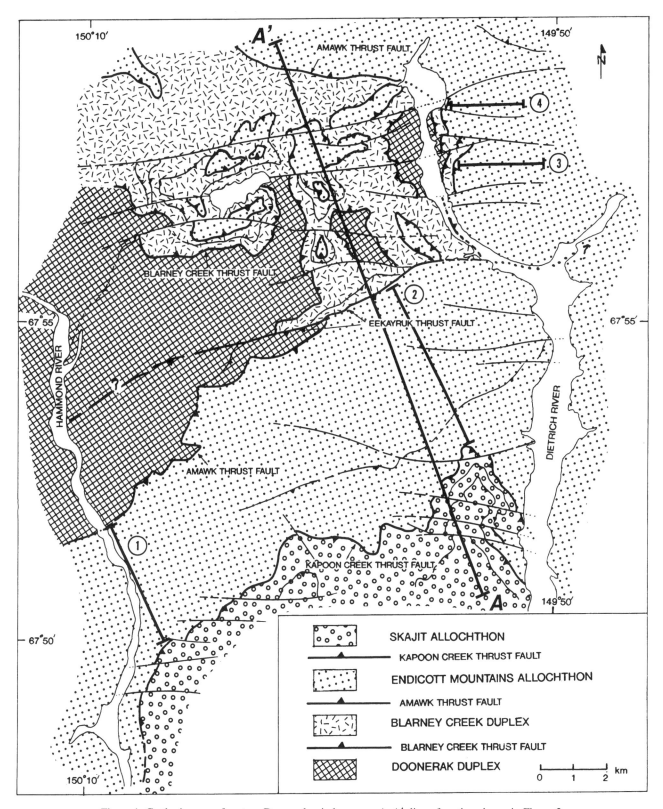

Figure 1. Geologic map of eastern Doonerak window area. A–A′: line of section shown in Figure 2; lines with numbers 1 to 4: stratigraphic sections shown in Figure 5.

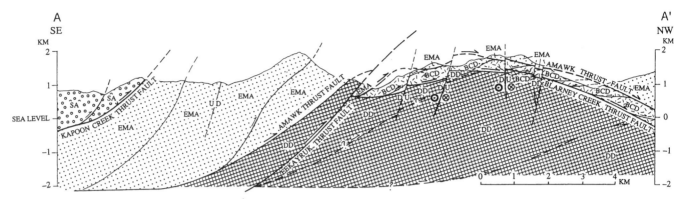

Figure 2. Cross section A–A′; location in Figure 1. Abbreviations: BCD, Blarney Creek duplex; DD, Doonerak duplex; EMA, Endicott Mountains allochthon; SA, Skajit allochthon.

Blarney Creek duplex

Overlying the Doonerak duplex is the Blarney Creek duplex; its upper contact is the Amawk thrust fault (see Seidensticker and Oldow, Chapter 6, this volume). The Blarney Creek duplex consists of several horses of Upper Devonian to Mississippian Kekiktuk Conglomerate, Mississippian Kayak Shale, and Mississippian to Pennsylvanian Lisburne Group carbonates, with rare slices of lower Paleozoic rocks (Figs. 1, 2, and 3; Seidensticker and Oldow, Chapter 6, this volume). The duplex has been traced for a short distance around the eastern terminus of the Doonerak window; it has been cut out in the south by the Eekayruk thrust fault.

Kekiktuk Conglomerate. The Kekiktuk Conglomerate has a maximum thickness of 35 m in the study area. It consists of about equal amounts of siltstone to sandstone and poorly to moderately sorted conglomerate in beds, 1 to 4 m thick. Clast size in the conglomerate is between 2 and 15 mm. The siltstone and sandstone consist primarily of quartz and chert grains; it is noteworthy that feldspar is absent. The conglomerate consists primarily of chert and argillite fragments; no volcanic fragments were encountered. Grains and pebbles are subangular to subrounded. Rare cross-beds have been observed.

Fossils were not found in the formation in the study area, nor elsewhere. Brosgé et al. (1962) assigned a Late Devonian to Mississippian age to the Kekiktuk Conglomerate on the basis of stratigraphic position. Nilsen and Moore (1984) and Melvin (1987) suggested a fluvial environment of deposition. It is important to note that the Apoon assemblage clearly is not the source for the Kekiktuk, as the Apoon is poor in chert and rich in volcanic rocks.

Kayak Shale. The original thickness of the Kayak Shale in the area is unknown because of tectonic disruption; tectonic thickness is between 100 and 300 m. The upper contact with the Lisburne Group limestones is generally sheared. Where not faulted, the lower contact of the Kayak Shale is gradational with the Kekiktuk Conglomerate. Generally, the Kayak Shale consists of dark gray to black slates and argillites with a distinctive

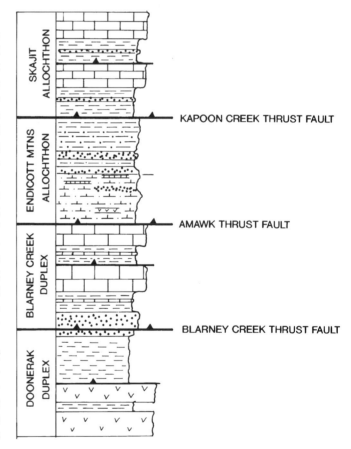

Figure 3. Simplified tectono-stratigraphic column of study area. Doonerak duplex consists of imbricated assemblage of phyllites, volcanic and volcaniclastic rocks, and, rarely, infolded slivers of Kekiktuk Conglomerate. Blarney Creek duplex consists of horses of Kekiktuk Conglomerate, Kayak Shale, and Lisburne Group limestone. Endicott Mountains allochthon consists of calcareous clastic rocks and limestone of the Beaucoup Formation, overlain by noncalcareous quartzite, argillite, and phyllite of the Hunt Fork Shale. Skajit allochthon consists of horses of marble, calcschist, and phyllite.

(2 to 3 m thick) red-weathering fossil-rich, bioclastic packstone near the upper contact.

Fossils in the packstone are the same as described elsewhere and are of Mississippian age (e.g., Bowsher and Dutro, 1957; Armstrong et al., 1976; Nilsen, 1981). Nilsen (1981) suggested that the Kayak Shale was deposited in a low-energy, deep-water environment or protected lagoon; the presence of well-preserved bryozoa in the study area fits better with a lagoonal environment of deposition, however.

Lisburne Group. The Lisburne in the study area is only 300 to 400 m thick; the upper depositional contact is not seen because of truncation by the Amawk thrust fault. The group consists of thick alternating beds (2 to 4 m) of light- to medium-gray packstone and wackestone rich in corals, crinoids, and bryozoa. At several horizons, irregular black chert nodules (5 to 10 cm long) and chert beds (2 to 6 cm thick) occur. The abundance of chert increases upward in the bottom half and decreases again in the upper half of the exposed section. Nodular chert and bedded chert are generally mutually exclusive. Dolostone horizons (between 1 and 3 m) and argillaceous horizons (2 to 5 m thick) occur rarely throughout the sequence. Stromatolites occur in the upper part of the section.

Elsewhere in the Brooks Range the Lisburne Group has been divided into three formations on the basis of microfossils (e.g., Armstrong, 1974; Armstrong and Mamet, 1974; Armstrong et al., 1976; Nilsen, 1981). No effort was spent in the study area to apply this division. Macrofossils were collected and were identified as Mississippian (Phelps, 1987). The Lisburne Group elsewhere has been dated as Mississippian and Pennsylvanian (e.g., Armstrong and Mamet, 1974). According to Armstrong and Mamet (1977), the Lisburne Group was deposited on a slowly subsiding shallow-water shelf; the environment of deposition alternated from open-marine to subtidal. The presence of stromatolites in the study area suggests very shallow conditions.

Endicott Mountains allochthon

The Endicott Mountains allochthon overlies the Blarney Creek duplex in the north, but on the southeast side of the Doonerak window it is directly in contact with the Apoon assemblage (Figs. 1, 2, and 3). Regionally, the allochthon comprises the Devonian Beaucoup Formation and the Devonian to Mississippian Endicott Group (Hunt Fork Shale, Kanayut Conglomerate, Kayak Shale, and Lisburne Group; Oldow et al., 1987), but only the Beaucoup Formation and Hunt Fork Shale occur in the study area.

Beaucoup Formation. A 1- to 2-km-thick sequence of mostly calcareous siltstone and sandstone overlies the southern and eastern flanks of the Doonerak window. The lower contact with the Apoon assemblage is a low-angle fault cutting out sections in both the lower and upper plates and is correlated with the Amawk thrust fault (Figs. 1 and 2). These rocks were mapped but unnamed by Brosgé and Reiser (1964, 1971), and

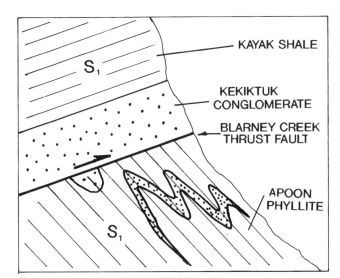

Figure 4. Schematic representation of contact between Apoon assemblage with infolded Kekiktuk Conglomerate and overlying Kekiktuk Conglomerate and Kayak Shale. Cleavage S_1 in Apoon formed at same time as S_1 in Kayak Shale but before juxtaposition along Blarney Creek thrust fault.

correlated by Dutro et al. (1979) with the Beaucoup Formation. Several low- and high-angle faults have been encountered within the unit; although displacements along these faults seem to be small (Phelps, 1987), the real thickness of the unit has not been established.

The Beaucoup Formation in the study area consists of two distinct facies. A "southern calcareous" facies (exposed along the southern and southeastern margin of the Doonerak window) is composed mainly of calcareous sandstone and siltstone, whereas the "eastern transitional" facies (exposed along the northeastern margin of the window) has a large component of interlayered noncalcareous clastic rocks (Fig. 5). The upper contact of the "southern calcareous" facies with the Hunt Fork Shale is sharp and conformable. The upper contact of the "eastern transitional" facies is gradational with the Hunt Fork Shale. The calcareous Beaucoup Formation appears to grade northeastward and upward into the noncalcareous Hunt Fork Shale. It may represent the transitional facies between the Hunt Fork Shale and a carbonate shelf to the south or southwest. This carbonate shelf may be represented by the Skajit Limestone (Oldow et al., Chapter 7, this volume).

The two facies of the Beaucoup Formation contain the same rock types, but in very different proportions. The main rock types are (1) calcareous sandstone and siltstone, (2) limestone, (3) argillite, (4) quartzite, (5) conglomerate, and (6) greenstone.

Calcareous sandstone and siltstone. About 80 to 90% of the "southern calcareous" facies consists of light gray, orange to brown weathering, rhythmically, thin- to thick-bedded (1 to 100 cm), calcareous sandstone and siltstone, whereas the proportion of this rock type is as low as 20% in the "eastern transitional" facies. The rock is poorly sorted and immature containing angu-

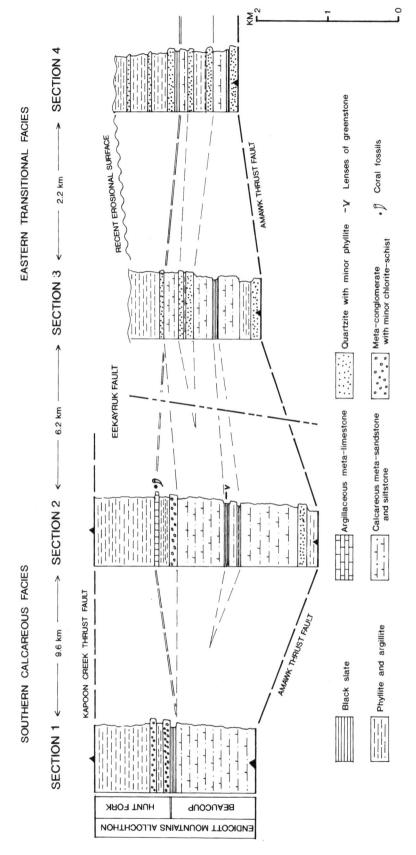

Figure 5. Litho-stratigraphic correlation of Beaucoup Formation and Hunt Fork Shale in study area; the four sections are from lines 1 to 4 in Figure 1.

lar to subangular grains of calcite, quartz, feldspar, and muscovite and fragments of chert, shale, and metamorphic rocks. Secondary chlorite is ubiquitous. The abundance of chlorite, feldspar, and metamorphic and shale fragments increases upward, whereas the calcite content decreases. Cross-beds, cut-and-fill structures, and other primary sedimentary features are rarely preserved.

Limestone. Lenses of fossiliferous, argillaceous, and graphitic limestone occurring in the upper part of the sections are 10 to 30 m thick and as much as 50 m long. The rock is generally a very fine grained mudstone. Interbedded carbonate sandstone contains brachiopod fragments, fenestrate bryozoan fonds, and corals. The corals have close affinity with *Diplophyllum caespitasum* and *Synatophyllum arundinaceum* (Phelps, 1987), both of Early to middle Late Devonian age (Hill, 1981). Locally, the limestone is metamorphosed to banded marble.

Argillite. Black argillite, slate, and phyllite occur as thin (2 to 10 cm) horizons in the calcareous sandstones. They are noncalcareous and show rhythmic layering.

Quartzite. The quartzite is fine to medium grained, moderately sorted, and thick bedded (1 to 2 m). It consists mainly of quartz grains and some chert and very few shale fragments; calcite grains were not observed. The quartzite contains little mud-sized matrix. Quartz and hematite cement is ubiquitous.

Conglomerate. Chert-pebble conglomerate occurs in the uppermost Beaucoup Formation. It is poorly sorted. Clast size ranges to as large as 1 cm. The clasts are subangular to subrounded and consist generally of chert, quartzite, and rare feldspar, mafic, and calcite grains. The matrix is a fine- to medium-grained, well-cemented quartzite. Secondary chlorite is ubiquitous.

Greenstone. Lenses of greenstone (1 to 3 m thick and 5 to 20 m long) occur in the argillite and conglomerate units. Chemical analyses indicate that they are of basaltic to andesitic composition (Phelps, 1987).

Hunt Fork Shale. The clastic sequence conformably overlying the Beaucoup Formation was correlated by Brosgé and Reiser (1964, 1971) and Dutro et al. (1976) with the Hunt Fork Shale. The upper contact with the Kanayut Conglomerate is not seen in the study area, because of structural removal along the Kapoon Creek thrust fault (Fig. 1). The rocks in the study area have undergone folding and faulting and, thus, thicknesses in Figure 5 are structural.

No fossils were recovered from the Hunt Fork Shale in the study area. Elsewhere, Late Devonian fossils have been found (Chapman et al., 1964; Dutro et al., 1979). Nilsen (1981) interpreted the Hunt Fork Shale as having been deposited in a slope or prodelta setting, the lower part having formed in a deep-marine and the upper in a shallow-marine environment.

The Hunt Fork Shale consists of three rock types: (1) conglomerate, (2) quartzite, and (3) argillite (Fig. 5). The main characteristic that distinguishes the Hunt Fork from the Beaucoup Formation is the scarcity of carbonate in the Hunt Fork Shale. The conglomerate is only found in the lowermost section of the Hunt Fork. Quartzite and argillite occur throughout the section, but quartzite is more abundant in the lower part,

whereas the upper part of the section contains more argillitic material.

Conglomerate. The conglomerate unit is an alternating sequence of pebble conglomerate/sandstone and purple/green slate with locally calcschist. The conglomerate and coarse-grained sandstone consist of quartz, chert, and shale fragments within a matrix of very fine grained chlorite and quartz. Maximum clast size is 8 cm. The pebbles are rounded. Individual beds are as much as 15 m thick. The purple and green slate is noncalcareous and occurs in 2- to 10-m-thick beds. The calcschist contains as much as 20% calcite; beds are also 2 to 10 m thick.

Quartzite. Quartzite is fine to medium grained, moderately well sorted with little or no clay matrix. It consists mostly of quartz grains; chert and shale fragments are rare. Grains are subangular to subrounded. In the lower part of the section, the quartzite is massive; up-section, the quartzite is interbedded with siltstone and mudstone. Beds have a thickness of between 1 and 5 m.

Argillite. Most of the upper Hunt Fork Shale in the study area consists of argillite, slate, and phyllite with interbeds of fine-grained sandstone and siltstone. Rhythmic cycles of graded sandstone to shale beds were locally observed. Individual sandstone beds range in thickness from 1 to 200 cm; siltstone beds range from 4 to 15 cm in thickness. The abundance and thickness of the argillite, slate, and phyllite increase up-section. This rock unit consists of quartz and minor albite grains, rare metamorphic and chert fragments, and variable amounts of phyllosilicates; the clastic grains are subangular.

Skajit allochthon

The Skajit allochthon overlies the "southern calcareous" facies of the Beaucoup Formation and Hunt Fork Shale along the Kapoon Creek thrust fault (Figs. 1, 2, and 3). The allochthon consists of chloritic and graphitic phyllite, slate, and schist, argillaceous to pure marble, and calcareous sandstone and siltstone (see Oldow et al., Chapter 7, this volume, for a more detailed description of these rocks).

No fossils were found in the Skajit during the present study. Elsewhere, Cambrian (Palmer et al., 1984), Silurian (Schrader, 1902; Smith and Mertie, 1930), and Middle to Late Devonian (Brosgé, 1960; Brosgé et al., 1962) fossils have been described. Oldow et al. (Chapter 7, this volume) propose that the Skajit was deposited on a carbonate platform, south of the more distal carbonates of the Beaucoup Formation.

STRUCTURAL GEOLOGY

The map and cross section of the study area (Figs. 1 and 2) demonstrate that the rocks have been folded into a long-wavelength, northeast-plunging antiform. They are disrupted by both low-angle and high-angle faults. Mesoscopic structural data, both ductile and brittle, were collected, and they show that the deformation is more complicated than the map and cross

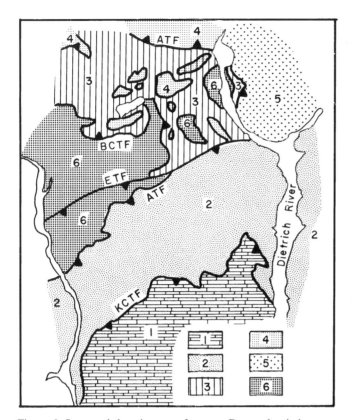

Figure 6. Structural domain map of eastern Doonerak window area. Domain 1: Skajit allochthon; 2: southern Endicott Mountains allochthon; 3: Blarney Creek duplex; 4: northern Endicott Mountains allochthon; 5: eastern Endicott Mountains allochthon; 6: Apoon assemblage. Abbreviations: ATF, Amawk thrust fault; BCTF, Blarney Creek thrust fault; ETF, Eekayruk thrust fault; KCTF, Kapoon Creek thrust fault.

sections indicate. Based on observed differences in the orientations of penetrative structures the area was divided into six distinct domains (Fig. 6). Ductile deformation data for each domain are shown in Figure 7. Brittle deformation data for the entire area are shown in Figure 8. At least three generations of mesoscopic folds and four generations of mesoscopic faults were recognized.

Folds

First-generation folds (D$_1$). The first phase of deformation resulted in the most penetrative ductile structures in the area. It is characterized by a very well developed, penetrative cleavage (S$_1$): an extremely fissile slaty cleavage in the fine-grained rocks and a phacoidal cleavage in the coarser lithologies. The cleavage is axial planar to relatively rare, mesoscopic, rootless, tight to isoclinal folds of "class 2" geometry (Ramsay, 1967). A very rare relict cleavage, which predates the regionally important S$_1$ cleavage, was observed in only one hand sample (collected from the Hunt Fork Shale). This older cleavage is much better developed in the southern belts of the Brooks Range (see Gottschalk, 1990; Oldow et al., Chapter 8, this volume).

The orientation of the bedding planes (S$_0$) and cleavages (S$_1$) in domains 1 to 5 (Figs. 6 and 7) clearly shows that, statistically, they are parallel to the Amawk and Blarney Creek thrust faults and wrap around the northeast-plunging Doonerak antiform (see also Fig. 1). In the Apoon assemblage (domain 6) though, the bedding and cleavage are discordant to the fault contacts: whereas the faults in the north dip to the north, the cleavage dips southeasterly.

The axes of the mesoscopic folds and intersection lineations trend northwest-southeast. Only a few stretching lineations were encountered in deformed pebble conglomerates in the Skajit allochthon just above the Kapoon Creek thrust fault; these plunge southeasterly. The vergence of the D$_1$ folds could not be established.

Second-generation folds (D$_2$). The second-generation folds are characteristically chevron- and kink-style folds and have "class 1b" geometry (Ramsay, 1967). Fold wavelengths vary from microns to hundreds of meters. Axial planar cleavage is locally well developed; some dissolution along the cleavage plane has occurred.

Most axes of the D$_2$ folds and D$_2$ lineations (intersection and crenulation lineations) plunge moderately to the northeast in all domains. Almost all folds are asymmetric; most folds are southeast vergent; some, however, verge to the northwest. Typically, the largest and most appressed south-vergent folds occur in the southern part of the study area, just underneath the Kapoon Creek thrust fault.

Third-generation folds (D$_3$). Gentle to open folds with a weakly developed axial planar cleavage characterize the third-phase folds. The folds are generally only of meso- to microscopic scale.

The fold axial planes strike northwest-southeast and are steep; fold axes and crenulation lineations trend northwest-southeast, but have variable plunge (Fig. 7). These structures are best developed near major, map-scale, high-angle faults and, thus, might be genetically related to them.

Faults

Numerous megascopic and mesoscopic faults occur in the area. Megascopic faults tend to strike east-west to northeast-southwest and are low-angle thrust faults and high-angle, mostly left-lateral strike-slip faults (Figs. 1 and 2). The amount of displacement along the low-angle thrust faults (Amawk, Blarney Creek, Eekayruk, and Kapoon Creek faults) could not be established but is believed to be quite large (Oldow et al., 1987; Seidensticker and Oldow, Chapter 6, this volume). Displacements along the high-angle faults seem to be minor. Mesoscopic faults of virtually every orientation and sense of displacement indicate a complex deformation history.

Amawk thrust fault. The Amawk thrust fault (Figs. 1 and 2) juxtaposes the Endicott Mountains allochthon against the Apoon and Blarney Creek duplexes. It has been traced along the northern flank of the Doonerak window for more than 20 km, from Mount Doonerak to the Hammond River (Brosgé and Reiser, 1964, 1971;

Mull et al., 1987; Seidensticker and Oldow, Chapter 6, this volume). During the present study, the fault was traced around the northeastern corner to the south flank of the window. The fault is not very well exposed and no kinematic indicators were found to establish sense of shear. Because the fault seems to cut up-section progressively from east to west (see map by Brosgé and Reiser, 1971) and because the transport direction is believed to be parallel to the fold axes of the mesoscopic isoclinal folds, we conclude that the Endicott Mountains allochthon was emplaced along the Amawk thrust fault from southeast to northwest.

Blarney Creek thrust fault. The Blarney Creek thrust fault (Figs. 1 and 2; Oldow et al., 1984, 1987; Seidensticker and Oldow, Chapter 6, this volume) separates the underlying Doonerak duplex from the overlying Blarney Creek duplex, which consists of several imbricate slices of Kekiktuk Conglomerate, Kayak Shale, and Lisburne Group limestone, and, rarely, of slices of the lower Paleozoic Apoon assemblage. Faults between adjacent thrust slices cut up-section toward the north, indicative of northward tectonic transport. For more details see Seidensticker and Oldow (Chapter 6, this volume). The Apoon assemblage underlying the Blarney Creek thrust fault consists of several imbricate sheets; Julian and Oldow (Chapter 5, this volume) show that they constitute a major duplex.

Eekayruk thrust fault. At the southeastern end of the Doonerak window (Figs. 1 and 2), a low-angle south-dipping fault juxtaposes rocks of the "southern" facies of the Beaucoup Formation in the hanging wall against Beaucoup rocks that are of the transitional "eastern" facies in the footwall. This fault, herein named the Eekayruk fault, has been traced to the west throughout the present study area and that of Julian and Oldow (Chapter 5, this volume) where it places rocks of the Endicott Mountains allochthon in the hanging wall against the Apoon assemblage in the footwall (Fig. 2). These relationships clearly indicate that it is a thrust fault. Because it breaches the Amawk and Blarney Creek thrust faults, the Eekayruk fault must be an out-of-sequence or an envelopment thrust fault. In their tectonic model for the Skajit allochthon, Oldow et al. (Chapter 8, this volume) prefer the Eekayruk to be an envelopment thrust fault.

Kapoon Creek thrust fault. The south-dipping Kapoon Creek thrust fault (Figs. 1 and 2) juxtaposes the Cambrian to Devonian Skajit assemblage in the hanging wall onto the Devonian Endicott Mountains assemblage. The fault is characterized by a severely deformed limestone pebble conglomerate, which indicates north-northwest or south-southeast tectonic transport.

High-angle faults. The nature of the major east- to northeast-trending high-angle faults is not completely understood. Fibrous calcite crystals decorating the fault planes invariably indicate that these faults are strike-slip faults (both left- and right-lateral) with a small component of dip-slip. The dip-slip component is from a few meters to as much as 500 m (Figs. 1 and 2); most dip-slip displacements are down to the north. The amount of strike-slip displacement could not be established. These observations do not preclude that these high-angle faults had an earlier history of mostly dip-slip displacements, but no trace of this has been found.

The timing of faulting is not well constrained; the faults cut across and, thus, are younger than D_1 and D_2 structures. D_3 folds are best developed adjacent to these strike-slip faults, which may indicate that the faults formed during D_3.

Mesoscopic faults. Numerous outcrop-scale faults occur throughout the area. Displacements along these faults are minor. Kinematic analysis indicates that there are four populations of faults: normal faults, reverse faults, and two sets of strike-slip faults (Fig. 8). The normal faults strike generally northwest-southeast approximately perpendicular to the tectonic transport direction of the thrust faults. Only a few reverse faults were observed; their orientations are variable. The orientations of two sets of conjugate strike-slip faults are compatible with northwest–southeast and northeast–southwest contraction. Whereas the relative timing of the various faults is generally unknown, crosscutting relationships reveal that the first set of strike-slip faults (northwest-southeast contraction) is older than the other set.

SUMMARY AND CONCLUSIONS

The Brooks Range fold and thrust belt consists of many east-west– to northeast-southwest–trending belts, most of which are allochthonous. Four allochthons are recognized in the present study area at the eastern termination of the Doonerak window in the central Brooks Range. They are from bottom to top: the Doonerak duplex, the Blarney Creek duplex, the Endicott Mountains allochthon, and the Skajit allochthon. The Doonerak window is a huge antiform, which in the study area plunges northeastward.

The Doonerak duplex consists of several thrust imbricates of the lower Paleozoic Apoon assemblage and is separated from the overlying Blarney Creek duplex by the Blarney Creek thrust fault. The Blarney Creek duplex consists of slices of the upper Paleozoic Kekiktuk Conglomerate, Kayak Shale, and Lisburne Group limestone and rare horses of the Apoon assemblage. Separating these allochthons from the overlying Endicott Mountain allochthon consisting of the Devonian Beaucoup Formation and Hunt Fork Shale is the Amawk thrust fault. The Endicott Mountain allochthon has been mapped as a continuous belt around the eastern terminus of the Doonerak window.

All four allochthons were deformed during three phases of folding and at least four phases of faulting. The hypothesis that rocks in the Doonerak window (Apoon assemblage) also underwent deformation in mid-Paleozoic time (e.g., Mull et al., 1976) could not be substantiated. The discordance of structures in the lower Paleozoic rocks with those in the overlying upper Paleozoic rocks (Endicott and Lisburne Groups) is not the result of an angular unconformity but of a thrust fault (Blarney Creek thrust fault). Locally, a depositional, conformable contact between the upper Paleozoic (Kekiktuk Conglomerate) and lower Paleozoic rocks (Apoon assemblage) has been observed.

The first-generation folds (D_1) are isoclinal and have a penetrative axial-planar cleavage and, locally, a well-developed stretching lineation. This lineation trends southeast-northwest and probably formed as the result of northwest-directed thrusting

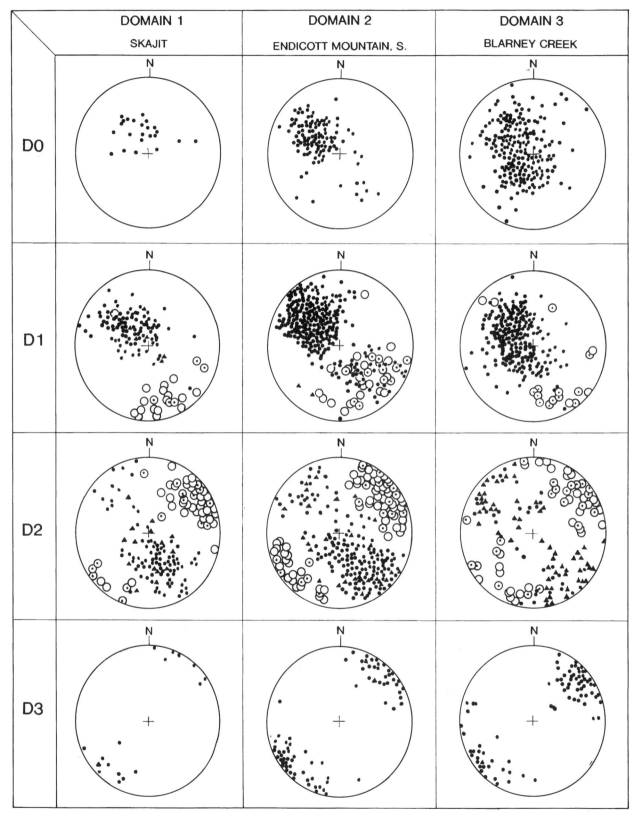

Figure 7 (on this and facing page). Lower hemisphere, equal-area projections of mesoscopic ductile fabric elements in six domains of Figure 6: A, Domains 1 to 3; B, Domains 4 to 6. Dots: poles to bedding (D_0) and cleavages (D_1 to D_3); triangles: poles to fold axial planes; dotted circles: fold axes; circles: intersection and crenulation lineations.

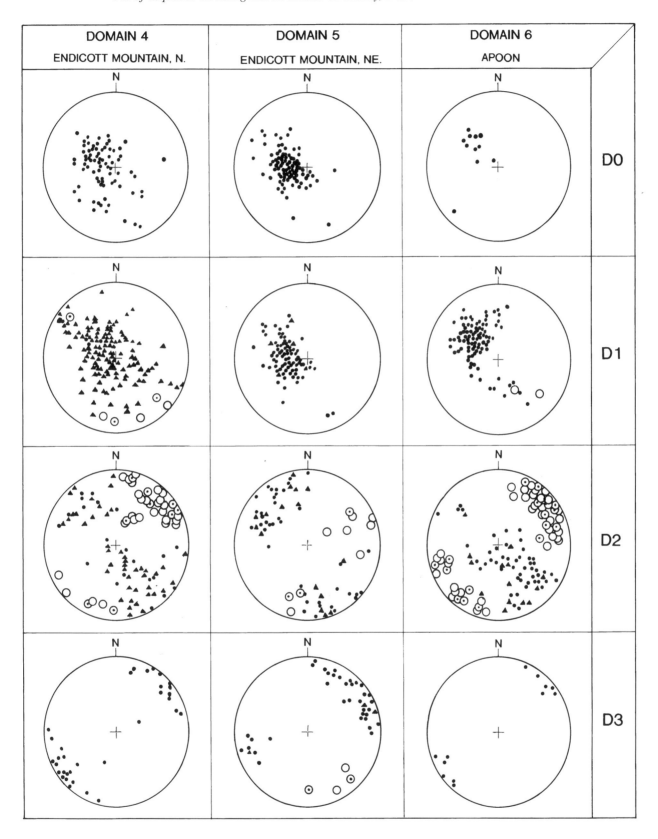

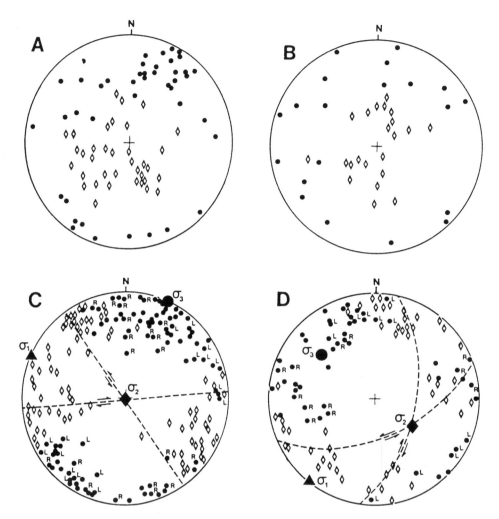

Figure 8. Lower hemisphere, equal-area projections of mesoscopic brittle fabric elements. Faults were separated into four groups: normal faults (A), reverse faults (B), and two sets of conjugate strike-slip fault systems (C and D). Set D faults are younger than set C faults. Dots: poles to faults (R: right lateral; L: left lateral); open diamonds: slickenlines; stippled planes in C and D: average orientation of left-lateral and right-lateral strike-slip faults; σ_1, σ_2, and σ_3: estimated orientations of the maximum, intermediate, and least principal compression axes, respectively.

along a basal thrust, which has to underlie the Apoon assemblage but is nowhere exposed in the area. Isoclinal fold axes are parallel to the stretching lineation and were probably rotated into alignment with the tectonic transport direction. At a late stage, the Amawk, the Blarney Creek, and the basal thrust faults were all three simultaneously active and the Apoon and the Blarney Creek duplexes formed (see Seidensticker and Oldow, Chapter 6, this volume) resulting in the discordant structures along the Blarney Creek thrust fault.

The second-generation folds (D_2) are generally southeast vergent (only a few folds are northwest vergent). They are best developed south of the Doonerak window and may be the result of accommodation to resolve a space problem related to the northwest-directed thrusting across the frontal ramp responsible for the formation of the Doonerak and Blarney Creek duplexes (see Avé Lallemant and Oldow, Chapter 14, this volume).

The north- to northwest-vergent Eekayruk thrust fault that breaches the Doonerak window appears to crosscut D_2 structures and, thus, is younger than D_2. It is an out-of-sequence or an envelopment thrust fault. Oldow et al. (Chapter 8, this volume) propose it to be the latter.

Mesoscopic reverse faults of variable orientation may be related to the thrust faults and the D_1 and D_2 folds. One conjugate set of mesoscopic strike-slip faults that resulted in northwest-southeast contraction may be related to D_1 as well as to D_2. Mesoscopic normal faults, oriented perpendicular to the Doonerak antiformal axis, may be related to bending stresses set up in a doubly plunging fold.

The deformational structures as discussed above are not thought to have formed in distinct phases of deformation separated by a time of quiescence. We believe that they are formed during continuous, progressive deformation that may have started in Late Jurassic time (Wirth et al., 1993) but continued through the Cretaceous (Blythe et al., Chapter 10, this volume; Gottschalk and Snee, Chapter 13, this volume), and probably the Tertiary (Blythe et al., Chapter 10, this volume; O'Sullivan et al., Chapter 11, this volume).

The third-generation folds (D_3) are north to northwest trending and are not penetrative. They are particularly well developed near major east- to northeast-trending strike-slip faults and, thus, may be genetically related to them. A conjugate set of mesoscopic strike-slip faults causing northeast-southwest contraction may be related to the formation of the megascopic, map-scale, left-lateral strike-slip faults and, thus, to the D_3 folds. The time of formation of the last structures is unknown. Whether they are related to the north-south contractional structures of the Brooks Range and resulted from displacement partitioning (Oldow et al., 1990) or are related to more recent changes in plate convergence, or both, is unknown.

ACKNOWLEDGMENTS

Field and laboratory work was supported by grants from the Department of Energy (DE-AS05-83ER13124), the National Science Foundation (EAR-8517384), and the Rice University Alaska Industrial Associates Group. The Alaska Industrial Associates were Amoco Production Company, Arco Exploration Company, Chevron U.S.A., Inc., Gulf Oil Exploration and Production Company, Mobil Exploration and Producing Services, Inc., and The Standard Oil Company. We are grateful to the late John T. Dillon of the Alaska Division of Geological and Geophysical Surveys for logistical support, geological discussions, and the use of his unpublished preliminary field maps of the northwest corner of the Chandalar Quadrangle (since published). We thank John S. Oldow for his advice and Wesley K. Wallace and Charles F. Kluth for their critical and very helpful reviews of this paper.

REFERENCES CITED

Armstrong, A. K., 1974, Carboniferous carbonate depositional models, preliminary lithofacies and paleotectonic maps, Arctic Alaska: American Association of Petroleum Geologists Bulletin, v. 58, p. 621–645.

Armstrong, A. K., and Mamet, B. L., 1974, Carboniferous biostratigraphy, Prudhoe Bay State 1 to northeastern Brooks Range, Arctic Alaska: American Association of Petroleum Geologists Bulletin, v. 58, p. 646–660.

Armstrong, A. K., and Mamet, B. L., 1977, Carboniferous microfacies, microfossils, and corals, Lisburne Group, Arctic Alaska: U.S. Geological Survey Professional Paper, v. 849, 144 p.

Armstrong, A. K., Mamet, B. L., Brosgé, W. P., and Reiser, H. N., 1976, Carboniferous section and unconformity at Mount Doonerak, Brooks Range, northern Alaska: American Association of Petroleum Geologists Bulletin, v. 60, p. 962–972.

Bowsher, A. L., and Dutro, J. T., Jr., 1957, The Paleozoic section in the Shainin Lake area, central Brooks Range, Alaska, *in* Exploration of Naval Petroleum Reserve No. 4 and adjacent areas northern Alaska, 1944–53, Part 3, Areal geology: U.S. Geological Survey Professional Paper, v. 303-A, p. 1–39.

Brosgé, W. P., 1960, Meta-sedimentary rocks in the south-central Brooks Range, Alaska: U.S. Geological Survey Professional Paper, v. 400-B, p. B351–B352.

Brosgé, W. P., and Patton, W. W., Jr., 1982, Regional bedrock geologic maps along the Dalton Highway, Yukon Crossing to Toolik, Alaska: U.S. Geological Survey Open-File Report 82-1071, 11 p.

Brosgé, W. P., and Reiser, H. N., 1964, Geologic map and section of the Chandalar Quadrangle, Alaska: U.S. Geological Survey Miscellaneous Geologic Investigations Map I-375, scale 1:250,000.

Brosgé, W. P., and Reiser, H. N., 1971, Preliminary bedrock geologic map: Wiseman and eastern Survey Pass Quadrangles, Alaska: U.S. Geological Survey Open-File Map 479, scale 1:250,000.

Brosgé, W. P., Dutro, J. T., Jr., Mangus, M. D., and Reiser, H. N., 1962, Paleozoic sequence in eastern Brooks Range, Alaska: American Association of Petroleum Geologists Bulletin, v. 46, p. 2174–2198.

Chapman, R. M., Detterman, R. L., and Mangus, M. D., 1964, Geology of the Killik-Etivluk Rivers region, Alaska: U.S. Geological Survey Professional Paper, v. 303-F, p. 325–407.

Churkin, M., Jr., and Trexler, J. H., Jr., 1981, Continental plates and accreted oceanic terranes in the Arctic, *in* Nairn, A. E. M., Churkin, M., Jr., and Stehli, F. G., eds., The Ocean basins and margins, v. 5, The Arctic Ocean: New York, Plenum Press, p. 1–20.

Crane, R. C., 1987, Arctic reconstruction from an Alaskan point of view, *in* Tailleur, I., and Weimer, P., eds., Alaskan North Slope geology: Society of Economic Paleontologists and Mineralogists, Pacific Section, Publication 50, p. 769–783.

Dillon, J. T., Harris, A. G., Dutro, J. T., Jr., Solie, D. N., Blum, J. D., Jones, D. L., and Howell, D. G., 1988, Preliminary geologic map and section of the Chandalar D-6 and parts of the Chandalar C-6 and Wiseman C-1 and D-1 Quadrangles, Alaska: Alaska Division of Geological and Geophysical Surveys, Report of Investigations 88-5, scale 1:63,360.

Dutro, J. T., Jr., Brosgé, W. P., Lanphere, M. A., and Reiser, H. N., 1976, Geologic significance of Doonerak structural high, central Brooks Range, Alaska: American Association of Petroleum Geologists Bulletin, v. 60, p. 952–961.

Dutro, J. T., Jr., Brosgé, W. P., Reiser, H. N., and Detterman, R. L., 1979, Beaucoup Formation, a new Upper Devonian stratigraphic unit in the central Brooks Range, northern Alaska, *in* Sohl, N. F., and Wright, W. B., eds., Changes in stratigraphic nomenclature by the U.S. Geological Survey: U.S. Geological Survey Bulletin, v. 1482-A, p. A63–A69.

Dutro, J. T., Jr., Palmer, A. R., Repetski, J. E., and Brosgé, W. P., 1984, Middle Cambrian fossils from the Doonerak anticlinorium, central Brooks Range, Alaska: Journal of Paleontology, v. 58, p. 1364–1371.

Grantz, A., Eittreim, S., and Whitney, O. T., 1981, Geology and physiography of the continental margin north of Alaska and implications for the origin of the Canada Basin, *in* Nairn, A. E. M, Churkin, M., Jr., and Stehli, F. G., eds., The Ocean basins and margins, v. 5, The Arctic Ocean: New York, Plenum Press, p. 439–492.

Hill, D., 1981, *Rugosa* and *Tabulata*, *in* Teichert, C., ed., Treatise on invertebrate paleontology, Volume F, Supplement 1: Lawrence, Kansas, Geological Society of America and University of Kansas Press, 762 p.

Jones, D. L., Silberling, N. J., Coney, P. J., and Plafker, G., 1987, Lithotectonic terrane map of Alaska (west of the 141st meridian): U.S. Geological Survey Miscellaneous Field Studies Map MF-1874-A, scale 1:250,000.

Melvin, J., 1987, Sedimentological evolution of Mississippian Kekiktuk Formation, Sagavanirktok Delta area, North Slope, Alaska, *in* Tailleur, I., and Weimer, P., eds., Alaskan North Slope geology: Society of Economic Paleontologists and Mineralogists, Pacific Section, Publication 50, p. 60.

Moore, T. E., and Churkin, M., Jr., 1984, Ordovician and Silurian graptolite discoveries from the Neruokpuk Formation (sensu lato), northeastern and central Brooks Range, Alaska: Paleozoic Geology of Alaska and Northwestern Canada Newsletter No. 1, May 1984, p. 21–23.

Mull, C. G., Tailleur, I. L., Mayfield, C. F., and Pessel, G. H., 1976, New structural and stratigraphic interpretations, central and western Brooks Range and Arctic Slope, *in* Cobb, E. H., ed., The U.S. Geological Survey in Alaska: Accomplishments during 1975: U.S. Geological Survey Circular, v. 733, p. 24–26.

Mull, C. G., Adams, K. E., and Dillon, J. T., 1987, Stratigraphy and structure of the Doonerak fenster and Endicott Mountains Allochthon, central Brooks

Range, Alaska, *in* Tailleur, I., and Weimer, P., eds., Alaskan North Slope geology: Society of Economic Paleontologists and Mineralogists, Pacific Section, Publication 50, p. 663–679.

Nilsen, T. H., 1981, Upper Devonian and Lower Mississippian redbeds, Brooks Range, Alaska, *in* Miall, A. D., ed., Sedimentation and tectonics in alluvial basins: Geological Association of Canada Special Paper 23, p. 187–219.

Nilsen, T. H., and Moore, T. E., 1984, Stratigraphic nomenclature for the Upper Devonian and Lower Mississippian (?) Kanayut Conglomerate, Brooks Range, Alaska: U.S. Geological Survey Bulletin, v. 1529A, 64 p.

Norris, D. K., and Yorath, C. J., 1981, The North American plate from the Arctic Archipelago to the Romanzof Mountains, *in* Nairn, A. E. M., Churkin, M., Jr., and Stehli, F. G., eds., The Ocean basins and margins: Volume 5, The Arctic Ocean: New York, Plenum Press, p. 37–103.

Oldow, J. S., Avé Lallemant, H. G., Julian, F. E., Seidensticker, C. M., and Phelps, J. C., 1984, The Doonerak window duplex: regional implications: Geological Society of America Abstracts with Programs, v. 16, p. 326.

Oldow, J. S., Seidensticker, C. M., Phelps, J. C., Julian, F. E., Gottschalk, R. R., Boler, K. W., Handschy, J. W., and Avé Lallemant, H. G., 1987, Balanced cross sections through the central Brooks Range and North Slope, Arctic Alaska: Tulsa, Oklahoma, American Association of Petroleum Geolo-

gists, Special Publication, 19 p., 8 pl.

Oldow, J. S., Bally, A. W., and Avé Lallemant, H. G., 1990, Transpression, orogenic float, and lithospheric balance: Geology, v. 18, p. 991–994.

Palmer, A. R., Dillon, J. T., and Dutro, J. T., Jr., 1984, Middle Cambrian trilobites with Siberian affinities from the central Brooks Range, northern Alaska: Paleozoic Geology of Alaska and Northwestern Canada Newsletter No. 1, May 1984, p. 29–30.

Phelps, J. C., 1987, Stratigraphy and structure of the northeastern Doonerak window area, central Brooks Range, northern Alaska [Ph.D. thesis]: Houston, Texas, Rice University, 293 p.

Ramsay, J. G., 1967, Folding and fracturing of rocks: New York, McGraw-Hill, 568 p.

Schrader, F. C., 1902, Geology of the Rocky Mountains in northern Alaska: Geological Society of America Bulletin, v. 13, p. 233–252.

Smith, P. S., and Mertie, J. B., Jr., 1930, Geology and mineral resources of northwestern Alaska: U.S. Geological Survey Bulletin, v. 815, 351 p.

Wirth, K. R., Bird, J. M., Blythe, A. E., and Harding, D. J., 1993, Age and evolution of western Brooks Range ophiolites Alaska: Results from $^{40}Ar/^{39}Ar$ thermochronometry: Tectonics, v. 12, p. 410–432.

MANUSCRIPT ACCEPTED BY THE SOCIETY SEPTEMBER 23, 1997

Geological Society of America
Special Paper 324
1998

Structure and lithology of the lower Paleozoic Apoon assemblage, eastern Doonerak window, central Brooks Range, Alaska

F. E. Julian
Department of Geological Sciences, University of Texas, El Paso, Texas 77968
J. S. Oldow
Department of Geology and Geological Engineering, University of Idaho, Moscow, Idaho 83844-3022

ABSTRACT

Deep structural levels of the central Brooks Range and a unique sequence of lower Paleozoic rocks are exposed in the Doonerak window. The lower Paleozoic rocks contain a sparse fossil population indicative of Cambrian to Silurian or Ordovician ages and reside in four thrust imbricates that constitute the lower horses of the Doonerak multiduplex. The thrust horses of the lower Doonerak multiduplex segregate the Apoon assemblage into four lithologic units: (1) a succession of fine-grained clastic rocks; (2) a coarse clastic sequence; (3) a mixed volcanic and volcaniclastic sequence; and (4) a succession of interbedded pyroclastic, volcanic flow, and sedimentary rocks. The coarse clastic rocks have a provenance suggestive of a magmatic-arc origin and the chemistry of the volcanic rocks is consistent with a continental arc source. The lower Paleozoic rocks exhibit three phases of folds: an early, penetrative fabric (D_1), overprinted by pervasive asymmetric chevron folds and crenulations (D_2), and sporadically developed crenulations and rare minor folds of the youngest generation of structures (D_3). The same generations of structures are developed in Carboniferous rocks that locally overlie the Apoon assemblage depositionally in exposures along the northern flank of the Doonerak window. The earliest recognizable penetrative structures in lower Paleozoic rocks of the Doonerak window are post-Mississippian in age. Penetrative structures associated with the Devonian Ellesmerian orogeny observed in coeval rocks of the northeastern Brooks Range and in the subsurface of the North Slope are not developed in the Apoon assemblage.

INTRODUCTION

The east-northeasterly trending Doonerak window (see Fig. 3, Preface, this volume) is a unique exposure in the central Brooks Range of the basal detachment of the Endicott Mountains allochthon (Amawk thrust) and of deeper structural levels of the thrust belt. The lower Paleozoic rocks exposed in the core of the Doonerak window compose the Apoon assemblage (Oldow et al., 1984) and have been interpreted as structural basement by numerous researchers (Dutro et al., 1976, 1984; Armstrong et al., 1976; Dillon et al., 1986; Mull, 1982; Mull et al. 1987). Typi-

cally, the lower Paleozoic rocks of the Doonerak window are correlated with the coeval rocks of the North Slope and northeastern Brooks Range. The rocks exposed in the northeastern Brooks Range and found in the subsurface of the North Slope constitute a basement (Franklinian sequence) of deformed lower Paleozoic and Proterozoic rocks (Carter and Laufeld, 1975) separated from overlying Carboniferous and younger units (Ellesmerian sequence) by an angular unconformity (Lerand, 1973; Grantz and May, 1982). The pre-Carboniferous unconformity is attributed to deposition following the Devonian Ellesmerian orogeny (Churkin, 1969, 1975; Brosgé et al., 1962; Reed, 1968; Dutro,

Julian, F. E., and Oldow, J. S., 1998, Structure and lithology of the lower Paleozoic Apoon assemblage, eastern Doonerak window, central Brooks Range, Alaska, *in* Oldow, J. S., and Avé Lallemant, H. G., eds., Architecture of the Central Brooks Range Fold and Thrust Belt, Arctic Alaska: Boulder, Colorado, Geological Society of America Special Paper 324.

1970; Sable, 1977; Oldow et al., 1987). Both Franklinian and Ellesmerian sequence rocks are involved in Brookian contraction in the northeastern Brooks Range (Oldow et al., 1987; Wallace and Hanks, 1990; Moore et al., 1994), but the Franklinian sequence contains an additional, pre-Mississippian penetrative phase of deformation represented by axial-planar cleavage and south-vergent isoclinal folds (Oldow et al., 1987). In contrast, the Ellesmerian sequence of the North Slope subsurface is not involved in Mesozoic or younger deformation and remains largely undeformed except for pre-Permian partially inverted half-grabens (Mauch, 1985) and normal faults associated with the passive margin that forms the northern coast of Alaska (Grantz and May, 1982; Moore et al., 1994). Numerous authors (Dutro et al., 1976, 1984; Armstrong et al., 1976; Mull, 1982; Mull et al., 1987; Moore et al., 1994) have suggested that the sub-Carboniferous angular unconformity underlying the North Slope extends south beneath allochthonous rocks of the northern Brooks Range and is exposed in the Doonerak window. The correlation of the Doonerak rocks with those along the Arctic margin to the north is a critical tie for regional stratigraphic and structural models of the Brooks Range orogen.

We investigated the structure of the rocks composing the Doonerak window to evaluate the validity of this pivotal paleogeographic link between allochthonous and autochthonous rocks of northern Alaska. In the context of Brookian deformation, the area around and including the Doonerak window may be divided into three primary structural units, which in structurally descending order are (1) the Endicott Mountains allochthon, a sequence of Devonian and Carboniferous rocks overlying the Amawk thrust; (2) the Blarney Creek allochthon, which contains Carboniferous Lisburne Limestone, Kayak Shale, and parts of the Devonian(?) to Carboniferous Kekiktuk Conglomerate separated for the most part from older rocks below by the underlying Blarney Creek thrust (Seidensticker and Oldow, this volume, Chapter 6); and (3) the Apoon assemblage composed of lower Paleozoic (Dutro et al., 1976; Repetski et al., 1987) volcanic and sedimentary rocks. On the southern flank of the Doonerak window the Blarney Creek allochthon is not present; the Endicott Mountains allochthon is in direct structural contact with the Apoon assemblage (Mull et al., 1987; Phelps and Avé Lallemant, this volume, Chapter 4). There, the southern extension of the Amawk thrust dips southeasterly, parallel to bedding and tectonic foliations in both the Apoon assemblage and the Endicott Mountains allochthon.

On the northern flank of the Doonerak window (Fig. 1), all three structural units outlined above are present. The Apoon assemblage lies below the shallowly north dipping Blarney Creek thrust (Seidensticker and Oldow, this volume, Chapter 6), which serves as the structural base of gently north-dipping Carboniferous rocks composing the Blarney Creek allochthon. The Blarney Creek allochthon is internally dismembered by north-dipping, generally bedding-parallel thrusts, which sole into the underlying Blarney Creek thrust. The Blarney Creek allochthon is itself structurally overlain by the north-dipping Amawk thrust, which carries Upper Devonian and Carboniferous rocks of the Endicott

Mountains allochthon north over the Doonerak window (Brosgé and Reiser, 1971; Dutro et al., 1976; Mull, 1982; Dillon et al., 1986; Handschy, this volume, Chapters 2 and 3).

One of the most regionally significant relations in the Doonerak window is the nature of the contact between Carboniferous rocks and the underlying Apoon assemblage. Most workers (Dutro et al., 1976, 1984; Armstrong et al., 1976; Dillon et al., 1986; Mull, 1982; Mull et al., 1987; Moore, 1987; Moore et al., 1994) interpret the contact as a major angular unconformity. This interpretation stems from the discordance of bedding across the contact, where the Apoon assemblage generally dips to the southeast and the Carboniferous rocks generally dip gently to the north. These authors attribute much of the structure in the Apoon assemblage to pre-Carboniferous deformation. The existence of the inferred angular unconformity is used as prima facie evidence for the correlation of the lower Paleozoic rocks of the Doonerak window (Apoon assemblage) with age-equivalent rocks of the North Slope and northeastern Brooks Range.

Although recognized as first-order point of paleogeographic control (Mull, 1982; Mull et al., 1987; Dutro et al., 1976), the rocks composing the Apoon assemblage received only cursory attention in previous investigations. Reconnaissance mapping (1:250,000) by Brosgé and Reiser (1971) identified a mixture of volcanic and phyllitic rocks in the Doonerak window, and a later study by Dutro et al. (1976) listed some of the various volcanic and sedimentary lithologies composing the Apoon assemblage. Neither study dealt with the relative abundance of the various lithologies nor their stratigraphic or structural relationships, and no detailed geologic maps were produced to show lithologic distribution within the window.

The primary objective of this project was to study the structure of the lower Paleozoic rocks. Detailed appraisal of the contact between Mississippian and lower Paleozoic rocks was a critical component of this work and was needed to document the origin for the disparity in dips of the lower Paleozoic (steep south dipping) and Carboniferous (shallow north dipping) lithologic successions. We were drawn to the Doonerak window as one of the only areas where so-called Ellesmerian structures were preserved in the central Brooks Range. Our intent was to unravel Brookian and Ellesmerian structures and to document the kinematics of the pre-Carboniferous structures in the lower Paleozoic section. To our great surprise, our work (outlined below) documented the lack of pre-Carboniferous penetrative structures in the lower Paleozoic rocks of the Doonerak window and called into question whether an unconformity exists below the Kekiktuk Conglomerate in this area.

AGE OF THE APOON ASSEMBLAGE

The age of the Apoon assemblage is poorly constrained. Dutro et al. (1976) dated five intrusive rocks with both K-Ar and $^{40}Ar/^{39}Ar$ techniques, yielding ages of 380 to 520 Ma (Fig. 1). The older ages came from dikes cross-cutting volcanic rocks and two of the younger ages came from dikes within

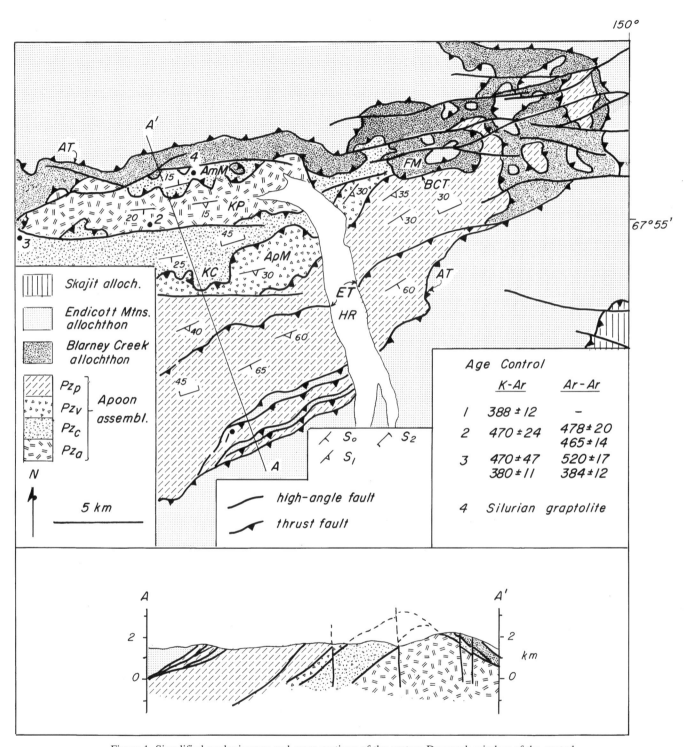

Figure 1. Simplified geologic map and cross sections of the eastern Doonerak window of the central Brooks Range. For detailed map refer to Figure 3 of the Preface (this volume). Radiometric age determinations given in Ma. Physiographic features are: Amawk Mountain (AmM); Apoon Mountain (ApM); Falsoola Mountain (FM); Hammond River (HR); Karilyukpuk Creek (KC); and Kinnorutin Pass (KP). Major fault systems are the Amawk thrust (AT), Blarney Creek thrust (BCT), and Eakayruk thrust (ET).

phyllitic units. Samples with both old and young ages come from the same area (Dutro et al., 1976) and cross-cutting relations were not established.

Fossils have been recovered by Dutro et al. (1984) and Moore and Churkin (1984) in the Doonerak window. Churkin and Moore (personal communication, 1984) identified a Silurian graptolite in phyllite from exposures along the northern flank of the duplex (Fig. 1), offering the only fossil age control for the Apoon assemblage in the eastern Doonerak window. They also report Ordovician-Silurian graptolites in phyllite of the western Doonerak window (Moore and Churkin , 1984), where Middle Cambrian brachiopods and trilobites with Siberian affinities were also found (Dutro et al., 1984). The Middle Cambrian fossils and most of the graptolites recovered by Churkin and Moore were found in phyllitic rocks of the western Doonerak window and, although lithologically similar to units in the east, specific correlations are not possible.

LITHOLOGIC UNITS

Detailed mapping of the eastern Doonerak window allows division of the Apoon assemblage into four units that are fault-bound successions with differing lithologic character. Two of these structural sheets contain predominantly clastic rocks (mapped together as Pzps by Brosgé and Reiser, 1971), and two contain mostly volcanic and volcaniclastic rocks (Pzvp and Pzv, respectively, of Brosgé and Reiser, 1971). The four lithologic units are easily recognized and mapped in the eastern Doonerak window, but the lack of age control results in stratigraphic ambiguity between structural sheets. It is conceivable that the rocks in different structural sheets are age-equivalent facies, telescoped by thrust faults. Similarities in two volcanic units residing in different thrust sheets lend support to this contention but do not supply compelling evidence.

The rocks have all been severely deformed and experienced low-grade metamorphism. Isoclinal folding has structurally thickened the units and often obscures original bedding, making it impossible to measure true stratigraphic thickness or to construct conventional stratigraphic columns. All thicknesses in this paper represent deformed state rather than original stratigraphic thickness. Because original rock textures and lithologies are recognizable in most of the rocks, sedimentary rock names are used without metamorphic modifiers in the following descriptions. Unlike the coarse-grained rocks, the depositional features in fine-grained clastic rocks are obscured by a penetrative tectonic foliation. The fine-grained sedimentary rocks were originally shale and mudstone, but are here referred to as argillite, slate, and phyllite in order of increasing degree of recrystallization and tectonic foliation development.

Lithologic descriptions of each of the four fault-bounded units are given in the following sections. The units are described in order of their exposure from south to north in descending structural position (Fig. 1): Phyllitic unit (Pzp), Volcaniclastic unit (Pzv), Clastic unit (Pzc), and Volcanic unit (Pza).

Phyllitic unit (Pzp)

The southernmost (and structurally highest) unit of the Apoon assemblage is also the thickest; containing ~3,000 m of fine-grained clastic rocks. The present thickness does not reflect depositional thickness because of known and suspected internal repetition and imbrication. The contact between the Phyllitic unit and the overlying Endicott Mountains allochthon is complicated by thrust faulting, with relatively thin (as much as 180 m) slices of Devonian rocks (Beaucoup Formation?) tectonically interleaved with slates of the Phyllitic unit in the complex fault zone that defines the southern flank of the Doonerak window (Phelps and Avé Lallemant, this volume, Chapter 4).

The rocks of the Phyllitic unit grade from black, iron-rich slate in the south to the north where both argillite and phyllite are common. The upper 1,500 m is dominated by black slate, which is monotonous and uniform lithologically. The slate is fissile with a penetrative tectonic foliation and contains occasional pyrite-rich layers that locally weather reddish brown. Within the upper 1,500 m of the unit, outcrops may be composed of as much as 5% siltstone with bedding thicknesses of less than 5 cm.

The upper part of the Phyllitic unit exposed along the southern margin of the Doonerak window contains two unique rock types: a marble layer and discontinuous outcrops of greenstone. The marble layer is unfossiliferous and occupies the top of the Phyllitic unit, west of the Hammond River (Fig. 1). The marble thickness varies from 1 to 15 m in a more-or-less constant structural position for more than 12 km along strike. The marble is generally recrystallized to a massive carbonate with few preserved sedimentary features, but a depositional base is occasionally exposed. In a few outcrops, the marble is composed of conglomerate, which contains rounded carbonate clasts (as much as 10 cm long) and angular shale rip-up clasts (as much as 0.5 cm long).

Thin (<2 m) aphanitic greenstone layers occur as isolated lenses (as much as 5 m long) throughout the upper part of the Phyllitic unit and are particularly abundant near the contact with the overlying Endicott Mountains allochthon. The greenstone does not form continuous layers, and individual outcrops are too small to be shown on the map. The greenstone is aphanitic but in thin section is a foliated aggregate of fine chlorite, calcite, quartz, white mica, and epidote.

The lower 1,500 m of the Phyllitic unit contains a wider variety of fine-grained clastic rocks than the upper section, but no marble or greenstone is present. Red-weathering black slate, like that comprising the upper sequence, makes up ~40% of the lower (northern) section. Interlayered with the slate is finely cleaved, gray, and green-gray phyllite. Phyllite layers range from 5 cm to 10 m thick and make up ~40% of the lower part of the Phyllitic unit. Also present (~20%) is weakly foliated siliceous argillite composing layers as much as 4 m thick. The argillite typically is banded (<1.5 cm) in hues of green, gray, and black, but no change in rock properties (hardness, fissility, weathering) are associated with color change. The argillite is the most weathering resistant lithology in the Phyllitic unit.

Volcaniclastic unit (Pzv)

North of and structurally underlying the Phyllitic unit is a wide variety of volcanic and volcaniclastic rocks, assigned to the Volcaniclastic unit. This unit has a laterally variable thickness reaching a maximum of ~400 m and has the greatest lithologic diversity of the Apoon assemblage. The Volcaniclastic unit consists of massive basaltic lava, agglomerate, tuff, volcanic litharenite, and fine-grained clastic rocks. Each of the major rock types represent a wide variety of composition and texture as described below. The spatial distribution of the rock types is discussed after the lithologic description, and the chemistry of the volcanic rocks is presented in a separate section.

Basaltic lava. The lava flows form massive featureless outcrops as much as 150 m thick (post-deformational thickness), and individual flows cannot be distinguished in the field. In thin section, the lava contains a homogeneous distribution of altered (usually sausseritized) euhedral feldspar (<5 mm; 10 to 80% of the total rock) and as much as 40% clinopyroxene phenocrysts (<7 mm). Accessory minerals consist of subhedral green amphibole, devitrified glass, epidote, and opaque minerals. The matrix comprises between 20 and 80% of the rock and is fine-grained albite, chlorite, sericite, epidote, and calcite.

Agglomerate. Agglomerate consists of rounded volcanic bombs within a subordinate matrix (<20%). The bombs typically are 5 to 12 cm in diameter with a maximum of 20 cm. The bombs are of two types: (1) scoria, sometimes bearing plagioclase microlites, with vesicles as much as 6 mm in diameter, and (2) nonvesicular fragments with 25% euhedral pyroxene porphyroclasts (<6 mm), 30% altered plagioclase laths (<2 mm), and glass shards (5%) in an ash matrix. The matrix is largely (~70%) coarse ash and lapilli (0.5 to 2 mm) containing fragmented plagioclase and pyroxene phenocrysts. Many of the phenocrysts exhibit some euhedral crystal faces and some jagged edges, suggesting that the crystals were broken during ejection or transport. Fine ash makes up the rest of the matrix.

Tuff. Like the agglomerate, the tuff layers contain porphyritic fragments, scoria, devitrified glass shards, and broken pyroxene and plagioclase crystals. The tuff is fine grained with most particles less than 2 cm in diameter. Most of the tuff contains a mixture of ash and lapilli but some of the tuff layers contain only ash (<2 mm). Matrix constitutes 30 to 70% of the rock and is composed of fine ash.

The tuff forms layers 15 m to 5 cm thick, and it is difficult to define individual tuff layers in the thick amalgamated sequences. Where tuff layers are interbedded with phyllitic rocks, both upper and lower contacts are sharp and well defined. Crude grading is present in a few of the tuff layers but most are poorly sorted with no internal texture.

Reworked tuff. Reworked tuff consists mainly of volcanic fragments composed of all the primary volcanic rocks described above (except agglomerates). Fine-grained volcanic fragments, either aphanitic or felted with plagioclase microlites, are the most common clast type, comprising as much as 70% of a given layer.

These fine-grained clasts are structurally flattened into the foliation plane and have long dimensions that vary from 1 mm to 20 cm. Plagioclase grains typically are less than 1 mm in diameter and make up about 10% of the coarse fraction, whereas pyroxene grains are rare (<1%). Subangular to subrounded monocrystalline quartz grains range from less than 0.5 mm to ~2 mm in diameter and comprise about 10% of the coarse fraction. Polycrystalline quartz (<8 mm) comprises 5 to 15% of the coarse fraction and sometimes shows good mylonitic ribbon texture, indicative of a metamorphic source. Immature sedimentary rock fragments, composed of subangular quartz grains in a mud matrix, were found in one sample and constitute ~5% of the clast population.

Volcanic litharenite is interbedded with the fine-grained conglomerate and is poorly sorted and grain supported. Litharentite never has greater than 15% fine-grained matrix (<0.1 mm) and exhibits no graded bedding or other sedimentary structures. Beds range from 30 cm to 2 m thick and all contacts are sharp. Individual beds were not traced between outcrops, but bedding thickness does not vary along strike over distances of as much as 30 m.

Fine-grained sedimentary rocks. Interlayered with the volcanic rocks of the Volcaniclastic unit are phyllite and argillite. The phyllite is black, gray, or green-gray with penetrative foliations and contains preserved beds 3 cm to 1.5 m thick. Argillite is massive with poorly developed foliations and usually exhibits color bands 2 mm to 4 cm thick in shades of green and gray. The banded argillite occurs in units as much as 8 m thick.

Lithologic distribution. The Volcaniclastic unit is divided by the Hammond River (Fig. 1) into two areas of different lithologic character. East of the Hammond River, the Volcaniclastic unit is dominantly basaltic lava flows with ~30% interbedded tuffs and 10% fine-grained sediments. The eastern region contains no agglomerates or volcanic litharenites. West of the Hammond River, however, basaltic lava flows are uncommon (~5%) and occur only near the top of the section in layers less than 12 m thick. Interbedded tuff and volcanic litharenite are the dominant lithologies, with fine-grained phyllite and argillite scattered throughout the sequence. Agglomerate is found in only three locations, each at the base of the Volcaniclastic unit.

Clastic unit (Pzc)

Lithology. This unit has a structural thickness of ~1,500 m and contains a variety of clastic rocks, ranging from argillite to conglomerate. Argillite, phyllite, and siltstone within the Clastic unit are indistinguishable from those in the Phyllitic unit but typically are associated with coarser clastic rocks: sandstone, pebbly mudstone, and fine-grained conglomerate. Volcanic-lithic fragments comprise most of the coarser clastic layers. Volcanic-lithic fragments are mostly fine grained with phenocrysts and microlites of altered plagioclase and rare remnant pyroxene. Some aphanitic-volcanic and vesicular-volcanic fragments also occur. Other major components of these clastic rocks are angular to subrounded monocrystalline quartz grains (often deformed and showing undulatory extinction), sausseritized and sericitized feldspar grains, and elongate shale clasts. Metamorphic (ribbon) quartz grains, immature sedimentary rock-fragments, detrital amphibole and epidote grains, and rare zircons are also present. The concentration of clast types varies between layers; some layers are nearly all volcanic, whereas others contain no identifiable volcanic fragments. Descriptions of the compositions of individual samples are given in Table 1, and show the variation within this unit.

The sandstone, pebbly mudstone, and fine-grained conglomerate layers generally are less than 2 m thick and contain clasts less than 2 cm in diameter. The fine-grained lithic fragments (volcanic and sedimentary) are flattened into the foliation plane and have long dimensions of as much as 2 cm. The quartz grains are more equant and are as much as 4 mm in diameter. Generally, the conglomerate and pebbly mudstone layers are poorly sorted and show no size grading, but some siltstone and sandstone layers exhibit graded bedding. Channel fill composed of sandstone is recognized in some interleaved shale-sandstone horizons. The coarse clastic rocks often contain a wide range of clast sizes dispersed throughout individual layers. Some of the conglomerate is clast supported but the pebbly mudstone is typified by a few large clasts isolated in a fine-grained matrix. The matrix composes from 5 to 50% of the various rock types and typically is composed of chlorite, white mica, clay minerals, carbonate, and fine-grained plagioclase and quartz. Occasionally, sand or conglomerate lenses (<20 cm long) appear in finer grained clastic rocks, usually siltstone.

Sedimentary petrology. Point counts (200 points per sample) of coarse silt and fine sand to fine conglomerate fractions for 14 samples are given in Table 1. In all point counts, the quartz component includes monocrystalline quartz grains; several samples also contain deformed, metamorphosed polycrystalline quartz.

The feldspar component is dominated by altered (sausseritized and sericitized) crystals, although three samples contain feldspar that was not extensively replaced. Lithic fragments are, with a few exceptions, volcanic clasts. Some sedimentary rock fragments are also present, and some of the fine-grained lithic fragments may have either sedimentary or volcanic origin.

The point-count data lie in or near the dissected magmatic arc field of the QFL triangular diagram (Dickinson and Suczek, 1979) presented in Figure 2, and one sample lies in the recycled orogen field. Applying the provenance fields of Dickinson et al. (1983), samples from this study are spread throughout the three divisions of the magmatic-arc regime: undissected arc, transitional arc, and dissected arc. The presence of a small but significant quartz component (>10%) in many of the samples shows that the arc was dissected and continental basement was exposed at the time of deposition. This conclusion is also supported by the presence of metamorphic clasts, sedimentary clasts, and sparse coarse-grained intrusive igneous clasts. In modern arc environments, these types of nonvolcanic clasts are seldom associated with stable arc systems (Marsaglia and Ingersoll, 1992). This compositional mixture from dissected and undissected arcs is often attributed to the uplift and unroofing of a volcanic arc, exposing increasingly large areas of its batholithic core (Dickinson, 1982). Recent work by Marsaglia (1991) shows that such provenance mixtures may be associated with complex tectonic environments, such as triple junctions, where transform motion or rifting has disrupted the simple subduction system, causing uplift and dissection of the arc. Alternatively, the composition of the coarse fraction may represent mixing of sediments from magmatic arc and continental source areas.

TABLE 1. POINT COUNTS (200 POINTS) FOR SAMPLES OF THE CLASTIC UNIT OF THE APOON ASSEMBLAGE*

Sample	Q	F	L	F_f	F_a	F_p	V_a	V_c	L_s	L_f
B5-107	37	46	17	19	27	0	4	11	2	0
B5-110	4	14	82	0	7	7	0	26	0	56
B5-133	40	34	82	13	20	1	8	14	2	2
B5-139	33	37	30	22	12	3	8	16	0	6
B5-145	38	32	30	0	32	0	8	4	18	0
B5-148	13	14	73	0	1	13	8	62	3	0
B5-149	52	26	22	0	26	0	8	11	0	3
B5-151a	20	23	57	0	15	8	18	29	4	6
B5-151b	14	48	38	1	36	11	20	16	1	1
B5-158	14	19	67	0	14	5	47	20	-	-
B5-159g	28	18	54	6	10	2	12	18	18	6
B5-162d	15	38	47	0	35	3	46	0	1	0
B5-164b	9	19	72	3	11	5	15	35	20	2
B5-164d	7	19	74	2	10	7	4	26	13	31

*Q = quartz; F = feldspar; L = lithic; F_f = feldspar with fresh appearance; F_a = altered feldspar (sericitized or sausseritized); F_p = feldspar phenocrysts in volcanic clasts; V_a = volcanic ash; V_c = volcanic fragments (w/microlites or phenocrysts); L_s = sedimentary rock fragments; L_f = fine-grained lithic fragment.

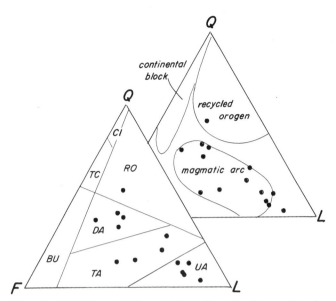

Figure 2. QFL (Quartz-Feldspar-Lithic) triangular diagram showing the composition of 14 samples from the Clastic unit of the Apoon assemblage. Provenance fields on right diagram are taken from Dickinson and Suczek (1979). Provenance fields on left diagram are from Dickinson (1982). CI, Cratonal interior; BU, Basement uplift; RO, Recycled orogen; DA, Dissected arc; TA, Transitional arc; TC, Transitional craton; UA, Undissected arc.

Volcanic unit (Pza)

The northernmost and structurally lowest unit in the Apoon assemblage has a thickness of ~500 m, but no bottom is observed. The unit contains predominately volcanic rocks with a systematic grain-size reduction from west to east. Western exposures are thick deposits of coarse-grained lapilli tuff and pyroclastic breccia, commonly clast supported, with abundant pumice and clasts ranging from ash to block size. Some agglomerate is present. There is very little fine-grained rock in this sequence. In the eastern part of the exposure belt, lapilli tuff is interlayered with laminated fine-grained volcanic and some phyllitic rocks. There is no abrupt transition between the coarse-grained rocks in the west and the finer grained rocks to the east.

The volcanic rocks typically are lapilli tuff or basaltic lava and have been metamorphosed to lowest greenschist facies. Plagioclase crystals are altered in several ways: (1) albitized (Na replacement of Ca is confirmed by EDP microprobe analyses and thin section staining), (2) sericitized, and (3) sausseritized. Vesicles are filled with calcite, chlorite, or epidote, and volcanic glass is devitrified. The groundmass of the basalt is commonly a fine-grained mixture of chlorite, albite, sericite, epidote, calcite, and Fe-Ti oxides.

The most common rock type is massive lapilli tuff with internally chaotic fragments composed of ash, lapilli, and block-size clasts in a fine-grained matrix (<0.25 mm). Most of the fragments are juvenile pyroclasts or broken feldspar crystals, but include angular cognate fragments locally. A variety of juvenile clasts are present: aphanitic basalt, glass shards (devitrified), fine-grained fragments with plagioclase microlites, plagioclase phenocrysts and microlites in a fine-grained matrix, rounded scoria, and vesicular clasts containing euhedral plagioclase phenocrysts. The number and proportion of these different types of clasts varies between layers as does the size. The layers also include different amounts of matrix; some are clast supported and others contain as much as 50% fine-grained matrix. Agglomerate constitutes less than 5% of the total sequence in the western exposures of the Volcanic unit.

Volcanic geochemistry

Many of the constituents of the Volcanic and Volcaniclastic units are juvenile volcanic rocks, and their chemical composition can provide information on the tectonic setting in which they formed. Twenty samples were analyzed for major and trace element compositions using an Induction Coupled Plasma Spectrometer (ICP) at the Rice University facility. Our work significantly expands on the geochemical characterization of three samples of the Apoon assemblage by Moore (1987), who compared the chemistry of volcanic rocks in the Doonerak window and northeastern Brooks Range.

Data from chemical analyses are shown in Table 2. The number labeled L.O.I. (loss on ignition) listed below the major element concentrations represents the volatile abundance driven off during sample fusion. Major element analyses have been recalculated to 100% without volatiles (H_2O or CO_2), placing all samples on the same anhydrous basis for comparison. Because these rocks are weathered and metamorphosed, interpretations are based on elements or element ratios; such as Ti, Zr, and Y (Cann, 1970), as well as Cr and Ni (Bloxam and Lewis, 1972). Rare-earth elements are stable but are difficult to measure accurately on an ICP and, as such, were not analyzed.

Samples from the Volcanic and Volcaniclastic units of the Apoon assemblage are plotted on discrimination diagrams that assign basalts of unknown origin to one of three main groups (Pearce and Cann, 1973). The differentiated tectonic setting of eruption are: (1) mid-ocean ridge basalts (MORB), (2) volcanic-arc basalts (VAB), and (3) within-plate basalts (WPB). Further subdivision within these groups is sometimes possible, but not all of the samples analyzed in this study have chemical compositions suitable for use on basalt discrimination diagrams (Fig. 3A). Five of the samples contain too much SiO_2 to be classified as basalts. Four of these samples also contain less than 12% CaO + MgO and do not qualify as basalts in the system devised by Pearce and Cann (1973) for use in areas where Si may have been mobilized by metamorphism. These five samples were not included in the discrimination diagrams and they are marked with an asterisk in Table 2.

The triangular diagram Ti/100-3Y-Zr of Pearce and Cann (1973) separates WPB from other basalts, and samples from the Apoon assemblage (Fig. 3B) clearly fall outside of the WPB field and within the VAB and MORB field. VAB can be distinguished from MORB by reduced values of Ti, Zr, and Y values for a given

TABLE 2. CHEMICAL ANALYSES OF VOLCANICLASTIC (Pzv) AND VOLCANIC (Pza) UNIT SAMPLES OF THE APOON ASSEMBLAGE

Elements	B5-7* Pzv	B5-10 Pzv	B5-14 Pzv	B5-17 Pzv	B5-18 Pzv	B5-30* Pzv	B5-33 Pzv	B5-152* Pzv	M3-37 Pzv
(%)									
SiO_2	57.48	49.49	48.72	50.05	50.88	56.08	46.36	55.72	48.88
Fe_2O_3	6.67	11.40	8.09	12.03	9.66	13.03	14.23	8.60	11.19
MnO	0.12	0.18	0.14	0.18	0.16	0.23	0.19	0.12	0.17
MgO	4.59	7.42	7.05	6.93	7.66	3.70	8.48	5.32	8.53
TiO_2	0.83	1.26	0.56	1.55	1.01	1.43	1.78	0.81	1.13
Al_2O_3	17.18	16.61	17.36	16.47	16.90	16.14	17.18	17.12	15.90
CaO	5.85	9.81	11.99	7.80	10.41	3.03	8.33	4.67	11.06
NaO	7.20	3.78	2.22	4.30	2.74	4.34	3.30	7.05	3.00
K_2O	0.03	0.03	3.60	0.54	0.51	0.25	0.06	0.36	0.07
P_2O_5	0.22	0.06	0.11	0.13	0.07	0.16	0.12	0.23	0.09
Total	100.17	100.04	99.84	99.98	100.00	98.39	96.73	100.00	100.02
L.O.I.	5.84	2.96	8.10	3.49	3.01	3.33	4.25	2.70	2.98
(ppm)									
Ni	28.3	42.82	113.63	40.61	60.88	8.43	49.17	22.09	55.13
V	183.8	299.17	258.21	326.61	232.35	312.00	368.69	293.96	284.03
Y	13.7	28.39	11.83	32.63	21.28	30.25	33.77	17.06	40.56
Sr	282.4	565.81	391.50	114.72	559.84	183.92	109.01	263.82	244.27
Ba	13.5	22.69	492.92	179.68	196.12	42.84	34.00	86.46	530.63
Zr	94.2	67.66	36.27	90.97	51.23	79.77	99.79	65.66	48.32
Cr	60.8	123.96	341.57	113.47	201.00	20.17	133.86	27.46	168.11
Sc	17.6	42.23	38.82	41.2	36.90	36.30	45.46	26.55	....

Elements	M4-52* Pza	M4-53 Pza	M4-54 Pza	M4-82* Pza	M4-97C Pza	M4-97E Pza	MJ-23 Pza	B4-64 Pza	B4-69B Pza	B4-120 Pza
(%)										
SiO_2	60.68	49.64	44.26	55.19	50.84	48.10	51.75	44.46	51.88	48.46
Fe_2O_3	7.66	7.23	9.78	8.92	9.44	11.95	9.09	10.07	10.08	10.60
MnO	0.14	0.22	0.24	0.13	0.20	0.17	0.05	0.25	0.16	0.19
MgO	3.47	3.78	4.55	5.37	6.89	6.45	11.20	4.03	5.93	5.75
TiO_2	0.85	0.87	1.14	0.73	0.85	0.96	1.06	1.16	0.96	0.98
Al_2O_3	14.84	19.30	19.38	19.87	19.09	20.31	19.46	20.51	18.75	18.28
CaO	9.51	13.09	16.07	2.86	8.36	7.33	1.80	14.81	9.06	12.38
NaO	2.34	4.86	4.01	4.44	2.22	4.15	2.26	3.43	1.61	2.46
K_2O	0.26	0.71	0.29	2.34	1.83	0.35	2.98	0.95	1.23	0.63
P_2O_5	0.26	0.31	0.28	0.15	0.29	0.17	0.35	0.33	0.34	0.27
Total	100.01	100.01	98.00	100.00	100.01	99.94	100.00	100.00	100.00	100.00
L.O.I.[†]	5.40	9.15	10.23	3.58	5.30	4.80	6.26	10.21	6.68	4.23
(ppm)										
Ni	42.37	26.94	60.50	93.26	71.25	73.77	230.96	15.80	30.61	43.75
V	208.98	213.11	246.06	191.01	273.00	278.06	218.65	255.95	261.81	283.74
Y	21.63	27.02	25.26	16.18	20.64	22.57	19.35	28.23	25.14	27.58
Sr	189.71	341.65	207.68	284.68	141.64	152.81	46.57	607.13	543.25	353.30
Ba	54.44	71.30	59.16	407.04	276.61	82.41	502.16	260.11	141.74	173.33
Zr	61.33	52.47	69.26	71.83	72.42	70.94	162.13	75.36	64.54	66.10
Cr	....	....	....	....	71.30	98.34	....	....	....	....
Sc	....	....	....	....	....	....	....	....	....	....

*Denotes samples not used in discriminant diagrams of Figure 3.
[†]L.O.I. = loss on ignition, the volatile abundance driven off during sample fusion.

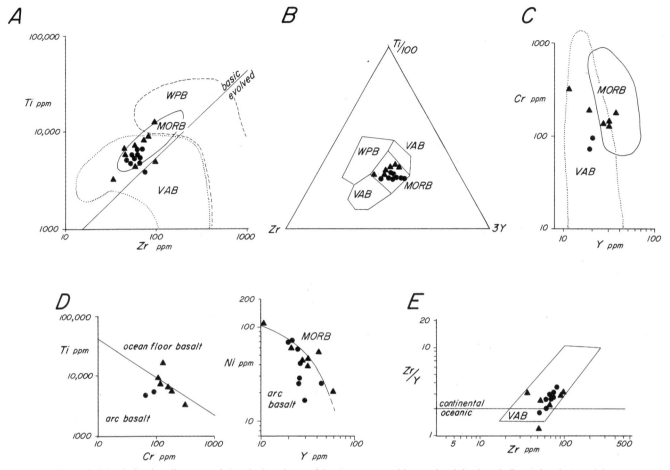

Figure 3. Discrimination diagrams of chemical analyses of the Apoon assemblage volcaniclastic unit, Pzv (triangles), and the volcanic unit, Pza (closed circles). A: Zr-Ti diagram of Winchester and Floyd (1977) separating basaltic rocks from metasediments and andesites. The two samples plotted to the right of the Zr/Ti = 0.014 line are not suitable for use on basalt discrimination diagrams. B: Zr-Ti-Y diagram from Pearce and Cann (1973). Diagram separates within-plate basalts (WPB) from mid-ocean ridge (MORB) and volcanic-arc (VAB) basalts. Samples fall within both the MORB and VAB field. C: Cr-Y diagram (Pearce, 1982) differentiates VAB from MORB. Samples tend toward the VAB field with overlap in the MORB field. D: Ti-Cr and Ni-Y diagrams (Crawford and Keays, 1978) differentiate arc basalts (VAB) from MORB. Samples generally show VAB characteristics. E: Zr/Y-Zr diagram (Pearce, 1983) discriminates VAB of this study has affinity with an active continental margin.

concentration of such stable element such as Cr (Pearce, 1975; Pearce and Norry, 1979), Ni, or V (Shervais, 1982; Crawford and Keays, 1978). The most effective discrimination diagram plots Cr against Y (Fig. 3C; Pearce, 1982), with most Apoon samples lying within the VAB field or within the overlap between VAB and MORB. Plots of Ti versus Cr and Ni versus Y shown in Figure 3D also give good separation of island-arc tholeiites and MORB, but allow some overlap between calc-alkaline arc basalts and MORB. This indicates that these are indeed volcanic-arc basalts as suggested by the pyroclastic texture and the composition of the coarse clastic sediments. The tendency of these samples to plot within the VAB field suggests that the volcanics may be calc-alkaline and erupted through continental lithosphere. The distinction between calc-alkaline and tholeiitic basalts cannot be made with certainty without comparison of rare-earth elements or the mobile elements Fe, Mg, Na, and K, however.

The final diagram Zr/Y-Zr (Fig. 3E) developed by Pearce (1983), differentiates between VAB erupted at ocean–ocean island arcs and those where ocean crust is subducted beneath an active continental margin. Our samples plot in the continental-arc field supporting the interpretation that Apoon assemblage is related to subduction beneath a magmatic arc on an active continental margin. The magmatic arc may actually be part of the continent, or it may be separated from the continent by a back-arc basin (Pearce, 1983).

STRUCTURE

High-angle faults, low-angle thrust faults, and three phases of folds, designated D_1, D_2, and D_3 in order of decreasing age, are recognized in the Apoon assemblage. The development of penetrative structures is dependent upon lithology and varies between the four lithotectonic units. The fine-grained rocks generally show more evidence of folding than conglomerate and coarse-grained volcanic rocks.

Faults

The Apoon assemblage is disrupted by both high-angle faults and low-angle thrusts. The high-angle faults trend east-west and are demonstrably younger than the thrusts. They truncate and offset thrust faults in three locations: Kinnorutin Pass, an unnamed area southwest of Apoon Mountain, and north of Amawk Mountain (Fig. 1). The high-angle faults are best recognized along the northern flank of the Doonerak multiduplex where the Carboniferous clastic and carbonate rocks overlie the Apoon assemblage. Elsewhere, where lithologic units are less distinct, the faults are recognized with difficulty and are not traced for distances greater than a few kilometers. Along the northern flank of the Doonerak window, the high-angle faults have trace lengths of 5 to 15 km and are arranged as a series of subparallel structures with overlapping terminations. Most faults have relatively small displacements with vertical off-sets on the range of 100 to 300 m. The greatest vertical displacements are found at the midpoints of the individual fault traces and the structures can be tracked along strike to their terminations. Phelps et al. (1987) record left-slip as the last motion of the fault system but the total lateral offset is small, probably on the order of a few hundred meters.

Low-angle faults internally disrupt the Apoon assemblage and define the major lithotectonic boundaries. The faults are generally subparallel to bedding and D_1 foliations and are recognized by local truncations of bedding and cleavage and by the truncation of section along strike. The faults consistently dip southeasterly and exhibit minor splays that sole southward (down-dip) into more extensive detachment surfaces which display rootless, north-vergent antiforms.

Thrust faults within the lower Paleozoic rocks are most easily observed where they separate different lithologic units of the Apoon assemblage. The best exposures lie on Amawk Mountain (Fig. 1), where shale, siltstone, and sandstone (Clastic unit) rest on massive volcanic rocks (Volcanic unit), and between Kinnorutin Pass and Karilyukpuk Creek (Fig. 1) where the Saint Patrick's Creek fault carries volcanogenic rocks (Volcaniclastic unit) over clastic rocks (Clastic unit). Fault breccia and orange iron staining associated with alteration along fault zones mark many structures. Elsewhere, thrust faults are present within the Phyllitic and Volcaniclastic units, but these faults are difficult to recognize and trace because of lithologic similarities of both plates.

Thrust faults within the Apoon assemblage typically are oriented subparallel to bedding and the penetrative D_1 foliation in the adjacent thrust sheets. Although the strike of the faults is consistently northeasterly, the dip varies; the southern faults dip more steeply than more northern faults. The southernmost, and structurally highest, thrust fault (Fig. 1) separates the Phyllitic unit from the underlying Volcaniclastic unit on a moderately southeast

dipping discontinuity (30°). The northernmost and structurally lowest exposed thrust separates the Clastic unit and underlying Volcanic unit on a subhorizontal fault surface. A spatially systematic change in dip is recorded on the thrust separating the Volcaniclastic and Clastic units of the Apoon assemblage. The fault dip changes from about 25° southeast to 15° northwest over a distance of approximately 3 km and forms a broad arch with an east-northeast axis. The thrust faults within the Apoon assemblage are laterally continuous and merge upward with the Blarney Creek thrust with the exception of the late-stage Eekayruk thrust (Phelps and Avé Lallemant, this volume, Chapter 4) that cross-cuts all structures but the high-angle faults. Thrust faults have undetermined displacements because of the lack of correlative stratigraphic markers on opposing sides of the structures. Individual displacements of at least several kilometers to tens of kilometers is required, however, by the lack of stratigraphic similarity of units on opposing sides of the structures.

There is a slight discordance between the orientations of thrusts within the Apoon assemblage and the overlying Blarney Creek thrust (Fig. 1). The thrusts within the Apoon trend northeasterly, whereas the Blarney Creek thrust trends east-west. These two systems of thrusts converge at the east end of the Doonerak window where the Blarney Creek allochthon overlies several different lithologic units of the Apoon assemblage. The relationship between these two thrust systems is clearly exposed in several locations on the steep flanks of mountains (Seidensticker and Oldow, this volume, Chapter 6).

Folds

Three generations of folds are observed in all of the rocks in the study area. The folds exhibit remarkably consistent orientations and the structural fabrics are shown separately for each tectonic slice of the Apoon assemblage (Fig. 4). Differences in data density reflect the lithologic control on structural development and style for each unit. The fine-grained clastic unit (Phyllitic unit) displays penetrative foliations (S_1) which are systematically folded by D_2 structures. Within the massive Volcanic unit, the S_1 foliation is poorly developed making subsequent D_2 folds indistinct. The Clastic and Volcaniclastic units show good development of S_1 and D_2 folds, but these structures are not as clearly defined as in the Phyllitic unit.

First-generation folds (D_1). The earliest generation of penetrative structures and the most obvious at the outcrop are foliations that are axial-planar to isoclinal folds. The foliation typically is oriented parallel to bedding (Fig. 4) and cross-cuts depositional surfaces only in the hinges of folds. The folds range in amplitude from 1 cm to 4 m and commonly occur as rootless structures. In the relatively homogeneous fine-grained slate, small-amplitude (1 to 5 cm) isoclinal folds of quartz veins are the best expressed mesoscopic feature. In lithologies such as banded argillite and interbedded siltstone where bedding ranges to 5 cm thick, the folds are defined by bedding surfaces and range from 20 cm to 4 m in amplitude.

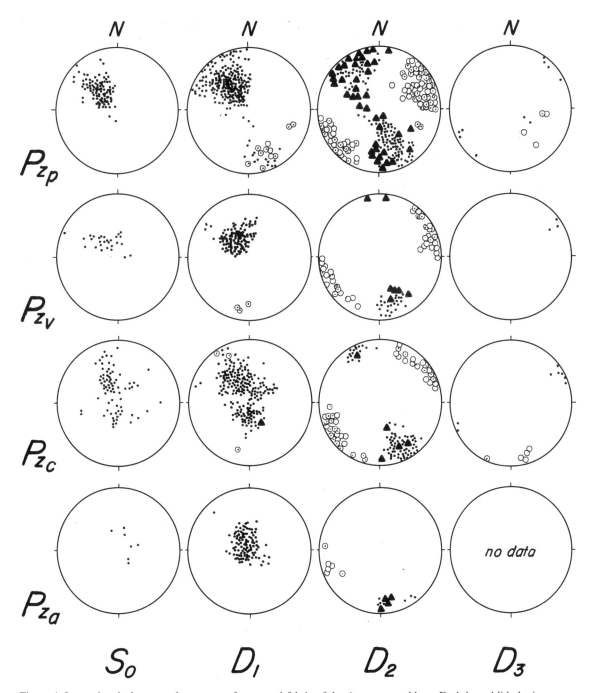

Figure 4. Lower hemisphere equal area nets of structural fabric of the Apoon assemblage. Fault-bound lithologic successions: P_{zp}, Phyllite unit; P_{zv}, Volcaniclastic unit; P_{zc}, Clastic unit; P_{za}, Volcanic unit. S_0, bedding; D_1, first-generation structures; D_2, second-generation structure; D_3, third-generation structures. Structural symbols: solid dot, pole to bedding or cleavage; solid triangle, pole to axial plane; open circle, lineation; and circle dot, fold axis.

There is no evidence of a penetrative foliation predating S_1 either at the outcrop or in thin sections. In thin section, S_1 is defined by alignment of phyllosilicate minerals and is enhanced by dissolution along the foliation plane. Lithic clasts are oblate (aspect ratio of 10:1) with their short axes perpendicular to the foliation plane and the long axis oriented downdip. Thin sections show evidence of shear strain associated with this foliation by the development of asymmetric pressure shadows, broken grains, and S-C relations.

S_1 generally strikes northeast and dips southeasterly (Fig. 4), but the dip decreases to nearly horizontal in the northern outcrops of the Doonerak window. Locally, where rocks of the Apoon assemblage are in close proximity to the Blarney Creek fault zone, they display shallow dips to the north. This south-to-north variation in S_1 is reflected in partial great circle distribution of poles to S_1 with a pole that corresponds to the maxima of D_2 fold axes (Fig. 4). The first-generation fold axes are rare but where observed are generally south to southeast trending with gentle plunges.

Shear-sense indicators studied in thin section (S-C orientations and broken grains) indicate a south-over-north transport of the upper plate. The similarity in orientation of the fold axes and the shear direction suggest that the fold axes experienced strain rotation into the tectonic transport direction.

Second-generation folds (D_2). Second-generation structures are dominated by crenulation and folding of the preexisting S_1. Axial-planar cleavage is often developed, but is not as pervasive as S_1. The D_2 crenulations are ubiquitous only in rocks with well-developed S_1. The crenulations often are accompanied by asymmetric chevron folds (amplitude less than 2 m; generally less than 30 cm) with southerly vergence and subhorizontal fold axes. The asymmetric chevron folds locally form orderly fold trains, but in other instances D_2 chevrons are replaced by disharmonic tight to close folds with rounded hinges and amplitudes of less than 6 m. The disharmonic folds are generally confined to outcrops containing interbedded lithologies.

Although the morphology of D_2 folds is variable, their orientation is quite consistent. D_2 axial planes strike about N60°E and generally dip steeply to the northwest (Fig. 4). Some southeast dips are observed and are associated with the sporadic development of box-folds showing conjugate axial planes. The orientation of D_2 folds is not dependent upon structural location; they have similar orientations in all four major thrust sheets and are independent of S_1 orientation. The vergence of asymmetric D_2 chevron folds indicates a northwest over southeast shear sense, which is further supported by microscopic offset along D_2 axial plane crenulations.

Third-generation folds (D_3). The last generation of structures (D_3) are sporadically developed as crenulations and minor folds. The structures are most commonly observed as a crenulation lineation that lies in S_1 and which cross-cuts D_2 lineations. A subvertical spaced cleavage is present in many outcrops and is parallel to S_3 where D_3 folds are present. Elsewhere the spaced cleavage is observed in exposures with no other indications of D_3 deformation.

AGE OF DEFORMATION

The Lower Mississippian Kekiktuk Conglomerate (Brosgé et al., 1962) is a thin, discontinuous unit of interbedded chert-pebble conglomerate, sandstone, and shale that separates the overlying Kayak Shale and superjacent Carboniferous Lisburne Group from the underlying lower Paleozoic rocks. The Kekiktuk Conglomerate is recognized throughout the North Slope, northeastern Brooks Range, and Doonerak window (Mull, 1982). In the Doonerak window, the Kekiktuk has a complex spatial distribution and appears in several structural levels (Seidensticker and Oldow, this volume, Chapter 6). Most commonly, the Kekiktuk forms laterally discontinuous slabs lying just above the Blarney Creek thrust but also has been recognized within the Blarney

Creek allochthon as isolated slices caught between thrust imbricates of Kayak Shale and Lisburne Limestone. The Kekiktuk Conglomerate is also found immediately below the Blarney Creek thrust and in a few instances as thin slices caught between the imbricated horses of Apoon assemblage. The detailed structure of the Kekiktuk is discussed elsewhere (Seidensticker, and Oldow, this volume, Chapter 6), but of critical importance here and to the understanding of the structural history of the lower Paleozoic rocks is the occurrence of the Kekiktuk below the Blarney Creek thrust.

The Kekiktuk Conglomerate is locally exposed in depositional contact with three of the Apoon assemblage units as infolded remnants of the largely detached conglomeratic layer overlying the Blarney Creek thrust (Seidensticker and Oldow, this volume, Chapter 6). These synclinal exposures exhibit all of the penetrative structures observed within the underlying Apoon assemblage. In fact, the infolds are D_1 structures characterized by an axial-planar cleavage that passes continuously from the Apoon into the overlying folded Kekiktuk. These relations are best preserved in two locations, the south flank of Falsoola Mountain and west of Amawk Mountain (Fig. 1).

At both locations D_1 folds below the Blarney Creek thrust deform the contact between the Kekiktuk and Apoon assemblage; the Kekiktuk is synformally infolded with the Apoon and contains south-dipping D_1 foliations concordant with S_1 in the older, underlying rocks. West of Amawk Mountain, the Kekiktuk is visibly folded and S_1 is parallel to the axial planes in both the Kekiktuk and the underlying Volcanic unit of the Apoon assemblage. On the southern flank of Falsoola Mountain, where D_1 folds are less visible, both the Kekiktuk and the Phyllitic unit of the Apoon contain southeast-dipping S_1. At these locations, bedding in the Kekiktuk defines large, close folds whereas bedding in the Apoon slate and siltstone is transposed into parallelism with S_1 during the formation of small isoclinal folds. Since S_1 passes continuously from the Apoon to the Kekiktuk, the bedding discordance between the Kekiktuk and the underlying Apoon is attributed to disharmonic folding due to ductility contrasts between the two lithologies.

The clear involvement of the Mississippian Kekiktuk in the earliest phase of penetrative deformation recognized in the Apoon assemblage requires that all of the observed structures within the Apoon assemblage formed after Mississippian deposition. Thus, contrary to many previous interpretations, the penetrative fabric of the Apoon assemblage is not related to the Devonian Ellesmerian orogeny but rather was developed together with structures in younger rocks during Brookian tectonism.

A distinct structural difference exists between the lower Paleozoic rocks of the Doonerak window, and coeval units in the northeastern Brooks Range and North Slope. In the northeastern Brooks Range an angular unconformity is confirmed by the existence of additional generations of structures within the Franklinian sequence rocks not found in overlying the Ellesmerian sequence (Oldow et al., 1987). In addition to the lack of structural evidence for an angular unconformity, Phelps and Avé Lallemant (this volume, Chapter 4) have documented depositional continuity between the Apoon assemblage and the Kekiktuk Conglomerate at the eastern end of the Doonerak window. The absence of an angular unconformity in the Doonerak window requires reassessment of the alleged structural continuity of the Apoon assemblage and the lower Paleozoic rocks of the North Slope.

TECTONIC RECONSTRUCTION

Stratigraphic uncertainties imparted by poor biostratigraphic control and complex structure precluded detailed assessment of the vertical and lateral variation of facies within the Apoon assemblage. Because all units of the Apoon reside in fault-bound sheets and no direct evidence exists to establish the sequence of movement on individual thrusts, several reconstructions are possible. In light of the consistent north-directed shear observed associated with D_1 structures, the lowest unit (Volcanic unit) was initially the northernmost facies and the upper unit (Phyllitic unit) the southernmost. However, details of the stacking sequence are nonunique. A general model of south-to-north imbrication of the Apoon horses beneath the Endicott Mountains and Blarney Creek allochthons is presented in Figure 5.

The least complicated model of lithologic distribution, in which the volcanic component systematically diminishes with

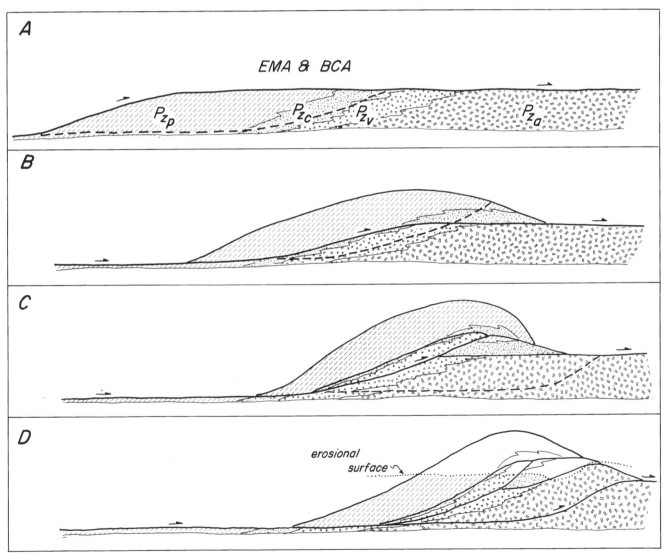

Figure 5. Imbrication model and stratigraphic reconstruction of Apoon assemblage rocks. Model assumes a simple north to south stratigraphic progression from volcanic rocks to fine-grained clastic rocks. Thrust history requires imbricate decapitation to preserve stratigraphic simplicity. Faults shown as heavy lines, incipient faults as dashed lines, current level of erosion by dotted line in last frame. BCA. Blarney Creek allochton; EMA, Endicott Mountains allochthon; P_{zp}, Phyllite unit; P_{zc}, Clastic unit; P_{zv}, Volcaniclastic unit; P_{za}, Volcanic unit.

distance from north to south from near-source volcanic deposits to distal fine-grained clastic, is supported by the existence of remnants of the Kekiktuk Conglomerate resting depositionally on three of four units of the Apoon assemblage. This stratigraphic scenario requires a complex imbrication history involving decapitation of older faults by younger. In this configuration (Fig. 5), the Clastic and Phyllitic units were initially imbricated and transported over the volcanic sequences, as a fault within the Apoon assemblage ramped upward to merge with the overlying Blarney Creek thrust. The lower detachment then propagated forward into the volcanic rocks before ramping up to the Blarney Creek roof thrust, cutting across the preexisting intra-Apoon thrust imbricates. This imbricate decapitation allows the emplacement of the Volcaniclastic unit and the structurally overlying Phyllitic unit to be carried together above the previously emplaced sheet composed of Clastic unit, which had been carried northward over the volcanic rocks. This imbrication scheme reproduces the observed distribution of duplex horses and preserves the depositional constraint imparted by the overlying Kekiktuk Conglomerate.

Regardless of the specific stratigraphic relations among the lithotectonic units composing the Apoon assemblage, their lithology, provenance, and geochemistry is typical of volcanic-arc assemblages. The high pyroclastic/flow volcanic ratio and the voluminous tephra interbedded with reworked tuff and mudstone are characteristic of magmatic-arc environments (Garcia, 1978). Provenance of the coarse clastic sediments also supports deposition in an environment supplied detritus from a magmatic arc. The presence of many detrital metamorphic and other nonvolcanic grains, however, suggests a continental sediment source. All these constituents fit in the framework of an Andean-type continental arc or an island arc separated from a continent by a marginal basin. This interpretation is further supported by the chemical composition of basaltic volcanic rocks in the Volcaniclastic and Volcanic units.

The paleogeographic significance of the lower Paleozoic Apoon assemblage is less clear. The lack of penetrative pre-Carboniferous structures sets it apart from age-equivalent rocks of the North Slope and northeastern Brooks Range. The position of the succession, lying to the south of coeval rocks involved in Ellesmerian structures, suggests that the Apoon assemblage may correspond to a foreland succession that was not directly involved in structures of the Devonian (Ellesmerian) contractional belt preserved farther north. This paleogeographic position is consistent with observations that Ellesmerian structures in the northeastern Brooks Range and elsewhere in northwestern Canada are south vergent (Oldow et al., 1987; Oldow and Avé Lallemant, 1989, 1993; Avé Lallemant and Oldow, 1993).

The fact that Kekiktuk Conglomerate depositionally overlies various lithologic constituents of the Apoon assemblage residing in different thrust-duplex horses points to the existence of a disconformity or possibly a low-angle unconformity preceding latest Devonian(?) or Carboniferous deposition of the regionally extensive conglomerate. Without better constraints on the magnitude of tectonic transport during imbrication of the Apoon horses, however, the contribution of regional tilt and erosion (possibly related to migration of a foredeep peripheral bulge) versus depositional overlap of relatively flat lying coeval facies of the lower Paleozoic Apoon assemblage cannot be differentiated. Be that as it may, the lack of penetrative structures of pre-Carboniferous age within the Apoon assemblage yields a vital point of control for any regional reconstruction of the Alaskan Arctic margin.

CONCLUSIONS

Four fault-bound lithologic units comprise the lower Paleozoic succession exposed in the Doonerak window and sparse fossil assemblages from fine-grained clastic rocks range in age from Cambrian through Siluro-Ordovician. The four lithologic units constitute the Apoon assemblage and range in composition from primary volcanic rocks, coarse volcaniclastic deposits, to fine-grained phyllites. The lithologies, provenance of clastic rocks, and chemistry of the volcanics are consistent with deposition in a continental magmatic arc complex. The Apoon assemblage comprises the lower horses of the Doonerak multiduplex that formed during late-stage Brookian deformation.

Imbrication of the rocks occurred during Brookian orogenesis and no evidence exists to suggest the involvement of the Apoon assemblage in earlier, pre-Mississippian tectonism. The lack of mid-Paleozoic structures severs the structural link between lower Paleozoic rocks of the Apoon assemblage and those exposed in the northeastern Brooks Range and in the subsurface of the North Slope. The southern limit of penetrative deformation during pre-Mississippian tectonism must have been north of the original site of deposition of the rocks now exposed in the Doonerak window. Stratigraphic linkage of the Apoon assemblage with rocks of the North Slope are still compelling, however. Thus, the structural evolution of the lower Paleozoic rocks of the Doonerak window clearly defines the need to reexamine models for the tectonic significance and spatial distribution of Ellesmerian structures in Arctic Alaska.

ACKNOWLEDGMENTS

The authors wish to acknowledge the careful reviews given earlier versions of this manuscript by T. E. Moore, C. L. Hanks, and R. Reifenstuhl. This research was partially supported by the Department of Energy (DE-AS05-83ER13124), the National Science Foundation (EAR-8517384, EAR-872017, EAR-9017835), and the Industrial Associates Program at Rice University.

REFERENCES CITED

Armstrong, A. K., Mamet, B. L., Brosgé, W. P., and Reiser, H. N., 1976, Carboniferous section and unconformity at Mount Doonerak, Brooks Range, Alaska: American Association of Petroleum Geologists Bulletin, v. 60, p. 962–972.

Avé Lallemant, H. G., and Oldow, J. S., 1993, Kinematics and areal extent of Devonian structures in northern Alaska and Yukon Territory: A link with the Innuitian fold and thrust belt of the Canadian Arctic Archipelago: Geological Society of America Abstracts with Programs, v. 25, p. A–171.

Brosgé, W. P., and Reiser, H. N., 1971, Preliminary bedrock geologic map, Wiseman and eastern Survey Pass Quadrangles, Alaska: U.S. Geological Survey Open-File Map 479, scale 1:250,000.

Brosgé, W. P., Dutro, J. T., Jr., Mangus, M. D., and Reiser, H. N., 1962, Paleozoic sequence in eastern Brooks Range, Alaska: American Association of Petroleum Geologists Bulletin, v. 26, p. 2174–2198.

Cann, J. R., 1970, Rb, Sr, Y, Zr, and Nb in some ocean floor basaltic rocks: Earth and Planetary Science Letters, v. 10, p. 7–11.

Carter, C., and Laufeld, S., 1975, Ordovician and Silurian fossils in well cores from North Slope Alaska: American Association of Petroleum Geologists Bulletin, v. 59, p. 457–464.

Churkin, M., Jr., 1969, Paleozoic tectonic history of the Arctic basin north of Alaska: Science, v. 165, p. 549–555.

Churkin, M., Jr., 1975, Basement rocks of Barrow Arch, Alaska, and circum-Arctic Paleozoic mobile belt: American Association of Petroleum Geologists Bulletin, v. 59, p. 451–456.

Crawford, A., and Keays, R. R., 1978, Cambrian greenstone belts in Victoria: marginal sea-crust slices in the Lachlan fold belt of southeastern Australia: Earth and Planetary Science Letters, v. 41, p. 197–208.

Dickinson, W. R., 1982, Compositions of sandstones in circum-Pacific subduction complexes and fore-arc basins: American Association of Petroleum Geologists Bulletin, v. 66, p. 121–137.

Dickinson, W. R., and Seely, D. R., 1979, Structure and stratigraphy of forearc regions: American Association of Petroleum Geologists Bulletin, v. 63, p. 2–31.

Dickinson, W. R., and Suczek, C. A., 1979, Plate tectonics and sandstone compositions: American Association of Petroleum Geologists Bulletin, v. 63, p. 2164–2182.

Dickinson, W. R., Beard, L. S., Brakenridge, G. R., Erjavec, J. L., Ferguson, R. C., Inman, K. F., Knepp, R. A., Lindberg, F. A., and Ryberg, P. T., 1983, Provenance of North American Phanerozoic sandstones in relation to tectonic setting: Geological Society of America Bulletin, v. 93, p. 222–235.

Dillon, J. T., Brosgé, W. P., and Dutro, J. T., Jr., 1986, Generalized geologic map of the Wiseman Quadrangle, Alaska: U.S. Geological Survey Open-File Map 86-219, scale 1:250,000.

Dutro, J. T., Jr., 1970, Pre-Carboniferous carbonate rocks, northeastern Alaska, *in* Adkinson, W. L., and Brosgé, W. P., eds., Proceedings, Geological Seminar on the North Slope of Alaska: Los Angeles, American Association of Petroleum Geologists, p. M1–M17.

Dutro, J. T., Jr., Brosgé, W. P., Lanphere, M. A., and Reiser, H. N., 1976, Geologic significance of Doonerak structural high, central Brooks Range, Alaska: American Association of Petroleum Geologists Bulletin, v. 60, p. 952–961.

Dutro, J. T., Jr., Palmer, A. R., Repetski, J. E., and Brosgé, W. P., 1984, Middle Cambrian fossils from the Doonerak Anticlinorium, central Brooks Range, Alaska: Journal of Paleontology, v. 58, p. 1364–1371.

Garcia, M. O., 1978, Criteria for the identification of ancient volcanic arcs: Earth-Science Reviews, v. 14, p. 147–165.

Grantz, A., and May, S. D., 1982, Rifting history and structural development of the continental margin north of Alaska, *in* Watkins, J. S., and Drake, C. L., eds., Studies in continental margin geology: American Association of

Petroleum Geologists Memoir 34, p. 77–100.

Lerand, M., 1973, Beaufort Sea, *in* McCrossam, R. G., ed., The future petroleum provinces of Canada—Their geology and potential: Canadian Society of Petroleum Geology Memoir 1, p. 315–386.

Marsaglia, K. V., 1991, Provenance of sands and sandstones from the Gulf of California, a rifted continental arc, *in* Fisher, R. V., and Smith, G. A., eds., Sedimentation in volcanic settings: Society of Economic Paleontologists and Mineralogists, Special Publication 45, p. 237–248.

Marsaglia, K. V., and Ingersoll, R. V., 1992, Compositional trends in arc-related, deep-marine sand and sandstone: A reassessment of magmatic-arc provenance: Geological Society of America Bulletin, v. 104, p. 1637–1649.

Mauch, E. A., 1985, A seismic stratigraphic and structural interpretation of the middle Paleozoic Ikpikpuk-Umiat basin, National Petroleum Reserve in Alaska [M.A. thesis]: Houston, Texas, Rice University, 220 p.

Moore, T. E., 1987, Geochemistry and tectonic setting of some volcanic rocks of the Franklinian assemblage, central and eastern Brooks Range, *in* Tailleur, I. L., and Weimer, P., eds., Alaska North Slope geology: Society of Economic Paleontologists and Mineralogists, Pacific Section, Publication 50, p. 691–710.

Moore, T. E., and Churkin, M., Jr., 1984, Ordovician and Silurian graptolite discoveries from the Neruokpuk Formation (*sensu lato*), northeastern and central Brooks Range, Alaska: Paleozoic Geology of Alaska and Northwestern Canada Newsletter, v. 1, p. 21–23.

Moore, T. E., Wallace, W. K., Bird, K. J., Karl, S. M., Mull, C. G., and Dillon, J. T., 1994, Geology of northern Alaska, *in* Plafker, G., and Berg, H. C., eds., The geology of Alaska: Boulder, Colorado, Geological Society of America, The Geology of North America, v. G-1, p. 49–140.

Mull, C. G., 1982, Tectonic evolution and structural style of the Brooks Range, Alaska: an illustrated summary, *in* Powers, R. B., ed., Geologic studies of the Cordilleran Thrust Belt, Volume 1: Denver, Colorado, Rocky Mountain Association of Geologists, p. 1–45.

Mull, C. G., Adams, K. E., and Dillon, J. T., 1987, Stratigraphy and structure of the Doonerak fenster and Endicott Mountains allochthon, central Brooks Range, Alaska, *in* Tailleur, I. L., and Weimer, P., eds., Alaskan North Slope geology: Society of Economic Paleontologists and Mineralogists, Pacific Section, Publication 50, p. 663–680.

Oldow, J. S., and Avé Lallemant, H. G., 1989, Tectonic elements of eastern Arctic Alaska and northwestern Canada: Eos (Transactions, American Geophysical Union), v. 70, p. 1337.

Oldow, J. S., and Avé Lallemant, H. G., 1993, Late Mesozoic and Cenozoic tectonism of the Arctic Margin of eastern Alaska and northwestern Canada: Implications for the evolution of the Canada basin: Geological Society of America Abstracts with Programs, v. 25, p. A–172.

Oldow, J. S., Avé Lallemant, H. G., Julian, F. E., Seidensticker, C. M., and Phelps, J. C., 1984, The Doonerak window duplex: regional implications: Geological Society of America Abstracts with Programs, v. 16, p. 326.

Oldow, J. S., Avé Lallemant, H. G., Julian, F. E., and Seidensticker, C. M., 1987, Ellesmerian(?) and Brookian deformation in the Franklin Mountains, northeast Brooks Range, Alaska and its bearing on the origin of the Canada basin: Geology, v. 15, p. 37–41.

Pearce, J. A., 1975, Basalt geochemistry used to investigate past tectonic environments on Cyprus: Tectonophysics, v. 25, p. 41–67.

Pearce, J. A., 1982, Trace element characteristics of lavas from destructive plate boundaries, *in* Thorpe, R. S., ed., Orogenic andesites: New York, John Wiley, p. 525–547.

Pearce, J. A., 1983, Role of the sub-continental lithosphere in magma genesis at active continental margins, *in* Hawkesworth, C. J., and Norry, M. J., eds., Continental basalts and mantle xenoliths: Nantwich, United Kingdom, Shiva, p. 230–249.

Pearce, J. A., and Cann, J. R., 1973, Tectonic setting of basic volcanic rocks determined using trace element analysis: Earth and Planetary Science Letters,

v. 19, p. 290–300.

Pearce, J. A., and Norry, M. J., 1979, Petrogenetic implications of Ti, Zr, Y, and Nb variations in volcanic rocks: Contributions to Mineralogy and Petrology, v. 69, p. 33–47.

Phelps, J. C., Avé Lallemant, H. G., Seidensticker, C. M., Julian, F. E., and Oldow, J. S., 1987, Late-stage high-angle faulting, eastern Doonerak Window, central Brooks Range, Alaska, *in* Tailleur, I. L., and Weimer, P., eds., Alaskan North Slope geology: Society of Economic Paleontologists and Mineralogists, Pacific Section, Publication 50, p. 685–690.

Reed, B. L., 1968, Geology of the Lake Peters area, northeastern Brooks Range, Alaska: U.S. Geological Survey Bulletin 1236, 132 p.

Repetski, J. E., Carter, C., Harris, A. G., and Dutro, J. T., Jr., 1987, Ordovician and Silurian fossils from the Doonerak anticlinorium, central Brooks Range, Alaska, *in* Hamilton, T. D., and Galloway, J. P., eds., Geologic

studies in Alaska by the U.S. Geological Survey during 1986: U.S. Geological Survey Circular 998, p. 40–42.

Sable, E. G., 1977, Geology of the western Romanzof Mountains, Brooks Range, northern Alaska: U.S. Geological Survey Professional Paper 897, 84 p.

Shervais, J. W., 1982, Ti-V plots and the petrogenesis of modern and ophiolitic lavas: Earth and Planetary Science Letters, v. 59, p. 101–118.

Wallace, W. K., and Hanks, C. L., 1990, Structural provinces of the northeastern Brooks Range, Arctic National Wildlife Refuge, Alaska: American Association of Petroleum Geologists Bulletin, v. 74, p. 1100–1118.

Winchester, J. A., and Floyd, P. A., 1977, Geochemical discrimination of different magma series and their differentiation products using immobile elements: Chemical Geology, v. 20, p. 325–343.

MANUSCRIPT ACCEPTED BY THE SOCIETY SEPTEMBER 23, 1997

Geological Society of America
Special Paper 324
1998

Structural development and kinematic history of ramp-footwall contraction in the Doonerak multiduplex, central Brooks Range, Arctic Alaska

C. M. Seidensticker
615 Baltimore Drive, El Paso, Texas 79902
J. S. Oldow
Department of Geology and Geological Engineering, University of Idaho, Moscow, Idaho 83844-3022

ABSTRACT

The Doonerak multiduplex of the central Brooks Range orogen uplifted and exposed deep structural levels of the fold and thrust belt. The duplex system was developed during footwall contraction as Devono-Carboniferous clastic and carbonate rocks of the Endicott Mountains allochthon were transported to the north on the Amawk thrust. Passage of the Endicott Mountains allochthon over a footwall ramp composed of lower Paleozoic clastic and volcanic rocks and uppermost Devonian to Carboniferous clastic and carbonate strata resulted in the simultaneous formation of two stacked duplexes in the lower plate assemblage. The upper Blarney Creek duplex and the lower Apoon duplex are separated by the Blarney Creek thrust, which served both as the floor thrust and as the roof thrust of the upper and lower duplexes, respectively. The Blarney Creek thrust varies in character along strike, from a sharp discontinuity to a diffuse zone of distributed shear as much as 250 m thick. In most locations, the fault zone is tens of meters thick and has a structural position controlled by a thin conglomerate that formed the depositional interface between lower Paleozoic strata and overlying Devono-Carboniferous rocks of the Doonerak footwall assemblage. The fault zone truncates structures at the bottom of the upper duplex and at the top of the lower duplex, and where it is broadest, the fault zone contains thin slices of both upper and lower duplex lithologies. The distribution of lithologic units and structures in lower plate rocks of the Doonerak window are inconsistent with a conventional serial development of the two duplexes and can be explained only by simultaneous formation of the upper and lower imbricate structures during footwall contraction.

INTRODUCTION

The Doonerak window (Brosgé and Reiser, 1971) is located in the central Brooks Range fold and thrust belt (see Fig. 3 of Preface, this volume) and exposes a thick succession of lower Paleozoic and Carboniferous strata and minor slices of Permian and Triassic rocks in an erosional breach of the overlying Endicott Mountains allochthon. The Endicott Mountains allochthon dominates exposures of the external part of the Brooks Range orogen and is composed largely of an imbricate stack of Devonian and Carboniferous clastic and carbonate rocks (Dutro et al., 1976; Mull, 1982). The rocks exposed in the Doonerak window differ in stratigraphy from coeval units in the overlying Endicott Mountains allochthon and share affinities with age equivalent units exposed in the northeastern Brooks Range and underlying the North Slope (Brosgé et al., 1974;

Seidensticker, C. M., and Oldow, J. S., 1998, Structural development and kinematic history of ramp-footwall contraction in the Doonerak multiduplex, central Brooks Range, Arctic Alaska, *in* Oldow, J. S., and Avé Lallemant, H. G., eds., Architecture of the Central Brooks Range Fold and Thrust Belt, Arctic Alaska: Boulder, Colorado, Geological Society of America Special Paper 324.

Armstrong et al., 1976; Dutro et al., 1976; Armstrong and Mamet, 1975; Mull, 1982; Mull et al., 1987; Moore et al., 1994). As such, the lower plate rocks of the Doonerak window have been recognized as a critical link between allochthonous units of the fold and thrust belt and autochthonous succession underlying the foreland to the north.

Recognition of the paleogeographic significance of the Doonerak window has proven to be less elusive than deciphering the structural history and the origin of the uplift. Although early workers identified thrust faults dipping away from both flanks of the east-northeast–trending antiformal structure (Brosgé and Reiser, 1971; Mull, 1982), the relationship between the structures on the northern and southern flanks of the window was uncertain. It was not clear whether the thrusts were parts of a single folded fault or separate structures with unrelated kinematic histories (Dutro et al., 1976). Mull (1982) was the first to recognize the tectonic significance of the Amawk thrust fault on the northern flank of the Doonerak window (Fig. 1) as the basal decollement of the Endicott Mountains allochthon, which was displaced tens of kilometers to the north (Mull et al., 1987; Oldow et al., 1987a). Mapping along the southern margin of the Doonerak window documented a south-dipping thrust fault between lower Paleozoic and Devonian strata (Julian et al., 1984; Oldow et al., 1984; Dillon et al., 1986; Mull et al., 1987) and established the continuity of the

Amawk thrust on both flanks of the Doonerak window. The folded Amawk thrust forms a nearly continuous tectonic surface around the eastern Doonerak window with only minor offset by the later, Eekayruk thrust fault and younger high-angle faults (Phelps, 1987; Phelps and Avé Lallemant, this volume, Chapter 4; Julian, 1989; Julian and Oldow, this volume, Chapter 5). The Doonerak window is framed by a relatively coherent upper plate assemblage composed of the Endicott Mountains allochthon.

Until recently, the structural relations between Carboniferous and lower Paleozoic rocks in the footwall assemblage exposed in the Doonerak window and their tectonic significance were poorly understood. The lower plate rocks were generally considered to be autochthonous or parautochthonous relative to Brookian contractional structures (Dutro et al., 1976), and their structurally high position within the orogen was attributed to isostatic uplift (Mull, 1982). Mapping and structural analysis (Julian and Oldow, this volume, Chapter 5; Phelps and Avé Lallemant, this volume, Chapter 4) and construction of regional balanced cross sections (Oldow et al., 1987a), however, demonstrated that the lower plate rocks were detached at depth and formed a regionally significant duplex system (Oldow et al., 1984). The duplex system developed during late-stage shortening and exhumation of the Brooks Range orogen in the Miocene (O'Sullivan et al., this volume, Chapter 11).

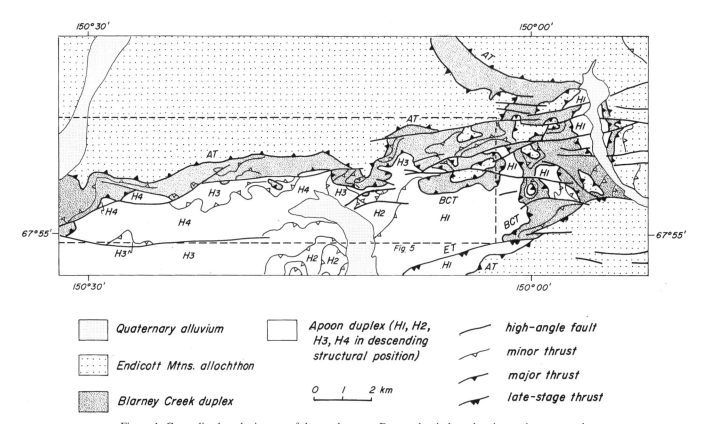

Figure 1. Generalized geologic map of the northeastern Doonerak window showing major structural assemblages forming the multiduplex. For regional setting, refer to Figure 3 of Preface, this volume. Major thrust faults are: AT-Amawk thrust; BCT-Blarney Creek thrust; ET-Eekayruk thrust.

Careful assessment of the structure and kinematics of lower plate rocks in the Doonerak window indicates a complex history of imbrication and duplex formation beneath the Endicott Mountains allochthon. The Endicott Mountains allochthon behaved as a relatively coherent upper plate unit, with the Amawk thrust serving as the roof thrust during imbrication of the underlying Devono-Carboniferous and lower Paleozoic rocks. Footwall shortening in the Doonerak multiduplex was accommodated by two subsidiary, stacked duplexes (Fig. 2). An upper duplex is structurally separated from a lower duplex by the Blarney Creek thrust (Seidensticker et al., 1987). The Blarney Creek thrust is a zone of shear that served both as the floor thrust of the overlying Blarney Creek duplex and simultaneously as the roof thrust that accommodated imbrication within the structurally deeper Apoon duplex. A remarkable feature of this fault zone is that it nucleated around a thin (5- to 20-m thick) conglomerate unit located at the pretectonic depositional contact between Devono-Carboniferous and lower Paleozoic rocks. The floor thrust of the overlying duplex developed immediately above the conglomerate unit, whereas the roof thrust of the underlying duplex developed immediately below the conglomerate. The conglomerate unit behaved as a boundary layer between and was itself shortened contemporaneously with the duplexes above and below. With few exceptions, rocks of the conglomerate boundary layer were not incorporated in the horses of either duplex, but rather "floated" in the intervening detachment zone.

The focus of this work is to document the kinematic evolution of footwall contraction in the upper and lower Doonerak footwall duplexes and to assess their implications for duplex formation in general. Emphasis is placed on the development of the stacked footwall duplexes and on the kinematic significance of the boundary-layer shear zone (Blarney Creek thrust) that separates them.

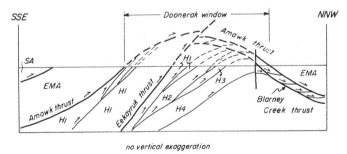

Figure 2. Schematic cross section through the eastern Doonerak window (oriented NNW–SSE). Beneath antiformily folded Amawk thrust imbricated horses of the Blarney Creek allochthon overlie the Blarney Creek thrust on the northern flank of the Doonerak window but are not found along the southern flank. Imbricate horses of the Apoon duplex merge with the Blarney Creek thrust on the north and with the Amawk thrust on the south. The Eekayruk thrust is a younger structure that cuts through the multiduplex. SA-Skajit allochthon; EMA-Endicott Mountains allochthon; H1, H2, H3, H4-horses of Apoon duplex.

DUPLEX KINEMATICS

Duplex structures are common architectural features of fold and thrust belts throughout the world (e.g., Willis, 1902; Bally et al., 1966; Dahlstrom, 1970; Elliott and Johnson, 1980; McClay and Price, 1981; Boyer and Elliot, 1982; Bell, 1983; Bosworth, 1984; Banks and Warburton, 1986; Fermor and Price, 1987; Yin et al., 1989; Wallace and Hanks, 1990; McClay, 1992; Skuce et al., 1992). The term "duplex" describes a plexus of imbricate thrust faults within which the trajectories of individual faults emerge from a common floor thrust and merge upward into a common roof thrust. Boyer and Elliott (1982) used balanced graphical techniques to model duplex formation as the progressive contraction of a thrust-ramp footwall. As a duplex develops, slip is transferred from the floor thrust to the roof thrust by way of a system of subsidiary ramping thrust faults. The ramping thrusts create imbricate horses that are plucked sequentially, in the direction of tectonic transport, from the footwall. Slip transfer and the creation of additional horses cause structural thickening; duplex growth is accompanied by folding of the overriding thrust sheet and addition of mass to the moving thrust complex.

Using forward models, several investigators have simulated a large variety of theoretical duplex structures arising from variations on the basic scheme proposed by Boyer and Elliott (1982). Variables include ramp height (horse thickness), ramp angle, spacing between horse-bounding thrusts (horse length), and the amount of displacement of individual horses (e.g., Butler, 1982; Boyer, 1986a; Banks and Warburton, 1986). Recognition of the geometric features that define a duplex, however, does not necessarily imply that the structure developed in the sequential manner. Boyer (1986b) suggested that some duplexes developed by simultaneous, rather than sequential, movement on multiple thrust faults. Other workers (Fermor and Price, 1987; Woodward et al., 1989; Yin et al., 1989) have also shown that the fundamental geometry of a duplex can arise through various kinematic paths.

In addition to complex kinematic histories, duplex morphology can also be quite complex. Fermor and Price (1987) used the term "multiduplex" to describe five or more stratigraphically distinct but structurally overlapping duplexes that occur together as nested, lenticular stacks of imbricated, sigmoidal thrust slices. They argued that even in light of unresolved complexities individual duplex structures within the multiduplex formed concurrently.

Within the Doonerak duplex system, the synchronous development of multiple levels of detachment during duplex formation can be documented on the basis of detailed structural and kinematic analysis. In fact, the Doonerak duplex can be described as a multiduplex consisting of two stacked, subsidiary duplexes that simultaneously accommodated footwall contraction beneath the Amawk thrust. Displacements that occurred on bounding fault surfaces between the two duplexes resulted in truncation of imbricate horses in both duplexes. The Doonerak multiduplex is an excellent example of nonsequential development of contractional structures, which may represent a more prevalent scheme of shortening than that generally attributed to fold and thrust belts.

STRUCTURAL FRAMEWORK OF THE DOONERAK MULTIDUPLEX

Rocks in and around the Doonerak window are complexly deformed and compose two fundamental structural units separated by the Amawk thrust: (1) an upper plate, or hanging-wall assemblage, consisting of the Endicott Mountains allochthon; and (2) a lower plate, or footwall assemblage, composed predominantly of Devono-Carboniferous and/or lower Paleozoic rocks (Figs. 2 and 3). This basic structural scheme is modified by a post-duplex thrust (the Eekayruk thrust) and by later high-angle faults. The Eekayruk thrust is a late-stage structure in the imbrication history of this part of the Brooks Range fold and thrust belt (Phelps, 1987; Phelps and Avé Lallemant, this volume, Chapter 4) and juxtaposes

different parts of the footwall and hanging-wall assemblages of the Doonerak window (Fig. 2).

Younger high-angle faults truncate thrust faults on the north flank of the Doonerak window and form an east-west–trending belt as much as 5 km wide (Fig. 1). Individual high-angle faults have trace lengths of as much as 15 km. Occasionally fault strands anastomose along strike, whereas in other cases, individual strands are discontinuous along strike and step, both left and right, in poorly defined en echelon patterns. Vertical displacement on the faults typically is between 100 to 300 m or less; the greatest throw observed does not exceed 1,500 m. Left-lateral motion is documented on several of the faults (Phelps et al., 1987; Phelps and Avé Lallemant, this volume, Chapter 4).

Although disrupted by the high-angle fault system (Fig. 1), continuity of the preexisting contractional structures is well preserved along the northeastern flank of the Doonerak window. We describe critical structural relations in the hanging-wall and footwall assemblages in the following sections.

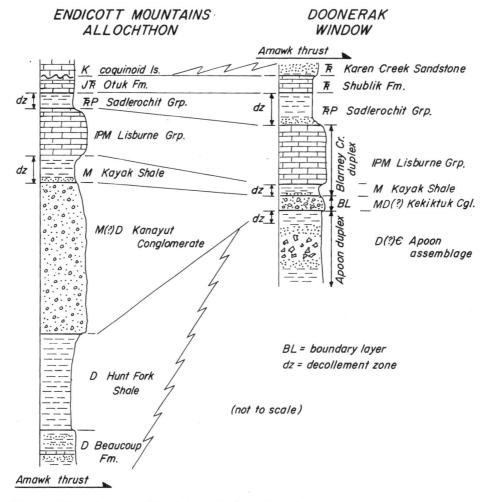

Figure 3. Schematic stratigraphic relations of Endicott Mountains allochthon and footwall duplexes of subjacent Doonerak window.

Hanging-wall assemblage: Endicott Mountains allochthon

The Endicott Mountains allochthon constitutes the hanging-wall assemblage of the Amawk thrust. Detailed stratigraphic and structural analyses of this assemblage are beyond the scope of this paper and we only address points important to the evolution of the Doonerak duplexes. For additional information about the Endicott Mountains allochthon, the reader is referred to Phelps and Avé Lallemant (volume, Chapter 4) and Handschy (this volume, Chapters 2 and 3).

Stratigraphy. In the vicinity of the Doonerak window, the Endicott Mountains allochthon is composed of Devonian clastic rocks that pass upward into Carboniferous and younger carbonate and clastic rocks along the northern flank of the Brooks Range (Fig. 3 of Preface, this volume). The Endicott Mountains allochthon and the footwall assemblage exposed in the Doonerak window share several stratigraphic units (Fig. 3). Nevertheless, the two successions are easily differentiated because of substantial differences in lithology, thickness, and depositional setting of the units within the footwall and hanging-wall assemblages.

The Endicott Mountains allochthon is dominated by the clastic and carbonate rocks of the Endicott and Lisburne Groups (Mull, 1982; Handschy, this volume, Chapters 1 and 2). In ascending stratigraphic order, the Upper Devonian and Lower Mississippian Endicott Group is composed of the Hunt Fork Shale, the Kanayut Conglomerate, and the Kayak Shale (Tailleur et al., 1967). Contacts between all three units are thought to be conformable but in many cases were strongly sheared during Brookian tectonism (Handschy, this volume, Chapter 3). Depositionally underlying the Hunt Fork Shale of the basal Endicott Group, the Middle(?) to lower Upper Devonian Beaucoup Formation is composed of fine-grained calcareous and siliciclastic rocks (Bowsher and Dutro, 1957; Brosgé et al., 1962; Chapman et al., 1964; Nilsen, 1981; Phelps and Avé Lallemant, this volume, Chapter 4; Handschy, this volume, Chapter 2). Conformably overlying the Kayak Shale are Lower Mississippian through Lower Pennsylvanian carbonate rocks of the Lisburne Group (Bowsher and Dutro, 1957).

The lithology and thickness of rocks composing the Endicott Group in the hanging-wall and footwall assemblages of the Amawk thrust differ greatly (Fig. 3). In the hanging-wall, the Endicott Group is composed of Hunt Fork Shale, Kanayut Conglomerate, and Kayak Shale, which have an aggregate thickness of greater than 8 km (Handschy, this volume, Chapters 2 and 3). In contrast, the Endicott Group of the footwall assemblage is composed of the Kekiktuk Conglomerate and Kayak Shale, which have a combined thickness of only several hundred meters.

Differences in the Kayak Shale and Lisburne Group carbonate rocks of the footwall and hanging-wall assemblages are subtle and for the Lisburne carbonates arise primarily from accumulation in different depositional settings. The Lisburne Group of the Endicott Mountains allochthon was deposited in more distal (farther from shore), open-marine conditions than coeval rocks of the footwall assemblage (Armstrong et al., 1976). Carbonate rocks of the Lisburne Group in the footwall assemblage represent a more proximal (closer to shore) facies and bear strong similarities with the Lisburne of the North Slope encountered in wells and exposed in the northeastern Brooks Range (Armstrong et al., 1976).

The Middle(?) to Upper Devonian Beaucoup Formation (Tailleur et al., 1967; Dutro et al., 1979) has a structural thickness of about 1.5 km in the southern Endicott Mountains allochthon (Handschy, this volume, Chapter 2; Phelps and Avé Lallemant, this volume, Chapter 4) and is not found within the footwall assemblage. The Beaucoup Formation is a sequence of interlayered shale, calcareous shale, sandstone, conglomerate, volcaniclastic rocks, and limestone reef deposits that underlies the Hunt Fork Shale, possibly disconformably (Phelps and Avé Lallemant, this volume, Chapter 4; Handschy, this volume, Chapter 2). On the south flank of the Doonerak window, the Beaucoup structurally overlies lower Paleozoic rocks of the footwall assemblage and can be traced around the eastern closure of the window to the north, where the Beaucoup rests on Carboniferous rocks. The Beaucoup Formation thins from a maximum 1.5 km in the south to less than a few hundred meters in the north. To the west, along the north flank of the window, the Beaucoup Formation is cut out by the underlying Amawk thrust.

Structure. As discussed in detail by Handschy (this volume, Chapter 3), the Endicott Mountains allochthon occupies most of the frontal part of the Brooks Range fold and thrust belt and consists of an east-west–striking stack of north-directed thrust sheets (see Fig. 3 of Preface, this volume). Substantial spatial variability exists in the structural character of the allochthon: in the north, the allochthon is composed of numerous imbricates, whereas in the south it is composed of a single thrust nappe. The thrust nappe composing the southern Endicott Mountains allochthon is about 8 km thick along the northern flank of the Doonerak multiduplex (Handschy, this volume, Chapter 3) but thins to the south, where the Endicott Mountains allochthon is cut out by the basal thrust of the overlying Skajit allochthon (e.g., Oldow et al., this volume, Chapters 7 and 8).

Three generations of structures are recognized in the basal part of the Endicott Mountains allochthon exposed around the eastern Doonerak window (Phelps and Avé Lallemant, this volume, Chapter 4; Handschy, this volume, Chapter 3). The relative ages of the structures are established by cross-cutting relationships, and they are designated D_1, D_2, and D_3 in order of decreasing age.

First-generation structures. First-generation structures (D_1) are most commonly expressed by a cleavage (S_1) that is axial planar to tight to isoclinal folds of bedding (S_0). Small first-phase folds are rare and where preserved generally are rootless structures, commonly with a sheath fold morphology. Fold axes generally lie in the down-dip orientation of S_1 cleavage (Fig. 4). In fine-grained clastic units, S_1 cleavage is penetrative, but in coarse clastic rocks and carbonate interbeds, S_1 cleavage is spaced and/or hackly and commonly is difficult to recognize. The orientation of S_1 is usually subparallel (within 10°) to the underlying

86   C. M. Seidensticker and J. S. Oldow

Amawk thrust. On the south flank of the Doonerak window, the Amawk thrust and S_1 cleavage in the hanging-wall dip steeply southeast, whereas north of the window both (Fig. 4) the thrust and the first cleavage dip gently north.

Second-generation structures. Second-generation structures (D_2) are expressed by gentle to open asymmetric folds and kink bands of S_1 cleavage. The kink bands and asymmetric D_2 folds are preferentially developed in fine-grained lithologies and have half-wavelengths ranging from millimeters to tens of meters. The D_2 folds and kinks are northeasterly trending and typically have subhorizontal fold axes (Fig. 4). D_2 axial planes generally dip moderately to steeply to the northwest, but occasional southeast dips are observed and are related to scarce box folds. The D_2 folds and kinks commonly have a crenulation cleavage parallel to the axial plane. Unlike D_1 structures, whose orientations vary from the north to the south side of the Doonerak window, D_2 structures within the Endicott Mountains allochthon maintain the same orientation throughout. The axial planes of D_2 folds and associated crenulation cleavage dip predominantly to the northwest, both in the gently north dipping northern exposure of the allochthon and in the steeply south dipping exposure on the southern side of the Doonerak window. The strike of D_2 axial planes and associated cleavage and the trend of D_2 fold axes are subparallel to the long axis of the Doonerak window. The development of D_2 structures is thought to be related to duplex development (Julian and Oldow, this volume, Chapter 5; Avé Lallemant and Oldow, this volume, Chapter 14).

Third-generation structures. Third-generation structures (D_3) are a northwest-striking, subvertical crenulation cleavage that overprints both S_1 and S_2 foliations. Rare, gentle D_3 folds are observed and they consistently have subvertical, northwest-striking axial planes that parallel S_3 cleavage. Fold axes vary according to the local orientation of S_0, S_1, and S_2 (Fig. 4). Like D_2 structures, D_3 structures have the same orientation on both flanks of the Doonerak window. The D_3 structures are related to late-stage deformation and may be associated with the young, high-angle en echelon faults observed on the north flank of the Doonerak window.

Footwall assemblage: Doonerak multiduplex

The Devono-Carboniferous and lower Paleozoic rocks of the footwall assemblage compose two stacked duplexes, which together constitute the Doonerak multiduplex. The upper, Blarney Creek duplex has a structural thickness of 180 to 800 m and, with the exception of thin, volumetrically insignificant slivers of Permian and Triassic clastic rocks and lower Paleozoic Apoon assemblage rocks, is composed of Carboniferous clastic and carbonate strata. Imbricate faults within the duplex merge upward into the Amawk thrust and downward into the Blarney Creek thrust, from which the duplex derives its name. The underlying Apoon duplex is composed almost entirely of lower Paleozoic clastic and volcanic rocks of the lower Paleozoic Apoon assemblage (Julian and Oldow, this volume, Chapter 5) but also contains minor exposures of Devono-Carboniferous clastic rocks (Kekiktuk Conglomerate and Kayak Shale). The Blarney Creek thrust serves as the roof thrust of the lower duplex, but its floor thrust is not exposed. The thickness of the Apoon duplex is uncertain but, based on regional balanced cross sections (Oldow et al., 1987a) and wide-angle seismic reflection data (Wissinger et al., this volume, Chapter 16), it is between 15 to 25 km thick. Only about 3 km of structural section of the Apoon duplex are exposed in the Doonerak window (Julian and Oldow, this volume, Chapter 5).

Stratigraphy. Although rocks within the two duplexes are highly deformed, they preserve their original, pretectonic stratigraphic order. The Apoon assemblage is overlain by a thin section of Devono-Carboniferous clastic rocks comprising the Endicott Group (Kekiktuk Conglomerate and Kayak Shale), which in turn is overlain by carbonate rocks of the Carboniferous Lisburne Group (Fig. 2). The stratigraphic succession is capped locally by minor remnants of fine-grained clastic rocks of the Permo-Triassic Sadlerochit Group, limestone and shale of the Triassic Shublik Formation, and the Triassic Karen Creek Sandstone. The lithologic units of the footwall assemblage are described below in ascending stratigraphic order.

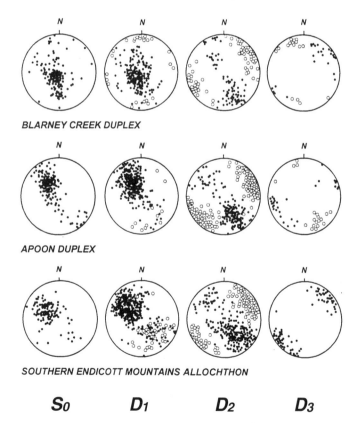

BLARNEY CREEK DUPLEX

APOON DUPLEX

SOUTHERN ENDICOTT MOUNTAINS ALLOCHTHON

S_0 D_1 D_2 D_3

Figure 4. Lower hemisphere, equal-area projections of structural data for eastern Doonerak window. Poles to bedding (S_0): solid dots. D_1, D_2, and D_3, structural elements: poles to cleavage and axial planes-solid dots; lineations and fold axes-open circles.

Apoon assemblage. Lower Paleozoic rocks of the Apoon assemblage are composed of volcanic, volcanogenic, and fine-grained clastic rocks (Julian and Oldow, this volume, Chapter 5). The rocks reside in four thrust-bound sheets that telescoped facies of a largely coeval assemblage (Julian, 1989). Palinspastic restoration (Julian and Oldow, this volume, Chapter 5) yields a series of east-west–trending sedimentary environments progressing from a northern volcanic arc, southward to a proximal volcaniclastic apron, to a facies of mixed volcanogenic and fine-grained clastic rocks that pass southward into a distal fine-grained clastic succession.

Age constraints for the Apoon assemblage are sparse and yield a wide range of dates. Conventional K/Ar and ^{40}Ar/^{39}Ar dates on mafic dikes containing hornblende (Dutro et al., 1976) define two clusters of ages, one at about 470 Ma (Middle Ordovician) and the other at about 380 Ma (Middle Devonian). Biostratigraphic age control is supplied by a few fossil assemblages found within the Apoon assemblage. Dutro et al. (1984) reported Middle Cambrian trilobites, brachiopods, and paraconodonts from a locality at the far western end of the Doonerak window. The trilobites have taxonomic affinities with Siberian forms, and they are similar to other trilobites found in the central Brooks Range (Dutro et al., 1984). Ordovician to Early Silurian graptolites have been recovered from Apoon assemblage exposures near the northwestern flank of the Doonerak window (Moore and Churkin, 1984). One graptolite location, near the summit of Amawk Mountain, is reported within the study area (Churkin, written communication, 1984) and indicates an Ordovician to Silurian age.

Kekiktuk Conglomerate. The Kekiktuk Conglomerate represents the basal unit of the Devono-Carboniferous Endicott Group (Brosgé et al., 1962) in the footwall assemblage (Fig. 3). Exposures of the Kekiktuk are concentrated along the northeastern flank of the Doonerak window, at the contact between the Apoon and Blarney Creek duplexes. The age of the Kekiktuk is poorly constrained. The only fossils that have been recovered are a few poorly preserved plant fragments of unknown age from a location in the northeastern Brooks Range (Brosgé et al., 1962). On the basis of its stratigraphic position conformably beneath the Lower Mississippian Kayak Shale, however, the Kekiktuk Conglomerate is assigned an age of Late(?) Devonian or Early Mississippian (Brosgé et al., 1962).

The Kekiktuk is a lithologically distinctive unit that consists of light gray, yellowish, or yellow-green weathering chert-pebble conglomerate, grit, and sandstone interbedded with siltstone and shale. Most of the coarse-clastic interbeds are lenticular and usually cannot be traced laterally in outcrop for more than a few meters. In the Doonerak window, the Kekiktuk Conglomerate typically is about 5 to 10 m thick, but the thickness is highly variable, increasing to 100 m in some places and decreasing to zero in others. Conglomerate clasts typically are flattened and lie parallel to S_1 cleavage that is penetratively developed in fine-grained lithologies of the unit. Conglomerate interbeds within the Kekiktuk achieve thicknesses of 5 m but generally are 2 m or

less. The conglomerates generally are clast supported and massive, but some display planar to trough cross-stratification. The subrounded to rounded clasts are almost entirely light gray to dark gray to black chert, with minor amounts of polycrystalline quartz clasts, and rare, dark gray to black mudstone clasts. A wide range of clast sizes is found, with a maximum of about 5 cm. The conglomerate matrix consists of subrounded to rounded quartz and chert grains of sand to silt size. Sandstone and siltstone interbeds are massive, with thickness of as much as 5 m but more typically of 2 m or less. The sand and silt fraction consists of subrounded to rounded, moderately well-sorted to well sorted quartz grains with subordinate chert. Frequently, the stratigraphically lowest unit of the Kekiktuk is a yellow-green phyllite.

The Kekiktuk Conglomerate is probably correlative, at least in part, with the upper Kanayut Conglomerate (Bowsher and Dutro, 1957) exposed in the Endicott Mountains allochthon. In addition to dramatic differences in thickness, the two units differ in the nature of underlying units, with the Kanayut conformably overlying the Devonian Hunt Fork Shale in the Endicott Mountains allochthon and the Kekiktuk resting on the lower Paleozoic Apoon assemblage in the Doonerak window.

In the Doonerak window, exposures of the Kekiktuk Conglomerate tend to be poor and discontinuous. The formation is thin and frequently is buried beneath talus derived from overlying units. These conditions make it difficult to find locations where either the upper or lower contact of the Kekiktuk is clearly exposed. Nevertheless, the Kekiktuk demonstrably overlies all four members of the Apoon assemblage and in some instances the contact is depositional. In a location just south of Amawk Creek, basal gray-green phyllite of the Kekiktuk is in depositional contact with volcaniclastic sandstone and conglomerate of the Apoon. As discussed below, dark gray phyllite of the Apoon assemblage exposed on the south flank of Falsoola Mountain is interbedded on the scale of centimeters to a meter with basal gray-green and yellow-green phyllite of the Kekiktuk that passes gradationally into the underlying Apoon assemblage. Elsewhere, in the eastern Doonerak window, Phelps and Avé Lallemant (this volume, Chapter 5) describe similar gradational contacts between the basal Kekiktuk and fine-grained clastic rocks of the underlying Apoon assemblage. In most cases, however, the contact between the Kekiktuk and the clastic and volcanic rocks of the Apoon assemblage is tectonic.

Kayak Shale. Gradationally overlying the Kekiktuk Conglomerate is the Mississippian Kayak Shale (Bowsher and Dutro, 1957; Armstrong et al., 1976). The Kayak Shale is found in all three structural units exposed in the Doonerak window area; it occurs within the Endicott Mountains allochthon of the hangingwall assemblage, and within both the Apoon and Blarney Creek duplexes of the footwall assemblage. At the base, the Kayak is composed of a black siltstone and/or very fine grained sandstone as much as 15 m thick. This basal unit is overlain abruptly by noncalcareous, black marine shale containing minor siltstone interbeds. The black shale constitutes the bulk of the formation and typically contains one or more distinctive interbeds of red-weathering crinoidal packstone or wackestone ranging from 20 cm to 1 m thick. The Kayak Shale is a mechanically weak horizon that provided an important decoupling surface during Brookian tectonism. As a result, the thickness of the unit varies dramatically, from very thin or absent in some places to a maximum of about 150 m. Generally, the Kayak thickness ranges from 50 to 100 m.

Lisburne Group. The carbonate rocks of the Lisburne Group (Bowsher and Dutro, 1957) conformably overlie the Kayak Shale. The Upper Mississippian to Lower Pennsylvanian Lisburne Group is composed of carbonate platform deposits (Armstrong et al., 1976) that achieve a stratigraphic thickness of about 300 m in the Doonerak window. This distinctive unit is an important regional stratigraphic marker consisting of light gray, usually massive, cliff-forming limestone. The Lisburne is abundantly fossiliferous and contains corals, crinoids, brachiopods, and bryozoans.

Sadlerochit Group (Shublik Formation and Karen Creek Sandstone). In foreland exposures, the Sadlerochit Group typically overlies the Lisburne Group disconformably (Detterman et al., 1975; Crowder, 1990). In the Doonerak window, the Sadlerochit Group is exposed only rarely, and only the basal Permian Echooka Formation is preserved in a few outcrops. One small exposure is located on the north wall of the canyon of Amawk Creek where the Echooka is about 15 m thick and is composed of reddish-brown weathering calcareous siltstone that contains Permian(?) brachiopods (Armstrong et al., 1976). The Echooka Formation overlies carbonate rocks of the Lisburne Group and is structurally decapitated by an intraduplex thrust that places the Kayak Shale above the Permian rocks. Elsewhere at Bombardment Creek (about 10 km west of the mouth of Amawk Creek), the Sadlerochit Group also is represented by the Echooka Formation (Mull, 1982; Mull et al., 1987), which has a thickness of about 125 m. In these exposures, a lower succession of calcareous sandstone and siltstone passes upward through a sharp contact into black, phyllitic, silty shale. The lower succession contains brachiopods dated as Early Permian (Mull et al., 1987). Here, the upper contact is sharp, and possibly represents a disconformity (Mull et al., 1987) between the Echooka and the overlying calcareous shale of the Shublik Formation. The Shublik consists of dark gray to black limestone, and carbonaceous and calcareous shale containing abundant phosphatic nodules. Pelecypods taken from several carbonate interbeds are interpreted to be late Middle to Late Triassic in age (cited in Mull et al., 1987, as written communication from N. J. Silberling). The Echooka represents a shallow marine transgressive succession (Detterman et al., 1975) that is overlain by basinal clastic and carbonate rocks of the Shublik. Also reported at Bombardment Creek is an exposure of Triassic Karen Creek Sandstone that is about 2 m thick (Mull, 1982; Mull et al., 1987).

Structure. Structures within the Blarney Creek and Apoon duplexes of the footwall assemblage (Doonerak multiduplex) are discussed separately below. In both duplexes, three generations of structures, comparable to those in the overlying Endicott Mountains allochthon, are observed.

Blarney Creek duplex. The Blarney Creek duplex is composed primarily of the Kayak Shale and carbonate rocks of the Lisburne Group. Minor slices of the Kekiktuk Conglomerate occur at the bottom of some horses, and one horse north of Amawk Creek contains a thin sliver of Permian Echooka Formation. The base of the duplex is occupied almost exclusively by the Kayak Shale, whereas the top of the duplex exposes mostly carbonate rocks of the Lisburne Group. The stratigraphic section of Kayak and Lisburne is repeated in a number of imbricate horses lying between the roof and floor thrusts. The horses generally merge smoothly with the roof and floor thrusts. In a few instances, however, horses containing Lisburne limestone and Kayak Shale are decapitated by the roof thrust (Amawk thrust). The bottoms of some of the duplex horses carry laterally discontinuous slivers of Kekiktuk Conglomerate beneath the Kayak Shale, but in general the Kekiktuk is conspicuously absent from duplex horses.

Duplex horses exhibit a wide range of dimensions. They have thicknesses ranging from 100 to 500 m and can be traced from south-southeast to north-northwest (parallel to the tectonic transport direction) for as much as 3 km. The horses are discontinuous along strike and have widths ranging from a few hundred meters to 5 km. Numerous lateral ramps are observed that dip shallowly to the east and west. Along the northeastern flank of the Doonerak window, the duplex horses have shallow north and south dips. Proceeding north to south around the eastern closure of the Doonerak window, however, the dips become consistently southerly.

The Blarney Creek duplex is preserved primarily along the northeastern flank of the Doonerak window, with only minor preservation on the southeastern flank (Fig. 1). On the north flank, the duplex is cut out to the west by the basal thrust (Amawk thrust) of the Endicott Mountains allochthon. Likewise, to the south, the characteristic carbonate rocks of the Blarney Creek duplex are absent and Devonian clastic rocks of the Beaucoup Formation (Endicott Mountains allochthon) rest directly on the fine-grained clastic rocks of the lower Paleozoic Apoon assemblage.

Within the Blarney Creek duplex, the limestone of the Lisburne Group rarely develops mesoscopic or microscopic structures. Map-scale structures are occasionally observed within the Lisburne, however, where they are open to tight inclined folds (axial planes dip to the southeast) with half-wavelengths of 400 to 800 m and amplitudes of 200 to 400 m. The development of minor structures is largely confined to the Kayak Shale. The geometry and superposition relations of the structures give important insight into the kinematic development of the duplex.

First-generation structures. The most prominent mesoscopic structure within the Blarney Creek duplex is a penetrative cleavage (S_1). S_1 is the dominant fabric of the Kayak Shale and in most instances transposes bedding. Rare D_1 folds consist either of poorly preserved rootless isoclines of thin sandstone interbeds or coherent isoclinal folds of crinoidal limestone interbeds. The S_1 cleavage consistently is axial planar to the minor folds. S_1 cleavage and bedding dip gently north on the northern flank of the Doonerak window (Fig. 4), but change to moderate to steep southeasterly dips around the eastern closure of the window. Sparse D_1 fold axis data illustrate a substantial variation in orientation. The variability is due, at least in part, to later deformation.

Second-generation structures. Second-generation structures (D_2) consist of gentle to open asymmetric folds and kink bands of bedding and S_1 cleavage. The folds have half-wavelengths ranging from millimeters to tens of meters and have subhorizontal, northeasterly trending fold axes (Fig. 4). The D_2 axial planes and kink bands (S_2) usually dip moderately to steeply to the northwest, but southeast dips are sometimes observed. Rarely observed D_2 box folds display both orientations of axial planes. Usually, D_2 folds and kinks have an axial-planar crenulation cleavage that is found even in areas where D_2 folds or kinks are not observed. The orientation of D_2 structures does not vary between the north and south flanks of the Doonerak window.

Third-generation structures. Third-generation (D_3) structures are sporadically developed. They consist of a subvertical crenulation cleavage (S_3) that strikes northwest (Fig. 4). The S_3 cleavage overprints both S_1 and S_2 cleavages and is associated with rare D_3 folds. The folds have subvertical, northwest-striking axial planes and northwest-trending, subhorizontal fold axes. Like the D_2 structures, D_3 structures are consistently oriented throughout the Blarney Creek duplex.

Apoon duplex. Most of the rocks exposed in the Doonerak window lie within the Apoon duplex. Along the northeastern margin of the window, the roof of the Apoon duplex is the Blarney Creek thrust, whereas along most of the southern margin of the window, where the Blarney Creek duplex is absent, the lower duplex is directly overlain by the Endicott Mountains allochthon (see Fig. 3 of Preface, this volume, and Fig. 1).

In the eastern Doonerak window, four major horses, designated H1, H2, H3, and H4 in structurally descending order from south to north, compose the Apoon duplex (Fig. 1). Each horse contains a different lithotectonic unit of the Apoon assemblage (Julian, 1989; Julian and Oldow, this volume, Chapter 5). At a few isolated locations, all of which are immediately below the Blarney Creek thrust, the top of some of the Apoon horses preserve small exposures of Kekiktuk Conglomerate that show gradational depositional contacts with the underlying Apoon rocks. In general, however, depositional contacts between the Kekiktuk and Apoon are not preserved.

For most of the Doonerak window, the horses and their bounding faults dip moderately to steeply to the southeast. As the Blarney Creek roof thrust is approached to the north, however, the three structurally lowest duplex horses roll over and become subhorizontal or dip shallowly to the north. The highest and lowest duplex horses (H1 and H4) are locally decapitated by the Blarney Creek thrust.

As in structurally higher units, three generations of superposed structures are recognized in the rocks of the Apoon duplex. Mesoscopic structures are best developed in fine-grained lithologies, but the structures are developed to some extent in all units.

First-generation structures. The oldest structures in the Apoon duplex form the most obvious fabric, which is a penetrative cleavage (S_1). The cleavage is parallel to the axial planes of rarely preserved isoclinal folds of bedding and quartz veins. The cleavage is penetrative in fine-grained rocks but is poorly developed in massive volcanic units of the Apoon assemblage (Julian and Oldow, this volume, Chapter 5). Bedding is generally transposed into parallelism with the S_1 cleavage except in the hinges of D_1 folds. D_1 folds of bedding range in amplitude from 20 cm to about 4 m; folds of quartz veins have amplitudes of 1 to 5 cm. The D_1 cleavage generally strikes northeast and dips southeast within the Apoon duplex (Fig. 4). In the vicinity of the Blarney Creek thrust, dips are nearly horizontal, and near the northern margin of the Doonerak window, some S_1 surfaces dip gently to the north. This variation in S_1 surfaces results in a partial great-circle distribution of poles with a normal that corresponds to the mode of D_2 fold axes. D_1 fold axes generally plunge to the southeast and lie in the down-dip position of the axial plane.

Second-generation structures. Second-generation structures (D_2) are asymmetric folds and kink bands of bedding and the D_1 foliation. An axial-planar crenulation cleavage (S_2) is commonly developed in rocks that display a well-developed S_1 cleavage. D_2 folds have both rounded and chevron morphologies with maximum amplitudes of about 2 m, although fold amplitudes measured on scales of less than 30 cm to millimeters are more prevalent. Axial planes of D_2 folds and S_2 crenulation cleavage generally dip steeply to the northwest, although occasionally they are found to dip southeast (Fig. 4). Rare box folds with conjugate axial planes have orientations consistent with the distribution of S_2 surfaces. D_2 fold axes are subhorizontal and northeast trending. The orientation of D_2 structures is consistent throughout all four horses of the Apoon duplex regardless of local S_1 orientations.

Third-generation structures. The youngest generation of structures (D_3) in the Apoon duplex is represented by sporadically developed folds with subvertical, northwest-striking axial planes and subhorizontal fold axes (Fig. 4). An S_3 crenulation cleavage parallels the axial planes of rare D_3 folds.

FOOTWALL BOUNDARY LAYER: BLARNEY CREEK THRUST

The Blarney Creek and Apoon duplexes are separated by the Blarney Creek thrust, which at the scale of mapping shown in Figure 5, appears as a single, well-defined discontinuity. In fact, in many areas the tectonic contact is sharp and only a few centimeters to meters wide. In contrast, over large areas the contact is a diffuse shear zone that reaches a thickness of 250 m. Within these broad zones of shear, slices of the Kekiktuk Conglomerate dominate, but occasionally thin sheets of Kayak Shale and rocks of the Apoon assemblage are also found. Where the Kekiktuk is absent, the Blarney Creek thrust generally juxtaposes Kayak Shale directly on the underlying Apoon assemblage, but in a few locations the Lisburne limestone forms the upper plate and rests directly on the Apoon assemblage.

Thrust faults separating imbricate horses within the over-

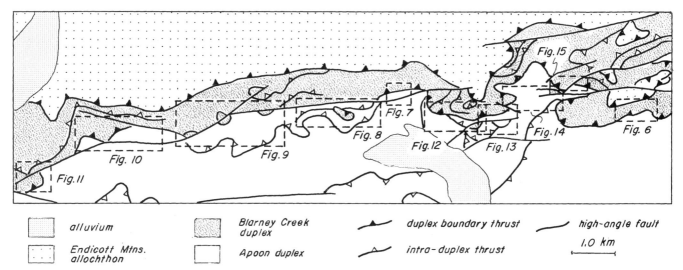

Figure 5. Index map showing the locations of Figures 6 through 15. Major thrusts indicated by solid teeth, minor thrusts by open teeth. Dashed lines in cross sections depict S_1 cleavage. Symbols for figures are: AT-Amawk thrust; BCT-Blarney Creek thrust; EMA-Endicott Mountains allochthon; H1, H2, H3, H4-Apoon duplex horse; Pe-Echooka Formation (basal unit of Sadlerochit Group); Ml-Lisburne Group; Mks-Kayak Shale; Mkc-Kekiktuk Conglomerate. Site locations for lower hemisphere, equal-area projections of structural data indicated by capital letters A–L as needed.

lying Blarney Creek duplex sole into a decollement zone (floor thrust) within the Kayak Shale, generally without affecting the underlying Kekiktuk Conglomerate. Similarly, thrust faults within the Apoon duplex merge upward into a roof thrust that is located immediately below the Kekiktuk Conglomerate. The Kekiktuk itself behaves as a boundary layer within the Blarney Creek thrust zone that is essentially detached from both the Apoon and Blarney Creek duplexes.

The geometry and structural history of the decollement zone separating the Blarney Creek and Apoon duplexes are quite complex and provide important constraints for developing kinematic models of footwall contraction during multiduplex formation. Along strike, the Blarney Creek thrust zone exhibits a number of complex structural and stratigraphic relationships that are divided into two variants: (1) those where the Kekiktuk is in depositional contact with the underlying Apoon assemblage, and (2) those where the Kekiktuk is either absent or, where present, is structurally detached from the Apoon. These relations are described below for individual locations along the fault zone.

Structurally attached Kekiktuk Conglomerate

Sites with preserved depositional contacts between the Kekiktuk Conglomerate and the Apoon assemblage are pivotal for both structural and stratigraphic assessments of the Doonerak multiduplex. At several locations the Kekiktuk depositionally overlies lower Paleozoic rocks at the upper contact of the Apoon duplex and is tectonically overlain by an original, although structurally modified, stratigraphic succession of Carboniferous rocks. Elsewhere, the Blarney Creek thrust zone contains slivers of the Apoon assemblage tectonically above exposures of Kekiktuk Conglomerate resting depositionally on

Apoon assemblage rocks. These older-over-younger structural contacts occur where the Blarney Creek thrust zone is a broad zone of distributed shear.

Southern Falsoola Mountain. Exposures on the southern flank of Falsoola Mountain (Fig. 6) preserve a depositional contact between the Kekiktuk Conglomerate and phyllite of the underlying Apoon assemblage (horse H1). The contact between the Apoon phyllite and the overlying Kekiktuk Conglomerate is sharp but gradational. A few meters below 1- to 2-m-thick layers of Kekiktuk Conglomerate, phyllite typical of the Apoon assemblage is interbedded with fine-grained and coarse-grained lithologies indistinguishable from those of the Kekiktuk. These transitional conglomerate and sandstone beds occur as channel-fill deposits with sharp lower contacts cut into phyllite. Beneath the channel-fill horizon is yellowish green phyllite, typical of the basal Kekiktuk, interbedded with black phyllite of the upper Apoon assemblage. The gradational nature of this depositional contact is difficult to reconcile with the contention of earlier workers that, in the Doonerak window, the Apoon assemblage and the overlying Kekiktuk Conglomerate are separated by an angular unconformity (Armstrong et al., 1976; Dutro et al., 1976; Mull, 1982; Mull et al., 1987). As pointed out by Julian and Oldow (this volume, Chapter 5), the lack of any pre-Brookian structures in the Apoon assemblage and the apparent gradational contact with the Kekiktuk is decidedly different from relations observed in the northeastern Brooks Range (e.g., Oldow et al., 1987b). In the Doonerak window, the original, pre-Brookian contact between the Kekiktuk and the lower Paleozoic rocks of the Apoon clearly is not an angular unconformity, but it may represent a disconformity associated with a glacially related lowstand in the latest Devonian (Diaz-Martinez and Isaacson, 1994).

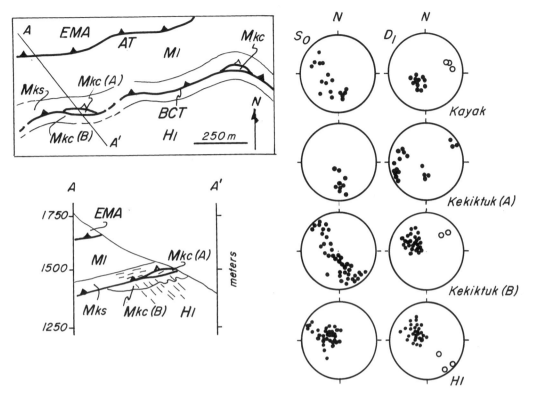

Figure 6. Southern Falsoola Mountain. Locally continuous outcrop of Kekiktuk Conglomerate (Mkc) is in depositional contact with horse H1 of the Apoon duplex; units share parallel S_1 cleavage. Discontinuous outcrops of Kekiktuk in boundary layer separate the Apoon and Blarney Creek duplexes underlying and overlying the Blarney Creek thrust (BCT), respectively. Phyllitic interbeds in Kekiktuk Conglomerate of boundary layer have S_1 cleavage parallel to that of rocks overlying the Blarney Creek thrust. Conglomerate beds of Kekiktuk have S_1 cleavage refracted to steeper dips. Structural data are presented in lower hemisphere, equal-area projections: solid dots represent poles to S_0 or S_1 cleavage, and open circles represent D_1 fold axes. See Figure 5 for definition of symbols.

Rocks of the Apoon assemblage and the Kekiktuk Conglomerate are intensely deformed. The dark gray phyllite of the Apoon assemblage contains a penetrative S_1 cleavage that is subparallel to bedding except in the hinges of rootless isoclinal folds of sandstone. S_1 cleavage is axial planar to these folds and consistently dips between 20 and 50° to the southeast (Fig. 6). Likewise, S_1 cleavage in the Kekiktuk Conglomerate dips southeast and generally lies at a large angle to shallowly dipping bedding. In several locations the cleavage demonstrably passes continuously from the Apoon phyllite into the overlying Kekiktuk.

Bedding in the Kekiktuk commonly dips shallowly north, but locally is folded in mesoscopic tight to isoclinal D_1 structures. D_1 folds are upright or overturned to the north with the S_1 foliations parallel to their axial planes. The D_1 fold axes in the Kekiktuk are subhorizontal and trend northeasterly and differ from D_1 fold axes in the underlying Apoon phyllite, which typically plunge to the southeast. During disharmonic folding, D_1 sheath folding in the fine-grained Apoon assemblage rotated fold axes downdip in S_1 cleavage, whereas the thicker and more competent bedding of the Kekiktuk Conglomerate did not experience comparable strain rotations. Nevertheless, the development of D_1 folds and S_1 cleavage in both lithologic successions

are related, clearly indicating that D_1 structures formed after deposition of the Kekiktuk.

Kayak Shale and Lisburne limestone structurally overlie most exposures of the lower plate Kekiktuk along the southern flank of Falsoola Mountain, and the Kayak Shale is the preferred basal decollement of the Blarney Creek duplex in this area. In two isolated occurrences, however, the Kekiktuk is incorporated in the upper plate of the Blarney Creek thrust near the base of the Blarney Creek duplex (Fig. 6). At these locations, detached Kekiktuk structurally overlies lower plate Kekiktuk, which is in depositional contact with the underlying Apoon. The upper plate Kekiktuk is deformed and has a structural fabric consistent with that of the Blarney Creek duplex. Within the Blarney Creek duplex, S_1 cleavage is a penetrative cleavage dipping to the north at about 30° in the Kayak Shale and phyllitic interbeds of the Kekiktuk. Locally S_1 is refracted to steeper dips in the coarser parts of the Kekiktuk, where it appears as a spaced cleavage. This refraction of cleavage between different lithologies within the Kekiktuk produces the two distinct populations of S_1 cleavage shown in the Kekiktuk fabric diagram of Figure 6. Nevertheless, S_1 in the upper plate Kekiktuk lies at a high angle to the S_1 in the underlying Kekiktuk of the Apoon duplex and documents decoupling of the two lithologic bodies.

Northern Hammond River. West of the northern Hammond River (Fig. 7), a belt of Kekiktuk quartzite is exposed between the overlying Kayak Shale and underlying Apoon assemblage (horse H4). Although obscured by poor exposures, the Kekiktuk ostensibly is in depositional contact with the underlying Apoon rocks. Bedding in the quartzite of the Kekiktuk dips 40 to 50° north-northwest and is cross-cut at a high angle by S_1 cleavage, which dips 20 to 45° southeast. The orientation of S_1 cleavage in the Kekiktuk compares favorably with S_1 in the Apoon. In contrast, the orientation of S_1 cleavage in the overlying Kayak Shale is north-northwest dipping at 20 to 45°. Thus, the Kekiktuk appears to be attached to the underlying Apoon assemblage and the Blarney Creek thrust is located between the Kekiktuk and the Kayak Shale.

Kinnorutin Mountain. Near Kinnorutin Mountain, which is capped by a structural sheet composed of Carboniferous clastic and carbonate rocks of the Kekiktuk, Kayak, and Lisburne, several laterally discontinuous outcrops of Kekiktuk Conglomerate reside at the boundaries of thrust imbricates within the Apoon duplex (Fig. 8). Of three isolated Kekiktuk exposures west of the peak, one lies at the contact between horses H3 and H4 of the Apoon duplex. The other two exposures (just north of map location D) occur along segments of minor thrusts contained within horse H3. One intra-horse thrust can be traced to the east where it passes beneath the structural sheet forming Kinnorutin peak (Fig. 8) and re-emerges on the eastern flank of the mountain.

West of Kinnorutin Mountain (Fig. 8, location D), the southernmost outcrop of conglomerate is about 20 m thick and ornaments the upper surface of horse H4. This exposure of Kekiktuk is structurally overlain by horse H3. The Kekiktuk is involved in open to tight D_1 folds that have subhorizontal axes trending about N75°W. S_1 cleavage parallels the axial planes of the folds and dips between 10 and 30° to the south and passes into the volcanic rocks of the underlying Apoon.

Farther to the north, two outcrops of Kekiktuk are surrounded by rocks of the horse H3. The more southerly of the two outcrops is small and poorly exposed. Due to the lack of demonstrably intact exposures at this locale, no structural observations were made. The northernmost exposure of Kekiktuk (Fig. 8, location B) has a maximum thickness of about 6 m and rests depositionally on the underlying Apoon. The Kekiktuk serves as a distinctive marker useful in tracing the intra-horse thrust to the east. The homoclinal Kekiktuk dips between 25 and 60° to the north-northeast and contains S_1 cleavage generally dipping less than 30° to the south. S_1 cleavage in adjacent Apoon assemblage rocks (Fig. 8, locations A and C) dips both to the north and south at less than 40°. The scatter in S_1 orientations in the Apoon is due, at least in part, to D_2 folds.

At location K (Fig. 8) east of the peak, a 10-m-thick layer of Kekiktuk Conglomerate underlies a minor thrust within horse H3 of the Apoon duplex. The Kekiktuk layer is internally folded and has an enveloping surface dipping 50° to the north. The Kekiktuk

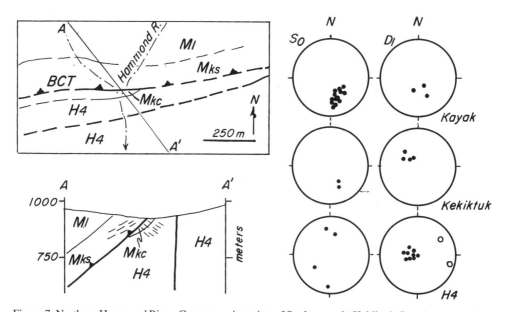

Figure 7. Northern Hammond River. Common orientation of S_1 cleavage in Kekiktuk Conglomerate (Mks) and underlying Apoon assemblage rocks (H4) indicates that conglomerate is in depositional contact with underlying Apoon substratum. Sharp contrast in S_1 orientation between Apoon assemblage rocks (H4 with south dipping S_1) and overlying Kayak Shale (Mks with north dipping S_1) indicates detachment of Kayak Shale from Kekiktuk Conglomerate and Apoon assemblage. Structural data are presented in lower hemisphere, equal-area projections: solid dots represent poles to S_0 or S_1 cleavage and open circles represent D_1 fold axes. See Figure 5 for definition of symbols.

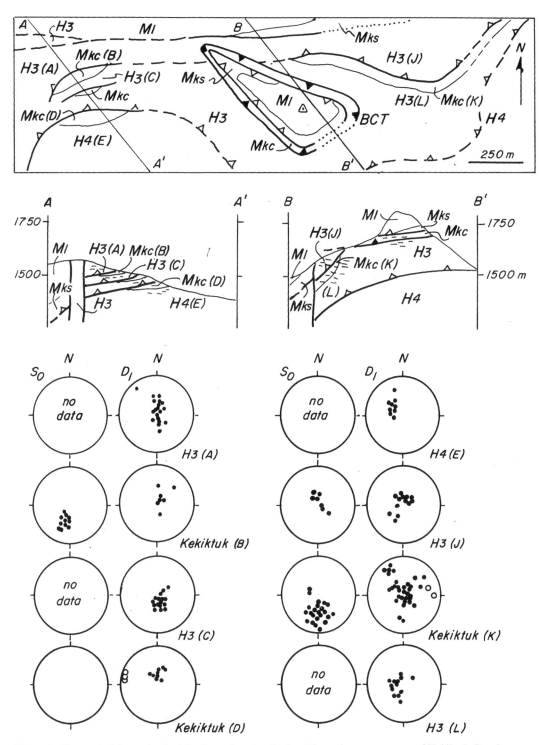

Figure 8. Kinnorutin Mountain. Peak is shown by triangle-dot. Discontinuous outcrops of Kekiktuk Conglomerate (Mkc) are both in depositional contact with underlying Apoon assemblage rocks (H3 and H4) and overlain structurally by imbricate horses of the Apoon. Near Kinnorutin Mountain, Kekiktuk Conglomerate is detached and separates Apoon and Blarney Creek duplexes below and above the Blarney Creek thrust (BCT), respectively. Structural data are presented in lower hemisphere, equal-area projections: solid dots represent poles to S_0 or S_1 cleavage, and open circles represent D_1 fold axes. See Figure 5 for definition of symbols.

contains S_1 cleavage that although generally subhorizontal, locally dips steeply to the southeast. Cleavage and bedding in the Apoon, both above and below the Kekiktuk (locations J and L, respectively), are subhorizontal. Although poorly exposed, the consistent orientation of S_1 cleavage suggests that the Kekiktuk rests depositionally on the underlying Apoon.

Northern Amawk Mountain. Kekiktuk Conglomerate is exposed at several locations along the steep north face of Amawk Mountain. Most of the exposures are cut on their northern side by east-west or northeast-striking high-angle faults, which appear to control the topography of this face of the mountain. Most of the exposures form north-dipping flatirons lying to the south of the high-angle faults (Fig. 9).

The flatirons exhibit minor differences in stratigraphic units exposed, but each has the Kekiktuk Conglomerate at its base. The easternmost flatiron (Fig. 9, B–B′) has Kayak Shale overlying the Kekiktuk as does the central flatiron (Fig. 9, A–A′), which is unique in being capped by Lisburne limestone. The westernmost flatiron (Fig. 9, area C) exposes only the Kekiktuk Conglomerate.

The three flatirons have similar structural fabrics (Fig. 9). Bedding in the Kekiktuk Conglomerate typically shows north dips of 30 to 90°, with the variability produced by mesoscopic D_1 folds. Where preserved, bedding in the Kayak and Lisburne also dips to the north. Below the flatirons, S_1 cleavage in fine-grained clastic rocks of the Apoon assemblage (horse H3) dips shallowly and passes into the overlying Kekiktuk Conglomerate. Bedding-cleavage intersections in the Kekiktuk are predominantly at a high angle, indicating that, although closure is not observed, the flatirons represent the hinge zone of a D_1 fold. Here as elsewhere, continuity between D_1 structures in the Apoon assemblage and lower plate Kekiktuk indicates that the structures must be post-Carboniferous in age. S_1 cleavage of the Kayak in the flatirons is discordant to the cleavage below and dips north at 10 to 45°. The Blarney Creek thrust is inferred to lie near the base of the Kayak separating it from the underlying Kekiktuk.

Amawk Creek. In the canyon containing Amawk Creek, the Blarney Creek thrust is down-dropped to the north and concealed by a high-angle fault that, in some areas, places rocks of the Apoon assemblage in direct contact with the Lisburne limestone (Fig. 10). On the south wall of the canyon and south of the high-angle fault, however, the Blarney Creek thrust is exposed. The thrust lies within the Kayak Shale and, in the area of section B–B′ (Fig. 10), isolates a small exposure of Kekiktuk Conglomerate and a thin layer of Kayak Shale in its footwall. In this restricted area, the Kekiktuk Conglomerate is in depositional contact with Apoon volcanic rocks of horse H4. The Kekiktuk and a thin interval of Kayak Shale are folded together with the underlying Apoon assemblage rocks in a D_1 syncline. The syncline has an axial plane dipping about 20° to the south-southeast and a west-southwest–trending fold axis plunging about 15°. S_1 cleavage within the Kekiktuk parallels the axial plane of the fold and passes continuously into the underlying Apoon volcanic rocks. Within the relatively incompetent Kayak Shale, the S_1 cleavage is reoriented by D_2 folds,

which do not affect the Kekiktuk Conglomerate or the volcanic rocks of H4.

The Blarney Creek thrust truncates the syncline and can be traced with confidence from the top to the bottom of the canyon wall (about 300 m of vertical relief). In the upper plate, both bedding and cleavage in the Kayak Shale have highly variable orientations, but generally dip to the northwest. Across the structural contact separating Kayak Shale in the upper and lower plates, bedding and S_1 orientations are clearly discordant.

Wien Mountain. At Wien Mountain (Fig. 11), Kekiktuk Conglomerate rests on volcanic and volcanogenic rocks (horse H4) of the Apoon assemblage and is exposed south of an east-northeast–striking, high-angle fault. The fault down-drops the Blarney Creek duplex to the north against the Apoon duplex to the south. Bedding in the Kekiktuk dips shallowly to moderately north-northwest (10 to 50°) and is cross-cut by poorly developed S_1 cleavage with a nearly horizontal dip (only one measurement was possible due to poor development of the cleavage). The underlying Apoon assemblage has shallowly west dipping bedding (up to 30°) and shallow, southeast-dipping S_1 cleavage (as much as 35°) that apparently is consistent in orientation with S_1 cleavage in the overlying Kekiktuk.

Bedding within the Kayak Shale in this location is transposed into S_1 cleavage and dips northwest at about 30°. S_1 cleavage in the Kayak Shale and underlying Apoon assemblage are clearly discordant and, although poorly developed, the S_1 cleavage in the Kekiktuk appears to be allied with that in the Apoon. Thus the Blarney Creek thrust is interpreted to lie at the base of the Kayak Shale above the Kekiktuk Conglomerate.

Structurally above the allochthonous Kayak Shale is a klippe of Apoon assemblage rocks (horse H3) that exhibit S_1 cleavage parallel to that of the underlying Kayak. The klippe is an example of the rare slices of Apoon assemblage rocks incorporated into the base of the Blarney Creek duplex.

Structurally detached or structurally removed Kekiktuk Conglomerate

The Kekiktuk Conglomerate has been stripped from the underlying Apoon assemblage at several locations along the Blarney Creek thrust. At these locations, where the upper and lower duplexes are in direct contact with no intervening boundary layer, the thrust is a sharp structural discontinuity. In several locations (Wien Creek, Kinnorutin Mountain, and northern Blarney Creek), the Kekiktuk demonstrably is stripped from the underlying Apoon but still is found at the contact between the upper and lower duplexes. At two of these locations, the detached Kekiktuk is readily differentiated from a depositional contact with the Apoon by the presence of (1) a shear zone between the Kekiktuk and underlying Apoon (Kinnorutin Mountain), or (2) a pronounced discordance in the orientation of S_1 cleavage across the contact (Wien Creek). At Lost Sheep Creek structural relations are more complex and numerous imbricates of Apoon assemblage and Carboniferous rocks lie within a broad shear

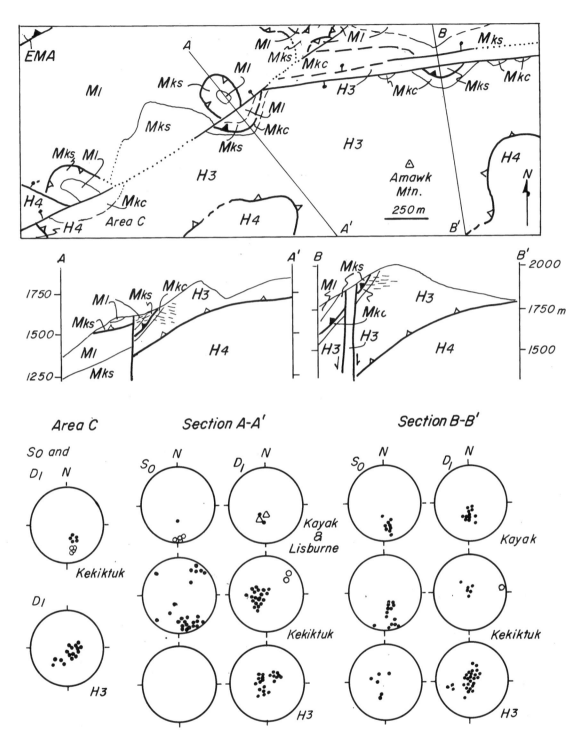

Figure 9. Northern Amawk Mountain. North-dipping flatirons of Kekiktuk Conglomerate (Mkc) are attached to the underlying Apoon duplex, and structurally detached from the Blarney Creek duplex overlying the Blarney Creek thrust (BCT). S_1 cleavage in Apoon duplex has variable orientations discordant with north-dipping S_1 cleavage in the overlying Blarney Creek duplex. Structural data are presented in lower hemisphere, equal-area projections. In Area C: poles to S_0 Kekiktuk represented by open circles, S_1 cleavage represented by solid dots; poles to S_1 cleavage in H3 Apoon represented by solid dots. For area of section A–A″, data for Kayak Shale and the Lisburne Group are combined: poles to S_0 and S_1 cleavage represented by solid dots for the Kayak, and open triangles for Lisburne Group. For Kekiktuk and Apoon assemblage rocks near section A–A′: solid dots represent poles to S_0 or S_1 cleavage, and open circles represent D_1 fold axes. For all units in area of cross section B–B′, poles to S_0 and S_1 cleavage represented by solid dots, and D_1 fold axes by open circles. See Figure 5 for definition of symbols.

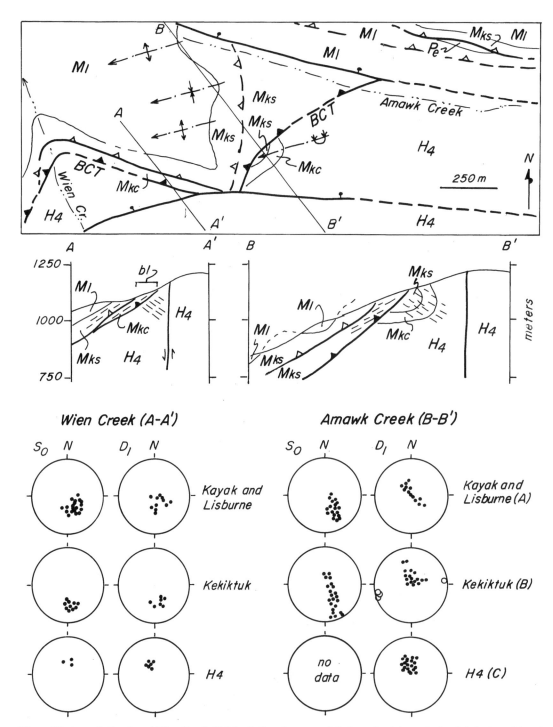

Figure 10. Amawk Creek and Wien Creek. Kekiktuk Conglomerate (Mkc) south of Amawk Creek is in depositional contact with Apoon assemblage (H4). Apoon, Kekiktuk, and Kayak Shale (Mks) are folded in D_1 syncline and truncated by overlying Blarney Creek thrust. At Wien Creek S_1 cleavage in Kekiktuk Conglomerate and Apoon assemblage is discordant indicating detachment of conglomerate from the underlying Apoon assemblage. Kekiktuk Conglomerate is part of boundary layer separating Blarney Creek and Apoon duplexes. Structural data are presented in lower hemisphere, equal-area projections: solid dots represent poles to S_0 or S_1 cleavage, and open circles represent D_1 fold axes. See Figure 5 for definition of symbols.

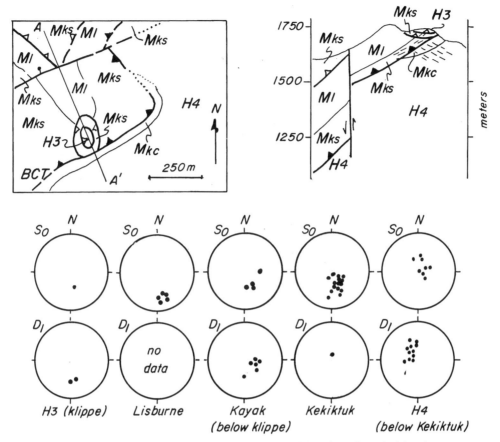

Figure 11. Wien Mountain. Kekiktuk Conglomerate (Mkc) and depositionally underlying Apoon assemblage (H4) share south-dipping S_1 cleavage. Kayak Shale (Mks) resides in Blarney Creek duplex overlying the Barney Creek Thrust (BCT) and has north-dipping S_1 cleavage. Klippe of Apoon horse H3 exhibits north-dipping S_1 cleavage and constitutes part of the Blarney Creek duplex. Structural data are presented in lower hemisphere, equal-area projections: solid dots represent poles to S_0 or S_1 cleavage. See Figure 5 for definition of symbols.

zone. In some areas (western Falsoola Mountain and western Blarney Mountain), evidence for structural detachment of the Kekiktuk from the substratum is not compelling, but the interpretation that a detachment exists is allowable and based on adjacent relations is preferred.

Kinnorutin Mountain. Underlying the peak at Kinnorutin Mountain (Fig. 8), the Kekiktuk forms a continuous horizon separating the Apoon and Blarney Creek duplexes. The upper duplex is dominated by an exposure of relatively flat lying Lisburne limestone that is underlain by the Kayak Shale and Kekiktuk Conglomerate. On the north side of the prominent peak, the Kayak is omitted and the Lisburne limestone rests directly on the Kekiktuk Conglomerate. The orientation of S_1 cleavage in all units, including the Apoon, shows substantial variability but essentially is subhorizontal. The contact between the Apoon and the Kekiktuk is locally well exposed and is sharp, without the gradational relations observed farther east.

The lack of structural discordance in the orientation of S_1 cleavage between the Apoon duplex, the overlying Kekiktuk, and the Blarney Creek duplex obscures the structural affinity of the

Kekiktuk and its relation to the Blarney Creek thrust. Using S_1 cleavage orientations alone, it cannot be demonstrated that the conglomerate is detached from the underlying Apoon. Similar relations elsewhere along the northern Doonerak window cause some ambiguity concerning the degree of detachment of the Kekiktuk Conglomerate. Nevertheless, at Kinnorutin Mountain the Kekiktuk clearly rests structurally above the Apoon duplex. This is shown by the continuity of an intra-horse thrust within the underlying Apoon duplex beneath the Kekiktuk underlying the structural stack forming Kinnorutin Mountain (Fig. 8).

Wien Creek. At Wien Creek (Fig. 10), discordant S_1 cleavage relations clearly document the detachment of the Kekiktuk Conglomerate from the underlying Apoon assemblage. The conglomerate is exposed along the east wall of the valley and lies above the lowest horse (H4) of the Apoon duplex. Above the Kekiktuk, two horses of the Blarney Creek duplex repeat the Kayak Shale and Lisburne limestone. Bedding and S_1 cleavage in the Apoon rocks (horse H4) dip 20 to 30° to the southeast. Bedding and S_1 cleavage in the Kekiktuk dip consistently to the north at 30 to 50° and parallel those of the overlying Kayak and Lisburne.

Western Blarney Mountain. For most of its exposed length along western Blarney Mountain (Fig. 12), the Blarney Creek thrust places Kayak Shale in direct contact with the Apoon assemblage, with no intervening Kekiktuk Conglomerate. Locally, however, at locations B and E (Fig. 12), a thin slice of Kekiktuk Conglomerate lies between the Apoon and Blarney Creek duplexes. Here, as seen locally elsewhere, the D_1 structural fabric in the Kekiktuk and the overlying and underlying duplexes are the same. In this instance, it is not possible to clearly document detachment of the Kekiktuk from the underlying Apoon. Only in areas of pronounced discontinuity of D_1 fabrics or where other structural relations are compelling, such as those described at Kinnorutin Mountain (Fig. 8), can detachment be proved. Where the Kekiktuk is discontinuous along the contact and not demonstrably forming the axis of a D_1 syncline, however, it is reasonable to attribute its absence to tectonic removal. Although equivocal, the correspondence between the structural fabric in the Kekiktuk and in the Blarney Creek duplex is interpreted as allowable for its detachment from the underlying Apoon.

Lost Sheep Creek. For part of its course, Lost Sheep Creek flows along the trace of an east-west–trending, high-angle fault that cuts and downdrops earlier contractional structures to the north. A complex imbricate stack of Apoon and Devono-Carboniferous rocks is preserved in the canyon (Fig. 13), but exposures on the low-relief hills south of the high-angle fault are limited due to vegetation cover. Within and near the canyon bottom south of the high-angle fault, however, Apoon volcanic rocks (H2; Fig. 13, location J) exhibit poorly developed S_1 cleavage that dips southeast at 20 to 40°. The volcanic rocks are structurally overlain by a thin, poorly exposed layer of Kayak Shale containing crinoidal packstone. The Kayak Shale contains S_1 cleavage dipping 45° to the east-southeast and is structurally overlain by a thin sheet of Kekiktuk Conglomerate. Based on graded bedding, the Kekiktuk Conglomerate is upright and contains bedding and S_1 cleavage dipping 20 to 40° to the east and southeast. The Kekiktuk sliver is structurally overlain by poorly exposed volcaniclastic rocks of horse H3 of the Apoon assemblage that also contains S_1 cleavage dipping shallowly to the south.

On the north side of the high-angle fault, volcaniclastic rocks of horse H3 are well exposed and contain a penetrative S_1 cleavage dipping south to southeast between 40 and 60° (location F). Above this exposure of Apoon (H3) is an unusually thick section (about 150 m) of Kekiktuk Conglomerate. At this location, the Kekiktuk is deformed in tight to isoclinal D_1 folds with a penetrative S_1 cleavage parallel to the axial planes. The S_1 cleavage orientation is variable due to superposed D_2 folds, but for the most part is consistent with the southeast-dipping S_1 cleavage in the underlying Apoon assemblage rocks. In sharp contrast, a 60-m section of Apoon volcanic rocks (H2) overlies the Kekiktuk and is penetratively deformed in a north 10- to 30°-dipping S_1 cleavage. The volcanic rocks of the Apoon are themselves structurally overlain by multiple, thin imbricates of the Kekiktuk, Kayak, and Lisburne that also exhibit shallow, north-dipping S_1 cleavage. The lack of preserved stratigraphic

contacts in the imbricate stack of Devono-Carboniferous rocks is documented along strike where each unit is seen to be a laterally discontinuous sliver. In the westernmost sliver, between locations C and A (Fig. 13), a double repetition of Kekiktuk, Kayak, and Lisburne is observed.

In the Lost Sheep Creek area, definition of the boundary between the Apoon and Blarney Creek duplexes becomes blurred. Nevertheless, using the orientation of D_1 structures as a guide, segregation is possible. The basal part of the structural stack is characterized by south to southeasterly dipping S_1 cleavage, consistent with cleavage orientations of the underlying Apoon duplex farther south. In contrast, the upper part of the imbricate stack exhibits D_1 structures with northerly dips, consistent with those of the overlying Blarney Creek duplex. Thus, based on associations of D_1 structures, the imbricate zone can be divided and assigned to the upper and lower duplexes.

Such a simple division of the imbricate stack obscures a critical structural relation, however. Of particular importance is the recognition that the imbricate stack is localized at the structural boundary between the Blarney Creek and Apoon duplexes. Specific constituents of the imbricate zone are not found in the superjacent and subjacent duplexes: Apoon assemblage rocks are rarely found within the Blarney Creek duplex, and then only at the base, and similarly, Carboniferous clastic and carbonate rocks seldom are found in the Apoon duplex except near its upper contact. The implication is that even though parts of the imbricate zone have structural affinities either with the upper or lower duplex, together they form a separate structural entity, the Blarney Creek thrust zone. Here, rocks of the underlying and overlying assemblages are complexly imbricated within the broad shear zone, but were not carried away from the shear zone into either adjacent structural unit.

Northern Blarney Creek. In two areas along northern Blarney Creek (Fig. 14), Lisburne limestone is in direct contact with underlying rocks of the Apoon duplex. The complete omission of Kayak Shale and the Kekiktuk Conglomerate from between the two duplexes demands a tectonic origin for the contact. The western exposure of this contact is poorly exposed and no mesoscopic structural data are available. Nevertheless, mapping in the western part of the area (Fig. 14) indicates that the Lisburne actually overlies two different sheets of Apoon assemblage rocks. Where the Kayak Shale is cut out to the north, the Lisburne overlies rocks of Apoon horse H3. Tracing the contact to the south, the Lisburne and horse H3 are separated by a thin wedge of Apoon volcanic rocks that are assigned to horse H2. Map relations suggest that this northward-thinning wedge of Apoon volcanic rocks is equivalent to one of the tectonic slices found within the diffuse Blarney Creek thrust zone exposed in Lost Sheep Creek 0.5 km to the south (Fig. 13).

Exposure of the Lisburne-Apoon contact east of Blarney Creek (Fig. 14) is more straightforward. There the orientation of S_1 cleavage found both in the Lisburne carbonate rocks and the Apoon assemblage essentially parallel the contact between the two lithologies and dips east at 10 to 50°. It is notable that this is

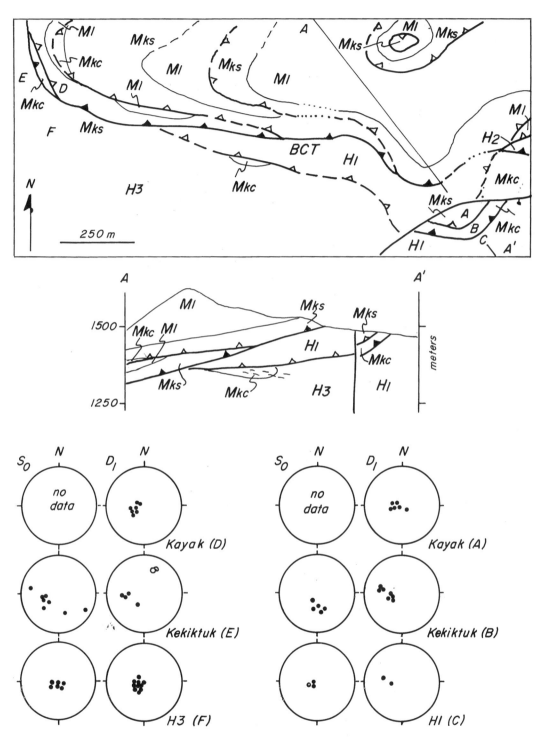

Figure 12. Western Blarney Mountain. Kekiktuk Conglomerate (Mkc) discontinuously outcrops along the boundary layer between Blarney Creek and Apoon duplexes overlying and underlying the Blarney Creek thrust (BCT), respectively. S_1 cleavage is parallel in both duplexes and within the boundary layer. Structural data are presented in lower hemisphere, equal-area projections: solid dots represent poles to S_0 or S_1 cleavage. See Figure 5 for definition of symbols.

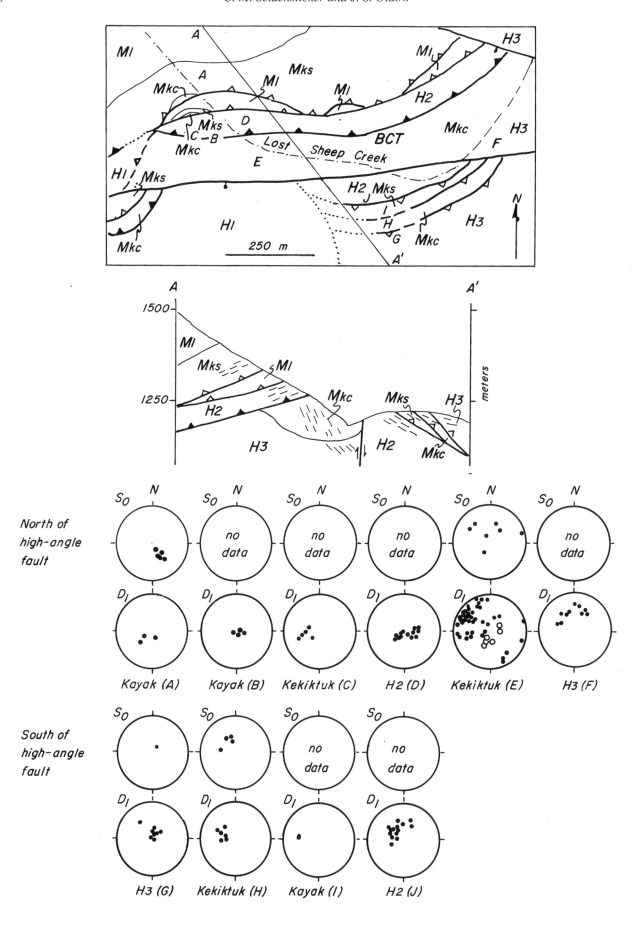

Figure 13. Lost Sheep Creek. South of the east-striking high-angle fault stretching across the center of map, two horses (H2 and H3) of the Apoon duplex are separated by thin imbricates of Kayak Shale (Mks) and Kekiktuk Conglomerate (Mkc) and all have south-dipping S_1 cleavage. North of the high-angle fault, the Apoon duplex exhibits south-dipping S_1 cleavage and is overlain depositionally by Kekiktuk Conglomerate. Tectonic slices of Apoon assemblage, Kekiktuk Conglomerate, Kayak Shale (Mks), and Lisburne Group (Ml) contain north-dipping S_1 cleavage and form part of the boundary layer beneath Blarney Creek duplex. Structural data are presented in lower hemisphere, equal-area projections: solid dots represent poles to S_0 and S_1 cleavage, open circles represent D_1 fold axes. See Figure 5 for definition of symbols.

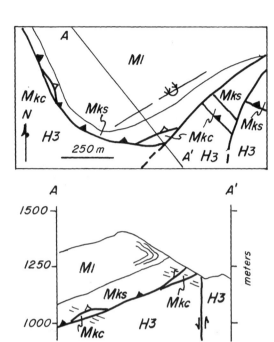

one of the few locations where S_1 cleavage is penetratively developed in Lisburne carbonate rocks, which may indicate that its development is related to shear on the Blarney Creek thrust. Normally, intense deformation is localized in the mechanically incompetent Kayak Shale, effectively insulating the Lisburne from the penetrative effects of shear associated with displacement on the Blarney Creek thrust.

Western Falsoola Mountain. Where the Kekiktuk is preserved both above and below the basal Blarney Creek thrust, as described elsewhere on southern Falsoola Mountain (Fig. 6), the differences in the orientation of S_1 cleavage conclusively documents detachment of the upper plate conglomerate from the underlying Apoon. Elsewhere, particularly on the western flank of Falsoola Mountain immediately east of Blarney Creek (Fig. 15), relationships are not as clear. In this area, the Kekiktuk Conglomerate occurs discontinuously at the boundary between the two footwall duplexes. The structural fabric, particularly the orientation of S_1 cleavage, within the isolated exposures of the Kekiktuk is consistent with that of both the upper and lower duplexes. S_1 cleavage and bedding are parallel within the Kekiktuk and, like cleavage in the overlying Kayak Shale and underlying Apoon, dip shallowly (10 to 30°) to the southeast. Conceivably, the Kekiktuk along the contact is in depositional contact with the underlying Apoon assemblage and is part of a preserved limb of a D_1 isoclinal fold, but no corresponding hinge zone is observed. Thus, the lower contact is interpreted to be tectonic, based on the discontinuous nature of the conglomerate and its similarity to allochthonous Kekiktuk elsewhere along the contact exposed on southern Falsoola Mountain.

KINEMATIC RECONSTRUCTION

The internal structural relations of the Blarney Creek and Apoon duplexes, when viewed individually, are consistent with simple models of duplex formation. The Blarney Creek duplex, in particular, has a classic duplex morphology. The floor and roof of the Blarney Creek duplex are respectively "tiled" and "shingled" almost entirely with rocks of a single stratigraphic unit: The floor of the duplex is composed mostly of Kayak Shale, and the roof is composed almost entirely of Lisburne Group carbonates. The Kekiktuk Conglomerate is sporadically present

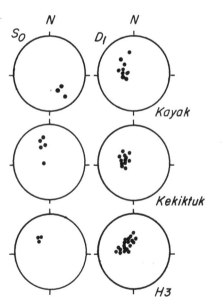

Figure 14. Western Falsoola Mountain. Kekiktuk Conglomerate (Mkc) discontinuously outcrops along the boundary layer between Blarney Creek and Apoon duplexes overlying and underlying the Blarney Creek thrust (BCT), respectively. S_1 cleavage is parallel in both duplexes and in the boundary layer. Structural data are presented in lower hemisphere, equal-area projections: solid dots represent poles to S_0 or S_1 cleavage. See Figure 5 for definition of symbols.

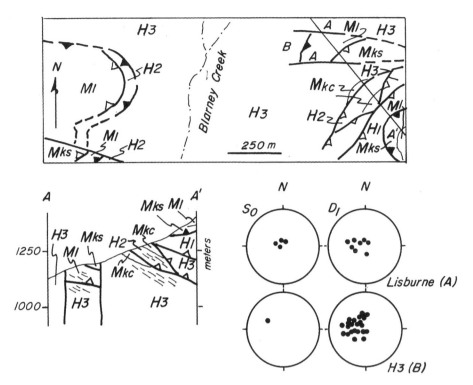

Figure 15. Northern Blarney Creek. Lisburne limestone (Ml) of Blarney Creek duplex is locally in direct structural contact with the underlying Apoon duplex (H2 and H3). Structural data are presented in lower hemisphere, equal-area projections: solid dots represent poles to S_0 or S_1 cleavage, and open circles represent D_1 fold axes. See Figure 5 for definition of symbols.

within the Blarney Creek duplex, at the base of some of the horses. In addition, rocks of the Sadlerochit Group and Shublik Formation are occasionally included at the top of a horse. These observations imply that during construction of the Blarney Creek duplex the floor thrust was located primarily within the Kayak Shale and that, in general, the Kayak Shale was detached from the underlying Kekiktuk Conglomerate during this kinematic episode. Occasionally, however, small slices of Kekiktuk remained attached to the Kayak and were thereby carried above the floor of the Blarney Creek duplex at the bottom of a horse. Similarly, the roof thrust was generally located either within or, more probably, at the top of the Lisburne Group carbonates during duplex formation. Since the Lisburne Group is a thick and structurally competent unit, it is no surprise that as the Endicott Mountains allochthon passed northward nucleation of the Amawk thrust, later to become the roof thrust of the Blarney Creek duplex, occurred at the top of the carbonates near their contact with overlying, less competent clastic rocks of the Sadlerochit Group. Only rarely were small sheets of the post-Lisburne strata left behind (that is, below the Amawk thrust), to be incorporated later within horses of the Blarney Creek duplex.

As discussed by Julian and Oldow (this volume, Chapter 5), the internal structure of the Apoon duplex requires a more complex imbrication history for its development than for that of the Blarney Creek duplex. Simple northward propagation of the

floor thrust does not account for the internal geometry of the Apoon duplex. Rather, horse imbrication involved out-of-sequence fault overstepping. Although complicated, the imbrication history of the Apoon duplex is a minor variant of existing models (e.g., Butler, 1982).

It is only when the entire footwall assemblage of the Doonerak multiduplex is viewed as a coherent unit that the complex behavior of interaction between its component structures becomes apparent. Attempts to unravel the kinematic history of the stacked footwall duplexes focus attention on the role of the boundary that separates them, the Blarney Creek thrust.

Boundary-layer kinematics: Blarney Creek thrust

Assessment of structural and stratigraphic relations exposed along the Blarney Creek thrust and within the Apoon and Blarney Creek duplexes indicates that the Kekiktuk Conglomerate deformed more or less independently of the bounding duplexes. The simple fact that the Kekiktuk Conglomerate is rarely found among the horses of the overlying and underlying duplexes indicates its predominant lack of direct involvement in their formation. For the overlying Blarney Creek duplex, this can be explained easily by the preferential development of its floor thrust in the mechanically weak Kayak Shale overlying the Kekiktuk. Although numerous examples of minor exposures of Kekiktuk Conglomerate lying between various horses of the Apoon duplex are described above, together they account for a small percentage of the total exposed length of intra-Apoon duplex horse contacts. This great scarcity of Kekiktuk Conglomerate exposures between horses of the Apoon duplex, however, cannot be explained so easily.

Evidence that the Kekiktuk Conglomerate originally was in depositional contact with the underlying Apoon assemblage is found in three of the four Apoon duplex horses. Small exposures of Kekiktuk that are in depositional contact with fine-grained clastic rocks of the Apoon horse H1 are found along the south flank of Falsoola Mountain (Fig. 6). Small exposures of Kekiktuk in depositional contact with the interbedded fine-grained clastic and volcanogenic rocks of horse H3 are preserved east and west of Kinnorutin Mountain (Fig. 8), and as flatirons on the north face of Amawk Mountain (Fig. 9). Similarly, small outcrops of the Kekiktuk depositionally overlie volcanic rocks of horse H4 just west of Kinnorutin Mountain (Fig. 8), in Amawk Creek (Fig. 10), and near the top of Wien Mountain (Fig. 11).

Given that the Kekiktuk Conglomerate was originally in depositional contact with the underlying Apoon assemblage but now, in general, is either structurally detached or completely removed from the Apoon, it can be concluded that in most cases the Kekiktuk must have been tectonically stripped from the Apoon assemblage rocks during imbrication of the Apoon duplex. Thus, the Kekiktuk constitutes an independent tectonic unit bound both above and below by faults. The Blarney Creek thrust, then, is not a discrete structural surface everywhere along its trace. Rather, the fault is a decollement zone, composed of an anastomosing system of faults localized near the stratigraphic horizon of the Kekiktuk Conglomerate. If this were not the case, one would expect to find the Kekiktuk at the base of horses within the Blarney Creek duplex and/or at the tops of horses in the Apoon duplex much more frequently than is observed.

The character of the Blarney Creek fault zone varies considerably along its trace. As shown in Lost Sheep Creek of eastern Blarney Mountain (Fig. 13), the fault zone does not exclusively contain rocks of the Kekiktuk, but also includes small slices of the underlying Apoon and overlying Kayak and Lisburne. Similar relations are observed on western Falsoola Mountain (Fig. 15) and on Wien Mountain (Fig. 11), as described above, and near Trembley Creek at the eastern closure of the Doonerak window (Phelps and Avé Lallemant, this volume, Chapter 4). In other segments of the fault zone, such as northern Hammond River (Fig. 7), northern Blarney Creek (Fig. 14), and western Blarney Mountain (Fig. 12), the tectonic boundary is sharp and well defined. Where it is well defined, the Blarney Creek thrust generally omits structural section and variably places rocks of the Kekiktuk, Kayak, or Lisburne on rocks of the underlying Apoon assemblage.

Mesoscopic structures both within and below the Blarney Creek thrust zone indicate that initiation of and continued motion along the fault occurred during D_1 deformation. In virtually all instances where the Kekiktuk is in depositional contact with the underlying Apoon assemblage, S_1 cleavage intersects bedding at a large angle. This is because the depositional contacts are preserved in the hinges of decapitated D_1 synclines involving both the Kekiktuk and the underlying Apoon. Where the Kekiktuk is exposed within the fault zone, S_1 cleavage and bedding are generally subparallel (except in the hinges of minor folds), representing detached limbs of large-scale tight to isoclinal D_1 folds.

Figure 16 illustrates a forward model of boundary-layer deformation that accounts for decapitation of D_1 synclines in the footwall, omission of units, and inclusion of minor slices of Apoon, Kayak, and Lisburne in the Blarney Creek thrust zone. North-vergent D_1 folds of the interface between the Kekiktuk Conglomerate and the underlying Apoon assemblage were formed in a north-directed (top-to-the-north) shear couple. The upper fault strand of the boundary-layer shear zone was localized in the Kayak Shale and it accommodated imbrication of the superjacent Blarney Creek duplex. The lower fault strand of the shear zone accommodated imbrication of the subjacent Apoon duplex and was nucleated within the Apoon assemblage below the upright lower limbs of major asymmetric tight to isoclinal folds of the Kekiktuk-Apoon interface. Displacement on anastomosing faults within the shear zone can result in decapitation of D_1 synclines and inclusion of small slices of Apoon rocks above the lower fault zones. The profound imbrication of units preserved in Lost Sheep Creek (Fig. 13) can be explained, at least in part, by such a system of anastomosing faults formed between the upper and lower boundaries of the shear zone.

Duplex horse decapitation, illustrated by the omission of structural section and the juxtaposition of units with highly discordant S_1 foliations, requires an additional component in this model. If imbricate faults within the underlying Apoon duplex merged sequentially, in a forward-propagating manner, with the lower strand of the Blarney Creek thrust zone, as predicted by simple duplex models, no major omission of section nor substantial discordance between S_1 cleavage orientations above and below the fault would be expected. Thus, it appears that imbricate horses of the lower duplex did not always merge smoothly with their roof thrust, but rather created local asperities in the fault zone (in a sequential model, such asperities would result in forward propagation of the ramp). Predictably, such irregularities in the boundary-layer fault zone would be truncated by new fault strands formed to maintain a smooth fault trajectory. Depending upon the magnitude of the asperity (Fig. 16), varying amounts of section can be removed in this manner, resulting in such dramatic juxtapositions as the Lisburne on the Apoon observed in northern Blarney Creek.

Formation of the Blarney Creek thrust zone followed development of tight to isoclinal D_1 folds involving the Kekiktuk Conglomerate and adjacent units. Discordance of D_1 structures and omission of section across the fault zone indicate that decapitation of horses resulted from continued displacement on the fault zone during Apoon duplex imbrication. Thus, the fault zone continued activity throughout D_1 and formation of the underlying Apoon duplex. Similarly, local truncation of horses at the top of the Blarney Creek duplex indicates continued motion on the Amawk thrust during subjacent duplex imbrication.

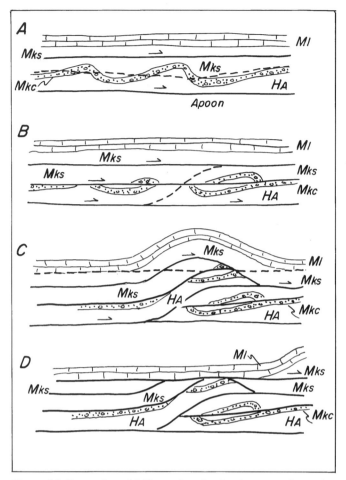

Figure 16. Forward model illustrating the development of complex structural relations observed at the boundary between Blarney Creek and Apoon duplexes. Bold solid lines represent thrust faults, dashed lines incipient faults. HA-detached Apoon assemblage; Mkc-Kekiktuk Conglomerate; Mks-Kayak Shale; Ml-Lisburne Group. A: Folding of the Kekiktuk Conglomerate and overlying Kayak Shale together with the underlying Apoon assemblage in the fault bound boundary layer; location of incipient fault within boundary layer indicated by dashed line. B: Displacement along intraboundary layer fault and decapitation of boundary layer folds. Incipient fault links upper and lower fault surfaces of boundary layer. C: Intraboundary layer duplexing and deformation of overlying Blarney Creek duplex. Incipient detachment below Lisburne Group cuts asperity formed by boundary layer duplex. D: Displacement of sub-Lisburne detachment and juxtaposition with underlying decapitated boundary layer duplex.

Relative timing of footwall contraction structures

An important consideration that arises from recognition of the stacked duplexes in the Doonerak window and their relation to the Blarney Creek thrust zone is an assessment of the relative timing of their structural development. It is of particular importance to address the alternative interpretations of serial versus simultaneous development of the duplexes and their intervening boundary layer during footwall contraction. Many of the first-

generation kinematic models for the evolution of duplexes employed a serial sequence of forward-propagating displacements (e.g., Boyer and Elliot, 1982). More recent models, however, either allow or require simultaneity of displacements within or among duplexes (Boyer, 1986b; Fermor and Price, 1987; Yin et al., 1989). Consequently, there is a growing awareness that contemporaneous motions on separate faults may be a common feature in the formation of duplexes, and probably also in the general scheme of fold and thrust belt evolution.

Mesoscopic structures. The geometric relations among the three phases of mesoscopic structures observed within the Blarney Creek duplex, the Blarney Creek thrust zone, and the Apoon duplex provide important relative timing constraints for the imbrication history. In each of these three structural units, as explained below, imbrication occurred during D_1 deformation and before the development of superposed D_2 and D_3 structures.

Along the northern margin of the Doonerak window, where most of the Blarney Creek duplex is exposed, the horses and constituent lithologic units dip gently to the north. Only on the southeastern flank of the Doonerak window, around the closure of the east–northeast trending antiform, do faults and bedding within the Blarney Creek duplex dip south. Strains associated with D_1 structures locally are large as indicated by nearly complete transposition of bedding into the orientation of S_1 cleavage. Planar D_1 structures are consistently oriented approximately parallel to the faults within the duplex. On the north flank of the Doonerak window, D_1 fabrics and bedding within the Blarney Creek duplex dip shallowly north, whereas on the southeast flank, the same fabric elements dip to the south.

In contrast, D_2 and D_3 structures within the Blarney Creek duplex have the same orientations on both flanks of the Doonerak window. The axial surfaces of D_2 folds and associated axial-planar cleavage consistently strike northeasterly and have steep northwesterly dips; subordinate southeast dips belong to conjugate D_2 folds. D_3 structures throughout the region have steeply dipping, north-northwest–striking axial surfaces and associated cleavage. D_3 structures clearly postdate development of the Doonerak multiduplex. D_2 structures, on the other hand, either postdate duplex formation or, considering their correspondence in orientation to the structural geometry of the multiduplex, may represent a final stage of duplex deformation (Avé Lallemant and Oldow, this volume, Chapter 14).

Similar structural relations are found in rocks composing the Blarney Creek thrust zone. Within the thrust zone, exposures of the Kekiktuk Conglomerate and less common slices of rocks of the Apoon assemblage, Kayak Shale, and Lisburne Group have the same generations of structures. Some of the coarse clastic, volcanic, and carbonate rocks do not develop the structures as well as fine-grained clastic units, however. In most rock units, a well-developed S_1 cleavage is found that, in places, is axial planar to tight to isoclinal folds. Bedding and S_1 cleavage are generally parallel, except in the hinges of rare D_1 folds. In many locations, S_1 is gently north dipping, coincident with S_1 orientations in the overlying Kayak Shale of the Blarney Creek duplex.

D_2 and D_3 structures have the same orientations as their counterparts in the Blarney Creek duplex.

Within the Apoon duplex, structural relations are a little more complex, but the same relations exist. D_1 structures do not show the same degree of variability between north and south dips as do those observed in the Blarney Creek duplex. For the most part, the axial surfaces of D_1 folds and their associated axial-planar cleavage have moderate to steep southeasterly dips throughout the Apoon duplex. Where it is recognizable, bedding is transposed into the orientation of S_1 foliation, and D_1 fold axes plunge consistently to the southeast (sheath-fold morphology). Only a few occurrences of north-dipping D_1 fabrics are preserved in the Apoon duplex. These are found along the northern flank of the Doonerak window, just below the Blarney Creek duplex, where the horses of the Apoon duplex merge smoothly into the overlying roof thrust (Blarney Creek thrust zone). Often, S_1 cleavage and D_1 folds in the Apoon duplex, including those synclines cored by the Kekiktuk Conglomerate, are truncated by the Blarney Creek thrust zone, resulting in a high-angle discordance between members of the same generation of structures. D_2 and D_3 structures within the Apoon duplex have the same orientations as those found in the overlying structural units.

As briefly outlined earlier, and discussed elsewhere (Handschy, this volume, Chapter 3; Phelps and Avé Lallemant, this volume, Chapter 4), rocks of the Endicott Mountains allochthon hanging-wall assemblage exhibit similar internal structural fabrics as those seen in the Doonerak footwall assemblage. D_1 structures in the Endicott Mountains allochthon vary in orientation between northerly and southerly dips, according to spatial position around the Doonerak window. The basal thrust of the Endicott Mountains allochthon, the Amawk thrust, is folded by the east-plunging, east-northeast–trending Doonerak antiform, which is a D_2 fold. D_2 and D_3 mesoscopic structures have the same orientations on both flanks of the Doonerak window.

Clearly, since its basal thrust was folded during D_2 deformation, emplacement of the Endicott Mountains allochthon preceded D_2 structures, which probably formed as the last stage of Doonerak multiduplex development (Avé Lallemant and Oldow, this volume, Chapter 14). Imbrication of the footwall duplexes and formation of the Blarney Creek thrust zone also predated D_2 deformation because D_2 structures are consistently oriented throughout the duplexes and have not been reoriented by later deformation. Thus, imbrication of both the Blarney Creek and Apoon duplexes of the footwall assemblage occurred during the first phase of deformation.

Thrust faults and duplex horses. Additional constraints for the relative timing of footwall deformation are derived from the relations between imbricate horses and bounding faults, and among D_1 structures in the Blarney Creek and Apoon duplexes and the Blarney Creek thrust zone. In a few rare occurrences, horses in the Blarney Creek duplex are decapitated by the Amawk thrust, implying displacement occurred on the roof thrust after subjacent horse imbrication. Likewise, the bottom of the Blarney Creek thrust zone truncates and removes substantial

sections of horses H1 and H4 of the Apoon duplex. The offset segments of decapitated Apoon horses are not observed in the field area, indicating significant displacement on the roof thrust after horse imbrication. The top of the Blarney Creek thrust zone also removes section from the bottom of the Blarney Creek duplex, placing Lisburne Group directly on the Apoon assemblage. Recurrent motion on roof and floor thrusts is also indicated by truncation of D_1 structures in upper and lower plate positions. Furthermore, the development of D_1 folds at the depositional interface of the Kekiktuk Conglomerate and the Apoon assemblage before or during displacement within and along the Blarney Creek thrust zone is documented by the few preserved synclinal hinges below the Blarney Creek thrust, and the presence of corresponding D_1 fold limbs within the fault zone in which sheets of Kekiktuk Conglomerate exhibit subparallel bedding and S_1 cleavage.

Imbrication models. Given the structural and stratigraphic constraints outlined above, two end-member models for footwall duplex formation are explored. In both models for the Doonerak multiduplex, imbrication is related to the emplacement of the overlying Endicott Mountains allochthon and subsequent imbrication of its footwall. In a serial model (Fig. 17A), the first step involves detachment between the Kayak Shale and autochthonous Kekiktuk Conglomerate to form the Blarney Creek duplex. During imbrication of the Blarney Creek duplex, Kayak Shale and Lisburne Group limestone are shortened to produce numerous horses. Younger clastic rocks (Sadlerochit Group, Shublik Formation, etc.) presumably were entrained at the base of the Endicott Mountains allochthon or were stripped from the top of the Lisburne during passage of the leading edge of the upper plate over the footwall ramp. During this stage of the serial model, all rocks beneath the floor of the Blarney Creek duplex remain autochthonous. The inception of D_1 folding of the Kekiktuk Conglomerate may have occurred during this stage, but large-scale development of recumbent folds of the Kekiktuk would not be likely, because such substantial deformation of the Kekiktuk would require its detachment from the underlying Apoon assemblage.

At this stage of development, the serial model encounters fatal difficulties. If the Kekiktuk Conglomerate and the underlying Apoon assemblage were essentially autochthonous during imbrication of the overlying Blarney Creek duplex, they would come to lie behind the advancing tail of the most distal horse of the Blarney Creek duplex. Subsequent involvement of the rocks composing the Blarney Creek thrust zone and the Apoon duplex would yield a footwall multiduplex geometry unlike that observed in the Doonerak window (Fig. 17A).

A model invoking simultaneous development of the footwall duplexes (Fig. 17B) better accounts for the relations observed in the Doonerak window. In this scheme, the Blarney Creek thrust zone acts simultaneously as an independent structural unit, and as the floor thrust to the Blarney Creek duplex, and as the roof thrust to the Apoon duplex. The omission of the Blarney Creek duplex on the southern flank of the Doonerak window and stripping of

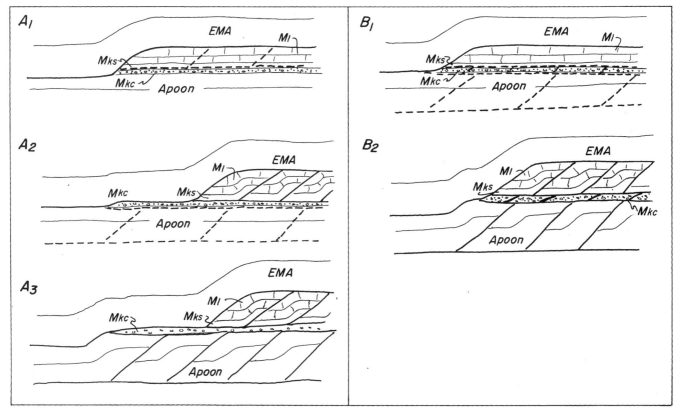

Figure 17. Comparison of serial (A) and simultaneous (B) kinematic models for Doonerak multi-duplex development. A₁: Emplacement of Endicott Mountains allochthon (EMA) above Amawk thrust ramp. A₂: Blarney Creek duplex (Ml and Mks) imbricated above Kekiktuk Conglomerate (Mkc) that remains attached to autochthonous Apoon assemblage. A₃: Apoon duplex imbricated below detached Kekiktuk Conglomerate. Sub-Kekiktuk detachment necessitated by lack of conglomerate involvement in Apoon duplex horses. Serial model fails because of unobserved "prong" of Kekiktuk Conglomerate between the Endicott Mountains allochthon and the Apoon duplex hindward of the Blarney Creek duplex. B₁: Emplacement of Endicott Mountains allochthon (EMA) above Amawk thrust ramp. B₂: Simultaneous imbrication of Blarney Creek duplex, boundary layer, and Apoon duplex. Geometry preserves overall balance and corresponds to observed relations in Doonerak window. Ml-Lisburne Group; Mkc-Kekiktuk Conglomerate; Mks-Kayak Shale.

post-Lisburne strata, nevertheless, require some degree of sequential development, at least in the early stages of the multiduplex history. Regardless, during D_1 deformation, shortening in the footwall was partitioned among three simultaneously developing decollement systems: the Amawk thrust, the Blarney Creek thrust zone, and the unexposed floor thrust of the Apoon duplex. Truncation of horses beneath the Amawk thrust and at the base of the Blarney Creek thrust zone indicates recurrent motion on both thrusts. Antiformal folding of the Blarney Creek thrust zone and the Blarney Creek duplex requires movement to have continued between horses within the Apoon duplex for some time after the Blarney Creek thrust zone became effectively locked. The rates of displacement on the different faults are unknown and could have been variable both temporally and spatially.

CONCLUSIONS

Typical models do not accommodate the geometry or kinematic development of the Doonerak window multiduplex beneath the Endicott Mountains allochthon of the central Brooks Range fold and thrust belt. The multiduplex does not contain a single roof and floor thrust system, but rather is composed of two stacked subsidiary duplexes (the Blarney Creek and Apoon duplexes) that are separated by a shear zone called the Blarney Creek thrust. The Doonerak multiduplex, whose stacked components developed simultaneously, formed during footwall imbrication beneath the Amawk thrust underlying the Endicott Mountains allochthon. Strict adherence to a serial kinematic model of deformation cannot account for observed structural relations. Rather, the stacked duplexes must have been imbricated at the same time during the earlier stages of their development. During the time that simultaneous development of the two stacked duplexes was ongoing, the Amawk thrust was employed as the roof thrust to the Blarney Creek duplex, while the Blarney Creek thrust zone served both as the floor thrust for the overlying Blarney Creek duplex and as the roof thrust of the underlying Apoon duplex.

Contemporaneous displacement on the Amawk and Blarney Creek thrusts is indicated by omission of structural section and truncation of horses of the Blarney Creek and Apoon duplexes. Therefore, the roof thrusts were not progressively locked during early stage duplex development, as suggested in typical serial models. However, during later stages of development of the Doonerak duplex, roof thrusts at higher structural levels did eventually become locked. This is indicated by folding of the Blarney Creek and Amawk thrusts above the Apoon duplex, and by the Eekayruk thrust, which ramps up through the Apoon duplex and cuts overlying duplexes.

The Blarney Creek thrust zone played a pivotal role in the development of the Doonerak multiduplex. The character of the fault zone varies along strike from a sharp structural contact between the upper and lower duplexes to a broad zone of distributed shear as much as 250 m thick. The fault zone typically contains deformed conglomerate and occasional slices of upper plate and/or lower plate rocks. The presence of the Kekiktuk Conglomerate, which originally was deposited as a thin layer (averaging 5 to 20 m thick) at the interface between Devono-Carboniferous rocks and underlying lower Paleozoic strata, determined the location of the Blarney Creek thrust zone. The floor thrust of the Blarney Creek duplex formed in shale overlying the conglomerate, concurrently with detachment of the conglomerate from the underlying lower Paleozoic rocks to form the roof thrust of the Apoon duplex. During shortening and development of the adjacent duplexes, the Blarney Creek thrust zone maintained its structural integrity and the conglomerate was not generally incorporated into overlying and underlying structural units.

ACKNOWLEDGMENTS

This research was supported by grants from the Department of Energy (DE-AS05-83ER13124), the National Science Foundation (EAR-8517384, EAR-8720171, EAR-9017835), and the Alaska Industrial Associates Program at Rice University. We thank Roy Kligfield and William J. Perry, Jr., for their careful reviews of the manuscript.

REFERENCES CITED

Armstrong, A. K., and Mamet, B. L., 1975, Carboniferous biostratigraphy, northeastern Brooks Range, Arctic Alaska: U.S. Geological Survey Professional Paper 884, 29 p.

Armstrong, A. K., Mamet, B. L., Brosgé, W. P., and Reiser, H. N., 1976, Carboniferous section and unconformity at Mount Doonerak, Brooks Range, northern Alaska: American Association of Petroleum Geologists Bulletin, v. 60, p. 962–972.

Bally, A. W., Gordy, P. L., and Stewart, G. A., 1966, Structure, seismic data, and orogenic evolution of southern Canadian Rocky Mountains: Bulletin of Canadian Petroleum Geology, v. 14, p. 337–381.

Banks, C. J., and Warburton, J., 1986, "Passive-roof" duplex geometry in the frontal structures of the Kirthar and Sulaiman Mountain belts, Pakistan: Journal of Structural Geology, v. 8, p. 229–237.

Bell, T. H., 1983, Thrusting and duplex formation at Mount Isa, Queensland, Australia: Nature, v. 304, p. 493–497.

Bosworth, W., 1984, Foreland deformation in the Appalachian Plateau, central New York: the role of small-scale detachment structures in regional overthrusting: Journal of Structural Geology, v. 6, p. 73–81.

Bowsher, A. L., and Dutro, J. T., Jr., 1957, The Paleozoic section in the Shainin Lake area, central Brooks Range, Alaska: U.S. Geological Survey Professional Paper 303-A, p. 1–39.

Boyer, S. E., 1986a, Styles of folding within thrust sheets: Some examples from the Appalachian and Rocky Mountains of the U.S.A. and Canada: Journal of Structural Geology, v. 8, p. 325–339.

Boyer, S. E., 1986b, Geometric evidence for simultaneous, rather than sequential, movement on major thrusts; implications for duplex development: Geological Society of America Abstracts with Programs, v. 18, p. 549.

Boyer, S. E., and Elliott, D., 1982, Thrust systems: American Association of Petroleum Geologists Bulletin, v. 66, p. 1196–1230.

Brosgé, W. P., and Reiser, H. N., 1971, Preliminary bedrock geologic map, Wiseman and eastern Survey Pass Quadrangles, Alaska: U.S. Geological Survey Open-File Map 71-56, scale 1:250,000.

Brosgé, W. P., Dutro, J. T., Jr., Mangus, M. D., and Reiser, H. N., 1962, Paleozoic sequence in eastern Brooks Range, Alaska: American Association of Petroleum Geologists Bulletin, v. 46, p. 2174–2198.

Brosgé, W. P., Reiser, H. N., and Tailleur, I. L., 1974, Pennsylvanian beds in Lisburne Group, south-central Brooks Range, Alaska, *in* U.S. Geological Survey Alaska Program, 1974: U.S. Geological Survey Circular 700, p. 41.

Butler, R. W. H., 1982, Hanging wall strain: a function of duplex shape and footwall topography: Tectonophysics, v. 88, p. 235–246.

Chapman, R. M., Detterman, R. L., and Mangus, M. D., 1964, Geology of the Killik-Etivluk Rivers region, Alaska: U.S. Geological Survey Professional Paper 303-F, p. 325–407.

Crowder, R. K., 1990, Permian and Triassic sedimentation in the northeastern Brooks Range, Alaska: deposition of the Sadlerochit Group: American Association of Petroleum Geologists Bulletin, v. 74, p. 1351–1370.

Dahlstrom, C. D. A., 1970, Structural geology in the eastern margin of the Canadian Rocky Mountains: Bulletin of Canadian Petroleum Geology, v. 18, p. 332–406.

Diaz-Martinez, E., and Isaacson, P. E., 1994, Late Devonian glacially-influenced marine sedimentation in western Gondwana: The Cumana Formation, Altiplano, Bolivia, *in* Embry, A. F., Beauchamp, B., and Glass, D. J., eds., Pangea; Global Environments and Resources: Canadian Society of Petroleum Geologists, Memoir 17, p. 511–522.

Detterman, R. L., Reiser, H. N., Brosgé, W. P., and Dutro, J. T., Jr., 1975, Post-Carboniferous stratigraphy, northeastern Alaska: U.S. Geological Survey Professional Paper 886, 46 p.

Dillon, J. T., Brosgé, W. P., and Dutro, J. T., Jr., 1986, Generalized geologic map of the Wiseman Quadrangle, Alaska: U.S. Geological Survey Open-File Report 86-219, scale 1:250,000.

Dutro, J. T., Jr., Brosgé, W. P., Lanphere, M. A., and Reiser, H. N., 1976, Geologic significance of Doonerak structural high, central Brooks Range, Alaska: American Association of Petroleum Geologists Bulletin, v. 60, p. 952–961.

Dutro, J. T., Jr., Brosgé, W. P., Detterman, R. L., and Reiser, H. N., 1979, Beaucoup Formation, a new Upper Devonian stratigraphic unit in the central Brooks Range, northern Alaska, *in* Sohl, N. F., and Wright, W. B., eds., Changes in stratigraphic nomenclature by the U.S. Geological Survey, 1978: U.S. Geological Survey Bulletin, v. 1482-A, p. A63–A69.

Dutro, J. T., Jr., Palmer, A. R., Repetski, J. E., and Brosgé, W. P., 1984, Middle Cambrian fossils from the Doonerak anticlinorium, central Brooks Range, Alaska: Journal of Paleontology, v. 58, p. 1364–1371.

Elliott, D., and Johnson, M. R. W., 1980, Structural evolution in the northern part of the Moine thrust belt, NW Scotland: Transactions of the Royal Society of Edinburgh: Earth Sciences, v. 71, p. 69–96.

Fermor, P. R., and Price, R. A., 1987, Multiduplex structure along the base of the Lewis thrust sheet in the southern Canadian Rockies: Bulletin of Canadian Petroleum Geology, v. 35, p. 159–185.

Julian, F. E., 1989, Structure and stratigraphy of lower Paleozoic rocks, Doonerak window, central Brooks Range, Alaska [Ph.D. thesis]: Houston,

Texas, Rice University, 127 p.

Julian, F. E., Phelps, J. C., Seidensticker, C. M., Oldow, J. S., and Avé Lallemant, H. G., 1984, Structural history of the Doonerak window, central Brooks Range, Alaska: Geological Society of America Abstracts with Programs, v. 16, p. 91.

McClay, K. R., ed., 1992, Thrust tectonics: London, Chapman and Hall, 447 p.

McClay, K. R., and Price, N. J., eds., 1981, Thrust and nappe tectonics: Geological Society of London, Special Publication 9, 539 p.

Moore, T. E., and Churkin, M., Jr., 1984, Ordovician and Silurian graptolite discoveries from the Neruokpuk Formation (sensu lato), northeastern and central Brooks Range, Alaska: Paleozoic Geology of Alaska and Northwestern Canada Newsletter, no. 1, p. 21–23.

Moore, T. E., Wallace, W. K., Bird, K. J., Karl, S. M., Mull, C. G., and Dillon, J. T., 1994, Geology of northern Alaska, *in* Plafker, G., and Berg, H. C., eds., The geology of Alaska: Boulder, Colorado, Geological Society of America, The Geology of North America, v. G-1, p. 49–140.

Mull, C. G., 1982, Tectonic evolution and structural style of the Brooks Range, Alaska: an illustrated summary, *in* Powers, R. B., ed., Geologic studies of the Cordilleran thrust belt, Volume 1: Denver, Colorado, Rocky Mountain Association of Geologists, p. 1–45.

Mull, C. G., Dillon, J. T., and Adams, K. E., 1987, The Doonerak fenster, central Brooks Range, Alaska, *in* Hill, M. L., ed., Cordilleran Section of the Geological Society of America Centennial Field Guide, Volume 1: Geological Society of America, p. 469–472.

Nilsen, T. H., 1981, Upper Devonian and Lower Mississippian redbeds, Brooks Range, Alaska, *in* Miall, A. D., ed., Sedimentation and tectonics in alluvial basins: Geological Association of Canada, Special Paper 23, p. 187–219.

Oldow, J. S., Avé Lallemant, H. G., Julian, F. E., and Seidensticker, C. M., 1984, The Doonerak window duplex: regional implications: Geological Society of America Abstracts with Programs, v. 16, p. 326.

Oldow, J. S., Seidensticker, C. M., Phelps, J. C., Julian, F. E., Gottschalk, R. R., Boler, K. W., Handschy, J. W., and Avé Lallemant, H. G., 1987a, Balanced cross sections through the central Brooks Range and North Slope, Arctic Alaska: Tulsa, Oklahoma, American Association of Petroleum Geologists, Special Publication, 19 p., 8 pl.

Oldow, J. S., Avé Lallemant, H. G., Julian, F. E., and Seidensticker, C. M., 1987b, Ellesmerian(?) and Brookian deformation in the Franklin Mountains,

northeastern Brooks Range, Alaska, and its bearing on the origin of the Canada Basin: Geology, v. 15, p. 37–41.

Phelps, J. C., 1987, Stratigraphy and structure of the northeastern Doonerak window area, central Brooks Range, northern Alaska [Ph.D. thesis]: Houston, Texas, Rice University, 171 p.

Phelps, J. C., Avé Lallemant, H. G., Seidensticker, C. M., Julian, F. E., and Oldow, J. S., 1987, Late-stage high-angle faulting, eastern Doonerak window, central Brooks Range, Alaska, *in* Tailleur, I. L., and Weimer, P., eds., Alaskan North Slope geology: Society of Economic Paleontologists and Mineralogists, Pacific Section, p. 685–690.

Seidensticker, C. M., Julian, F. E., Phelps, J. C., Oldow, J. S., and Avé Lallemant, H. G., 1987, Lateral continuity of the Blarney Creek thrust, Doonerak window, central Brooks Range, Alaska, *in* Tailleur, I. L., and Weimer, P., eds., Alaskan North Slope geology: Society of Economic Paleontologists and Mineralogists, Pacific Section, Publication 50, p. 681–683.

Skuce, A. G., Goody, N. P., and Maloney, J., 1992, Passive-roof duplexes under the Rocky Mountain foreland basin, Alberta: American Association of Petroleum Geologists Bulletin, v. 76, p. 67–80.

Tailleur, I. L., Brosgé, W. P., and Reiser, H. N., 1967, Palinspastic analysis of Devonian rocks in northwestern Alaska, *in* Oswald, D. H., ed., International Symposium on the Devonian System: Alberta Society of Petroleum Geologists Proceedings, v. 2, p. 1345–1361.

Wallace, W. K., and Hanks, C. L., 1990, Structural provinces of the northeastern Brooks Range, Arctic National Wildlife Refuge, Alaska: American Association of Petroleum Geologists Bulletin, v. 74, p. 1100–1118.

Willis, B., 1902, Stratigraphy and structure, Lewis and Livingston Ranges, Montana: Geological Society of America Bulletin, v. 13, p. 305–352.

Woodward, N. B., Boyer, S. E., and Suppe, J., 1989, Balanced geological cross-sections: An essential technique in geological research and exploration: American Geophysical Union, Short Course in Geology, v. 6, 132 p.

Yin, A., Kelty, T. K., and Davis, G. A., 1989, Duplex development and abandonment during evolution of the Lewis thrust system, southern Glacier National Park, Montana: Geology, v. 17, p. 806–810.

MANUSCRIPT ACCEPTED BY THE SOCIETY SEPTEMBER 23, 1997

Geological Society of America
Special Paper 324
1998

Stratigraphy and paleogeographic setting of the eastern Skajit allochthon, central Brooks Range, Arctic Alaska

J. S. Oldow
Department of Geology and Geological Engineering, University of Idaho, Moscow, Idaho 83844-3022
K. W. Boler and H. G. Avé Lallemant
Department of Geology and Geophysics, Rice University, Houston, Texas 77005-1892
R. R. Gottschalk
Exxon Production Research Company, P.O. Box 2189, Houston, Texas 77252
F. E. Julian
Department of Geological Sciences, University of Texas, El Paso, Texas 77968
C. M. Seidensticker
615 Baltimore Drive, El Paso, Texas 79902
J. C. Phelps
Chevron Petroleum Technology Center, P.O. Box 446, La Habra, California 90633-0446

ABSTRACT

The Skajit allochthon of the south-central Brooks Range is one of the structurally highest units in the late Mesozoic and Cenozoic fold and thrust belt and overlies metasedimentary rocks of the Endicott Mountains allochthon on the north and metamorphic tectonites of the Schist belt to the south. The Skajit allochthon is deformed internally in a complex array of folds and thrusts and is composed of two, partially coeval, stratigraphic sequences. The stratigraphic sequences are composed of metamorphosed fine- to coarse-grained siliciclastic, volcanogenic, and volcanic rocks and thin-bedded to massive carbonate units. The two stratigraphic sequences, which constitute numerous imbricate sheets, are spatially segregated into a lower and upper assemblage of thrust nappes. The structurally lower sequence is composed of intercalated shallow-marine to platformal carbonates and fine- to coarse-grained siliciclastic rocks of Middle Cambrian to early Late Devonian age. The clastic and carbonate rocks are conformably overlain by a post-lower Upper Devonian succession of fine-grained siliciclastic and thin-bedded carbonates. The stratigraphic sequence of the structurally higher nappes is composed of fine- to coarse-grained siliciclastic and volcanogenic rocks and subordinate thin-bedded carbonate rocks of Early or Middle Ordovician to early Late Devonian age. The two lower Paleozoic sequences substantially overlap in age but were deposited in distinctly different depositional environments. The Cambrian to Devonian carbonate succession was deposited in a carbonate platform or shelf environment, whereas the partially coeval siliciclastic, carbonate, and volcanic succession is of basin affinity. Paleogeographic restoration of the imbricate thrust sheets places the carbonate succession to the north of the siliciclastic sequence. The carbonate rocks of the shelf succession are the apparent source of fine-grained carbonate mud and occasional carbonate olistoliths found within the basin sequence. Siliciclastic detritus intercalated within the shallow-marine carbonates of the shelf sequence was derived from a continental source that also supplied most of the detritus for the basinal assemblage. Volcanic and volcanogenic clastic rocks within the basin sequence were derived from a magmatic source located to the south of the basin and are unrelated to other lower Paleozoic volcanic and volcanogenic rocks underlying the North Slope of Arctic Alaska and northwestern Canada.

INTRODUCTION

Unlike the North Slope and the frontal parts of the Brooks Range fold and thrust belt, the stratigraphy of low-grade metamorphic rocks within the interior of the mountain range is poorly understood. Formulation of a stratigraphic framework for this part of the orogenic belt has been hampered by poor age control, the lack of distinctive stratigraphic markers, metamorphic recrystallization, and structural complexities. Previous attempts to reconstruct the stratigraphy of parts of the interior of the fold and thrust belt have been based on local mapping and/or stratigraphic studies (e.g., Dillon et al., 1987; Dumoulin and Harris, 1987; Moore et al., 1992, 1994) but had limited success due to unresolved structural complexity.

We have expanded on earlier work (Brosgé and Reiser, 1964, 1971; Dillon et al., 1986, 1987, 1988; Oldow et al., 1987) and have identified several fundamental structural units or allochthons that compose the internal part of the fold and thrust belt. As used here, an allochthon is viewed as a coherent structural unit composed of numerous thrust sheets. Recognition of far-traveled allochthons, such as the Endicott Mountains allochthon (Mull, 1982; Oldow et al., 1987; Handschy, this volume, Chapter 3) underlying much of the frontal part of the thrust belt and the Skajit allochthon in the interior of the belt (Oldow et al., 1987: 1989) allows a more objective assessment of the stratigraphic relations in this part of the orogen.

Characterization of the internal stratigraphy of the Skajit allochthon necessarily must precede regional correlation of units and focuses attention on the need to exercise caution when stratigraphic relations established in the frontal and foreland regions of the contractional belt (Handschy, this volume, Chapter 1) are imported to the interior. Here, we outline in considerable detail our current understanding of the stratigraphy of the eastern Skajit allochthon.

STRUCTURAL SETTING

The Skajit allochthon is more than 10 km thick and is composed of numerous thrust imbricates (Oldow et al., this volume, Chapter 8) containing lower Paleozoic and possibly Precambrian rocks. The allochthon is one of the highest structural units within the central Brooks Range (Oldow et al., 1987) and on the north overlies Devonian siliciclastic rocks of the Endicott Mountains allochthon (Mull, 1982; Boler, 1989; Phelps and Avé Lallemant, this volume, Chapter 4). The northern contact is a well located basal thrust, named the Table Mountain thrust by Dillon et al. (1987), traced from south of the exposures of the Apoon assem-

blage in the Doonerak window duplex eastward to Table Mountain (Fig. 1). West of the Dalton Highway, the east-northeast–striking thrust dips moderately to the south-southeast, whereas east of the highway, the fault strikes north-northeast and dips shallowly to the east-southeast. South of Table Mountain (Fig. 1), the basal thrust is cut by the Foggytop thrust that carries rocks of the underlying Endicott Mountains allochthon over part of the northern Skajit allochthon. In the Chandalar River drainage, the Skajit allochthon is underlain by a subhorizontal basal thrust and structurally overlies the metasedimentary rocks of the Endicott Mountains allochthon in the north and in the south the amphibolite facies metamorphic rocks of the Schist belt. Several klippen of the Skajit allochthon are found tens of kilometers to the east of the Chandalar River, where they overlie the Endicott Mountains allochthon. Southwest of the Chandalar River, the Skajit allochthon overlies the Schist belt on a northeast-striking thrust fault originally recognized by Brosgé and Reiser (1964) that dips moderately to the northwest. To the south, the basal thrust of the allochthon is truncated by the south-dipping Minnie Creek thrust (Fig. 1), which carries metamorphic rocks of the Schist belt over the southern flank of the Skajit allochthon (Brosgé and Reiser, 1964, 1971; Dillon et al., 1986; Oldow et al., 1987; Gottschalk, 1990). Thus, the Skajit allochthon structurally overlies both the Endicott Mountains allochthon and the Schist belt and, on the south, is itself overlain by the Schist belt.

LITHOTECTONIC UNITS

Stratigraphic thicknesses and the depositional settings of lithologic units within the Skajit allochthon are obscured by the general lack of age control, metamorphic recrystallization, and severe deformation. Nevertheless, broad aspects of the stratigraphy have been established within individual thrust nappes composing the allochthon, and in this context, several lithotectonic units are identified. The lithotectonic units are composed variably of (1) coarse- to fine-grained volcaniclastic and terrigenous clastic rocks that commonly contain thin interbeds of carbonate, (2) thick units of massive and well-foliated marble, and (3) volcanic and intrusive greenstones.

Lithologic descriptions are based on our work along a transect stretching from the Endicott Mountains allochthon in the north to the Minnie Creek thrust in the south. Our mapping was confined to a corridor west of the Dalton Highway (Fig. 1) but relations determined there are extrapolated to the east and west of the transect on the basis of our reconnaissance studies and by using earlier reconnaissance and detailed maps of the region (Brosgé and Reiser, 1964, 1971; Dillon et al., 1986, 1988). All

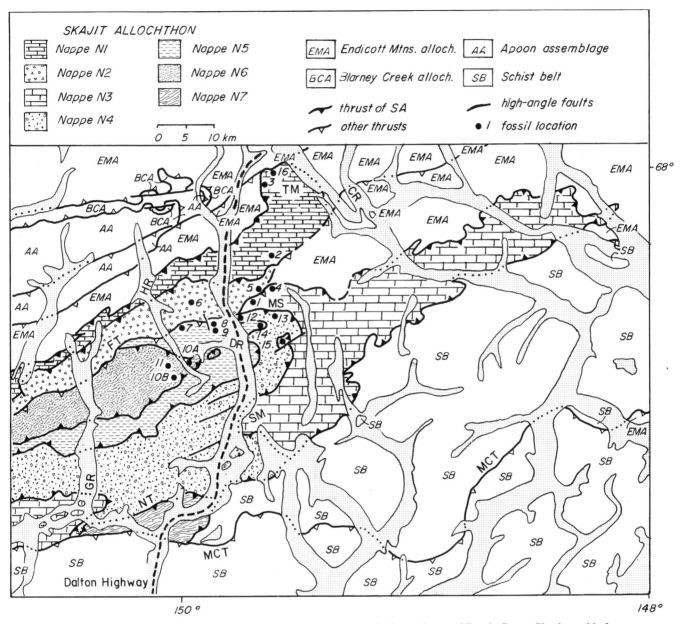

Figure 1. Generalized geologic map of the eastern Skajit allochthon in the south-central Brooks Range. Physiographic features mentioned in the text include, Chandalar River (CR), Dietrich River (DR), Glacier River (GR), Hammond River (HR), Mount Snowden (MS), Sukakpak Mountain (SM), and Table Mountain (TM). Major structures are Foggytop thrust (FT), Minnie Creek thrust (MCT), and Nolan thrust (NT). Location numbers correspond to fossil sites summarized in Figure 7 and discussed in text. Heavy dashed line marks track of Dalton Highway. Cross sections for the eastern Skajit allochthon are presented in Chapter 8 (Oldow et al., this volume).

lithologic units within the Skajit allochthon have undergone syn-metamorphic recrystallization to some degree, and in this discussion of metasedimentary and metaigneous rocks, we drop the prefix "meta."

In the eastern Skajit allochthon, seven lithotectonic units are recognized and correspond to the thrust nappes designated N1 through N7 in Figure 1. Composite columns for each of the lithotectonic units are presented in Figures 2 through 6. In all cases the columns are observed structural thicknesses and do not repre-

sent true stratigraphic thicknesses within each thrust nappe. In some lithotectonic units, stratigraphic facing is ambiguous whereas in others it is relatively well constrained. Considering the degree of internal deformation, sedimentary tops indicators are of questionable utility but stratigraphic facing can be established where sufficient fossil age control exists.

Thrust nappes composed of the same lithotectonic unit commonly are repeated internally by thrusts with relatively small displacement. In the central Brooks Range, the Skajit allochthon is divided into northern and southern domains by the Foggytop thrust (Oldow et al., this volume, Chapter 8) that cuts the previously imbricated thrust nappes. Similarities in the age, lithology, and structural relations indicate that several nappes are structurally repeated by the thrust. Nappes N1 and N2 occupy the northern domain, whereas thrust nappes N3 through N7 reside in the southern domain.

Thrust nappe N1

The northern thrust nappe, N1, structurally overlies the Endicott Mountains allochthon and consists of about 400 m of marble and chloritic calc-schist overlain by about 850 m of varicolored phyllite and thin-bedded carbonate (Fig. 2). The thick marble unit is part of the Skajit marble, from which the allochthon derives its name, and on the basis of sparse fossil age control, forms the base of the lithotectonic unit composing nappe N1.

Skajit marble unit. The Skajit marble contains two lithofacies in this area: massive marble, and calcareous and chloritic schist and phyllite. The two siliciclastic intervals lie at the base and near the middle of the succession (Fig. 2) and both pass upward into massive marble. The basal 90 m of the succession consist of chloritic, graphitic, and slightly calcareous schist and phyllite containing intrafolial lenses of stretched chert and quartz pebble conglomerate. The siliciclastic rocks pass upward into 60 m of fine-grained crystalline, banded marble that contains occasional interbeds of calcareous and chloritic schist. Rare, laterally discontinuous conglomerate lenses occur within the carbonate and include quartz, chert, shale, and quartzite pebbles. The lower carbonate interval is overlain by 110 m of calcareous and chloritic phyllite and schist, with 1- to 2-m-thick intrafolial lenses of sheared chert pebble conglomerate. The transition between the siliciclastic rocks and the overlying carbonate rocks is poorly exposed. The contact largely is covered by marble talus derived from the upper carbonate, which is composed of massive, sparsely fossiliferous crystalline marble. The upper carbonate is 140 m thick and forms the distinctive mottled gray and tan weathering "Skajit" cliffs in the study region. The marble is typically massive, although indistinct zones of banding and lamination are preserved locally. The marble varies in composition from pure white crystalline calcite to black argillaceous marble and a few calcareous phyllite interbeds occur in the lower 30 m.

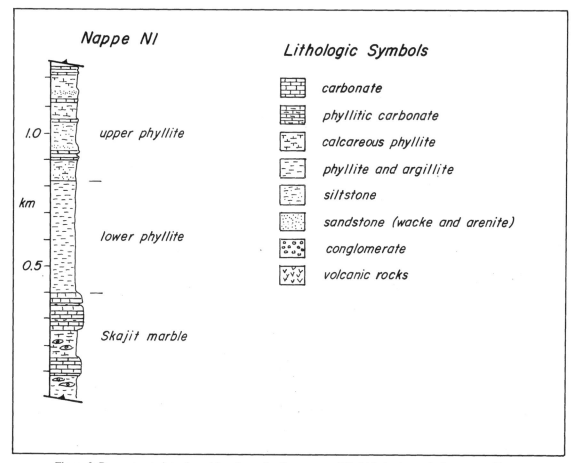

Figure 2. Reconstructed stratigraphic column for thrust nappe N1. Lithologic symbols presented here are used in all stratigraphic sections. The thickness shown is structural and does not represent true stratigraphic thickness.

Poorly preserved colonial rugose corals, crinoid columnals, gastropods, and brachiopods were found in the carbonate rocks along the transect but could not be dated. Along strike to the east, however, Middle to early Late Devonian fossil assemblages (Fig. 1; location 16) are reported near the top of the upper Skajit marble (Dillon et al., 1988). Elsewhere, marble exposed east of the Dalton Highway (Fig. 1) and interpreted by us to be part of the Skajit carbonate rocks (Fig. 1; location 2) yield conodonts, corals, and ostracodes of Middle Ordovician to Early Devonian age (Dillon et al., 1987, 1988). Brachiopods derived from the lateral extension of the upper phyllitic interval yield brachiopods of Middle(?) Devonian age (Fig. 1; location 3). Collections from the uppermost Skajit marble in the Wiseman and Chandalar Quadrangles indicate a Frasnian (early Late Devonian) age (Fig. 1; location 4) that may serve as an upper age limit for the unit (Brosgé and Reiser, 1964; Dillon et al., 1987).

Phyllite unit. Overlying the Skajit marble with apparent conformity is a sequence of fine-grained siliciclastic rocks with subordinate carbonate intercalations. The unit has a total thickness of about 850 m and is divided into upper and lower members by the presence or lack of interbedded carbonate rocks. The lower member consists of about 425 m of purple and green phyllite, slate, and siltstone overlain by 425 m of interbedded green phyllite, quartz siltstone, and thin buff marble of the upper member.

The lower member is laterally homogeneous and is composed of noncalcareous green, purple, gray, and black phyllite exposed in low-relief outcrops. The lowest and highest thirds of the unit are dominated by green phyllite and slate and the middle third by purple and black slate. Green and gray, graphitic silty phyllite, with accessory quartz silt, and very fine grained quartz sand is interspersed throughout the unit. Sparse, thin beds of quartzose, sandy siltstone are also observed. The unit is not calcareous except for rare, tan-weathering, fine crystalline marble lenses. No fossils were observed along the transect, but by their apparent stratigraphic position, the rocks are post-early Late Devonian in age.

The upper member is a poorly exposed sequence of phyllite, quartz siltstone, and thin marble interbeds. The basal contact of the upper member is defined by the first appearance of bedded carbonate and is conformable with the underlying clastic rocks. The upper contact is a south-dipping thrust that cuts down section to the west. The lower 375 m of the upper member consist of thin to medium-bedded phyllite and sandy siltstone that generally become more calcareous higher in the section. The dominant lithology is a green weathering, silty phyllite. Infrequent dark phyllite interbeds contain pyrite horizons. Interbeds of laminated, buff-weathering micritic marble are found throughout the section but are most common in the uppermost 50 m. Argillaceous, often sandy, quartz siltstone occurs throughout the section and is often finely intercalated with thin phyllite beds. Fine-grained, graded, and cross-bedded quartzite interbeds are also present.

A few fossils were recovered from the upper member. Several marble lenses found near the base of the upper member contain trilobite fragments and have yielded one 5-mm cephalon of post-Cambrian morphology and several indeterminate Stauriid rugose corals (L. Gore, personal communication, 1986). Based on the occurrence of the cephalon and the rugose corals, the unit certainly is younger than Early Ordovician (Moore et al., 1952) and by virtue of its stratigraphic position is interpreted as Late Devonian or younger.

Thrust nappe N2

Thrust nappe N2 consists of two major thrust sheets, each with its own lithologic assemblage (Fig. 3). The upper thrust sheet of nappe N2 contains several small displacement imbricates. Although structurally disrupted in the study area, the rocks are thought to be part of the same lithologic succession because stratigraphic continuity is preserved to the east (Dillon et al., 1988). Sedimentary tops indicators suggest that the entire succession is rightway-up, but sufficient biostratigraphic control does not exist to confirm this conclusion.

Rocks of the lower thrust sheet constitute three laterally continuous, south-dipping units, which in ascending order are phyllitic marble, argillite containing carbonate olistoliths, and black slate. The units form a structurally continuous succession about 2 km thick and are structurally overlain by an internally imbricated sequence of volcanogenic conglomerate, wacke, and phyllite. The upper structural sheet has a maximum thickness of about 2.2 km and is divided into two coeval units that interfinger laterally: volcanic-pebble conglomerate and volcanogenic wacke and phyllite.

Phyllitic marble unit. The lower unit rests on a basal thrust and consists of about 500 m of thin-bedded, phyllitic marble and calcareous phyllite containing channels filled with sheared quartz and lithic pebble conglomerate. The dominant lithology is argillaceous, black marble with phyllitic partings, thin phyllite interbeds, and amalgamated phyllite intervals. Marble intervals are separated by 1- to 10-m intercalations of dark green phyllite and phyllitic siltstone, and subordinate lithic quartz wacke. The uppermost 100 m of the member become increasingly phyllitic and pyritic toward the top.

Conglomerate lenses and channels within the phyllitic marble reach thicknesses greater than 7 m and are both matrix- and class-supported with thick indistinct bedding and no evidence of grading. The dominant clast size is 0.5 to 1 cm and in order of relative abundance are: rounded to angular quartz pebbles, angular black marble pebbles and cobbles, subrounded quartzite and siltstone pebbles, argillite pebbles, and rare coral fragments. The marble clasts are lithologically similar to those of the enclosing rocks and tend to be larger and more angular than the siliciclastic clasts. Matrix composition is approximately equivalent to that of the quartz wacke found as interbeds throughout the lower unit. One fossil suite containing a gastropod mold and several rugose corals was recovered but could not be dated. Correlative units east of the Dalton Highway (Fig. 1; location 5) have yielded an Early to middle Middle Ordovician conodont (Dillon et al., 1987, 1988).

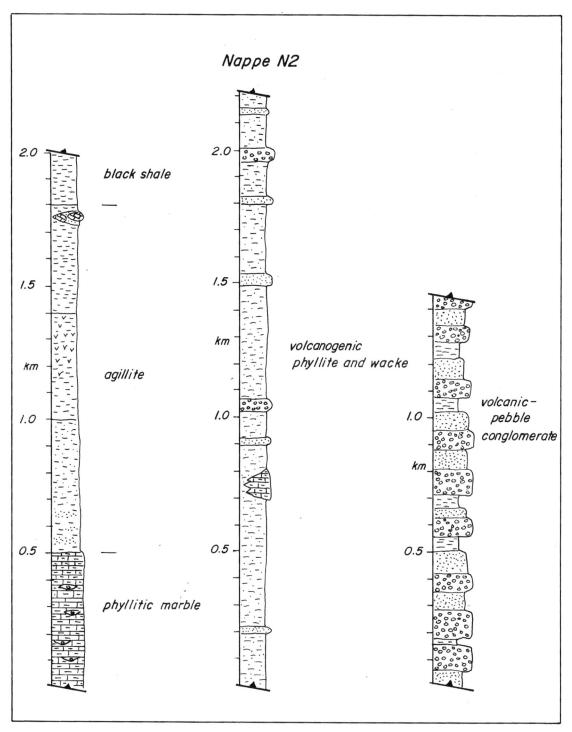

Figure 3. Reconstructed columns for thrust nappe N2. Lithologic symbols are shown on Figure 2.

Argillite unit. The argillite unit consists of about 1.3 km of a poorly exposed, fining upward sequence of argillaceous quartz siltstone, phyllite, and volcanogenic argillite. Dolomite olistoliths several meters in diameter are found in channelized carbonate conglomerate bodies in the upper third of the unit.

Thin-bedded black siltstone becomes increasingly fine grained and argillaceous upward, eventually grading into black silty phyllite and porcelaneous argillite. Relict cross-bedding, graded beds, and laminations were observed in outcrop and in thin section. The lower 500 m are dominantly phyllitic siltstone composed of quartz silt, fine-grained quartz sand, pyrite, feldspar, and phyllosilicates. Laminations, climbing ripples, tabular cross-beds, and graded beds are locally preserved. The middle 400 m are silty, locally pyritic phyllites exhibiting laminations and thin-bedded, fining-upward sequences. The uppermost 400 m consist of black slaty phyllite containing distinctive intervals of porcelaneous siliceous argillite whose constituents are suggestive of a volcanogenic protolith.

Two dolomite olistoliths occur in a zone of channelized carbonate lithoclastic conglomerate about 80 m from the top of the unit. The olistoliths are about 25 m in diameter and protrude prominently from the surrounding phyllite. They are composed of laminated, patchy microcrystalline dolomite containing abundant soft sediment structures. The black, siliceous dolomite weathers mottled orange-brown and contains a fauna of gastropods, brachiopods, and rugose corals. Internal bedding is contorted and discordant to that of the surrounding phyllites and lenticular bodies of carbonate lithoclastic conglomerate that envelope the olistoliths and pinch out laterally over a distance of about 50 m. The conglomerate is composed of poorly sorted subangular dolomite and micritic marble clasts ranging in size from pebbles to boulders. No fossils were found in the argillite, but the fossil suite within the olistoliths at location 6 (Fig. 1) is post–Early Ordovician in age (Moore et al., 1952).

Slate unit. The slate unit consists of about 180 m of black phyllitic slate. The basal contact is gradational with the underlying argillite and the slate is noncalcareous and contains as much as 15% quartz silt. No fossils were recovered from this unit.

Volcanogenic phyllite and wacke unit. This unit has a tectonic thickness of as much as 2.2 km and consists of volcanogenic wacke, siltstone, and phyllite with subordinate arenite and conglomerate interbeds. A laterally discontinuous horizon of argillaceous, thin-bedded marble is exposed in the eastern parts of the unit where the unit becomes coarser and interfingers with and grades into the volcanic pebble conglomerate unit (described below).

Lithologically, the succession is comprised of interbedded, green weathering, thin- to medium-bedded siltstone and phyllite containing relict graded bedding and cross-beds. Lithic and quartz arenite, siltstone, and wacke are often rhythmically interbedded with phyllite, and frequently show relict cross stratification, grading, and laminations. Fine-grained lithic conglomerate with a matrix composed of volcanogenic, lithic, and quartz sand and silt is present but decreases in frequency westward. A laterally restricted carbonate interval of as much as 100 m of orange-brown weathering black marble apparently interfingers with siltstone and phyllite. The marble is laminated to medium bedded, argillaceous and fine crystalline, and has phyllitic partings.

Volcanic-pebble conglomerate unit. The volcanic-pebble conglomerate unit consists of polymictic and monomictic volcanic and lithic pebble conglomerate interbedded with volcanogenic arenite, wacke, and phyllite. This lithofacies is exposed in two thrust sheets and has a tectonic thickness of at least 1.45 km. The conglomerate is medium to thick bedded and contains lenses and interbeds of medium- to thick-bedded lithic wacke and phyllite. Both matrix- and class-supported conglomerate grade laterally and vertically into wacke. Most conglomerate is polymictic although occasional monomictic conglomerate beds occur. Abundant cut and fill surfaces, cross-stratification, and graded, nongraded, and reverse-graded sequences are preserved. Internally chaotic matrix- and clast-supported debris flows as much as 3 m thick are most common. Graded and nongraded matrix- and clast-supported conglomerate is observed with the percentage of matrix increasing upward in the section. The conglomerate has sharp and erosive lower contacts but channels are not well developed and occasional fining-upward sequences are found.

The conglomerate clast composition varies between exposures in the two thrust imbricates. The northern imbricate has a structural thickness of at least 900 m and is characterized by volcanic and quartz pebble conglomerate and wacke. Clasts range in size from coarse sand (1 mm) to medium boulders (30 cm) in diameter. More than half of the clast population is composed of intermediate volcanic and volcanogenic rock fragments, quartz, and quartzite pebbles, with the remainder made up of lithic wacke, quartz wacke, quartz siltstone, and phyllite clasts. Carbonate and chert clasts occur locally and angular phyllite flags form the largest angular blocks, with the volcanic and lithic clasts being smaller and more rounded.

Volcanic sandstone, wacke, and phyllite are interbedded with the conglomerate. The sandstone comprises about 20 to 30% of the section. Bedding is variable, but 1- to 3-cm-thick beds are most common. Some ripples and normal grading are observed. Sand composition is similar to the matrix composition of the conglomerates, and grains are angular and generally poorly sorted. Crudely graded, argillaceous, silty, and sandy wacke is poorly sorted and occurs in thin to medium beds. Thin beds and partings of silty or sandy phyllite occur throughout the sections and increase in frequency and thickness toward the west

No fossils have been found in the transect area, but similar polymictic volcanic conglomerate 5 km to the east is assigned an Ordovician(?) to Cambrian(?) age by Dillon et al. (1987, 1988). The age assignment is based on the inferred stratigraphic position of the conglomerate below an Lower to middle Middle Ordovician carbonate horizon (Fig. 1; location 5). Elsewhere in the region (nappe N4), lithologically similar carbonate units contain conodonts with ages as old as late Early to Middle Ordovician (Dillon et al., 1987).

Thrust nappe N3.

Thrust nappe N3 underlies much of the eastern Skajit allochthon from the Chandalar River south to the vicinity of the Minnie Creek thrust and is exposed west of the Dalton Highway in the Glacier River drainage (Fig. 1). Nappe N3 and overlying nappes N4, N5, and N6 are folded in a post-imbrication, east-northeast–trending synform whose northern flank is cut by the Foggytop thrust that divides the Skajit allochthon into two structural domains.

Nappe N3 is composed of a complex imbricate stack dominated by exposures of the Skajit marble with subordinate siliciclastic units (Brosgé and Reiser, 1964; Dillon et al., 1988). The structure and lithology of nappe N3 was not studied in detail but based on the results of previous work (Brosgé and Reiser, 1964; Dillon et al., 1988) and our reconnaissance, clearly contains the same lithologies as those in nappe N1 (Fig. 2). The prominent marble exposures, forming the precipitous cliffs that characterize the eastern Skajit allochthon, consist of white, gray, and black massive and foliated marble that weathers a mottled gray and tan. The upper part of the marble is composed in part of black fine-crystalline, thick-bedded limestone and dolomite. The foliated marble commonly contains thin interbeds of phyllite and in some locations, schist. Siliciclastic intervals several tens of meters thick, consisting of phyllite and schist, occur within the predominantly marble succession, but their stratigraphic position is unclear. A thick succession of interbedded siliciclastic rocks, which locally contains thin carbonate beds, overlies the Skajit marble. The siliciclastic rocks are composed of calcareous and noncalcareous phyllite, siltstone, and sandstone.

Fossil suites collected from the marble of nappe N3 consist of solitary corals, brachiopods, pelecypods, stromatoporoids, and conodonts (Brosgé and Reiser, 1964; Dillon et al., 1988; Oldow, unpublished data). Recovered conodont fragments yield little useful age constraint and have an allowable range of Ordovician to Triassic (Dillon et al., 1988). Megafossil assemblages from the top of the carbonate unit (e.g., Fig. 1; location 4) have ages of Middle to early Late Devonian (Dillon et al., 1988; Brosgé and Reiser, 1964). The lower age bond for the Skajit marble, however, is given by an early Middle Cambrian trilobite of Siberian affinity (Palmer et al., 1984) collected in a thin carbonate bed lying immediately below more massive marble units forming the precipitous northern flank of Mount Snowden (Fig. 1, location 1). Our interpretation of the stratigraphic significance of the Cambrian fauna differs from that of Dillon et al. (1987, 1988) who assigned it to a different map unit. We argue that the faunal assemblage was collected from the basal member of the Skajit marble. We base this interpretation on the lithologic similarity of the clastic and carbonate rocks exposed at Mount Snowden, where the fossils were collected, and those near the base of the Skajit marble in Nappe N1 along our map transect (Figs. 1 and 2).

Thrust nappe N4

The rocks of nappe N4 correspond lithologically to and have the same structural position as those of nappe N2 to the north and are interpreted as a structural repetition of the lower nappe (N2). The tectonic contact between nappes N4 and N3 is traced from near the center of the Skajit allochthon south to the vicinity of the Minnie Creek thrust (Fig. 1). The bounding fault is deformed in a major west-southwest–plunging synform-antiform pair with a wavelength greater than 25 km. Nappe N4 underlies most of the south-central Skajit allochthon and generally is poorly exposed except along ridge tops. Due to subdued topography and the east-

northeast–trending folds, only about 1.8 km of structural section is exposed in nappe N4.

In the line of the transect, the lithologic succession is composed of three units of uncertain thickness. The lowest is composed of interbedded thin carbonate rocks and phyllite that is overlain by the middle unit, consisting of intercalated phyllite, siltstone, and sandstone. The highest lithologic unit is a black siltstone. The stratigraphic succession is pieced together from several blocks bound by steeply dipping, east-west–striking faults. Reconstructed minimum thicknesses are 650 m for the lower unit, whose base is not exposed, 900 m for the middle unit, which is not exposed in a single structural block containing upper and lower contacts, and 250 m for the upper unit (Fig. 4). By comparison with the lithologic succession of nappe N2, nappe N4 is inferred to be stratigraphically upright. No direct evidence exists to support this contention, however. No fossils were found in the succession, but by correlation with units immediately east of the Dalton Highway (Fig. 1; locations 12 and 13), they overlie carbonate rocks dated by conodonts as late Early to Middle Ordovician (Dillon et al., 1987, 1988) and, as such, the rocks are interpreted to be Middle Ordovician or younger.

Phyllite and carbonate unit. The lower unit is composed of interbedded gray phyllite, finely crystalline gray limestone, and micaceous limestone. The unit generally consists of the three lithologies interbedded on a scale of between 2 to 20 cm, with the crystalline limestone beds being slightly thicker than the other lithologies. The crystalline limestone often has light and dark bands less than 1 cm thick and weathers to a light tan. The proportion of the three lithologies varies greatly between isolated outcrops, commonly with one or two of the lithologies absent. There is no distinct pattern to the variation, but to the north phyllite appears to increase at the expense of crystalline carbonate.

Phyllite and sandstone unit. The middle unit of nappe N4 is composed entirely of siliciclastic rocks. The contact with the underlying calcareous unit is indistinct because of the abundance of phyllite in both units and the poor exposures. The middle unit appears to vary laterally as well as vertically but in general coarsens upward in two or more cycles. The cycles are characterized by an increase in sandstone and ultimately the appearance of conglomerate beds. Near the top of the unit, the grain size diminishes and forms a fining-upward succession as the contact with the overlying upper black siltstone unit (Fig. 4) is approached.

In the north, the lower part of the middle unit is composed almost entirely of phyllite with less than 5% siltstone. To the south, successive outcrops of the lower middle unit contain progressively greater amounts of coarser grained clastic rocks with as much as 50% siltstone and fine-grained sandstone, with subordinate medium- to coarse-grained sandstone beds. Most beds are between 0.5 and 5 cm thick with occasional coarse micaceous sandstone beds as much as 1 m thick. In the southern part of nappe N4, the lowest exposures of the middle member, which probably represent stratigraphically higher levels than those exposed in the north, are composed of rhythmically bedded (0.5 to 2 cm) gray, green, and black shales. The transition is gradational but the shales

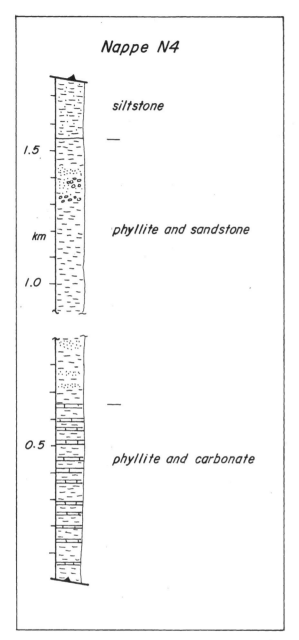

Figure 4. Reconstructed column for thrust nappe N4. Lithologic symbols are shown on Figure 2.

Thrust nappe N5

Nappe N5 is exposed in the core of the west-southwest–plunging synform that dominates the structure of the central Skajit allochthon (Fig. 1). The nappe contains 1.7 km of Devonian rocks and an undated structural sheet composed of quartzite (Fig. 5). The Devonian rocks are divided into four units: (1) a coarse clastic succession, (2) a lower calcareous phyllite that passes upward into, (3) a massive carbonate, and (4) an upper unit of calcareous phyllite. Biostratigraphic age control (Fig. 1; locations 6, 7, and 8) is not sufficient to establish the stratigraphic facing of the Devonian succession. Without contrary data, however, we assume that the lithotectonic unit is upright. Faunal assemblages and sedimentary structures are consistent with carbonate and siliciclastic shelf depositional environments characterized by variable terrigenous clastic input.

Coarse clastic unit. The coarse clastic unit consists of about 500 m sandstone and conglomerate that are capped by about 100 m of carbonate. The contact is obscured by talus but is thought to be conformable. The lower third of the carbonate is composed of banded (0.5 mm to 3 cm), finely crystalline marble intercalated with clean white calcarenite in beds as much as 3 m thick. The calcarenite is gradually replaced upsection by the banded marble, which passes upward into black fissile carbonate mudstone. The carbonate mudstone is about 60 m thick and passes upward into interlayered noncalcareous gray phyllite and fine-grained carbonate sandstone.

Lower calcareous phyllite unit. The transition between the coarse clastic unit and the overlying lower calcareous phyllite is obscured by cover, but the lithology above and below the contact zone is consistent with a gradational transition. The lower calcareous phyllite unit is comprised of about 800 m of poorly exposed calcareous phyllite, sandstone, fossiliferous lenses of argillaceous marble, and igneous sills, dikes, and flows. The igneous rocks are concentrated in the lower third of the unit and locally preserve relict flow banding. These rocks have been interpreted as felsic flows, tuff, lahar, and intrusives (Brosgé and Reiser, 1964; Dillon et al., 1986, 1987). Interbedded phyllite constitutes the bulk of the section but is poorly exposed. More commonly, a thin to medium bedded quartzose and calcareous argillite is exposed. Calcareous and noncalcareous, silty gray phyllite composes about 20% of the section. Thin to medium interbeds of cross-bedded calcareous quartz sandstone and calcarenite are interspersed throughout the section. Fossiliferous interlayers and biohermal lenses of argillaceous marble occur throughout but are most common in the upper third of the unit. Fossils are rugose corals, crinoid columnals, brachiopods, gastropods, and platyceratids with varying degrees of preservation.

Thin carbonate beds are fossiliferous, but the preservation varies substantially due to the degree of recrystallization. Dated fossils were not broken or abraded and apparently were autochthonous. Elsewhere in the unit, corals were observed to be resedimented and were not collected for dating. Brosgé and Reiser (1964) dated the rugose corals in one biohermal lens as of probable

are overlain by coarser grained clastics consisting of siltstone (60%), sandstone (20%), and gray phyllite (20%). Rare conglomerate beds as much as 1.0 m thick are found and contain highly flattened siltstone and phyllite clasts as much as 6 cm long. The succession fines upward, over a stratigraphic interval of about 250 m, to phyllite (35%) and fine-grained sandstone (15%).

Siltstone unit. The upper member is a black siltstone about 150 m thick. The black siltstone is siliceous and has a regularly spaced foliation of 0.5 to 3 cm and occasionally weathers reddish yellow. The unit is lithologically correlated with the black shale in nappe N2.

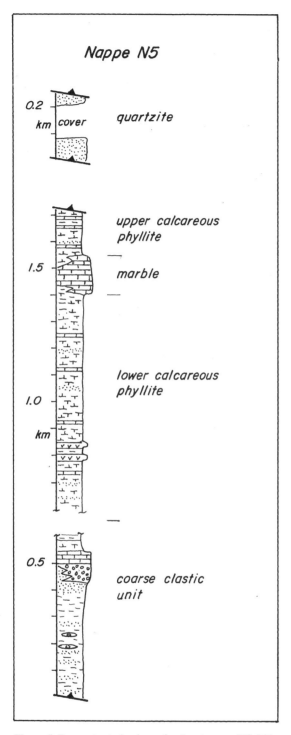

Figure 5. Reconstructed column for thrust nappe N5. Lithologic symbols are shown on Figure 2.

Frasnian age. One conodont assemblage is dated as Givetian to early Frasnian (Dillon et al., 1987). Fossil suites of brachiopods and gastropods collected in the lower member (Fig. 1; location 7 and 8) constrain the age to post-Siegenian to pre-Frasnian but are strongly suggestive of Eifelian to Givetian (Middle Devonian) age (R. Blodgett, written communication, 1985).

Marble unit. The marble unit forms a prominent exposure of mottled gray cliffs with blocky talus slopes. The massive marble is fossiliferous and varies in thickness between 40 to 150 m. Bedding is generally thick but variable and commonly indistinct. Abundant brachiopods, bryozoans, rugose corals, gastropods, and crinoid columnals are observed in the less recrystallized zones. Generally, fossils are poorly preserved in beds of packstone and wackestone interbedded with cross-bedded calcarenite.

Fossil suites collected from the unit (Fig. 1; location 9) include gastropods, solitary and colonial rugose corals, and brachiopods indicative of Eifelian to Givetian (Middle Devonian) age (R. Blodgett, written communication, 1985). Elsewhere in the map unit, Brosgé and Reiser (1971) recognized Devonian to early late Devonian corals, brachiopods, and stromatoporoids.

Upper calcareous phyllite unit. The upper calcareous phyllite unit is composed of interbedded calcareous phyllite, wacke, and thin marble lenses. The maximum preserved thickness beneath the basal thrust of nappe N6 is 150 m. Interbedded calcareous and siliceous, silty phyllite constitutes most of the section. Fossiliferous, fine-crystalline, argillaceous marble lenses often contain coral debris, relict upright burrows, and cross-beds. Carbonate buildups contain colonial and solitary rugose corals and gastropods. No age-diagnostic fossils were recovered from this unit, but by its stratigraphic position, it is Middle or possibly Late Devonian in age.

Quartzite unit. The poorly exposed quartzite is composed of about 250 m of interbedded quartzite, sandy siltstone, and quartz wacke. The unit contains no fossils and is undated. The sands commonly contain tabular and small trough stratification, laminae, and climbing ripples. Infrequent graded bedding and pinch and swell structures are also observed. In thin section, the quartzite is a sucrosic quartz arenite with accessory feldspar.

Thrust nappe N6

Rocks of nappe N6 form the klippe exposed as the highest lithotectonic unit in the eastern Skajit allochthon. Approximately 1.8 km of structural section was mapped and the stratigraphic facing of the lithologic succession is unknown. The lithotectonic assemblage is divided into three map units: a basal calcareous phyllite, a phyllitic marble, and an upper section of argillite (Fig. 6). The rocks of the klippe are more strongly deformed than the underlying nappe N5 and primary sedimentary structures are almost entirely obliterated by the tectonic fabric.

The rocks of the klippe (Fig. 1) were previously mapped as Devonian(?) although no fossils were reported (Brosgé and Reiser, 1964, 1971; Dillon et al., 1986, 1987). Argillaceous rocks, roughly corresponding to the upper unit described here, were assigned a Cambrian age and were suggested by Brosgé and Patton (1982) to be lithologically similar to some of the lower Paleozoic rocks exposed in the Doonerak window farther north. The correlation served as an important stratigraphic link between the North Slope assemblages and the southern Brooks

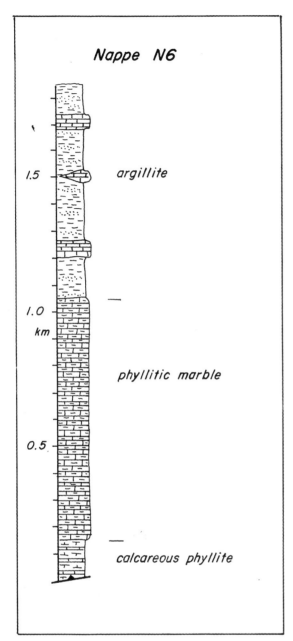

Figure 6. Reconstructed column for thrust nappe N6. Lithologic symbols are shown on Figure 2.

Range. Work in the Doonerak window (Julian and Oldow, this volume, Chapter 5) and in the klippe cast doubt on this correlation, but meager fossil evidence does suggest a lower Paleozoic age for the rocks.

Calcareous phyllite unit. The lower units consists of 80 to 150 m of argillaceous, fine-crystalline marble, intercalated with schistose calcwacke, and calcareous phyllite. This lower unit is cut out to the north by the basal thrust, and the upper contact is bedding parallel and lithologically gradational with the overlying succession. The dominant lithology of the lower unit is argillaceous marble with subordinate phyllitic intercalations. Calcare-

ous, silty phyllite is found throughout the section. No fossils were observed in the lower unit.

Phyllitic marble unit. The phyllitic marble is 900 m thick and crops out as a north-dipping sequence on the south limb of the folded klippe. The unit consists of a sequence of dark, finely crystalline argillaceous marble with phyllitic partings. Thin interbeds of calcareous quartz siltstone and calcareous phyllite comprise less than 5% of the section. The unit is regularly bedded with a maximum thickness of 10 cm. Relict, graded sequences are preserved sporadically but in general sedimentary structures are obscured by recrystallization. Several conodont forms and fragments (Fig. 1; location 10A and 10B) of pre-Devonian morphology were recovered (R. Tipnis, written communication, 1985) and on this basis the rocks are interpreted to be of lower Paleozoic age.

Argillite unit. The upper unit is composed of more than 800 m of siliceous argillite, quartzite, and marble. The lower contact occurs at a largely covered interval 30 m thick where the lithology changes from carbonate to siliciclastic. The contact is bedding parallel and is interpreted to be stratigraphic. The lithology of the upper unit is characterized by a heterogeneous sequence of intercalated quartzite, quartz wacke, siliceous argillite, phyllite, and argillaceous marble, which becomes more abundant upward. Strongly deformed 20- to 50-m-thick intervals of silty, carbonaceous and quartzose marble form mappable horizons within the otherwise poorly exposed sequence. Bedding, where recognizable, is generally thinner than 5 cm in the lower 600 m of section. The upper 200 m of section is a thick to thinly bedded sequence of calcareous clastics, fine-grained quartzite, and coarse quartzose siltstone. The argillite unit sporadically exhibits relict cross-beds in quartzite layers and laminations, and small-scale cross-stratification are locally distinguishable.

Several brachiopods and coral fragments were collected from an argillaceous quartzite near the base of the unit and constitute the only known megafossils from the klippe. Conodont forms and fragments of pre-Devonian morphology were recovered from carbonate rocks (R. Tipnis, written communication, 1985). Poorly preserved atrypid(?) brachiopods recovered from the basal part of the member (Fig. 1; location 11) suggest a post–Early Ordovician age (Moore et al., 1952). The upper member is tentatively interpreted to be Ordovician(?) and/or Silurian(?) in age.

Thrust nappe N7

Thrust nappe N7 is exposed at the southern margin of the Skajit allochthon (Fig. 1) where the Minnie Creek thrust carries amphibolite-facies metamorphic rocks of the Schist belt over the lower Paleozoic carbonate and siliciclastic succession. The rocks of nappe N7 are composed of siliciclastic and subordinate carbonate metasedimentary units that, unlike units to the north, are strongly schistose. They are substantially higher grade than rocks to the north, and mineral thermometry indicates that they have been metamorphosed at temperatures of as much as 500° C (Gottschalk et al., this volume, Chapter 12). Segregation of nappe N7 rocks from those of the overlying Schist belt is based on the following observations: (1) where the schistose rocks of nappe

N7 have coarse-grained clastic protoliths, sedimentary grains and pebbles are clearly preserved in sharp contrast to the wholly recrystallized nature of the Schist belt rocks; (2) the metasedimentary rocks of nappe N7 show little metamorphic segregation of quartz and mica-rich layers and synmetamorphic quartz and/or carbonate veinlets are rare, whereas in the Schist belt highly deformed quartz veinlets are ubiquitous and quartz-mica schists have undergone extensive metamorphic segregation.

Lithologically, the schists of nappe N7 have protoliths similar to those exposed in nappe N4 (Fig. 4) and differ only in degree of metamorphic recrystallization. Three units are recognized in two thrust sheets comprising nappe N7. The lower sheet contains a quartz-mica schist, overlain by banded chloritic quartzite and quartz-mica phyllite. The upper thrust sheet is composed of quartz-mica schist and marble.

The quartz-mica schist is a heterogeneous unit composed of a variety of metaclastic rocks that provide few continuous marker horizons. The main rock type is a brown-weathering, calcareous gray quartz + muscovite + chlorite + albite + garnet + biotite schist. The schist locally contains tiny lenses of orange-brown ferruginous calcite and/or dolomite, and in many areas the schist itself is also calcite bearing. Thin gray to black, or white and black mottled marble horizons as much as 2.5 cm thick are occasionally present, as are thin layers (as much as 3 cm) of black to dark-gray micaceous marble.

Fine-grained, limy schistose rocks are locally interlayered with immature sandstone and grit intervals ranging from 5 to 10 cm thick. Coarse clastic rocks are markedly less mica-rich than the associated quartz-mica schists, and weather readily, producing a distinctively banded outcrop pattern. They are made up primarily of well-sorted, angular to subangular quartz grains, feldspar, and lithic grains, with occasional carbonate grains or clasts. Thin (less than 1 m thick), discontinuous conglomeratic layers are exposed at several locations. Conglomerates are matrix supported and are made up of subrounded quartz (chert?) and minor lithic pebbles as much as 7 mm in diameter.

Banded quartzite forms a distinctive map unit and contains several lithologies. The most widespread lithology consists of quartz + white mica + albite + chlorite and a low birefringent mineral (quartz or albite) forming pseudomorphs after amphibole. This rock type forms a massive, blocky weathering interval within the quartz-mica schist ranging from 200 to 260 m thick. A mineral lineation defined by pseudomorphs after amphibole is characteristic of this unit. Less common is a banded pink quartzite that differs from the main body in that it does not contain chlorite and is dominated by quartz and muscovite. This unit has a variable thickness as much as 250 m and may be gradational with the banded quartzite.

Thinly layered (5 to 15 cm) white-, gray-, and orange-weathering marble, locally interfoliated with a phyllitic or schistose mica-rich interval, forms a unit as much as 270 m thick. The marble probably has gradational contacts above and below with the more calcareous parts of the quartz-mica schist. The marble is often spotted in appearance, consisting of dark gray nodules in a light gray or white matrix. A number of small conglomerate lenses 5 to 10 cm thick are present within the marble and consist of angular to subrounded lithic pebbles 2 to 20 mm in diameter.

No fossils were found in nappe N7. The protolith of the metamorphic rocks are lithologically similar to those exposed in the southern part of nappe N4 and we correlate the two successions.

STRATIGRAPHY AND DEPOSITIONAL SETTING

The eastern Skajit allochthon is composed of two partially coeval stratigraphic sequences of Cambrian through Devonian age. The structurally lower sequence, exposed in the northern and central part of the allochthon, consists of the thick, cliff-forming carbonate and interbedded and overlying siliciclastic rocks of thrust nappes N1 and N3. This sequence is dominated by carbonate rocks deposited in relatively shallow water conditions typical of shelf environments and accordingly is referred to as the shelf sequence. The structurally higher stratigraphic assemblage composing nappes N2, and N4 to N7 is composed of fine- to coarse-grained siliciclastic rocks interbedded with thin carbonate rocks and locally contains thin greenstones of volcanic and intrusive origin. These rocks form a lithologic succession typical of distal shelf and substorm wave-base environments of deposition and are termed the basin sequence in the following discussion.

Shelf sequence

The carbonate and siliciclastic rocks of the shelf sequence are divided into a lower assemblage characterized by carbonate and thin interbeds of siliciclastic rocks and an upper assemblage of fine-grained siliciclastic and subordinate thin-bedded carbonate rocks. The upper age limit of the lower carbonate assemblage is well established as early Late Devonian (Brosgé and Reiser, 1964, 1971; Dillon et al., 1987). The lower age limit is not well constrained but appears to be at least as old as early Middle Cambrian, but the possibility that the succession passes into the Proterozoic cannot be discounted. Thick carbonate units that we include in the Skajit marble contain Late Ordovician to Silurian conodonts (Dillon et al., 1987).

Although largely obscured by metamorphic recrystallization, sedimentary structures and fossil assemblages are consistent with deposition in a shallow-marine, platformal environment (Brosgé and Reiser, 1964; Henning, 1982; Boler, 1989). Siliciclastic rocks composing thin interbeds and discrete stratigraphic horizons (Fig. 2) within the carbonate assemblage are dominantly fine-grained but locally contain channels filled with lithic, quartz, and chert pebble conglomerate. Few diagnostic sedimentary structures have survived deformation and metamorphism, but the clastic rocks are typical of deposition in shelf or deltaic environments interspersed with the dominant carbonate sedimentation.

The thick siliciclastic succession overlying the carbonate rocks is poorly dated. A few poorly preserved fossil fragments indicate a post–Early Ordovician age for the siliciclastic rocks and corals are dated variously as Devonian and Devonian to Silurian (Brosgé and Reiser, 1964). The siliciclastic succession overlies lower Upper Devonian rocks at the top of the subjacent carbonate assemblage conformably and as such are thought to be Late Devonian and younger.

The depositional setting of the upper siliciclastic succession is poorly understood. The predominance of phyllite suggests deposition below storm wave base conditions, possibly in an outer shelf or basin environment, but water depth cannot be determined conclusively. The abrupt transition from the shallow-marine carbonate conditions of the underlying rocks to deposition of non-calcareous phyllite points to substantial increase in relative water depth, possibly related to a major sequence boundary. The siliciclastic rocks become more calcareous in the upper half of the succession and near the top of the exposed section thin interbeds of carbonate appear. The upward increase in carbonate content may indicate either shoaling through time or increase of carbonate supply via mass transport.

The sparse fossil assemblages in nappe N1 and N3 suggest that the rocks represent the entire lower Paleozoic section. The succession contains no evidence of significant internal unconformities but probably contains numerous unrecognized sequence boundaries. The Cambrian through early Late Devonian sequence is characterized by shallow-marine, probably platformal, carbonate environments periodically interrupted by siliciclastic deposition. The carbonate environment apparently was drowned in the Late Devonian when more basinal, siliciclastic conditions prevailed. The upper age limit of the clastic rocks is unknown.

Basin sequence

Within the limits of the sparse fossil age control, the basin sequence (nappes N2, N4, N5, N6, and N7) contains rocks of Cambrian(?) through early Late Devonian age. The succession is composed predominantly of fine- to coarse-grained siliciclastic and volcanogenic rocks that are variably calcareous and intercalated with thin-bedded carbonate rocks.

No pre-Ordovician fossils have been recovered from the basin sequence, but Cambrian(?) and Ordovician(?) ages for many of the units are permissible based on stratigraphic position. Within the line of our transect, few age-diagnostic fossils were recovered, but fortunately, equivalent rocks along strike have been dated in several locations (Dillon et al., 1987, 1988). Where age controls are lacking, provisional lithologic correlations are made (Fig. 7).

A Middle to early Late Devonian age is well established for most of the rocks composing nappe N5. The dated succession (Fig. 7) is composed of fine-grained calcareous clastic rocks, thin-bedded limestone, and local interbeds of intermediate volcanic rocks near the base (Fig. 5). The rocks are abundantly fossiliferous with shallow water corals, brachiopods, and gastropods. Sedimentary structures within both the siliciclastic and carbonate rocks indicate sedimentary transport and substantial reworking before lithification. In light of the faunal assemblage and sedimentary structures, a shelf environment of deposition is visualized for the upper part of this lithotectonic assemblage.

Undated fine- to coarse-clastic rocks and resedimented carbonates are interpreted to underlie the dated clastic and carbonate assemblage of nappe N5. The lower unit is composed of a coarsening upward succession of argillite, sandstone, and conglomerate, which is overlain conformably by a marble member composed of resedimented carbonate clastics. Here, as elsewhere, water depth determination is equivocal, but based on the few preserved sedimentary structures, the siliciclastic component is thought to represent mass-flow deposition. The lower succession may represent either part of a shoaling-upward section in transition with the overlying carbonate and siliciclastic rocks or alternatively nonmarine to shallow-marine siliciclastic environments passing into dominantly carbonate conditions.

Within nappe N5, a structural sheet composed of an undated quartzite is unique in the basin sequence. Its relations to other lithologic units is not known. Sedimentary structures within the sandstone and sandy siltstone indicate tractive and suspension transport mechanisms during sedimentation (Boler, 1989). Deposition apparently occurred in relatively high-energy conditions.

Most of the siliciclastic rocks in exposures of nappes N2 and N4 are assigned Cambrian(?) and Ordovician(?) and younger ages. No age-diagnostic fossils have been recovered from these rocks (along the transect) but a trilobite cephalon of post–Early Ordovician age was recovered from near the top of one section in nappe N2 (Fig. 7). East of the Dalton Highway, a succession of thinly interbedded carbonaceous, calcareous phyllite, minor dolomitic mudstone, and subordinate thin-bedded carbonate rocks has yielded late Early to Middle Ordovician conodonts (Dillon et al., 1987, 1988) in several locations (Fig. 1; locations 11, 12, and 13) both in nappes N2 and N4. The dated rocks overlie the volcaniclastic rocks of nappe N2 and locally intertongue with and underlie the fine-grained siliciclastic and carbonate rocks in nappes N2 and N4.

The lithologic assemblages in nappes N2 and N4 are similar and are interpreted as lateral equivalents, at least in part. Good lithologic correspondence exists between the northern succession exposed in nappe N2 and the reconstructed stratigraphic section exposed in nappe N4 (Fig. 7). In nappe N4, a vertical succession of noncalcareous phyllite and phyllitic marble passes upward into noncalcareous argillite containing interbeds of siltstone, sandstone, and conglomerate, which is overlain by black siltstone. Similar rocks are exposed in the northern part of nappe N2, where a lower phyllitic marble unit, containing interbeds of quartz wacke and siltstone and a few lenses of conglomerate, grades upward into argillite with thin interbeds of siltstone and porcelaneous siliceous argillite with a probable volcanogenic protolith. Near the top of the argillite unit, olistoliths of shallow-water carbonate are found. The argillite is overlain by black silty shale.

Sedimentary structures in fine- to coarse-grained siliciclastics indicate both tractive and suspension mechanisms during sedimentary transport. Conglomerate is found in different parts of the two successions and may represent debris flows and channelized mass-flow deposits. The lithologies are suggestive of upper or mid-fan facies in a basin and the presence of shallow-water olistoliths suggests relatively steep topographic gradients, at least locally, between basin and carbonate shelf environments.

122 J. S. Oldow and others

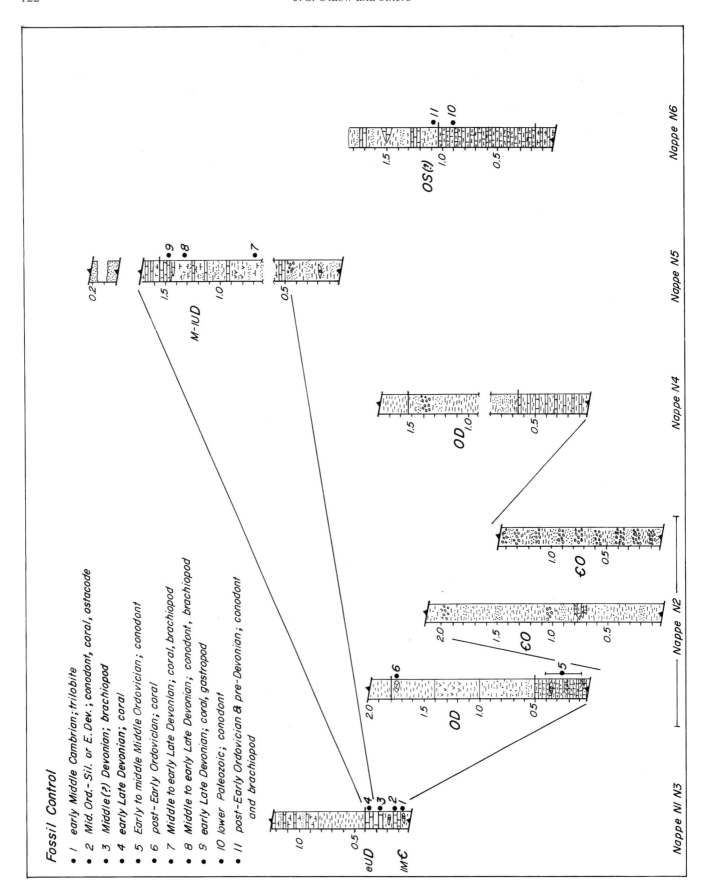

Figure 7. Stratigraphic correlation of lithotectonic units of the eastern Skajit allochthon. Section thicknesses measured in kilometers. Age ranges for lithologic sections: eUD-early Upper Devonian; lM-C-lower Middle Cambrian; OD-Ordovician-Devonian; CO-Cambrian-Ordovician; M-lUD-Middle-lower Upper Devonian; OS-Ordovician-Silurian.

In nappe N2, a thick assemblage of volcanogenic conglomerate and volcanogenic phyllite and wacke structurally overlies the fine-grained siliciclastic and carbonate rocks. No fossils have been recovered from this unit, but north of Mount Snowden (Fig. 1) the unit is interbedded with fine-grained siliciclastic and carbonate rocks overlain by upper Lower to Middle Ordovician rocks (Dillon et al., 1988). In the transect area, the thick volcanogenic conglomerate is observed to pass laterally (east to west) into the finer grained phyllite and wacke. The coarse- and fine-grained successions are thought to represent a proximal subaqueous debris apron associated with a volcanic highland and its more distal equivalent, respectively. The relationship between the volcanogenic rocks and the structurally underlying fine-grained clastic and carbonate succession is obscured by thrust faults. Based on mapping farther east by Dillon et al. (1988), however, the volcanogenic clastics lie near the bottom of the stratigraphic section and may represent some of the oldest parts of the Paleozoic section.

The fine-grained siliciclastic and carbonate rocks in nappe N6 are dated only as lower Paleozoic on the basis of poorly preserved conodont fragments (Boler, 1989). Lithologically, it is conceivable that the upper argillite and underlying phyllitic marble correspond stratigraphically to the two lower units of nappe N4 and the northern succession in nappe N2 (Fig. 1). The sedimentary structures, lithology, and fossil fragments are all consistent with basinal environments of deposition.

PALEOGEOGRAPHIC RECONSTRUCTION

Restoration of the original spatial distribution of the upper and lower stratigraphic sequences composing the Skajit allochthon requires removal of the effects of Brookian contraction. This is accomplished in light of the structural relations outline in Chapter 8 (Oldow et al., this volume).

The reconstruction, presented in Figure 8, places the carbonate and siliciclastic rocks of the shelf sequence in a position north of the basin sequence. For much of the lower Paleozoic (Cambrian through Devonian) the shelf and basin had separate but seemingly related depositional histories. The original proximity of the two sequences is not known, but they may not have been separated by more than a few tens of kilometers, as suggested by the occurrence of shallow-water carbonate olistoliths within basinal rocks of nappe N2.

The carbonate shelf lay to the north and supplied carbonate detritus to the basin as indicated by the abundance of thin-bedded carbonate and marl intercalations within the basin sequence. For most of the lower Paleozoic, volcanic and terrigenous debris was supplied to the basin, possibly during sea-level lowstands, from an undetermined direction. The continentally derived component

of the siliciclastic rocks in the basin sequence is similar in composition to the thin terrigenous clastic interbeds in the marbles of the shelf sequence and suggests a related source region, probably to the north. The prevalent volcanic and volcanogenic component of the basin sequence is subdued or lacking in the shelf sequence, however. In the Cambrian(?) and Ordovician, a nearby volcanic highland existed and supplied coarse clastic sediments now residing in nappe N2. Stratigraphically overlying siliciclastic rocks are finer grained than the volcanogenic succession and contain abundant continentally derived components, but volcanic constituents still exist as pointed out by Anderson (1987). It is unlikely that the source region for the volcanic debris lay to the north where carbonate shelf conditions prevailed. Rather, the inferred magmatic source probably was on the southern flank of the basin.

A shallowing upward transition is suggested in the upper part of the basin sequence. In nappe N5, undated coarse- and fine-grained siliciclastic rocks of inferred basinal affinity are interpreted to depositionally underlie Middle to lower Upper Devonian shallow water carbonate and siliciclastic strata (Fig. 5). During Middle to early Late Devonian time, shallow marine carbonate conditions persisted in the shelf sequence. Conceivably, the shoaling-upward succession in the basin reflects southerly progradation of the carbonate shelf over the old basin-shelf transition.

The late Devonian and younger history of the basin sequence is not preserved. In the shelf sequence, on the other hand, a thick succession of fine-grained siliciclastic deposition suggests a significant relative rise in sea level during the Late Devonian. The increase in water depth may correspond to global eustatic fluctuations for the Late Devonian (Harland et al., 1989) or could be the product of subsidence associated with the onset of Devonian tectonism.

The upper siliciclastic succession of the shelf sequence is lithologically similar to rocks of the middle to late Late Devonian Beaucoup Formation (Dutro et al., 1979) found in the Endicott Mountains allochthon (Phelps and Avé Lallemant, this volume, Chapter 4; Handschy, this volume, Chapter 2) which structurally underlies the Skajit allochthon. The post-lower Upper Devonian siliciclastic rocks are tentatively correlated with the Beaucoup Formation and as such supply a possible stratigraphic link between the Endicott Mountains allochthon and the Skajit allochthon.

CONCLUSIONS

The eastern Skajit allochthon is comprised of numerous thrust sheets containing rocks of at least Cambrian through early Late Devonian age. Reconstruction of the major structural units allows recognition of two stratigraphic sequences. The two sequences are at least partially age equivalent and were juxtaposed during imbrication and assembly of the Skajit allochthon. The structurally lower sequence, exposed in the northern and central parts of the allochthon, is composed of a lower Middle Cambrian through lower Upper Devonian carbonate shelf or platform assemblage overlain conformably by deeper water,

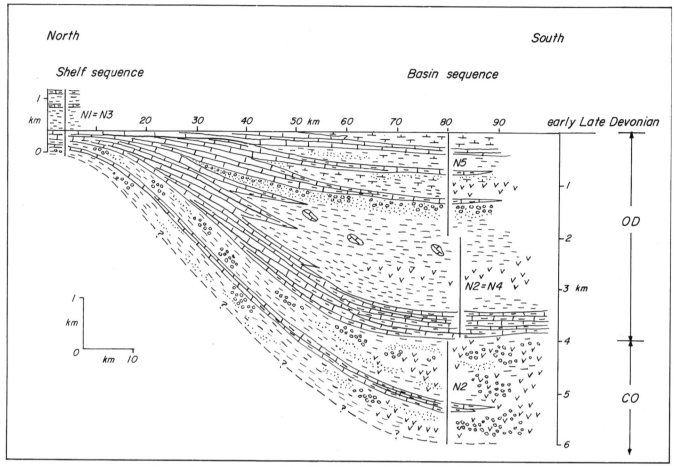

Figure 8. Schematic representation of the depositional setting envisioned for the shelf and basin
sequences of the eastern Skajit allochthon. Heavy vertical lines indicate the position of structural sections (Figs. 2 through 6) used in reconstruction.

fine-grained siliciclastic rocks of Late Devonian and possibly
younger ages. Structurally above the lower stratigraphic
sequence are Cambrian(?) through Middle to early Late Devonian age fine- to coarse-grained siliciclastic and volcanogenic
rocks, locally interbedded with thin carbonate units. The structurally higher sequence is interpreted as the product of a basin
deposition that shoaled to shelf conditions in the mid-Devonian.

Paleogeographic reconstruction of the stratigraphy suggests
that the lower Paleozoic rocks of the Skajit allochthon are part of
a linked shelf-basin system. The carbonate shelf, which lay to the
north, supplied carbonate detritus to the basin. Continentally
derived terrigenous clastic rocks, both in the basin and shelf,
were probably shed from the same source region. Volcanic and
volcanogenic clastic rocks within the basin are not found in the
shelf sequence and probably had a source to the south.

ACKNOWLEDGMENTS

The authors are indebted to R. Dunbar and R. Nilsen for
their careful reviews of this manuscript. This research was
partially funded by the Department of Energy (DE-AS05-
83ER13124), the National Science Foundation (EAR-8517384,
EAR-872017, EAR-9017835), and the Industrial Associate Program at Rice University.

REFERENCES CITED

Anderson, A. V., 1987, Provenance and petrofacies of the Endicott and Hammond
        terranes: Philip Smith Mountains and Arctic Quadrangles, Brooks Range,
        Alaska: Geological Society of America Abstracts with Programs, v. 19,
        p. 354.
Boler, K. W., 1989, Stratigraphy, structure, and tectonics of the central Brooks
        Range, near Dietrich Camp, Alaska [M.A. thesis]: Houston, Texas, Rice
        University, 163 p.
Brosgé, W. P., and Dutro, J. T., Jr., 1973, Paleozoic rocks of northern and central
        Alaska, *in* Pitcher, M. G., ed., Arctic geology: American Association of
        Petroleum Geologists Memoir 19, p. 361–375.
Brosgé, W. P., and Patton, W. W., Jr., 1982, Regional bedrock geologic maps
        along the Dalton Highway, Yukon Crossing to Toolik, Alaska: U.S. Geo-
        logical Survey Open-File Report 82-1071, 18 p., 2 sheets, scale
        1:500,000.
Brosgé, W. P., and Reiser, H. N., 1964, Geologic map and section of the
        Chandalar Quadrangle, Alaska: U.S. Geological Survey Miscellaneous
        Geologic Investigations Map 1375, 1 sheet, scale 1:250,000.
Brosgé, W. P., and Reiser, H. N., 1971, Preliminary bedrock geologic map:
        Wiseman and eastern Survey Pass Quadrangles, Alaska: U.S. Geological

Survey Open-File Map 479, 2 sheets, scale 1:250,000.

Dillon, J. T., Harris, A. G., Dutro, J. T., Jr., Solie, D. N., Blum, J. D., Jones, D. L., and Howell, D. G., 1988, Preliminary geologic map and section of the Chandalar D-6 and parts of the Chandalar C-6 and Wiseman C-1 and D-1 Quadrangles, Alaska: Alaska State Division of Geological and Geophysical Survey, Report of Investigations 88-5, scale 1:63,360.

Dillon, J. T., Harris, A. G., and Dutro, J. T., Jr., 1987, Preliminary description and correlation of lower Paleozoic fossil-bearing strata in the Snowden Mountain area of the south-central Brooks Range, Alaska, *in* Tailleur, I., and Weimer, P., eds., Alaska North Slope geology: Society of Economic Paleontologists and Mineralogists, Pacific Section, Publication 50, p. 337–345.

Dillon, J. T., Brosgé, W. P., and Dutro, J. T., Jr., 1986, Generalized geologic map of the Wiseman Quadrangle, Alaska: U.S. Geological Survey Open-File Report 86-19, scale 1:250,000.

Dumoulin, J. A., and Harris, A. G., 1987, Lower Paleozoic carbonate rocks of the Baird Mountains Quadrangle, western Brooks Range, Alaska, *in* Tailleur, I., and Weimer, P., eds., Alaska North Slope geology: Society of Economic Paleontologists and Mineralogists, Pacific Section, Publication 50, p. 311–336.

Dutro, J. T., Jr., Brosgé, W. P., Detterman, R. L., and Reiser, H. N., 1979, Beaucoup Formation, a new Devonian stratigraphic unit in the central Brooks Range, northern Alaska, *in* Sohl, N. F., and Wright, W. B., eds., Changes in stratigraphic nomenclature by the U.S. Geological Survey: U.S. Geological Survey Bulletin 1482-A, p. A63–A69.

Gottschalk, R. R., 1990, Structural evolution of the Schist belt, south-central Brooks Range fold and thrust belt, Alaska: Journal of Structural Geology, v. 12, p. 453–470.

Harland, W. B., Armstrong, R. L., Cox A. V., Craig, L. E., Smith, A. G., and Smith, D. G., 1989, A geologic time scale: Cambridge, United Kingdom, Cambridge University Press, 263 p.

Henning, M. W., 1982, Reconnaissance geology and stratigraphy of the Skajit Formation, Wiseman B-5 Quadrangle, Alaska: Alaska State Division of Geological and Geophysical Surveys Open-File Report AOF-147, scale 1:63,360.

Moore, R. C., Lalicker, C. G., and Fischer, A. G., 1952, Invertebrate fossils: New York, McGraw-Hill Book Company, 766 p.

Moore, T. E., Wallace, W. K., Bird, K. J., Karl, S. M., Mull, C. G., and Dillon, J. T., 1992, Stratigraphy, structure, and geologic synthesis of northern Alaska: U.S. Geological Survey Open-File Report 92-330, 182 p.

Moore, T. E., Wallace, W. K., Bird, K. J., Karl, S. M., Mull, C. G., and Dillon, J. T., 1994, Geology of northern Alaska, *in* Plafker, G., and Berg, H. C., eds., The geology of Alaska: Boulder, Colorado, Geological Society of America, The Geology of North America, v. G-1, p. 49–140.

Mull, C. G., 1982, Tectonic evolution and structural style of the Brooks Range, Alaska: An illustrated summary, *in* Powers, R. B., ed., Geologic studies of the Cordilleran Thrust belt: Denver, Colorado, Rocky Mountain Association of Geologists, p. 1–45.

Oldow, J. S., Seidensticker, C. M., Phelps, J. C., Julian, F. E., Gottschalk, R. R., Bowler, K. W., Handschy, J. W., and Avé Lallemant, H. G., 1987, Balanced cross sections through the central Brooks Range and North Slope, Arctic Alaska: American Association of Petroleum Geologists, Special Publication, 19 p., 8 pl., scale 1:450,000.

Oldow, J. S., Avé Lallemant, H. G., and Gottschalk, R. R., 1989, Envelopment thrusting in the central Brooks Range fold and thrust belt, Arctic Alaska: Geological Society of America Abstracts with Programs, v. 21, p. 125.

Palmer, A. R., Dillon, J. T., and Dutro, J. T., Jr., 1984, Middle Cambrian trilobites with Siberian affinities from the central Brooks Range, Alaska: Paleozoic Geology of Alaska and northwest Canada Newsletter No. 1, May 1984, p. 29–30.

Manuscript Accepted by the Society September 23, 1997

Geological Society of America
Special Paper 324
1998

Envelopment thrusting and the structure of the eastern Skajit allochthon, central Brooks Range, Arctic Alaska

J. S. Oldow
Department of Geology and Geological Engineering, University of Idaho, Moscow, Idaho 83844-3022
R. R. Gottschalk
Exxon Production Research Company, P.O. Box 2189, Houston, Texas 77252
H. G. Avé Lallemant and K. W. Boler
Department of Geology and Geophysics, Rice University, Houston, Texas 77005-1892
F. E. Julian
Department of Geological Sciences, University of Texas, El Paso, Texas 77968
C. M. Seidensticker
615 Baltimore Drive, El Paso, Texas 79902

ABSTRACT

The eastern Skajit allochthon forms one of the highest structural sheets of the central Brooks Range fold and thrust belt and contains complexly deformed imbricates of metasedimentary rocks with early Paleozoic and possibly Proterozoic protolith ages. The Skajit allochthon presents an outstanding example of envelopment thrusting, in which structurally underlying rocks are carried over previously imbricated rocks of the superjacent roof-complex by a younger thrust fault that emanated from a deeper decollement level. The Skajit allochthon has a structurally complex configuration and both overlies units of the external and internal belts of the Brooks Range orogen, and itself is overlain by a thrust sheet composed of rocks of the internal belt. The metamorphic rocks of the internal belt overlie the southern flank of the Skajit allochthon and were emplaced during north-directed displacement on the Minnie Creek envelopment thrust. The Minnie Creek thrust carried amphibolite-facies metamorphic rocks from deep structural levels over the previously imbricated nappes of the Skajit allochthon and was accompanied by penetrative deformation and recrystallization of the footwall assemblage. Rocks within the Skajit allochthon are deformed in three and locally five generations of ductile structures. First structures exhibit a gradient, manifest by decreasing strain and associated generations of superposed structures (D_{1a}, D_{1b}, and D_{1c}) from south to north. The earliest first-generation structures (D_{1a}) formed during internal imbrication and emplacement of the allochthon before movement on the envelopment thrust. Subsequent first-generation structures (D_{1b} and D_{1c}) formed during movement on the Minnie Creek thrust and emplacement of metamorphic tectonites of the Schist belt over the southern part of the Skajit allochthon. The multiple generations of D_1 structures are associated with diminishing bulk strain from south to north. Second- and third-generation structures are found throughout the Skajit allochthon and are similar to younger structures elsewhere in the central Brooks Range. The second- and third-generation structures postdate emplacement of the Minnie Creek thrust but have unclear relations with a late stage out-of-sequence or envelopment

Oldow, J. S., Gottschalk, R. R., Avé Lallemant, H. G., Boler, K. W., Julian, F. E., and Seidensticker, C. M., 1998, Envelopment thrusting and the structure of the eastern Skajit allochthon, central Brooks Range, Arctic Alaska, *in* Oldow, J. S., and Avé Lallemant, H. G., eds., Architecture of the Central Brooks Range Fold and Thrust Belt, Arctic Alaska: Boulder, Colorado, Geological Society of America Special Paper 324.

thrust that cuts the northern part of the Skajit allochthon. The younger Foggytop thrust repeats the internal structure of the Skajit allochthon and locally carries clastic rocks of the underlying external belt in the hanging wall.

INTRODUCTION

Lower Paleozoic and possibly Proterozoic metasedimentary rocks are widely recognized in the south-central Brooks Range and form an ill-defined, east-west–trending belt of exposures. For years, the structural and stratigraphic relations between the belt of metasedimentary rocks and better understood carbonate and clastic units of the external and metamorphic rocks of the internal parts of the Brooks Range have been enigmatic. In their pioneering studies of the central Brooks Range, Brosgé and Reiser (1964, 1971) recognized the structural complexity of these metasedimentary rocks, but a poorly understood stratigraphy arising from sparse age control and the lack of easily recognized stratigraphic markers, coupled with intense mesoscopic deformation and accompanying metamorphic recrystallization, obscured regional structural relations. The uncertainty of the structure and stratigraphy of the succession also was recognized by Jones et al. (1987) in their assignment of these rocks to the Hammond terrane.

As an outgrowth of local mapping and regional compilation by Dillon and his colleagues (Dillon et al., 1980, 1986, 1987, 1988) and our mapping and structural analysis along the central Brooks Range transect (Oldow et al., 1987), we have recognized that the highly folded and imbricated metasedimentary rocks compose a structurally coherent unit of regional extent. The Skajit allochthon as defined by Oldow et al. (1987), and refined here, forms one of the structurally highest sheets in the Brooks Range orogen and is composed of two stratigraphic sequences of intensely deformed and variably recrystallized carbonate and clastic rocks (Oldow et al., this volume, Chapter 7). The tectonic history of this allochthon and its relation to structurally underlying and overlying units plays a pivotal role in deciphering the imbrication history of the central Brooks Range fold and thrust belt. Significantly, the southern boundary of the Skajit allochthon preserves an excellent example of envelopment thrusting (refer to Fig. 6 in Bally et al., 1966) in which a thrust sheet derived from a deep structural level preserves in-sequence imbrication with respect to the deeper decollement but cross-cuts imbricates of a shallower detachment level. This structure differs from the more familiar out-of-sequence thrust, where younger thrusts cut older structures derived from the same detachment level.

STRUCTURAL SETTING

The Skajit allochthon is more than 10 km thick and is composed of numerous thrust imbricates containing lower Paleozoic and possibly Proterozoic rocks. In the north, the Skajit allochthon overlies the Endicott Mountains allochthon, which is composed of far traveled and internally imbricated Devonian clastic and Carboniferous carbonate rocks (Mull, 1982; Boler, 1989; Phelps and

Avé Lallemant, this volume, Chapter 4). The contact between the Skajit and Endicott Mountains allochthons is a well-located basal thrust, traced from south of the Doonerak window duplex (Fig. 1) eastward to Table Mountain (Fig. 2). West of the Dalton Highway (Fig. 1), the east-northeast–striking thrust dips moderately to the south-southeast, whereas east of the highway, the fault strikes north-northeast and dips shallowly to the east-southeast. South of Table Mountain (Fig. 2), the basal thrust is cut by a younger thrust, named here the Foggytop thrust. In the Chandalar River drainage (Fig. 2), the allochthon is underlain by a subhorizontal basal thrust and structurally overlies the metasedimentary rocks of the southern Endicott Mountains allochthon in the north and, in the south, metamorphic rocks of the Schist belt. The contact between the Endicott Mountains allochthon and Schist belt clearly underlies the basal thrust of the Skajit allochthon and juxtaposes amphibolite facies rocks in the Schist belt with lower greenschist facies rocks in the Endicott Mountains allochthon.

Southwest of the Chandalar River, the Skajit allochthon overlies the Schist belt on a northeast-striking thrust fault that dips moderately to the northwest. This segment of the basal thrust was identified earlier (Brosgé and Reiser, 1964), but its regional significance was not recognized until recently (Oldow et al., 1987, 1989). Underlying the thrust, the Schist belt is composed of metamorphic tectonites derived from sedimentary and volcanic protoliths and highly deformed Devonian granitoids (Dillon et al., 1980) recrystallized during Brookian tectonism at amphibolite-facies metamorphic conditions (Gottschalk, 1990; Patrick, 1995). The overlying metasedimentary rocks of the Skajit allochthon did not experience conditions greater than greenschist conditions, with the exception of a narrow belt along the southern margin of the allochthon as discussed later.

To the south (Fig. 2), the basal thrust of the allochthon is truncated and overlain by the south-dipping Minnie Creek thrust (Oldow et al., 1987; Gottschalk, 1990). The Minnie Creek thrust carries the highest grade rocks of the Schist belt over the southern flank of the Skajit allochthon and contains subsidiary structures, such as the Wiseman thrust (Dillon et al., 1986), in the hanging-wall assemblage. Thus the Skajit allochthon structurally overlies both the Endicott Mountains allochthon and the Schist belt and, in the south, is itself overlain by rocks of the Schist belt.

The Minnie Creek thrust is traced east and west from the Dalton Highway for over 100 km. The western continuation of the fault is marked by the truncation of older thrusts and folds along the southern margin of the Skajit allochthon and by juxta-

Figure 1. General geologic map of the central Brooks Range fold and thrust belt showing the distribution of major structural assemblages. Heavy line (north-south) is the track of the Dalton Highway.

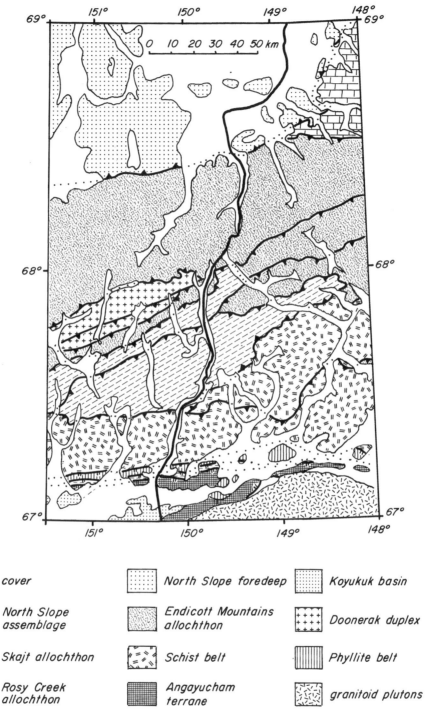

	cover		North Slope foredeep		Koyukuk basin
	North Slope assemblage		Endicott Mountains allochthon		Doonerak duplex
	Skajit allochthon		Schist belt		Phyllite belt
	Rosy Creek allochthon		Angayucham terrane		granitoid plutons

thrust fault bounding or cutting lithotectonic assemblages, teeth on upper plate

moderate- to low-angle normal fault, tick marks on upper plate

Dalton Highway

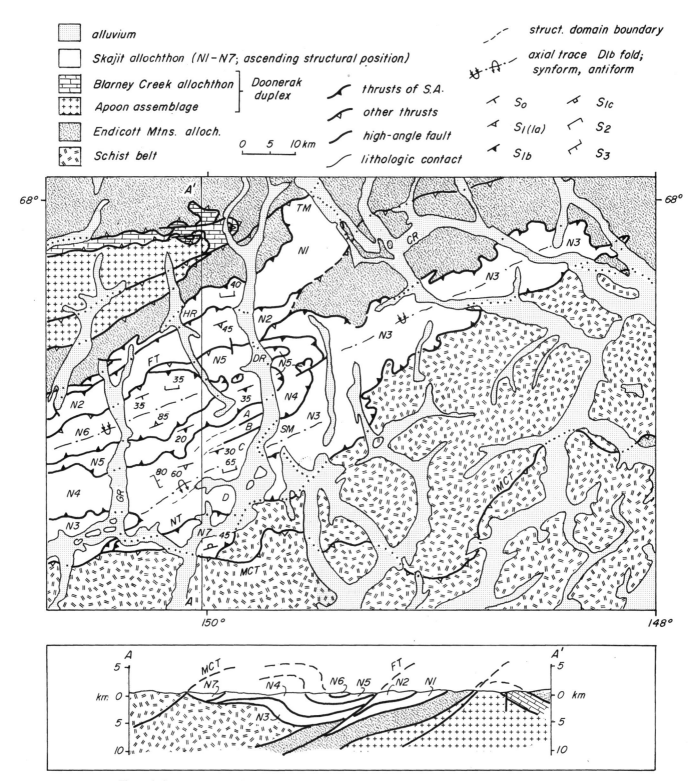

Figure 2. Structure map and cross section of the eastern Skajit allochthon. Thrust nappes are designated N1 through N7 in structurally ascending order. Structural subdomains in nappe N4 (A, B, C, and D) are separated by bold lines where known and dashed lines where suspected. Abbreviations: CR, Chandalar River; DR, Dietrich River; FT, Foggytop thrust; GR, Glacier River; HR, Hammond River; MCT, Minnie Creek thrust; NT, Nolan thrust; SM, Sukakpak Mountain; TM, Table Mountain.

position of rocks of different metamorphic facies. In the east, the fault is located with more difficulty because rocks of the Schist belt compose both the hanging-wall and footwall assemblages (Fig. 1). Nevertheless, the fault is traced to the east where a relatively small slice of the Endicott Mountains allochthon is found in the hanging wall (Fig. 1). The western extent of the Skajit allochthon, beyond the Glacier River, is poorly understood, but it appears to correspond, at least in part, with rocks of the Central Metamorphic belt (Till et al., 1988) of the western Brooks Range.

INTERNAL STRUCTURE

The internal structure of the eastern Skajit allochthon is complex and records a protracted history of progressive deformation. Polyphase folding at various scales (millimeter to kilometer) resulted in multiple generations of penetrative slaty and crenulation cleavage. Early generations of structures are associated with the highest strains, and bedding is transposed into tectonic foliations. The number of fold generations varies spatially between three and five and records a north-to-south increase in strain and associated metamorphic recrystallization. Penetrative structures associated with the imbrication of the thrust nappes composing the allochthon are deformed together with the nappes by later superposed structures. Map-scale folds involving the nappes are cut by the Foggytop thrust (Fig. 2) that divides the allochthon into northern and southern structural domains.

Thrust nappes

Seven thrust nappes, designated N1 through N7, are recognized in the eastern Skajit allochthon (Fig. 2). Two nappes, N1 and N2, form a northern structural domain and are interpreted as structural repetitions of nappes N3 and N4 of a southern domain lying in the hanging wall of the Foggytop thrust. For the most part, nappe-bounding thrust faults, which are well defined in high-relief exposures and cliff faces, cannot be traced laterally with certainty due to widespread vegetation covering lower relief topography. The inferred continuity of the thrusts shown in Figure 2 is based on recognition and lateral continuity of specific lithologic assemblages lying above and below exposed segments of the faults. In several instances, relatively small displacement faults are recognized within individual thrust nappes. Within the limits set by the nature of the exposures, these faults appear to be laterally discontinuous and do not juxtapose distinctly different lithologic units as do the nappe-bounding structures.

As discussed elsewhere (Oldow et al., this volume, Chapter 7), the imbricates of the eastern Skajit allochthon are composed of rocks derived from partially coeval basin and platform successions of early Paleozoic and possibly Precambrian ages. Nappes N1 and N3 are the structurally lowest units in the northern and southern structural domains, respectively, and are composed of a Cambro-Devonian carbonate shelf or platform assemblage overlain by Devonian and younger fine-grained siliciclastic rocks. In the northern structural domain, nappe N2 overlies nappe N1 and is lithologically similar to the siliciclastic, volcanogenic, and thin-bedded carbonate rocks of nappe N4, which overlies nappe N3 of the southern domain. Nappe N4 is overlain by two klippen, nappes N5 and N6, which themselves are composed of thin-bedded carbonate and siliciclastic rocks. The lithology and poorly preserved sedimentary structures suggest a basinal origin for the siliciclastic and carbonate rocks of nappes N2, N4, N5, and N6. Sparse fossil age control suggests a Cambro-Devonian age for the basin succession, similar to the shelf or platform carbonate assemblage of nappes N1 and N3, but a Proterozoic age for some of the rocks cannot be discounted (Oldow et al., this volume, Chapter 7).

The map-scale structure of the northern domain is relatively straightforward. The two thrust nappes strike east-northeast and dip shallowly to the southeast. The thick carbonate and clastic succession of nappe N1 overlies fine-grained siliciclastic rocks (Beaucoup Formation) of the Endicott Mountains allochthon on an easily identified thrust fault. In the Dietrich River drainage and to the west, nappe N1 is structurally overlain by nappe N2, which rests on a moderately southeast dipping basal thrust. In the Chandalar River drainage to the east, nappe N2 and N1 in turn are structurally overlain by rocks of the Endicott Mountains allochthon in the hanging wall of the Foggytop thrust (Fig. 2).

In the southern domain, structural relations are more complex. Nappes N3 and N4 constitute the bulk of the exposures and their contact serves as an important form surface in the identification of map-scale structures. In this domain, the Skajit allochthon is folded in a regionally extensive synform-antiform pair exposed in a plunge depression between the Glacier and Dietrich Rivers.

The only exposures of nappes N5 and N6 are preserved as klippen in the core of an east-northeast–trending synform. The axial trace of the synform continues to the east (Fig. 2), where the structure clearly involves the basal thrust of nappe N4 and also deforms the basal thrust of the Skajit allochthon. The northern limb of the regional synform is truncated by the Foggytop thrust, which carries nappes N3 through N6 over nappes N1 and N2 in the Hammond and Dietrich River drainages and carries Devonian clastic rocks of the underlying Endicott Mountains allochthon over nappes N1 and N2 in the Chandalar River drainage farther to the east.

South of the synform, the trace of the fault separating nappes N3 and N4 defines a map-scale antiform. Along the antiform axis, carbonate and clastic rocks of nappe N3 plunge beneath the clastic rocks of nappe N4 west of Sukakpak Mountain (Fig. 2) and reemerge in the Glacier River drainage to the southwest. There the folded fault between nappes N3 and N4 is truncated at a high angle by the Minnie Creek thrust.

The Minnie Creek thrust carries some of the highest grade metamorphic rocks exposed along the transect over the southern flank of the Skajit allochthon. Tracing the Minnie Creek thrust from the Glacier River to the east, the footwall is composed of nappes N3, N4, and N7. East of Dietrich River, both the hanging-wall and footwall assemblages of the Minnie Creek thrust are composed of metamorphic rocks of the Schist belt. Nappe N7 contains the most intensely deformed and recrystallized rocks of the Skajit allochthon and is separated from nappe N4 on the north by the steeply south dipping Nolan thrust (Fig. 2). At its western extent, the Nolan thrust is truncated by the Minnie Creek thrust, whereas in the east, the fault merges with the upturned basal thrust of the Skajit allochthon.

The protoliths of the metamorphic rocks in nappe N7 are lithologically similar to the units of nappe N4 and differ only in degree of deformation and recrystallization. Although metamorphosed, the schist of nappe N7 is of significantly lower pressure than structurally overlying metamorphic tectonites of the Schist belt but in the immediate vicinity of the Minnie Creek thrust the schist of nappe N7 is of comparable temperature to rocks of the Schist belt (Gottschalk, 1990; Gottschalk et al., this volume, Chapter 12). Metamorphic temperatures decrease north of the Minnie Creek thrust from approximately 500°C to less than 300°C in only a few kilometers (Gottschalk and Snee, this volume, Chapter 13). Consequently, the metamorphic rocks of nappe N7 are viewed as part of the Skajit allochthon.

Mesoscopic structures

Analysis of mesoscopic structures is essential in assessing the kinematic history of the eastern Skajit allochthon. Limited markers within the stratigraphic section accompanied by subdued topography and associated vegetation obscure most map-scale structures. The distribution of small outcrops and their attendant mesoscopic structures, however, yields adequate coverage over the entire study area and serves to illuminate the internal structure of the allochthon. Three generations of superposed ductile structures are recognized and designated D_1, D_2, and D_3 in order of decreasing age.

First-generation structures (D_1). First-generation structures (D_1) are found throughout the Skajit allochthon and record an increase in strain from north to south. The increase in bulk strain is manifested by the development of additional generations of D_1 penetrative structures and increased metamorphic recrystallization. The additional structures and associated metamorphism demonstrably predate formation of D_2 structures, and where developed, the earlier structures are sequentially designated D_{1a}, D_{1b}, and D_{1c}. The continuity of the D_1 strain gradient is disrupted by two thrust faults, the Foggytop thrust in the north and the Nolan thrust in the south (Fig. 2).

In the Skajit allochthon north of the Foggytop thrust, D_1 structures are represented by a single generation of folds and associated axial-planar cleavage. The sparse D_1 folds typically are of centimeter scale and have rootless tight to isoclinal hinges with a class 1c to 2 morphology. The hinge lines generally are arcuate, occasionally with a sheath-fold morphology. Bedding, where preserved, is transposed into the penetrative axial-planar cleavage, which throughout the area disrupts bedding and in many instances obliterates internal sedimentary structures. Where observed, the shear sense of the D_1 folds, determined from their symmetry, is consistently top-to-the-north.

Poles to bedding (S_0) correspond in orientation and distribution to those of the D_1 cleavage (S_1). As shown in Figure 3, the poles to cleavage and to a lesser degree bedding define weak north-south girdles, the normal to which corresponds to the locus of D_2 fold axes. First-generation fold axes consistently

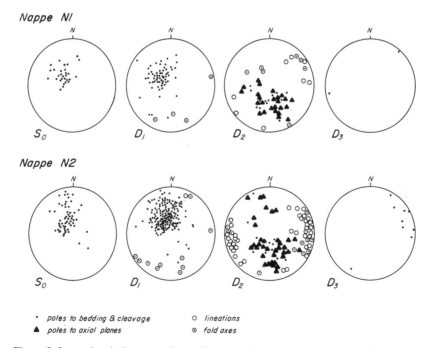

Figure 3. Lower hemisphere, equal-area diagrams of mesoscopic structures of nappes N1 and N2 of the northern structural domain.

have moderate to shallow plunges and, although their azimuths are widespread, suggest a dominant northeast-southwest or north-south trend.

South of the Foggytop thrust and north of the Nolan thrust, first-phase structures consist of two fold and cleavage generations (D_{1a} and D_{1b}). D_{1a} structures are like those in the area north of the Foggytop thrust and consist of tight to isoclinal rootless folds with a penetrative axial-planar cleavage. Bedding is strongly transposed and the poles to bedding and cleavage yield north-south girdles (Fig. 4A and B). D_{1a} fold axes are sparse but consistently have shallow plunges.

D_{1b} structures range in scale from millimeters to tens of meters and are overturned to recumbent, close to tight folds with a class 1b to 1c morphology. The folds typically have a spaced to slaty crenulation cleavage parallel to the axial plane. Fold axes and crenulation lineations are subhorizontal (Fig. 4A and B). S_{1b} is rarely penetrative and is generally localized in fold hinges. New mica growth is oriented parallel to S_{1b}, but for the most part the foliation is defined by crenulations of S_{1a}. Sheath-fold geometries are locally developed, particularly in the southern exposures of nappe N4 and in nappe N7 (Fig. 5).

In the core of the map-scale synform south of Foggytop thrust, D_{1b} axes of mesoscopic folds in nappes N5 and N6 trending east-northeast and west-southwest are tightly clustered (Fig. 4A). In the underlying nappe N4 southward to the Nolan and Minnie Creek thrusts, the distribution of D_{1b} fold axes varies systematically (Fig. 4B). Immediately below nappe N5, D_{1b} fold axes in nappe N4 define a well-clustered, west-southwest point maximum similar to counterparts in the overlying thrust sheets. From north to south, however, the shallowly plunging D_{1b} axes define a more diffuse girdle pattern with the mean orientation systematically changing to a more north-south orientation. The spread and change in orientation of fold axes is accompanied by a decrease in the dihedral angle of the folds from close-to-tight to tight-to-isoclinal. Decrease in the dihedral angle (increase in limb appression) is not accompanied by a corresponding change in orientation of the poles to D_{1b} axial planes and axial-planar cleavage. The reorientation of fold axes and decrease in dihedral angle of folds is the product of increased strain during D_{1b} deformation and is not caused by superposed structures.

Approximately equal numbers of mesoscopic D_{1b} structures display top-to-the-north and top-to-the-south shear, with eleven and seven determinations, respectively. Although not directly observed, we suspect that the measured folds are parasitic structures to poorly defined major folds with gently dipping axial surfaces. The top-to-the-north mesoscopic structures are thought to reside on upright limbs and the top-to-the-south structures lie in overturned limbs of the cryptic major folds. In nappe N4, recognition of major D_{1b} folds is confounded by the monotonous lithology and the degree of cover.

The doubly plunging synform and antiform pair that deforms nappes N3, N4, N5, and N6 are interpreted to be D_{1b} folds. The nappes and their bounding thrusts serve as useful form surfaces and define the strongly eroded map-scale folds.

Although somewhat obscured by their large size, the map-scale folds are interpreted as having close dihedral-angles and to be strongly vergent to the north (Fig. 2). The relationship between the map-scale folds and the mesoscopic D_{1b} structures is drawn from the consistent orientation of the D_{1b} axial planes and axial-planar cleavage throughout the region. Bedding, D_{1a} cleavage, and nappe-bounding faults are clearly folded at outcrop- and map-scales and vary substantially in strike and dip from north to south. D_{1b} fabric elements, which only are moderately reoriented by later D_2 and D_3 structures, do not exhibit the same degree of variability. This is consistent with the interpretation of their development as parasitic structures to the map-scale folds.

In nappe N7 along the southern margin of the Skajit allochthon, an additional generation of structures is assigned to D_1 deformation (Fig. 5). Three generations of structures, D_{1a}, D_{1b}, and D_{1c} are recognized. The primary manifestation of D_{1a} deformation is a penetrative metamorphic foliation (S_{1a}), which in all cases is parallel to compositional layering; S_{1a} is the main penetrative fabric developed. Locally, the S_{1a} foliation is axial-planar to isoclinal to very tight folds of bedding and quartz veins; the hinges of which are rarely preserved. Poles to S_{1a} form a girdle distribution around a roughly east-west trending, subhorizontal axis. Although few mesoscopic D_{1a} folds were noted, the parallelism of lithologic layering and the foliation support the interpretation of extensive transposition of bedding by isoclinal D_{1a} folds.

D_{1b} folds are sporadically developed throughout nappe N7 and in many cases are difficult to differentiate from D_{1c} folds with small dihedral angles. In most outcrops where they are present, D_{1b} folds are tight to isoclinal with axial planes and the associated penetrative axial-planar cleavage (S_{1b}) distributed in a weak girdle about an east-west–trending, horizontal axis (Fig. 5). Here, as for D_{1a} structures, the geometry is due to D_{1c} folds.

The primary manifestation of D_{1b} deformation is a penetrative crenulation cleavage that cross-cuts S_{1a} at a small angle. In the field, it is not always clear to what extent S_{1b} is metamorphically differentiated, but the presence of an L_{1b} amphibole lineation in nappe N7 clearly indicates that D_{1b} is synmetamorphic locally. S_{1b} is not everywhere detectable, but an L_{1b} intersection lineation is commonly (although not ubiquitously) developed on the S_{1a} foliation. As with D_{1b} fold axes, the trend of the associated lineation is variable, forming a girdle distribution as shown in Figure 5. In several outcrops north of the Minnie Creek thrust, syn-D_{1a} ferruginous calcite segregations are boudinaged and stretched parallel to L_{1b}.

Map-scale D_{1c} folds are the dominant structures in nappe N7. Attendant mesoscopic D_{1c} folds are asymmetric, with north-vergent folds occurring on the north-dipping limbs of large-scale D_{1c} folds, and south-vergent folds on the south-dipping limbs. The axial planes of D_{1c} folds dip variably to the north and south, with fold axes clustering in an east-northeast to west-southwest subhorizontal orientation (Fig. 5). A penetrative axial-planar crenulation cleavage (S_{1c}) is present in some out-

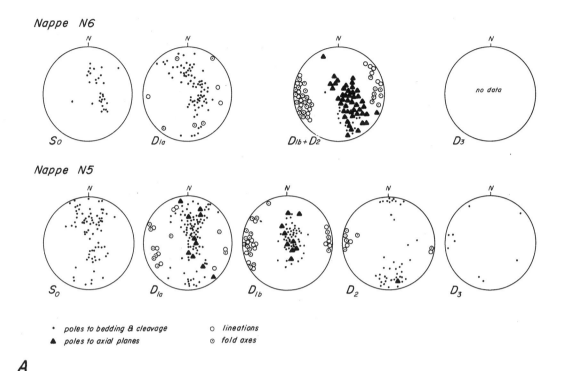

Figure 4 (on this and facing page). A, Lower hemisphere, equal-area diagrams of mesoscopic structures in nappes N5 and N6 of the southern structural domain. B, Lower hemisphere, equal-area diagrams of mesoscopic structures in various internal structural domains within nappe N4.

crops, and related intersection lineations (chiefly S_{1a}/S_{1c}) are oriented in a northeast-trending, subhorizontal orientation similar to D_{1c} fold axes.

In nappe N7, numerous ductile shear zones and small-scale faults are associated with D_1 folds. Several south-southeast–dipping ductile shear zones occur in the sheared limbs of D_{1b} folds immediately north of the Minnie Creek thrust. Northerly trending mineral stretching-lineations in the shear zones, and the sweeping of D_{1a} foliation planes into the shear zone indicates top-to-the-north transport. Fold axes of adjacent and nearby D_{1b} folds were not measurable due to the lack of three-dimensional exposure. The girdle distribution of D_{1b} axes probably results from strain rotation during north-vergent shear.

A number of small-scale faults with minimal offset are present in the exposures of nappe N7 north of the Minnie Creek thrust. Both D_{1c} folds and associated shear zones have a dominant south-over-north shear sense. Nearly all of the thrust faults dip gently north, at a steeper angle than the predominantly north-dipping D_{1c} axial planes (Fig. 5). Slickenside lineations are scarce but where present trend northerly. A few south-directed faults are also present; these occur on the south-dipping limbs of large-scale D_{1c} folds and result from intrafolial slip, producing a series of small fault splays terminating into south-vergent D_{1c} folds. Locally, spectacular ductile or semibrittle shear zones are developed in the sheared limbs of mesoscopic D_{1c} folds. As with the thrust faults, shear zones dip gently to the north and consis-

tently indicate south-over-north shear. No lineations were noted on the shear zone surfaces.

On the ridge crests north of the Minnie Creek thrust, a series of small, steeply to shallowly north-dipping faults with normal offset were observed. A single south-dipping normal fault with listric geometry was also observed in south-dipping tectonites. Normal faults offset lithologic banding and metamorphic foliations (S_{1a}), but the relationship of normal faults to D_{1b} and D_{1c} folds is unknown. Normal faults are not abundant and offsets are on the order of centimeters. Extension in nappe N7 is a limited, small-scale phenomenon.

Second-generation structures (D_2). D_2 structures, although common, are not as pervasive as first-generation structures. No map-scale D_2 folds are known to exist and all structures are of mesoscopic scale. D_2 folds are gentle to close, characteristically with an asymmetric kink geometry. D_2 folds rarely exceed 10 cm in amplitude and only locally are as large as 5 m. The folds have an axial-planar crenulation cleavage that rarely is penetrative and usually is spaced on millimeter to centimeter scales. Generally, the shear sense of D_2 folds is top-to-the-south and folds characteristically have axial planes dipping steeply to the north-northwest (Figs. 3, 4, and 5). Infrequently, a conjugate set of D_2 folds or cleavage dips steeply to the south-southeast or south. Second-phase fold axes and intersection lineations in either case are subhorizontal and trend northeast to southwest, parallel to the strike of the axial surface.

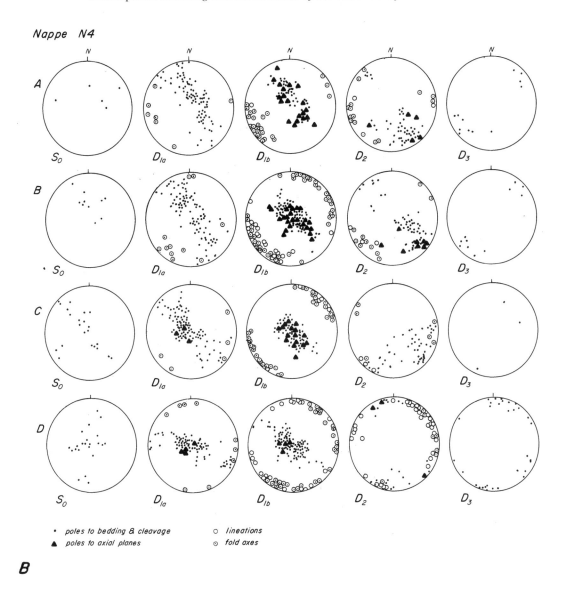

Nappe N4

- poles to bedding & cleavage   ∘ lineations
▲ poles to axial planes   ⊙ fold axes

B

Third-generation structures (D₃). Structures designated D_3 are rare and nonpenetrative. Most of the surfaces designated as S_3 are the axial surfaces of crenulations that occasionally appear to be dilational and are filled with calcite veins. A few folds with S_3 parallel to the axial plane are small kinks, generally with less than 2 cm amplitude with northwesterly strikes (Figs. 3 and 4). D_3 structures had little or no effect on the orientation of earlier structures and accommodate only minor shortening.

IMBRICATION HISTORY

To completely develop the deformational history of the eastern Skajit allochthon a regional view must be taken and the structural and geochronological relations with the Schist belt (Gottschalk and Snee, this volume, Chapter 13), the Endicott Mountains allochthon (Handschy, this volume, Chapter 3; Phelps and Avé Lallemant, this volume, Chapter 4), and the

Doonerak duplex (Julian and Oldow, this volume, Chapter 5; Seidensticker and Oldow, this volume, Chapter 6) must be invoked. This is beyond the scope of this chapter. Here we focus only on the relative timing of structures and internal imbrication of the eastern Skajit allochthon.

As outlined below, thrust nappe imbrication and development of first-generation structures within the eastern Skajit allochthon both predate and are related to the envelopment of the southern part of the allochthon by the schists of the hanging wall of the Minnie Creek thrust (Oldow et al., 1987; Gottschalk et al., this volume, Chapter 12). Most thrusts in the Skajit allochthon formed during the early stages of first-generation deformation. Penetrative D_1 (northern domain) and D_{1a} fabrics parallel mapped thrust surfaces that are folded into D_{1b} folds (southern domain). Displacement on the Nolan thrust and Foggytop thrust is less well constrained. The Nolan thrust juxtaposes nappes N4 and N7, which are differentiated by the degree of metamorphic

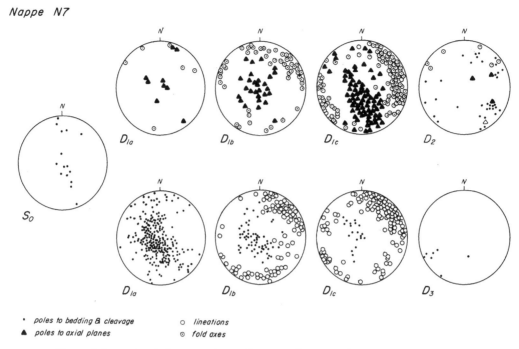

Figure 5. Lower hemisphere, equal-area diagrams of mesoscopic structures in nappe N7.

recrystallization associated with D_{1a} and D_{1b} structures and the presence or lack of D_{1c} folds and foliation. The Nolan thrust certainly experienced displacement after or during development of D_{1c} structures and may have been active during D_{1b} deformation. Based on its contact relations with the overlying Minnie Creek thrust, however, motion on the Nolan thrust ceased before that on the Minnie Creek thrust. The timing of the Foggytop thrust can only be established as postdating D_{1b} major folds from relations within the Skajit allochthon and the relation of the fault with D_2 and D_3 structures is unknown. Within the Skajit allochthon, second- and third-generation structures accommodate only minor shortening and formed after Schist belt envelopment.

The initial thrust-nappe imbrication within the Skajit allochthon occurred during the early stages of D_1 deformation. Where multiple phases of D_1 structures exist, thrust imbrication coincided with the formation of D_{1a} folds and associated penetrative foliations. D_1 folds of the northern domain, together with D_{1a} folds in the southern domain, do not involve thrust nappe boundaries. Where exposed, thrust faults are subparallel to D_1 and D_{1a} axial planes and axial-planar cleavage. Sparse shear sense indicators from mesoscopic D_1 and D_{1a} structures indicate top-to-the-north shear, suggesting north-directed tectonic transport for the thrust nappes. Thus, displacement on thrust nappe boundaries formed contemporaneously with D_1 folds in the northern domain (nappes N1 and N2) and D_{1a} folds in the southern domain (nappes N3–N7).

D_{1b} folds and associated penetrative cleavage (S_{1b}), found south of the Foggytop thrust, demonstrably postdate imbrication

of the thrust nappes. The map-scale synform and antiform dominating the structure of the central and southern Skajit allochthon are D_{1b} folds. On the south limb of the synform involving nappes N3, N4, N5, and N6, S_{1b} predominantly dips shallowly to the south and cross-cuts steeply dipping bedding, S_{1a}, and nappe-bounding thrusts. Shear-sense determinations from mesoscopic D_{1b} structures are ambiguous due to uncertainty in the locations of poorly expressed D_{1b} major folds. The map-scale structures depicted in the cross-sections of Figure 2, however, are north-vergent to overturned folds, and like D_1 and D_{1a} structures are consistent with a top-to-the-north shear sense.

A genetic relation between D_{1b} structures and emplacement of the Minnie Creek thrust is suggested by the north to south increase in strain associated with D_{1b} folds as the trace of the thrust is approached. South from the Foggytop thrust, D_{1b} fold axes progressively rotate from strike parallel to nearly dip parallel. As discussed previously, this change in orientation of the fold axes is associated with a decrease in dihedral angle of folds and is not accompanied by a reorientation of axial planes. This geometry is interpreted as the product of strain rotation like that originally documented in the Apuane Alps of northern Italy (Carmignani et al., 1978).

In the southernmost nappe, N7, lying immediately below the Minnie Creek thrust, greenschist to amphibolite facies metamorphism accompanied D_{1b} deformation. Here, D_{1b} structures are superposed on D_{1a} structures, which themselves exhibit the highest grade metamorphic fabrics of the Skajit allochthon. D_{1b} fold axes are variably oriented but have north-northeast– to south-southeast–trending modes, roughly parallel to locally developed stretching lineations in ductile shear zones associated with D_{1b} minor folds. The shear zone geometries typically indicate top-to-the-north displacement and are consistent with the interpretation of D_{1b} tectonic transport in the lower strain structures observed to the north.

Nappe N7 is the only part of the eastern Skajit allochthon known to exhibit D_{1c} structures and differs from the other parts of the allochthon in the degree of associated metamorphic recrystallization during D_1 deformation. The ductile character of D_{1a} and D_{1b} structures and the degree of syntectonic recrystallization suggest that deformation occurred at significantly higher pressures and temperatures than parts of the Skajit allochthon exposed farther north. Furthermore, greater D_1 strain is manifest in nappe N7 as D_{1c} structures, which are not found within nappe N4. Based on lithologic similarity, nappes N7 and N4 are thought to be related (Oldow et al., this volume, Chapter 7) and parts of a formerly continuous sheet imbricated internally by the Nolan thrust. The difference in D_1 structure between the nappes ostensibly is related to displacement on the Nolan thrust, which foreshortened the interpreted strain gradient represented by D_{1b} and D_{1c} structures.

The association between formation of D_1 structures and continued displacement on the Minnie Creek thrust is supported by structural relations in both the hanging wall and footwall of the thrust system (Gottschalk, 1990; Gottschalk et al., this volume, Chapter 12). Within the hanging-wall assemblage, the intensity of D_{1c} folds progressively increases from south to north as the fault is approached, in a relationship comparable to that within the Skajit allochthon. Thus, the development of D_{1c} structures in the Schist belt and in the Skajit allochthon exhibits a crude bilateral symmetry about the Minnie Creek thrust. The implication is that displacement on the thrust was long lived and was accompanied by development of D_{1b} and D_{1c} structures in the footwall assemblage and D_{1c} structures in hanging-wall assemblage.

D_2 and D_3 structures contribute little shortening in the Skajit allochthon. Superposition relations clearly indicate that the structures postdate imbrication of the Skajit allochthon and envelopment by the Minnie Creek thrust. The origin of D_2 structures, which differ from D_1 folds in being predominantly south vergent, is interpreted to be the product of antithetic shear associated with development of the Doonerak duplex (Seidensticker and Oldow, this volume, Chapter 6) and is discussed elsewhere (Avé Lallemant and Oldow, this volume, Chapter 14). The origin of D_3 structures is less clear. As outlined by Phelps et al. (1987) for the Doonerak window area to the north, D_3 structures may be related to poorly understood left-lateral transcurrent faults in the central Brooks Range.

The relations outlined above constrain the forward model presented in Figure 6. Initial imbrication of the thrust nappes within the Skajit allochthon occurred during the early stages of D_1 deformation (Fig. 6). In the northern structural domain, a single generation of D_1 structures is associated with thrust imbrication. To the south of the Foggytop thrust, additional generations of D_1

structures exist but clearly postdate nappe imbrication, since thrusts are folded into D_{1b} structures. D_1 and D_{1a} structures associated with thrust imbrication consistently are north vergent and, with the exception of nappe N7, are associated with only moderate metamorphic recrystallization. D_{1a} structures in nappe N7 formed at elevated pressure and temperature conditions indicative of greater depths of burial.

D_{1b} structures, found south of the Foggytop thrust, preserve a disrupted strain gradient. Strain associated with D_{1b} folds progressively increase from north to south in nappe N4. The relatively homogeneous composition of nappe N4 precludes attribution of the difference in strain to the mechanical properties of different lithologies. D_{1b} structures in nappe N7, unlike their

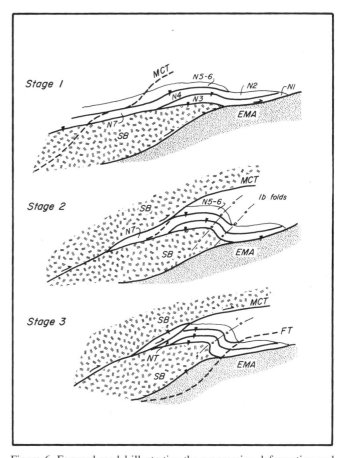

Figure 6. Forward model illustrating the progressive deformation and imbrication of the eastern Skajit allochthon during D_1 deformation. Stage 1 depicts structural geometry after emplacement of previously imbricated Skajit allochthon and underlying Schist belt (SB) over the southern margin of the Endicott Mountains allochthon (EMA). Dashed line indicates position of future Minnie Creek thrust (MCT). Stage 2 illustrates envelopment of Schist belt over southern flank of Skajit allochthon by the Minnie Creek thrust. During envelopment, Skajit allochthon nappes and the basal decollement are deformed in D_{1b} structures. Stage 3 depicts emplacement of Nolan thrust (NT), a splay of the Minnie Creek thrust. Development of D_{1b} and D_{1c} structures are related to continued displacement on the Minnie Creek thrust. Dashed line shows future position of the Foggytop thrust (FT).

counterparts to the north, are associated with metamorphic recrystallization at depth. The development of D$_{1b}$ structures is attributed to the initial uplift and emplacement of the Schist belt over the southern Skajit allochthon. D$_{1b}$ structures are best formed and exhibit greatest strain in closer proximity to the Minnie Creek thrust (Fig. 6).

Displacement of the Minnie Creek thrust continued through formation of D$_{1c}$ structures found both within the hanging-wall assemblage and in nappe N7 of the footwall. Foreshortening of the footwall strain-gradient was accommodated by the Nolan thrust, which carried rocks deformed in deeper parts of the Skajit allochthon over shallower portions of the sheet. The Nolan thrust is viewed as a subsidiary splay of the Minnie Creek thrust that was ultimately deactivated during progressive deformation (Fig. 6).

The timing of the Foggytop thrust and its relation to the Minnie Creek and Nolan thrusts cannot be determined with data derived solely from the Skajit allochthon. The Foggytop thrust clearly postdates formation of the D$_{1b}$ synform of thrust nappes but its relation with D$_2$ structures is unclear. North-directed thrusting recorded by the Foggytop thrust, however, is incompatible with east-west contraction recorded by D$_3$ folds and thus predates the third phase of deformation.

CONCLUSIONS

The history of imbrication and deformation of the eastern Skajit allochthon presents an outstanding example of envelopment thrusting (Oldow et al., 1987) in which previously imbricated rocks overlying a shallow decollement system (Skajit allochthon) are overthrust by part of the underlying basement complex (Schist belt). Recognition of this structural configuration represents a key in deciphering the tectonic evolution of the central Brooks Range fold and thrust belt. As such, several structural relations must be reconciled with any forward model of imbrication in the Brooks Range:

1. The Skajit allochthon is a structurally coherent unit composed of numerous thrust nappes. The allochthon is one of the highest structural units in the central Brooks Range and structurally overlies metasedimentary rocks of the Endicott Mountains allochthon on the north and rocks of the Schist belt on the south. The southern boundary of the Skajit allochthon is defined by the Minnie Creek thrust, which carries rocks of the Schist belt over the southern flank of the allochthon.

2. Three generations of folds are recognized in the Skajit allochthon. First-generation structures represent the most pervasive deformation and record a north-to-south strain gradient manifest by an increasing number of superposed folds and cleavage. Subsequent phases of folds (second- and third-generation) accommodate only minor shortening.

3. For the most part, imbrication of thrust nappes composing the allochthon occurred during the early stages of D$_1$ deformation at relatively shallow structural levels. An exception exists in the southernmost nappe, which underwent substantial recrystal-

lization during deformation prior to being carried to the surface by late-stage D$_1$ deformation.

4. Later generations of D$_1$ folds define a north-to-south increase in strain associated with the emplacement of the Minnie Creek thrust. The strain gradient is disrupted by two thrust faults. The Foggytop thrust in the north is an envelopment or out-of-sequence thrust that clearly postdates D$_{1b}$ folds but its relation to the Minnie Creek thrust is unclear. The Nolan thrust in the southern part of the allochthon juxtaposes rocks with different structural and metamorphic histories, but formed during D$_1$ deformation as a splay active during part of the displacement on the Minnie Creek thrust.

5. Second- and third-generation structures accommodate only minor shortening and postdate envelopment thrusting.

ACKNOWLEDGMENTS

We thank R. Kligfield and W. Wallace for careful reviews of the manuscript. This research was partially supported by the Department of Energy (DE-AS05-83ER13124), the National Science Foundation (EAR-8517384, EAR-872017, EAR-9017835), and the Industrial Associates Program at Rice University.

REFERENCES CITED

Bally, A. W., Gordy, P. L., and Stewart, G. A., 1966, Structure, seismic data, and orogenic evolution of southern Canadian Rockies: Canadian Society of Petroleum Geologists Bulletin, v. 14, p. 337–381.
Boler, K. W., 1989, Stratigraphy, structure, and tectonics of the central Brooks Range, near Dietrich Camp, Alaska [M.A. thesis]: Houston, Texas, Rice University, 163 p.
Brosgé, W. P., and Reiser, H. N., 1964, Geologic map and section of the Chandalar Quadrangle, Alaska: U.S. Geological Survey Miscellaneous Geologic Investigations Map I-375, 1 sheet, scale 1:250,000.
Brosgé, W. P., and Reiser, H. N., 1971, Preliminary bedrock geologic map: Wiseman and eastern Survey Pass Quadrangles, Alaska: U.S. Geological Survey Open-File Map 479, 2 sheets, scale 1:250,000.
Carmignani, L., Giglia, G., and Kligfield, R., 1978, Structural evolution of the Apuane Alps: An example of continental deformation in the northern Apennines, Italy: Journal of Geology, v. 86, p. 487–504.
Dillon, J. T., Pessel, G. H., Chen, J. H., and Veach, N. C., 1980, Middle Paleozoic magmatism and orogenesis in the Brooks Range, Alaska: Geology, v. 8, p. 338–343.
Dillon, J. T., Brosgé, W. P., and Dutro, J. T., 1986, Generalized geologic map of the Wiseman Quadrangle, Alaska: U.S. Geological Survey Open-File Report 86-219, scale 1:250,000.
Dillon, J. T., Harris, A. G., and Dutro, J. T., Jr., 1987, Preliminary description and correlation of lower Paleozoic fossil-bearing strata in the Snowden Mountain area of the south-central Brooks Range, Alaska, in Tailleur, I., and Weimer, P., eds., Alaska North Slope geology: Society of Economic Paleontologists and Mineralogists, Pacific Section, Publication 50, p. 337–345.
Dillon, J. T., Harris, A. G., Dutro, J. T., Jr., Solie, D. N., Blum, J. D., Jones, D. L., and Howell, D. G., 1988, Preliminary geologic map and section of the Chandalar D-6 and parts of the Chandalar C-6 and Wiseman C-1 and D-1 Quadrangles, Alaska: Alaska State Division of Geological and Geophysical Survey, Report of Investigations 88-5, scale 1:63,360.
Gottschalk, R. R., 1990, Structural evolution of the schist belt, south-central

Brooks Range fold and thrust belt, Alaska: Journal of Structural Geology, v. 12, p. 453–470.

Jones, D. L., Silberling, N. J., Coney, P. J., and Plafker, G., 1987, Lithotectonic terrane map of Alaska (west of the 141st Meridian): U.S. Geological Survey Miscellaneous Field Studies Map, MF 1874-A, scale 1:2,500,000.

Mull, C. G., 1982, Tectonic evolution and structural style of the Brooks Range, Alaska: An illustrated summary, *in* Powers, R. B., ed., Geologic studies of the Cordilleran Thrust belt: Denver, Colorado, Rocky Mountain Association of Geologists, p. 1–45.

Oldow, J. S., Seidensticker, C. M., Phelps, J. C., Julian, F. E., Gottschalk, R. R., Boler, K. W., Handschy, J. W., and Avé Lallemant, H. G., 1987, Balanced cross sections through the central Brooks Range and North Slope, Arctic Alaska (eight plates and text): American Association of Petroleum Geologists, Special Publication, 19 p., 8 pl., scale 1:450,000.

Oldow, J. S., Avé Lallemant, H. G., and Gottschalk, R. R., 1989, Envelopment thrusting in the central Brooks Range fold and thrust belt, Arctic Alaska: Geological Society of America Abstract with Programs, v. 21, p. 125.

Patrick, B. E., 1995, High-pressure low-temperature metamorphism of granitic orthogneiss in the Brooks Range, northern Alaska: Journal of Metamorphic Petrology, v. 13, p. 111–124.

Phelps, J. C., Avé Lallemant, H. G., Seidensticker, C. M., Julian, F. E., and Oldow, J. S., 1987, Late-stage high-angle faulting, eastern Doonerak window, central Brooks Range, *in* Tailleur, I. L., and Weimer, P., eds., Alaskan North Slope geology: Society of Economic Paleontologists and Mineralogists, Pacific Section, Publication 50, p. 685–690.

Till, A. B., Schmidt, J. M., and Nelson, S. W., 1988, Thrust involvement of metamorphic rocks, southwestern Brooks Range, Alaska: Geology, v. 16, p. 930–933.

MANUSCRIPT ACCEPTED BY THE SOCIETY SEPTEMBER 23, 1997

Geological Society of America
Special Paper 324
1998

Petrology of eclogite and associated high-pressure metamorphic rocks, south-central Brooks Range, Alaska

Richard R. Gottschalk
Exxon Production Research Company, P.O. Box 2189, Houston, Texas 77252

ABSTRACT

Eclogite containing the prograde assemblage garnet + sodic-augite + glaucophane + actinolitic-hornblende + epidote + rutile + apatite (+ sphene) occurs within the Schist belt of the south-central Brooks Range along the Middle Fork Koyukuk River near Wiseman, Alaska. Prograde minerals in eclogite are overprinted and partially replaced by a secondary assemblage of zoned amphibole (actinolite to hornblende, core to rim), epidote, chlorite, garnet, albite, and sphene in diffuse veins and in pervasively altered patches. Veins and patches of secondary mineralization are crosscut, in turn, by mono-minerallic actinolite veins.

Host rocks consist mainly of pelitic and semipelitic quartz-mica schists with subordinate intercalated layers of mafic schist, albite schist, massive metabasite, metagabbro, marble, calc-schist, and granitic orthogneiss. These rocks have been severely deformed and recrystallized under lower amphibolite- to greenschist-facies conditions, but an older high pressure/low temperature (*HP/LT*) metamorphic event is indicated by the presence of relict crossite in metabasite, and the common occurrence of pseudomorphs after glaucophane in quartz-mica schists. Textural relations indicate that the high-pressure assemblage quartz + phengite + paragonite + chlorite + glaucophane + chloritoid was widespread in quartz-mica schists prior to retrograde metamorphism. Mineral geothermometry, phase equilibria, and microstructural and textural relations indicate that eclogite and the enclosing rocks of the Schist belt both experienced peak metamorphic conditions within the blueschist facies at temperatures between 350 and 550°C and pressures above the lower stability limit of glaucophane (greater than approximately 7 kbar); retrograde metamorphism occurred within the lower-amphibolite to greenschist facies at temperatures between approximately 380 to 480°C and at pressures below the lower stability limit of glaucophane.

The occurrence of the assemblage garnet + Na-pyroxene + rutile in metabasites appears to be compositionally controlled. Whereas eclogite was derived from a Fe-Ti-rich basaltic protolith, mafic schists, metabasites, and metagabbros from the surrounding schists were derived from rocks of normal basaltic composition and contain lower amphibolite- or greenschist-facies mineral assemblages.

Gottschalk, R. R., 1998, Petrology of eclogite and associated high-pressure metamorphic rocks, south-central Brooks Range, Alaska, *in* Oldow, J. S., and Avé Lallemant, H. G., eds., Architecture of the Central Brooks Range Fold and Thrust Belt, Arctic Alaska: Boulder, Colorado, Geological Society of America Special Paper 324.

INTRODUCTION

The Schist belt is a regionally extensive, east-west–trending belt of high pressure/low temperature (*HP/LT*) metamorphic rocks that crop out along the length of the southern Brooks Range. Although many aspects of the structural and metamorphic history of the Schist belt are still poorly known, it has long been recognized that the Schist belt and correlative rocks in the Seward Peninsula (the Nome Group schists) were affected by deformation under blueschist-facies metamorphic conditions, followed by recrystallization under amphibolite- to greenschist-facies metamorphic conditions (Sainsbury et al., 1970; Gilbert et al., 1977; Carden, 1978; Turner et al., 1979; Nelsen, 1979; Nelson and Grybeck, 1982; Thurston, 1985; Hitzman et al., 1986; Zayatz, 1987; Gottschalk, 1987, 1990; Patrick, 1988; Evans and Patrick, 1987; Till et al., 1988; Till, 1988; Till and Patrick, 1991; Patrick and Lieberman, 1988; Patrick and Evans, 1989; Dusel-Bacon et al., 1989; Little et al., 1994; Patrick, 1995). The absolute ages of these two deformational/metamorphic events has been a persistent topic of debate, but recent K-Ar, Rb-Sr, and $^{40}Ar/^{39}Ar$ geochronologic studies indicate that blueschist-facies metamorphism in the Schist belt is Middle to Late Jurassic or older in age (Armstrong et al., 1986; Till and Patrick, 1991; Christiansen and Snee, 1994; Gottschalk and Snee, this volume, Chapter 13). Lower amphibolite- to greenschist-facies metamorphism and associated deformation is significantly younger and is interpreted to have occurred in the Early Cretaceous (135 to 110 Ma; Till and Patrick, 1991; Christiansen and Snee, 1994; Blythe et al., this volume, Chapter 10; Gottschalk and Snee, this volume, Chapter 13).

In the south-central Brooks Range along the Middle Fork Koyukuk River, the Schist belt is dominated by rocks that contain lower amphibolite- to greenschist-facies metamorphic assemblages. Evidence of early *HP/LT* metamorphism is largely indirect, and consists of mainly pseudomorphs after high-pressure metamorphic minerals and rare occurrences of relict sodic-amphibole. One of the best preserved pieces of evidence of *HP/LT* metamorphism is a partially altered body of eclogite that crops out along the Middle Fork Koyukuk River near its confluence with Clara Creek. Although eclogite is present at a number of localities on the Seward Peninsula in the Nome Group schist (Forbes et al., 1984; Thurston, 1985; Patrick and Evans, 1989), eclogite is rare in the Brooks Range Schist belt; the only other known eclogite occurrence in the Schist belt is a small body of omphacite-bearing metabasite in the Ruby Ridge area of the Ambler mining district, described by Carden (1978; Fig 1).

Because of the generally poor preservation of *HP/LT* metamorphic assemblages in the south-central Brooks Range, the Clara Creek eclogite provides a rare opportunity to study the early metamorphic history of the Schist belt in this region. It is the purpose of this paper to describe the general geology of Clara Creek eclogite, to summarize new data on the mineralogy and petrology of the eclogite body, and to discuss the implications of the Clara Creek eclogite for the metamorphic history of the

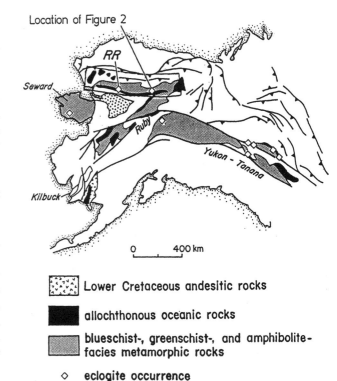

Location of Figure 2

Figure 1. Generalized distribution of metamorphic rocks, northern and central Alaska and Yukon Territory. RR = Ruby Ridge eclogite occurrence.

Schist belt. Because disequilibrium metamorphic textures are common in the Clara Creek eclogite, textural criteria such as those outlined in Spry (1969) were used to determine the relative timing of mineral growth; from these relations it was possible to (1) separate equilibrium assemblages in the Clara Creek eclogite; (2) identify critical reactions that occurred in the Clara Creek eclogite in response to changing pressure (*P*), and temperature (*T*) conditions; and (3) to delineate a partial *P,T* path using constraints supplied by phase equilibria and mineral geothermometry. The schistose host rocks of Clara Creek eclogite were also studied to determine whether the Clara Creek eclogite is relatively high grade, tectonically emplaced block similar to those described in subduction melanges (e.g., Coleman and Lee, 1963; Sorensen, 1986; Moore, 1986), or whether eclogite was metamorphosed in situ and is isofacial with its host rocks.

GENERAL GEOLOGY

Along the Middle Fork Koyukuk River, the Schist belt is exposed in a 7-km-thick section composed mainly of pelitic and semipelitic quartz-mica schist with subordinate intercalated layers of mafic schist, albite schist, massive metabasite, metagabbro, marble, calc-schist, metachert, and granitic orthogneiss. These rocks were derived from latest Devonian and older marine sedimentary rocks that were deposited on continental crust, and were locally intruded by Devonian granitic plutons (Turner et al., 1979; Dillon et al., 1980, 1986; Hitzman et al., 1986; Gottschalk, 1990; Moore et al., 1994; Gottschalk et al., this volume, Chapter 12). Moore et al. (1994) assign rocks exposed along the Middle Fork Koyukuk River to the quartz-mica schist unit of the Coldfoot Subterrane. This study does not include the sodic-amphibole-bearing schistose rocks north of the Minnie Creek thrust assigned by Till and Moore (1991) to the Schist belt.

The Clara Creek eclogite occurs as an isolated knob at the base of the hills north of Clara Creek, about 0.8 km east of the Dalton Highway, near Wiseman, Alaska (67° 17′ 02″ N, 150° 08′ 20″ W; Fig. 2). Although the outcrop has been largely reduced to rubble by frost heave, several large blocks of eclogite are in situ, defining what was probably a continuous outcrop more than 150 m in length and as much as 75 m in width. Contacts with the surrounding schists are obscured by cover; however, the northeasterly trending distribution of outcrops runs parallel to the regional trend of the foliation in the Schist belt, suggesting that eclogite exists as a conformable, lenticular body within the enclosing quartz-mica schist. Where relatively fresh, the eclogite body is massive and structureless to weakly foliated, with the foliation defined by elongate clinopyroxene and amphibole grains and by garnet segregation bands. The foliation is parallel to the long dimension of the eclogite body and to the foliation in the surrounding schists.

Eclogite is crosscut by numerous veins and overprinted by diffuse patches of secondary mineralization. Two generations of veins with distinct morphologies and mineralogies occur: (1) Irregular veins composed of epidote + amphibole with accessory sphene, chlorite, plagioclase, pyrite, and apatite are the oldest veins in the Clara Creek eclogite. Wall-rock boundaries are generally diffuse, and selvages composed of dark green amphibole, epidote, sphene, and to a lesser extent, chlorite and albite are well developed. Patches of secondary mineralization, dominantly amphibole + epidote + sphene + garnet + chlorite, are present throughout the eclogite body and occur most prominently in areas of high vein density; patches of secondary mineralization are laterally contiguous with vein selvages. (2) The youngest veins are composed of fibrous actinolite in brittle fractures that crosscut diffuse polymineralllic veins and patches of secondary mineralization. Fibrous actinolite veins differ markedly from older veins in that they are essentially planar features with sharply defined boundaries and no discernible selvages.

MINERALOGY AND PETROGRAPHY OF THE CLARA CREEK ECLOGITE

The mineralogy and metamorphic textures of the Clara Creek eclogite have been studied in detail and are described below. Microprobe analyses presented in this paper were collected on a JEOL 733 Superprobe at Southern Methodist University, Dallas, Texas, and on an ETEC Autoprobe electron microprobe analyzer at Rice University, Houston, Texas. Details on microprobe analysis and calculation of mineral formulae are given in Appendix 1.

Clinopyroxene

Clinopyroxene is the most abundant mineral in relatively unaltered portions of the eclogite body, occurring as colorless, subidioblastic, lathlike grains ranging from 0.15 to 0.3 mm in length. Microprobe analysis indicates that clinopyroxene is sodic-augite according to the classification of Essene and Fyfe (1967); sodic-augite grains exhibit a weak compositional zoning characterized by a slight decrease in Fe^{+2}/Mg from core to rim (Table 1; Appendix 2). Sodic-augite hosts tiny grains of rutile, garnet, and pyrite, and contains numerous fluid inclusions; the negative crystal shape and solitary aspect of fluid inclusions suggest that they are primary in nature. Where present in altered portions of the eclogite body, sodic-augite is epitaxially overgrown and partially replaced by hornblende.

Garnet

At least three compositionally and morphologically distinct generations of garnet are recognizable in the Clara Creek eclogite. From oldest to youngest, these are designated garnet$_I$, garnet$_{II}$, and garnet$_{III}$.

Garnet$_I$ and garnet$_{II}$ are present mainly in relatively unaltered portions of the eclogite body, and are in textural equilibrium with sodic-augite, amphibole, and epidote. Garnet$_I$ forms large xenoblastic poikiloblasts 2 to 5 mm in diameter that often contain an internal foliation made up of tiny aligned amphibole, epidote, sphene, and rutile grains; inclusion trails are usually linear but occasionally form snowball-type trails indicative of synkinematic

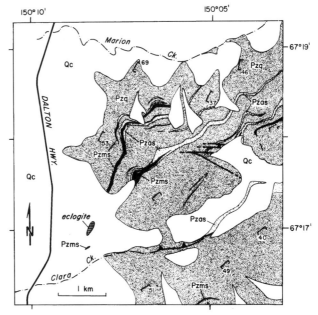

Figure 2. Detail of geology in vicinity of eclogite, after Gottschalk (1987). Pzq = quartz-mica schist; Pzas = albite schist; Pzms = mafic schist; Qc = Quaternary cover; brackets = orientation of D$_{1b}$ foliation.

TABLE 1. REPRESENTATIVE MICROPROBE ANALYSES OF GARNETS AND CLINOPYROXENE*

	Garnet										Clinopyroxene Eclogite	
	Eclogite					Mafic Schist 6-22-84-1		Pelitic Schist 8-9-84-5	Metachert 7-21-85-1			
	6-14-83-3				8-4-84-3a						6-14-83-3	
	Gt$_I$ Core	Gt$_I$ Rim	Gt$_{II}$ Core	Gt$_{II}$ Rim	Gt$_{III}$ Rim	Core	Rim	Rim	Rim		Core	Rim
SiO$_2$	38.37	39.34	39.01	39.03	37.14	37.41	37.46	37.41	37.02		54.17	54.03
TiO$_2$	0.20	0.07	0.04	0.06	0.08	0.12	b.d.	0.08	0.05		0.12	0.16
Al$_2$O$_3$	21.32	21.68	21.58	21.58	20.52	19.24	19.40	20.63	20.17		4.15	4.40
FeO†	24.41	22.61	23.85	23.58	32.55	21.19	30.80	29.94	35.81		6.20	5.18
MnO	1.14	0.84	0.84	0.85	0.79	9.20	0.64	3.11	0.46		0.09	0.05
MgO	3.78	6.48	6.57	7.04	1.06	0.29	1.19	0.64	0.54		12.67	13.06
CaO	10.76	9.77	8.51	8.25	7.73	12.66	10.68	8.34	6.09		20.48	20.47
Na$_2$O	n.a.	n.a.	n.a.	n.a.	n.a.	n.a.	n.a.	n.a.	n.a.		2.38	2.57
Total	99.98	100.79	100.41	100.39	99.87	100.11	100.17	100.15	100.14		100.26	99.92
	Formulae based on 12 oxygens										On 6 oxygens	
Si	3.002	3.009	3.002	3.000	3.001	3.022	3.022	3.013	3.010	Si	1.979	1.971
										Al^{iv}	0.021	0.029
Al	1.966	1.954	1.957	1.954	1.955	1.833	1.845	1.957	1.933			
Ti	0.011	0.002	0.002	0.002	0.004	0.007	....	0.004	0.002	Al^{vi}	0.157	0.159
Fe^{+3}	0.023§	0.044§	0.041§	0.044§	0.041§	0.160§	0.155§	0.039§	0.065§	Ti	0.003	0.004
										Fe^{+3}	0.012**	0.019**
Fe^{+2}	1.573	1.402	1.493	1.472	2.158	1.272	1.923	1.977	2.370	Fe^{+2}	0.176	0.138
Mn	0.073	0.054	0.054	0.054	0.052	0.630	0.044	0.210	0.031	Mn	0.002	0.001
Mg	0.441	0.739	0.752	0.805	0.125	0.035	0.143	0.075	0.064	Mg	0.689	0.710
Ca	0.900	0.799	0.700	0.679	0.669	1.096	0.923	0.718	0.530	Ca	0.800	0.800
Σ	2.987	2.994	2.999	3.010	3.004	3.033	3.033	2.980	2.995	Na	0.168	0.181
										Σ	2.007	2.012
	Endmembers in mole percent											
Alm	52.7	46.8	49.7	48.9	71.8	41.9	63.4	66.3	79.1	*Ac*	1.2	1.9
Gr	28.4	24.4	21.3	20.3	20.0	27.6	22.5	22.0	14.4	*Jd*	15.5	16.0
And	1.7	2.3	2.1	2.3	2.3	8.5	7.9	2.1	3.3	*CaTs*	1.1	1.3
Py	14.8	24.7	25.1	26.7	4.2	1.2	4.7	2.6	2.2	*Hd*	17.5	13.6
Sp	2.4	1.8	1.8	1.8	1.7	20.8	1.5	7.0	1.0	*Di*	60.9	64.1
										En	3.8	3.1

*Abbreviations listed in Appendix 2.
†Total Fe = FeO.
§Fe^{+3} = 2-Al-Ti.
**Fe^{+3} = 12-total cation charge assuming all Fe = Fe^{+2}.
n.a. = not analyzed. b.d. = below detection limit.

garnet$_I$ growth. Compositionally, garnet$_I$ is zoned (Fig. 3; Table 1), with a continuous increase in pyrope content from core to rim, and a corresponding decrease in almandine and grossular content. Garnet$_I$ poikiloblasts are subsidiary in numbers to smaller (0.4 to 1.1 mm in diameter), relatively inclusion poor garnet$_{II}$ grains. Garnet$_{II}$ is pyrope-rich, and displays a more subdued compositional zoning than garnet$_I$ (Table 1). Because garnet$_{II}$ grains often form a mantle around garnet$_I$ poikiloblasts, and because garnet$_{II}$ is similar in composition to garnet$_I$ rims, garnet$_{II}$ growth is interpreted to postdate, or to have occurred during the latest stages of garnet$_I$ growth. Although garnet$_{II}$ is relatively inclusion poor in comparison to garnet$_I$, it often contains tiny inclusions of rutile, epidote, and pyrite.

A third generation of compositionally and texturally distinct garnet occurs only in pervasively altered domains of the Clara Creek eclogite. Garnet$_{III}$ is distinguished from earlier generations of garnet on the basis of the following criteria. (1) Garnet$_{III}$ exists in apparent equilibrium with minerals of secondary origin such as hornblende, epidote, chlorite, sphene, albite, ilmenite, and pyrite. (2) Garnet$_{III}$, hornblende, epidote, and sphene occur as nonreactant idioblastic inclusions within large weblike grains of late-stage albite. (3) Garnet$_{III}$ grains are occasionally poikiloblastic, and contain inclusions of hornblende, epidote, chlorite, and ilmenite. As discussed later in this paper, chlorite, albite, and ilmenite postdate the growth of garnet$_I$ and garnet$_{II}$, as does growth of much of the epidote and amphibole

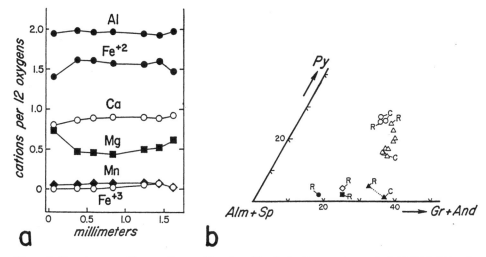

Figure 3. Garnet compositions. a, Compositional profile of eclogite garnet$_I$, sample 6-14-83. b, Triangles = garnet$_I$, eclogite; circles = garnet$_{II}$, eclogite; diamond = garnet$_{III}$, eclogite; filled square = garnet from metapelite; filled circle = garnet from metachert; filled triangles = garnet from mafic schist. R = rim analysis, C = core analysis. Abbreviations listed in Appendix 3.

present in eclogite; these relations suggest that garnet$_{III}$ represents a distinctly later phase of garnet growth. Garnet$_{III}$ is significantly more almandine-rich than garnet$_I$ and garnet$_{II}$ and is compositionally similar to almandine-rich garnets in mafic schists of the Schist belt (Fig. 3).

Amphibole

Amphibole is abundant in the Clara Creek eclogite, and is divided into three textural types: Amphibole$_I$ occurs as tiny (<0.05 mm in length) inclusions in garnet$_I$ poikiloblasts; because of their small size, reliable microprobe analyses of amphibole$_I$ grains were not obtained. Amphibole$_{II}$ is the most abundant amphibole in the Clara Creek eclogite, and is discussed in detail below. Amphibole$_{III}$ is the youngest amphibole present, occurring as fibrous actinolite fracture-fill in late-stage veins; a representative microprobe analysis of amphibole$_{III}$ is presented in Table 2, analysis 7, and its composition is shown graphically in Figure 4a.

Amphibole$_{II}$. Amphibole$_{II}$ occurs in textural configurations indicating that it grew throughout the metamorphic history of the Clara Creek eclogite. It is intergrown with other silicate phases in relatively unaltered as well as pervasively altered portions of the eclogite body. Amphibole$_{II}$ is also an important component of polyminerallic veins, vein selvages, and diffuse patches of secondary alteration. Amphibole$_{II}$ occurs throughout the eclogite body in a variety of textural configurations, and exhibits a consistent zoning pattern regardless of whether it occurs in relatively pristine eclogite, in veins, or in pervasively altered regions. A typical compositional zoning profile is shown in Figure 5, and representative microprobe analyses are presented in Table 2. Cores of zoned amphibole$_{II}$ are composed mainly of colorless to light green pleochroic actinolitic-hornblende with occasional patches of glaucophane (Figs. 5a and 6).

Where actinolitic-hornblende and glaucophane are both present, they comprise single, crystallographically continuous amphibole grains; contacts between the two amphiboles are optically and chemically sharp. Similar textures have been described by other workers (e.g., Himmelberg and Papike, 1969; Black, 1973; Robinson et al., 1982; Thurston, 1985) who interpret them as an equilibrium intergrowth phenomenon, rather than a syntaxial replacement texture. Actinolitic-hornblende and glaucophane from the Clara Creek eclogite are similar in composition to coexisting sodic and calcic amphiboles from other relatively high temperature (T>400°C) blueschist terranes, supporting an equilibrium intergrowth interpretation for textures in amphibole$_{II}$ (Fig. 7). Representative microprobe analyses of coexisting actinolitic-hornblende and glaucophane are presented in Table 2, analyses 1 through 3. Amphibole$_{II}$ cores are usually inclusion-free, but occasionally contain tiny grains of unreacted garnet, rutile, and sphene.

Actinolitic-hornblende ± glaucophane cores are mantled by amphibole that exhibits a rimward color gradation from light green to dark green (Fig. 5b) reflecting a rimward change in composition from actinolite to hornblende, characterized by the coupled substitution:

$$(\square)_A (Ca)_{M4} (Mg, Fe^{+2})_C (Si)_T = (Na)_A (Na)_{M4} (Al^{VI}, Fe^{+3})_C (Al^{IV}).$$

An abrupt compositional break between zoned actinolite to hornblende rims and actinolitic-hornblende ± glaucophane cores is indicated by a chemical inflection on composition profiles (e.g., at about 225 µm, Fig. 5a). These relations suggest that zoned amphibole rims may represent a distinctly later phase of amphibole growth on cores of actinolitic-hornblende ± glaucophane. Representative microprobe analyses of zoned amphibole rims are presented in Table 2, analyses 4 through 6.

TABLE 2. REPRESENTATIVE MICROPROBE ANALYSIS OF AMPHIBOLES

	Eclogite						Vein		Metagabbro	Sodic-	Massive Metabasite		Mafic Schist		
	Glaucophane	Glaucophane	Actinolitic Hornblende	Actinolite	Hornblende	Hornblende	Actinolite	Crossite	Barroisite	Hornblende	Hornblende	Hornblende	Actinolite	Actinolite	Actinolite
	6-14-83-3	8-4-84-3a	6-14-83-3	6-14-83-3	6-14-83-3	6-20-84-2	6-20-84-3	SBR-88-56	7-23-84-1	SBR-88-56	7-22-85-7	7-22-85-6	7-24-83-1	6-12-84-2	6-12-85-9
	1	2	3	4	5	6	7	8	9	10	11	12	13	14	15
SiO_2	56.30	57.50	51.55	54.74	45.34	47.89	54.75	54.50	48.18	47.61	45.94	48.99	55.32	55.20	52.81
TiO_2	b.d.	0.06	0.35	0.10	0.15	0.10	0.03	0.04	0.12	0.20	0.14	0.04	b.d.	0.03	0.05
Al_2O_3	8.30	9.30	7.48	1.40	11.84	7.07	1.88	4.98	6.01	5.73	10.73	9.69	1.25	1.95	3.51
Fe_2O_3*	3.91	2.55	1.65	2.04	2.44	4.90	2.61	8.49	5.26	4.01	3.56	3.48	0.05	1.32	1.55
FeO	12.04	10.69	5.35	9.53	13.21	18.62	11.57	16.22	18.92	19.18	20.15	16.29	12.95	11.05	13.79
MnO	0.09	0.07	0.05	0.10	0.39	0.55	0.22	0.21	0.41	0.57	0.18	0.06	0.22	0.22	0.16
MgO	8.52	9.70	17.72	16.88	10.64	8.24	15.56	5.62	8.04	8.66	7.44	8.82	15.53	16.15	13.93
CaO	1.05	1.01	10.64	11.09	10.02	8.77	10.66	0.89	8.05	9.22	8.65	8.78	12.40	11.05	11.26
Na_2O	6.68	6.95	2.18	1.11	3.11	2.77	1.38	6.39	2.76	2.74	1.26	2.36	0.25	1.00	0.70
K_2O	0.04	0.03	0.63	0.05	0.07	0.04	0.13	0.03	0.33	0.32	0.21	0.15	0.03	0.04	0.10
H_2O†	2.13	2.16	2.14	2.11	2.04	1.99	2.08	2.06	1.99	1.98	2.00	2.04	2.09	2.11	2.07
Total	99.06	100.02	99.74	99.15	99.25	100.94	100.87	99.43	100.07	100.22	100.26	100.70	100.09	100.12	99.93
Formulae based on 23 oxygens															
Si	7.986	7.981	7.250	7.839	6.701	7.134	7.787	7.998	7.253	7.188	6.870	7.144	7.922	7.851	7.645
Al^{iv}	0.014	0.019	0.750	0.161	1.299	0.866	0.213	0.002	0.747	0.812	1.130	0.856	0.078	0.149	0.355
Al^{vi}	1.374	1.503	0.490	0.075	0.764	0.376	0.102	0.859	0.320	0.207	0.762	0.810	0.133	0.178	0.244
Ti	......	0.006	0.037	0.011	0.017	0.011	0.003	0.004	0.014	0.023	0.016	0.004	......	0.003	0.005
Fe^{3}†	0.417	2.66	0.175	0.220	0.272	0.549	0.280	0.938	0.596	0.456	0.400	0.382	0.006	0.141	0.169
Fe^{2}	1.428	1.241	0.629	1.141	1.633	2.320	1.376	1.991	2.383	2.422	2.520	1.987	1.550	1.314	1.670
Mn	0.011	0.008	0.006	0.012	0.049	0.069	0.027	0.026	0.052	0.073	0.023	0.007	0.027	0.027	0.020
Mg	1.801	2.007	3.714	3.603	2.343	1.829	3.298	1.229	1.804	1.948	1.658	1.917	3.314	3.423	3.005
Σ	13.031	13.031	13.050	13.062	13.078	13.154	13.086	13.047	13.169	13.129	13.379	13.107	13.030	13.086	13.133
Ca	0.160	0.150	0.603	1.702	1.587	1.400	1.625	0.140	1.299	1.492	1.386	1.372	1.903	1.684	1.747
Na(M4)	1.837	1.850	0.397	0.298	0.413	0.600	0.375	1.818	0.701	0.508	0.365	0.628	0.069	0.276	0.197
Na(A)	......	0.020	0.197	0.010	0.478	0.200	0.006	......	0.105	0.294	......	0.039	......	......	......
K	0.007	0.005	0.113	0.009	0.013	0.008	0.024	0.006	0.063	0.062	0.040	0.028	0.006	0.007	0.019

*Mean calculated Fe^{+3} values, using algorithm of Spear and Kimball, 1984.
†Stoichiometric H_2O content.
b.d. = below detection limit.

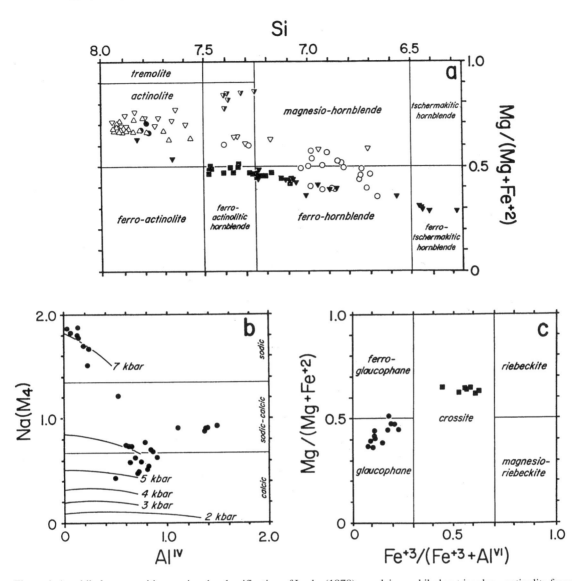

Figure 4. Amphibole compositions, using the classification of Leake (1978). a, calcic amphiboles; triangles = actinolite from mafic schists; inverted triangles = rims of zoned amphiboles from eclogite; half-filled inverted triangles = actinolitic-hornblende intergrown with glaucophane in cores of zoned amphiboles from eclogite; filled inverted triangles = zoned amphiboles from pervasively altered eclogite; circles = amphibole from massive metabasites; filled squares = calcic amphibole from metagabbro. b, amphiboles from metagabbro, plotted with NaM_4 versus tetrahedral Al; isobars from Brown (1977). c, sodic amphiboles; circles = glaucophane from eclogite; squares = crossite from metagabbro.

Amphibole$_{II}$ in polymineralic veins and vein selvages exhibit identical zoning to that observed in matrix amphibole$_{II}$. Although actinolitic-hornblende ± glaucophane cores are rarely present, most vein-hosted amphibole$_{II}$ displays only the continuous actinolite to hornblende zoning described above. The similarity in the zoning patterns suggests that growth of vein-hosted and matrix amphibole$_{II}$ was, at least in part, contemporaneous.

Amphibole$_{II}$ in pervasively altered eclogite is also zoned, and displays a compositional zoning similar to that in vein-hosted and matrix amphibole$_{II}$ (Fig. 5b). Individual grains exhibit a gradational color change from light green cores to blue-green rims, reflecting a change in composition from actinolite to ferro-

hornblende or ferro-tschermakitic hornblende (Fig. 4a). Unzoned ferro-hornblende grains are also common. Glaucophane rarely occurs as isolated patches within light green cores.

Textural and microstructural relations in eclogite indicate that the zoned actinolite to hornblende rims of amphibole$_{II}$ represent a relatively late stage of mineral growth that, in addition to post-dating the growth of clinopyroxene, glaucophane, and actinolitic-hornblende, also postdated the growth of garnet$_I$ and garnet$_{II}$. Hornblende, along with epidote, sphene, apatite, and pyrite, often occurs in fractures that transect garnet$_I$ and garnet$_{II}$ grains. Hornblende grains, some exhibiting the actinolite to hornblende compositional zoning, often occur along with epidote + sphene +

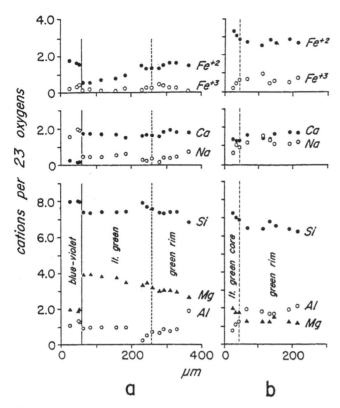

Figure 5. Compositional profile of amphiboles from eclogite, core to rim = left to right. a, profile of matrix amphibole from sample 6-23-84-1b; includes glaucophane patch (blue-violet zone) in core. b, profile of amphibole from pervasively altered eclogite, sample 6-20-84-4. Constructed from microprobe analyses normalized to median calculated Fe+3 using the algorithm of Spear and Kimball (1984).

pyrite ± apatite in irregular clots and masses that occur along grain boundaries between garnet$_I$, garnet$_{II}$, and sodic-augite. Zoned amphibole$_{II}$ is abundant in pervasively altered eclogite, and is commonly intergrown with garnet$_{III}$, chlorite, sphene, epidote, albite, and ilmenite; it occasionally contains inclusions of sphene, epidote, ilmenite, and pyrite.

Epidote

Epidote occurs in textural configurations indicating that it remained stable over much of the metamorphic history of the Clara Creek eclogite. It occurs in apparent textural equilibrium with garnet$_I$, garnet$_{II}$, sodic-augite, amphibole$_{II}$, rutile, and apatite in relatively unaltered portions of the eclogite body, and exists as inclusion within garnet$_I$ and garnet$_{II}$ grains. In altered domains of the eclogite body where garnet$_I$ and garnet$_{II}$ are partially preserved, epidote occurs as tiny veins in fractured garnets, and forms coronas around partially resorbed garnet grains. Epidote is also the chief constituent of polymineralic (epidote + amphibole$_{II}$ + sphene ± apatite) veins. Where eclogite is pervasively altered, epidote occurs as idioblastic to subidioblastic grains that are intergrown with zoned amphibole$_{II}$ (actinolite to hornblende, core to rim), sphene, albite, chlorite, garnet$_{III}$, pyrite, and ilmenite; epidote occasionally contains inclusions of hornblende, sphene, and rarely, glaucophane and pyrite.

Plagioclase

Plagioclase is nearly pure end-member albite (An$_1$ to An$_3$), and is restricted in its occurrence to pervasively altered eclogite. Most often, albite occurs as large, weblike grains that surround idioblastic to subidioblastic grains of hornblende, epidote, sphene, garnet, pyrite, and apatite. Albite also occurs as fracture-fill within broken garnet$_I$, garnet$_{II}$, and amphibole$_{II}$ grains, and

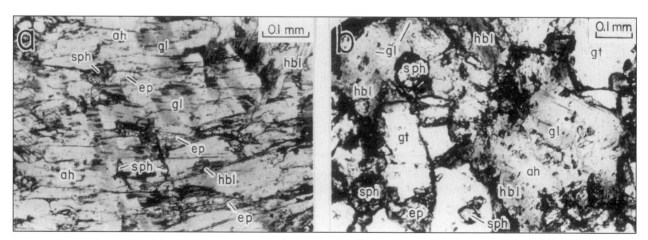

Figure 6. Photomicrographs of zoned amphiboles from eclogite. a, patchy intergrowth of actinolitic-hornblende (ah) and glaucophane (gl) in cores of zoned amphiboles; note partial replacement of glaucophane by hornblende. b, zoned amphiboles with cores of glaucophane + actinolitic-hornblende, and rims of continuously zoned amphibole, grading outward from actinolite to hornblende; color gradation in rim corresponds to composition gradient; ep = epidote, sph = sphene, gt = garnet, hbl = hornblende.

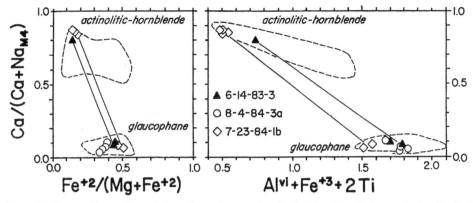

Figure 7. Compositions of coexisting glaucophane and actinolitic-hornblende from eclogite. Dashed lines outline fields of coexisting sodic and calcic amphiboles from relatively high temperature (T > 400°C blueschist terranes, from Robinson et al. (1982).

forms monomminerallic veins that crosscut epidote + amphibole$_{II}$ + sphene veins. These relations indicate that albite formed at a relatively late stage in the metamorphic history of the Clara Creek eclogite.

Chlorite

Chlorite is also a late-stage mineral phase. It occurs mainly within fractures in garnet$_I$ and garnet$_{II}$, as coronas on partially resorbed garnet$_I$ and garnet$_{II}$ grains, and in pervasively retrograded domains where it occurs as part of an equilibrium intergrowth of epidote, amphibole$_{II}$ (displaying actinolite to hornblende rimward zoning), albite, sphene, and garnet$_{III}$, ilmenite, and pyrite.

Ti-bearing phases

Rutile is a ubiquitous accessory mineral in eclogite, occurring as grains as large as 1.5 mm in diameter that are mantled by coronas of sphene. Rutile also exists as unreacted inclusions where armored by sodic-augite, garnet$_I$, garnet$_{II}$, and actinolitic-hornblende ± glaucophane cores of amphibole$_{II}$.

Sphene is also a common accessory mineral, occurring in textural configurations indicating a long history of growth: It occurs as inclusions in garnet$_I$ and garnet$_{II}$, in cores and rims of zoned amphibole$_{II}$, and in epidote and albite. Sphene is a constituent of epidote + amphibole$_{II}$ + sphene veins, and rarely forms monomineralic veins that crosscut, and are themselves transected by epidote + amphibole$_{II}$ + sphene veins. Sphene also occurs in intergranular regions between sodic-augite, garnet$_I$, and garnet$_{II}$, where it forms an equilibrium intergrowth with epidote, hornblende, chlorite, and pyrite. Sphene grains often contain tiny inclusions of relict rutile; rarely, sphene is intergrown with fine-grained ilmenite.

Ilmenite also occurs in pervasively altered domains of eclogite where it occurs as needlelike inclusions within garnet$_{III}$ and zoned amphibole$_{II}$.

Other accessory minerals

Apatite, pyrite, and rarely, chalcopyrite are minor constituents of the Clara Creek eclogite. Apatite is intergrown with sodic-augite, garnet$_I$, garnet$_{II}$, epidote, and amphibole in relatively unaltered domains, occurs as a minor constituent of epidote + amphibole$_{II}$ + sphene veins, and coexists with epidote, amphibole$_{II}$, sphene, and albite in pervasively altered portions of the eclogite body.

Xenoblastic opaque masses as large as 2.0 mm in diameter are common in relatively fresh and pervasively altered portions of the Clara Creek eclogite. These large, ragged grains are composed mainly of pyrite, but tiny blebs of chalcopyrite are also rarely present. Pyrite also occurs as rare inclusions in amphibole$_I$, amphibole$_{II}$, epidote, albite, garnet$_{II}$, and garnet$_{III}$.

SUMMARY OF TEXTURAL RELATIONS IN ECLOGITE

Based on the textural relations described in the previous section, the relative ages of mineral growth in the Clara Creek eclogite have been determined (Fig. 8). These relations, listed below, indicate that sodic-augite, garnet$_I$, garnet$_{II}$, epidote, actinolitic-hornblende, glaucophane, rutile, apatite, and possibly sphene comprise the prograde equilibrium assemblage. Textural relations also indicate that the retrograde assemblage consists of continuously zoned actinolite-to-hornblende (as overgrowths on older actinolitic-hornblende ± glaucophane cores), epidote, chlorite, garnet, albite, sphene, pyrite ± chalcopyrite, ilmenite, and apatite. (1) Sodic-augite, garnet$_I$, garnet$_{II}$, epidote, actinolitic-hornblende, glaucophane, rutile, and apatite comprise an equilibrium intergrowth in unaltered portions of the eclogite body. (2) Inclusions of amphibole$_I$, epidote, rutile, sphene, and pyrite exist within minerals of the prograde assemblage, indicating that they either predated, or grew contemporaneously with their host minerals. (3) Textural relations indicate that zoned actinolite-to-hornblende overgrowths on actinolitic-hornblende ± glaucophane cores also postdate the growth of sodic-augite, garnet$_I$, and garnet$_{II}$; there-

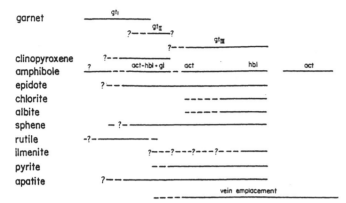

Figure 8. Relative timing of mineral growth in eclogite; act = actinolite; gt = garnet; hbl = hornblende.

fore zoned amphibole rims and equilibrium assemblages that include zoned amphibole postdate the assemblage sodic-augite + garnet$_I$/garnet$_{II}$ + actinolitic-hornblende + glaucophane. (4) In domains of the eclogite affected by retrograde metamorphism, sodic-augite, garnet$_I$, garnet$_{II}$, epidote, actinolitic-hornblende, glaucophane, rutile, and apatite are overprinted and partially to completely replaced by an assemblage that includes zoned amphibole, epidote, chlorite, garnet, albite, sphene, pyrite (± chalcopyrite), ilmenite, and apatite. This late-stage assemblage is in turn postdated by the emplacement of fibrous actinolite in brittle fractures.

Field, microstructural, and textural relations indicate that both prograde and retrograde metamorphism of eclogite involved a metamorphic fluid phase. The presence of hydrous mineral phases such as glaucophane, actinolitic-hornblende, and epidote, as well as the abundance of primary fluid inclusions in sodic-augite attest to the importance of metamorphic fluids during prograde metamorphism. These relations support the conclusions of other workers (e.g., Newton, 1986; Koons et al., 1987; Wayte et al., 1989; Klemd, 1989) that fluids play an essential role in the eclogitization of mafic rocks. The abundance of veins containing secondary minerals indicates that retrograde metamorphism of eclogite was accommodated by the influx of fluids into fractures in eclogite, and aided by diffusion along grain boundaries and microscopic intragranular cracks. Hydrous retrograde metamorphism of eclogite to blueschist-, amphibolite-, or greenschist-facies assemblages is common, and has been described in other *HP/LT* metamorphic terranes (e.g., Maresch and Abraham, 1981; Klemd, 1989).

GEOLOGY OF HOST ROCKS

Quartz-mica schist

Pelitic and semipelitic quartz-mica schists are the dominant rock-type in Schist belt along the Middle Fork Koyukuk River, and are inferred to enclose the Clara Creek eclogite (Fig. 2). Quartz-mica schists are complexly deformed and affected by two syn-

metamorphic phases of penetrative deformation. The oldest structures (referred to herein as D_{1a}) are poorly preserved isoclines and sheath folds of probable Middle to Late Jurassic age that formed under *HP/LT* metamorphic conditions (Gottschalk et al., this volume, Chapter 12). In the Early Cretaceous, D_{1a} folds and fabrics were refolded by a second set of close to isoclinal folds (D_{1b} folds); D_{1b} folding was accompanied by dynamic recrystallization and metamorphic mineral growth under lower amphibolite- to greenschist-facies conditions (Gottschalk, 1987, 1990; Little et al., 1994; Gottschalk et al., this volume, Chapter 12; Gottschalk and Snee, this volume, Chapter 13).

Pelitic and semipelitic quartz-mica schists are composed of quartz + phengite + paragonite + chlorite ± albite ± chloritoid, with accessory ilmenite, sphene, pyrite, apatite, tourmaline, and epidote; calcite, and very rarely, almandine-rich garnet may also be present in minor amounts. Pelitic and semipelitic varieties differ mainly in their quartz content, with quartz being the dominant mineral in semipelitic schists, and white mica being the main constituent of pelitic schists. Representative whole-rock chemical analyses and modal mineralogies of quartz-mica schists are presented in Table 3.

Phengite from quartz-mica schists exhibit a range of composition from 3.25 to 3.48 Si per formula unit (normalized to 11 oxygens) and Mg/(Mg + Fe) between 0.61 to 0.77; paragonite exhibits little compositional variation and is nearly pure end-member paragonite (Fig. 9). Quartz-mica schists commonly contain both white micas in textural equilibrium, occurring either as discrete grains, or as subgrain-scale intergrowths. Chloritoid in quartz-mica schist is Fe-rich ((Fe/Fe+Mg) = 0.81 to 0.87 in analyzed samples) and plagioclase is essentially pure end-member albite. Representative microprobe analyses of phengite, paragonite, and chloritoid in quartz-mica schists are presented in Table 4. Garnet is rare in quartz-mica schists, occurring in only three out of several hundred quartz-mica schist samples examined in this study (Fig. 3, Table 1).

Although glaucophane has been recognized as an important minor constituent in quartz-mica schists in other portions of the Schist belt (Gilbert et al., 1977; Nelson and Grybeck, 1982; Hitzman et al., 1986; Till, 1988), no glaucophane was identified in quartz-mica schists along the Middle Fork Koyukuk River. However, diamond- and lath-shaped mats of finely intergrown albite + chlorite ± quartz after glaucophane are common in pelitic and semipelitic schists (Fig. 10), and have been interpreted as pseudomorphs after glaucophane (Gottschalk, 1987, 1990). The long axes of pseudomorphs are oriented parallel to the D_{1a} foliation, and are locally folded around D_{1b} fold hinges. These relations indicate that glaucophane was part of the syn-D_{1a} assemblage in quartz-mica schists, which also includes quartz + phengite + paragonite + chlorite ± chloritoid. Along the Middle Fork Koyukuk River, glaucophane growth appears to be restricted to the D_{1a} folding phase (Gottschalk, 1987, 1990) but elsewhere in the Schist belt glaucophane growth may have occurred during the D_{1b} folding phase (A. B. Till, personal communication, 1993).

TABLE 3. REPRESENTATIVE WHOLE-ROCK CHEMICAL ANALYSES AND MODAL MINERALOGIES OF QUARTZ-MICA SCHIST AND ALBITE SCHIST*

	Quartz-Mica Schist						Albite Schist		
	7-8-83-13	7-24-84-3	6-14-83-1	7-8-83-7	7-28-85-1	6-10-83-2	6-24-85-1	7-12-84-7	6-19-85-4
(wt. %)									
SiO_2	73.60	76.76	69.71	62.18	70.85	65.32	72.53	61.32	59.04
TiO_2	0.65	0.77	0.94	1.16	0.79	1.12	1.11	0.94	0.68
Al_2O_3	13.37	11.25	16.20	19.08	14.62	18.58	14.18	16.66	11.57
$FeO^†$	6.03	5.94	5.57	6.18	4.54	6.37	5.32	8.48	4.99
MnO	0.05	0.05	0.04	0.06	0.04	0.05	0.05	0.06	0.09
MgO	1.89	1.74	2.01	2.01	1.68	2.25	1.69	3.09	3.04
CaO	0.17	0.67	0.09	0.13	0.16	0.16	0.13	0.15	8.85
Na_2O	1.22	1.07	1.18	1.56	1.68	0.74	0.71	1.52	1.17
K_2O	1.44	1.24	2.47	2.50	2.64	3.51	2.66	2.51	1.17
P_2O_5	0.11	0.08	0.15	0.14	0.14	0.14	0.06	0.22	0.09
H_2O	2.98	2.02	3.21	3.94	3.66	3.21	2.69	3.85	8.78
Total	101.51	101.59	101.57	98.94	100.80	101.45	101.13	98.80	99.47
Qtz	x	x	x	x	x	x	x	x	x
Wt mica	x	x	x	x	x	x	x	x	x
Chl	x	x	x	x	x	x	x	x	x
Ctd	-	x	-	x	x	x	-	-	-
Ab	ag	ag	x	-	-	-	-	x	x
(Gl)	x	x	-	-	-	-	-	-	-
Ap	x	x	x	x	x	x	x	x	x
Tour	x	x	-	x	x	x	-	x	x
Op	x	x	x	x	x	x	x	x	x
Cc	-	-	-	-	-	-	-	-	x
Dol	-	-	-	-	-	-	-	-	x

*All elements determined by induction-coupled plasma spectroscopy (ICP) except H_2O, determined by loss on ignition. Abbreviations are listed in Appendix 2. x = mineral present; (gl) = pseudomorphs after glaucophane; ag = after glaucophane.
†Total Fe = FeO.

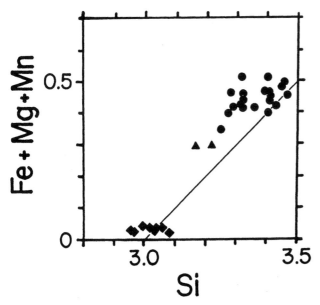

Figure 9. White mica compositions, normalized to 11 oxygens per formula unit. Diamonds = paragonite from quartz-mica schist; circles = phengite from quartz-mica schist; triangles = phengite from marble.

Quartz-mica schists were extensively recrystallized during D_{1b} folding, resulting in the dynamic recrystallization of quartz, and the syndeformational growth of phengite, paragonite, chlorite, and chloritoid along microlithon boundaries. Syn-D_{1b} albite porphyroblasts with snowball-type inclusion trails are also common in pelitic and semipeltitc quartz-mica schists; there is no unequivocal evidence for pre-D_{1b} growth of albite in quartz-mica schists. Following D_{1b}, quartz-mica schists experienced a period of recrystallization under "static" conditions, resulting in the growth of randomly oriented white mica and chlorite grains, and the appearance of albite porphyroblasts that helicitically enclose D_{1b} structures.

Mafic schist and massive metabasite

Mafic schist and massive metabasite occur as intrafolial layers within the quartz-mica schist that range from a few centimeters to as much as 140 m in thickness. Although mafic schist and metabasite are minor components of the Schist belt along the Middle Fork Koyukuk River, they are particularly abundant in the area between Marion and Clara Creeks (Fig. 2). Mafic schists are weakly foliated rocks composed of tiny (>1.0 mm in length)

TABLE 4. REPRESENTATIVE MICROPROBE ANALYSES OF MICAS AND CHLORITOID*

	Pelitic Schist		Metachert	Pelitic Schist
	Phengite 8-6-85-1	Paragonite 8-9-84-5	Biotite 7-21-85-1	Chloritoid 7-24-84-1
SiO_2	51.32	47.09	36.93	24.57
TiO_2	0.28	0.04	1.72	b.d.
Al_2O_3	28.78	39.53	17.72	40.34
$FeO^†$	1.63	0.35	24.17	25.04
MnO	b.d.	0.01	0.17	0.35
MgO	2.46	0.13	6.02	2.26
CaO	0.07	0.10	0.02	b.d.
Na_2O	0.50	7.78	0.04	b.d.
K_2O	10.15	0.47	9.29	b.d.
$H_2O^§$	4.54	4.69	3.90	7.26
Total	99.73	100.19	99.98	99.82

	Formulae based on 11 oxygens			On 12 oxygens	
Si	3.400	3.001	2.839	Si	2.032
Al^{iv}	0.600	0.999	1.161	Al	3.934
Al^{vi}	1.648	1.971	0.445	Fe^{+3}	0.066**
Ti	0.014	0.002	0.099	Fe^{+2}	1.666
Fe	0.090	0.019	1.554	Mn	0.025
Mn	....	0.001	0.011	Mg	0.279
Mg	0.243	0.012	0.690	Ca	....
Σ	1.995	2.005	2.799	Σ	1.970
Ca	0.005	0.007	0.002		
Na	0.064	0.961	0.006		
K	0.858	0.038	0.911		
Σ	0.927	1.006	0.919		

*Procedure for estimation of Fe^{+3} in chloritoid from Schliestedt, 1986.
†Total Fe = FeO.
§Stoichiometric H_2O content.
**Fe^{+3} = 4 - Al^{vi}.
b.d. = below detection limit.

actinolite. Representative microprobe analyses of hornblende in massive metabasite are presented in Table 2, analyses 11 and 12.

Metagabbro

Metagabbros are massive metabasites with relict igneous textures that consist of metamorphic albite, chlorite, amphibole, epidote, calcite, stilpnomelane, and sphene with accessory magnetite, pyrite, and apatite. Metagabbro is rare along the Middle Fork Koyukuk River, and occurs only in the southernmost and lowest grade exposures of the Schist belt near the settlement of Coldfoot. Amphiboles from metagabbro straddle the compositional boundary between sodic-hornblende and barroisite, and contain rare cores and patches of crossite (Fig. 4b and c; Table 2, analyses 8, 9, and 10). The absence of a sharp chemical boundary between crossite and barroisite and the lack of a distinct core-to-rim zoning suggests that this is a replacement texture in which barroisite replaces older crossite grains. Metagabbro outcrops are transected by late-stage brittle fractures filled by calcite, chlorite, albite, stilpnomelane, and epidote.

Albite schist

Albite schist is an important minor component of the Schist belt in the immediate area surrounding the Clara Creek eclogite (Fig. 2); here, it is interlayered with quartz-mica schist on a scale of centimeters to meters, and is laterally and vertically gradational into mafic schist. Albite schist contains the assemblage albite + chlorite + white mica + quartz + calcite ± dolomite ± sphene ± epidote ± garnet with subsidiary graphite, tourmaline, apatite, and pyrite. Albite porphyroblasts are the main constituent of albite schist, comprising between 50 to 60% of the rock, and occurring in configurations indicating that albite growth is syn- to

grains of chlorite, actinolite, albite, epidote, garnet, and sphene, enclosed in porphyroblasts of albite, epidote, garnet, and calcite. Amphibole in mafic schist is actinolite (Fig. 4a; and Table 2, analyses 13, 14, and 15), and garnets are almandine-rich with spessartine-rich cores. Mafic schists commonly possess a single penetrative fabric and rarely display the polyphase structures characteristic of quartz-mica schists. Random growth and replacement textures indicate postkinematic growth of albite, epidote, calcite, chlorite, stilpnomelane, and biotite.

Massive metabasites are weakly foliated to massive aggregates of amphibole, chlorite, epidote, sphene, calcite, and rare white mica. Although massive mafic schists are mineralogically similar to mafic schists, the composition of amphibole from these rocks differ significantly. In thin section, amphiboles from massive mafic schist are deep blue-green in color and range from Fe-rich hornblende to actinolitic-hornblende in composition (Fig. 4a). In several samples, hornblende is partially to completely replaced by a fine-grained intergrowth of albite and

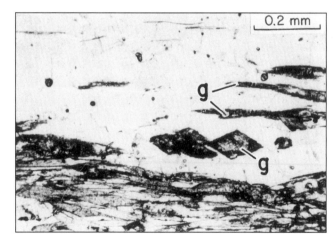

Figure 10. Photomicrograph of pseudomorphs after syn-D_{1a} glaucophane (g) composed of randomly oriented intergrowth of albite + chlorite ± quartz, from quartz-mica schist. Other syn-D_{1a} foliation-forming minerals in photomicrograph include white mica, chlorite, and quartz.

post-D_{1b}. Albite is essentially pure Na-feldspar (An_1 to An_3). Representative whole-rock chemical analyses and modal mineralogies of albite schist are presented in Table 3.

Marble

Marble is a minor but conspicuous constituent of the Schist belt along Middle Fork Koyukuk River, forming massive gray outcrops that are readily distinguishable from the surrounding quartz-mica schists. Marbles are mineralogically simple and consist of mainly of calcite ± dolomite with subsidiary quartz, phengite, and graphite. Outcrops are generally gray in color, with boudinaged, isoclinally folded layers of white marble suggesting a refolded early metamorphic fabric. In thin section, however, calcite forms a polygonal-granoblastic aggregate of slightly elongate grains suggesting complete dynamic recrystallization during D_{1b}.

Metachert

Metachert was found at only one locality along the Middle Koyukuk River and is noteworthy because it contains biotite, a relatively rare mineral in the Brooks Range Schist belt. Metachert occurs as a 0.5-m-thick layer within quartz-mica schist about 1 km northeast of the Clara Creek eclogite, and contains the assemblage quartz + stilpnomelane + garnet + biotite.

COMPARATIVE CHEMISTRY OF METABASITES FROM THE SCHIST BELT

Samples of mafic schist, metagabbro, and relatively fresh eclogite were selected for major and trace element analysis by inductively coupled plasma spectroscopy (ICP) and instrumental neutron activation analysis (INAA). Details on ICP and INAA analysis are presented in Appendix 3. For eclogite, care was taken to eliminate secondary vein material, but inclusion of minor amounts (<2%) of microscopically intergranular secondary material was unavoidable. Analyses are presented in Table 5.

Eclogite is a silica-poor, Fe-Ti-rich metabasite, in contrast to mafic schists and metagabbros that are uniformly basaltic in composition. Compositionally, eclogite differs markedly from other metabasites in the Schist belt because it contains less SiO_2 and Na_2O, and is relatively enriched in CaO, MgO, and particularly, FeO (as total Fe).

Since it is likely that metamorphism has caused significant changes in the major-element chemistry of metabasites in the Schist belt, the primary nature of metabasite protoliths was characterized using discrimination diagrams utilizing trace and minor elements that are relatively resistant to mobilization during secondary processes (e.g., Hughes, 1982). In the Schist belt, Zr, Ti, Y, P, Cr, and Ni all retain good correlations (Gottschalk, 1987) and are considered to have remained relatively immobile with respect to one another during metamorphism. Also, the relatively smooth trends exhibited by the rare-earth elements (REE) are suggestive of primary igneous values rather than secondary processes. Data plotted on discrimination diagrams (Fig. 11) show that metabasites from the Schist belt, including eclogite,

have high-field-strength element compositions that are similar to mid-ocean ridge basalts (MORB).

As in other blueschist/greenschist facies metamorphic terranes (e.g., Dungan et al., 1983; Sorensen, 1986; Owen, 1989; Biino and Pognante, 1990), metabasites from the Schist belt retain REE patterns comparable to unmetamorphosed basaltic rocks (Fig. 11d). Eclogite and mafic schist both exhibit the flat to light rare-earth element (LREE) depleted patterns typical of MORB (Sun et al., 1979; Le Roux et al., 1983; Vierek et al., 1989). Chondrite-normalized La/Sm ratios, $(La/Sm)_N$, of mafic schist show a slight LREE enrichment compared to normal MORB $(La/Sm)_N = 0.76$ and 0.90 versus 0.4 to 0.7 for normal MORB), and patterns are similar to those of the transitional MORB of Sun et al. (1979). Metagabbro sample 6-23-84-1 exhibits a marked LREE enrichment pattern, with $(La/Sm)_N = 1.59$. LREE enrichment in the mafic schist and metagabbro is consistent with derivation of mafic schist from a mantle-derived N-MORB that has undergone significant interaction with continental crust (Thompson et al., 1982; Dupuy and Dostal, 1984). This explanation best satisfies geologic relations in the Schist belt, where metabasites exist as thin layers within a thick metasedimentary sequence dominated by metamorphosed shales, siliciclastic rocks, and limestones deposited on continental crust. It should be noted, however, that LREE enrichments may also result from elemental mobility accompanying metamorphism or hydrothermal alteration processes (Hellman et al., 1979).

Major- and trace-element geochemical data indicate that eclogite was probably derived from a Fe- and Ti-rich N-MORB protolith. The consistency of trace-element data with an N-MORB protolith suggests that the low SiO_2 content of eclogite may have resulted from depletion during metamorphism, rather than reflecting an ultramafic composition for the protolith. It is not clear whether the contrasting major-element compositions of eclogite and other metabasites are a reflection of original compositional differences, or whether they result from varying degrees of major-element mobility during metamorphism. The depleted LREE pattern in eclogite suggests that (1) in contrast to mafic schists and metagabbros, its protolith was relatively uncontaminated by continental crust, or (2) LREE was removed from eclogite and incorporated in vein epidote during retrograde metamorphism.

If the compositional differences between the Clara Creek eclogite and other metabasites are primary, then chemical composition may have played a critical role in determining which rocks formed eclogitic assemblages and which did not; such compositional effects have been noted in the Nome Group schist of the Seward Peninsula (Thurston, 1985; Evans et al., 1987), as well as in other blueschist metamorphic belts (e.g., Schleistedt, 1986; Black et al., 1988).

GEOTHERMOMETRY

Garnet-Clinopyroxene

Temperatures of prograde metamorphism in the Clara Creek eclogite have been estimated using the garnet-

R. R. Gottschalk

TABLE 5. REPRESENTATIVE WHOLE-ROCK CHEMICAL ANALYSES
AND MODAL MINERALOGIES OF METABASITES*

	Mafic Schist		Massive Metabasite		Metagabbro		Eclogite	
	6-12-84-2	6-17-84-1	7-22-85-6	6-29-85-4	8-3-83-1	6-23-84-1	6-14-83-3	6-20-84-3
(wt. %)								
SiO_2	49.73	45.90	42.85	48.02	46.24	48.84	41.15	44.36
TiO_2	1.29	1.83	2.27	2.47	2.50	2.08	2.33	2.01
Al_2O_3	13.99	14.56	16.85	14.45	14.15	15.34	14.43	13.30
Fe_2O_3†	13.00	9.60	9.26	11.65	10.82	12.17	19.80	17.02
MnO	0.27	0.18	0.16	0.21	0.16	0.19	0.57	0.39
MgO	7.09	7.53	3.68	5.74	4.31	4.63	8.43	8.80
CaO	12.08	8.13	9.21	8.09	10.22	5.68	11.65	14.20
Na_2O	1.68	3.88	5.41	4.30	4.55	5.85	1.00	1.16
K_2O	0.07	0.01	0.13	0.03	0.20	0.15	0.06	0.01
P_2O_5	0.09	0.18	0.43	0.40	0.45	0.56	0.40	0.12
H_2O	1.44	8.11	7.97	4.05	4.28	2.15	1.81	0.58
Total	100.73	99.91	99.25	99.41	97.88	97.64	101.63	101.95
(ppm)								
Sr	166	237	377	188	762	243	34	34
Ba	83	4	331	30	144	40	14	13
Zr	58	103	157	164	231	178	172	102
Y	28.7	29.9	31.7	48.2	42.3	53.5	73.0	47.3
Sc	54.7	36.4	20.4	40.0	37.3	24.5	58.4	64.2
V	361	239	198	317	369	132	428	494
Cr	123	224	139	158	95	125	237	271
Ni	58.2	93.4	50.6	40.1	42.3	8.4	207.1	33.0
Co	40.3	37.9	n.d.	n.d.	n.d.	30.7	n.d.	26.0
Cs	0.24	n.d.	n.d.	n.d.	n.d.	1.70	n.d.	n.d.
La	3.8	4.5	n.d.	n.d.	n.d.	20.2	n.d.	1.3
Ce	10.4	13.5	n.d.	n.d.	n.d.	48.1	n.d.	4.3
Nd	9.0	8.0	n.d.	n.d.	n.d.	28.0	n.d.	16.8
Sm	2.70	3.72	n.d.	n.d.	n.d.	8.02	n.d.	3.14
Eu	1.03	1.29	n.d.	n.d.	n.d.	2.71	n.d.	1.30
Tb	0.72	0.78	n.d.	n.d.	n.d.	1.37	n.d.	0.92
Yb	3.48	2.81	n.d.	n.d.	n.d.	4.38	n.d.	4.10
Lu	0.53	0.42	n.d.	n.d.	n.d.	0.66	n.d.	0.66
Hf	1.87	3.03	n.d.	n.d.	n.d.	6.63	n.d.	3.20
Ta	0.30	0.29	n.d.	n.d.	n.d.	2.19	n.d.	0.70
U	b.d.	0.27	n.d.	n.d.	n.d.	0.43	n.d.	n.d.
Th	0.18	0.31	n.d.	n.d.	n.d.	1.81	n.d.	n.d.
Ab	x	x	x	x	x	x	x	-
Chl	x	x	x	x	x	x	-	-
Am	act	act	hbl	(hbl)act	cr-hbl	cr-hbl	gl-act-hbl	gl-act-hbl
Ep	x	x	x	x	x	x	x	x
Cc	x	-	-	-	x	x	-	-
Sph	x	x	-	x	x	x	rr	rr
Gt	-	x	x	x	-	-	x	x
Op	x	x	x	x	x	x	x	x
Qtz	x	-	-	-	-	-	-	-
Wt. mica	-	-	-	-	x	-	-	-
Ap	x	-	x	-	x	x	-	-
Stp	-	-	-	-	x	x	-	-
Bi	-	x	-	-	-	-	-	-
Cpx	-	-	-	-	-	-	x	x

*All elements determined by induction-coupled plasma spectroscopy (ICP) except rare-earth elements, determined by instrumental neutron activation analysis (INAA), and H_2O, determined by loss on ignition. Abbreviations listed in Appendix 2.
†Total Fe as Fe_2O_3.
n.d. = no data; x = mineral present ( ) = pseudomorphed mineral; rr = rims on rutile grains.

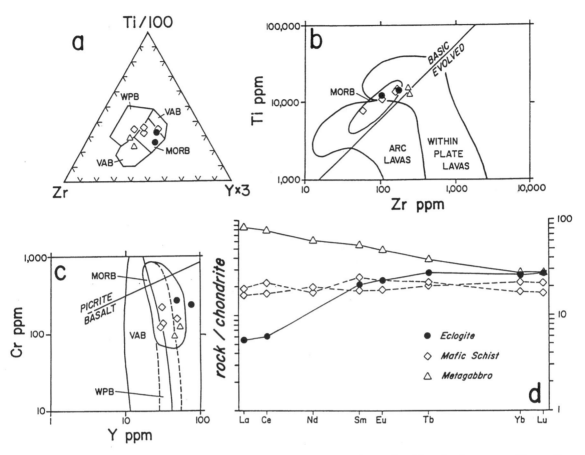

Figure 11. Trace-element contents of metabasites. Circles = eclogite; diamonds = mafic schist, triangles = metagabbro. a, discrimination diagram of Pearce and Cann (1973); b, discrimination diagram of Pearce and Norry (1979); c, discrimination diagram of Pearce (1982); WPB = within-plate basalt; MORB = mid-ocean ridge basalt; VAB = volcanic arc basalt; d, rare-earth element (REE) contents of metabasites normalized to CI (using values of Ebihara et al., 1982).

clinopyroxene Fe-Mg exchange geothermometers of Ellis and Green (1979), Krogh (1988), and Pattison and Newton (1989). Whereas the Pattison and Newton (1989) geothermometer does not account for Fe^{+3} in garnet and clinopyroxene, the calibrations of Ellis and Green (1979) and Krogh (1988) do, and are therefore dependent upon the method in which Fe^{+2}/Fe^{+3} ratios are determined from microprobe analyses. Because the calculated Fe^{+3} contents of sodic-augite and garnet from the Clara Creek eclogite are small (less than 0.046 Fe^{+3} per 12 oxygens in garnet, and less than 0.022 Fe^{+3} per 6 oxygens in sodic-augite) temperatures determined by the thermometers of Ellis and Green (1979) and Krogh (1988) should be relatively insensitive to the method of Fe^{+3} determination. Similarly, calculated temperatures should not be significantly affected by ignoring Fe^{+3}, as in the Pattison and Newton (1989) calibration. The Fe^{+3} content of sodic-augite and garnet was estimated from assumed charge balance according to the procedures outlined in Laird and Albee (1981; Appendix 1).

All of the calibrations applied in this study take into account the effect of cations other than Fe and Mg in garnet

and clinopyroxene. Krogh (1988) and Pattison and Newton (1989) incorporate a nonlinear dependence of K_D on X_{Ca}^{gt} in their calibrations, while Ellis and Green (1979) assume that the dependence is linear. The Pattison and Newton (1989) thermometer also incorporates a series of factors that account for the variation of K_D with (Mg/Mg+Fe) in garnet. None of the geothermometers employed in this study accounts for the jadeite-acmite component of clinopyroxene, which at the relatively low concentrations found in the sodic-augites of this study, probably has no significant effect on K_D (Koons, 1984).

Rim compositions of sodic-augite and garnet$_{II}$ grains were used in temperature calculations; results are summarized in Table 6. At an assumed pressure of 10 kbar, the thermometers of Ellis and Green (1979), Krogh (1988), and Pattison and Newton (1989) yield temperatures of 657, 606, and 490°C, respectively. Error analysis based on reproducibility of microprobe analyses indicates that calculated temperatures are precise to ± 25°C. Markedly lower temperatures determined by the Pattison and Newton (1989) geothermometer result mainly from corrections to ln K_D for (Mg/Mg+Fe) in garnet.

TABLE 6. SUMMARY OF MINERAL GEOTHERMOMETRY

Garnet-clinopyroxene

X_{Ca}	$(Fe^{+2}/Mg)_{gt}$	$(Fe^{+2}/Mg)_{cpx}$	K_D	$T_1(°C)$	$T_2(°C)$	$T_3(°C)$
0.222	1.823	0.201	9.11	657	606	490

T_1 - Ellis and Green, 1979; T_2 - Krogh, 1988; T_3 - Pattison and Newton, 1989; temperatures estimated at P = 10 kbar.

Garnet-hornblende - Graham and Powell, 1984.

X_{Ca}	$(Fe^{+2}/Mg)_{gt}$	$(Fe^{+2}/Mg)_{hbl}$	K_D	$T(°C)$
0.223	17.000	1.574	10.80	478

Calcite-dolomite

Sample	X_{Mg}	X_{Fe}	$T_1(°C)$	$T_2(°C)$	$T_3(°C)$
8-2-84-1	0.035	0.002	462	451	466
8-5-84-4	0.039	0.002	479	467	482
7-28-85-7	0.037	0.011	483	465	474

T_1 - Anovitz and Essene, 1987; T_2 - Powell et al., 1984; T_3 - Rice, 1977.

Garnet-biotite

X_{Ca}	$(Mg/Fe)_{gt}$	$(Mg/Fe)_{bi}$	K_D	$T_1(°C)$	$T_2(°C)$	$T_3(°C)$	$T_4(°C)$
0.182	0.026	0.442	0.059	381	410	421	385

T_1 - Hodges and Spear, 1982; T_2 - Ganguly and Saxena, 1984; T_3 - Perchuk and Lavrent' eva, 1983; T_4 - Hoinkes, 1986, temperatures estimated at P = 6 kbar.

Garnet-hornblende

Temperatures of retrograde metamorphism in eclogite were estimated by applying the geothermometer of Graham and Powell (1984) to garnet and hornblende in pervasively retrograded eclogite. Rim compositions of adjacent garnet$_{III}$ and amphibole$_{II}$ grains yielded a temperature of 478°C (Table 6). Error analysis based on reproducibility of microprobe analyses indicates that calculated temperatures are precise to ± 25°C.

Calcite-dolomite

Calcite and dolomite in marble from several localities were analyzed for calcite-dolomite solvus geothermometry (sample 7-28-85-7, 67° 24' 43" N, 150° 02' 15" W; sample 8-2-84-1, 67° 23' 00" N, 150° 05' 39" W; sample 8-5-84-4, 67° 22' 43" N, 150° 06' 11" W). Geothermometers employed in this study include the thermometer of Rice (1977), and the Fe-corrected thermometers of Powell et al. (1984), and Anovitz and Essene (1987). All samples produced consistent results regardless of which thermometer is used, and yielded metamorphic temperatures ranging from 451 to 483°C (Table 6). These temperatures are consistent with the absence of talc, tremolite, or diopside in marble.

Garnet-biotite

Garnet-biotite geothermometry was employed to estimate metamorphic temperatures for metachert (sample 7-21-85-1, 67° 18' 14" N, 150° 06' 27" W) cropping out about 1.5 km east-northeast of the Clara Creek eclogite. Rims of biotite and garnet were analyzed, and temperatures were calculated for an assumed pressure of 6 kbar using the thermometers of Hodges and Spear (1982), Ganguly and Saxena (1984), Perchuk and Lavrent' eva (1983), and Hoinkes (1986). Whereas the Perchuk and Lavrent' eva (1983) geothermometer accounts only for mixing of Fe and Mg, the calibrations of Hodges and Spear (1982), Ganguly and Saxena (1984), and Hoinkes (1986) incorporate the effects of other species (i.e., Ca and Mn in garnet). Estimated temperatures are in good agreement, ranging between 381 and 421°C (Table 6).

METAMORPHIC HISTORY

The structural and textural relationships described in this paper indicate that the Clara Creek eclogite and its host rocks have experienced a common metamorphic history characterized by early *HP/LT* metamorphism followed by retrograde metamorphism under lower amphibolite- to greenschist-facies conditions. Using constraints supplied by phase equilibria and mineral geothermometry, it will be shown that the prograde assemblage in the Clara Creek eclogite formed under pressure and temperature conditions that are comparable to those necessary to produce the inferred *HP/LT* assemblages in the surrounding rocks. In addition, it will be demonstrated that retrograde metamorphism of eclogite and syn-D$_{1b}$ mineral assemblages in the enclosing schists occurred under similar pressure and temperature conditions. Results are used to delineate an approximate pressure-temperature path for the Schist belt in the south-central Brooks Range.

High pressure–low temperature metamorphism

Eclogite. The prograde assemblage in the Clara Creek eclogite (garnet + sodic-augite + glaucophane + actinolitic-hornblende + epidote + rutile + apatite) is characteristic of the epidote-blueschist facies (as defined by Evans, 1990) except that it lacks SiO_2 as a free phase. Minimum metamorphic temperatures are constrained by the reaction clinozoisite + tremolite + H_2O = pumpellyite + chlorite + quartz, which indicates that eclogite formed at temperatures greater than approximately 350 to 400°C (Fig. 12). Maximum temperatures are delimited by the reaction glaucophane + epidote = Na-pyroxene (jadeite + diopside) + garnet + quartz + H_2O, which for the composition of glaucophane in the Clara Creek eclogite indicates maximum metamorphic temperatures of about 550 to 600°C. Minimum metamorphic pressures are constrained by a reaction involving the breakdown of glaucophane + epidote (+ quartz) to albite + calcic-amphibole + chlorite (Maruyama et al., 1986; Evans, 1990); albite, hornblende, and chlorite are important secondary minerals in eclogite, indicating that this reaction played an important role in producing the retrograde assemblage. Note that this reaction required the introduction of SiO_2 into the quartz-free eclogite. The reaction glaucophane + epidote = omphacite + lawsonite defines the low temperature, high pressure limit to the stability field of glaucophane + epidote.

Reaction curves define a polygon in P,T space indicating that the prograde assemblage in the Clara Creek eclogite formed at temperatures between approximately 350 and 600°C and pressures in excess of 7 kbar. Precise estimates of prograde metamorphic pressure in the Clara Creek eclogite were prohibited due to the lack of quartz in the Clara Creek eclogite, thus preventing application of the jadeite = albite + quartz geobarometer. Attempts to establish reliable estimates of prograde metamorphic temperature in the Clara Creek eclogite were hampered by garnet-clinopyroxene geothermometers that produced inconsistent results. Inspection of Figure 12 indicates that the calibration of Pattison and Newton (1989) yielded temperatures that are the most compatible with phase relations in the Clara Creek eclogite. These results are consistent with analyses by Powell (1985), Krogh (1988), Pattison and Newton (1989), and Green and Adam (1991) who argue that the Ellis and Green (1979) geothermometer may overestimate temperatures by as much 150°C at crustal pressures.

Host rocks. Although severely recrystallized during retrograde metamorphism, evidence for *HP/LT* metamorphism in the Schist belt is locally preserved:

1. Albite + chlorite ± quartz pseudomorphs after glaucophane are common in pelitic and semipelitic quartz mica schists, and occur in textural configurations indicating that glaucophane growth was restricted to the oldest (D$_{1a}$) deforma-

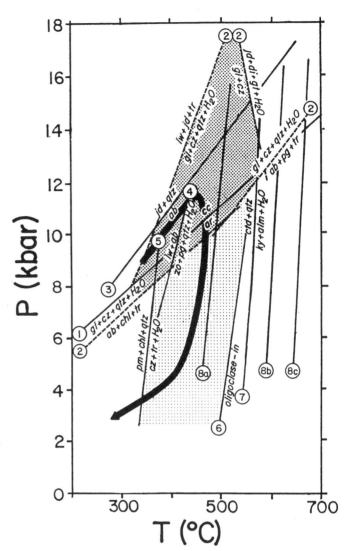

Figure 12. Pertinent phase equilibria and approximate pressure-temperature path. (1) From Boettcher and Wyllie (1968); (2) approximate location of reaction curves from Evans (1990); (3) from Holland (1979); (4) from Heinrich and Althaus (1980); (5) from Liou et al. (1985); (6) from Winkler (1979); (7) from Ghent et al. (1987); (8) K_D isopleths from from garnet-clinopyroxene geothermometry; a, Pattison and Newton (1989) calibration; b, Krogh (1988) calibration; c, Ellis and Green (1979) calibration. Heavily shaded area: pressure-temperature field for prograde (syn-D$_{1a}$) metamorphic conditions. Lightly shaded area: pressure-temperature field for retrograde (syn-D$_{1b}$) metamorphic conditions. Abbreviations listed in Appendix 3.

tional phase (Gottschalk, 1987, 1990; Little et al., 1994). The implied syn-D$_{1a}$ assemblage (quartz + phengite + paragonite + chlorite + glaucophane + chloritoid in the absence of jadeite, garnet, and albite) occurs over a stability field that is limited to relatively high metamorphic pressures and moderate temperatures (Guiraud et al., 1990). Temperatures were sufficiently low (less than about 550°C to maintain chloritoid + quartz assemblages in lieu of kyanite + garnet; Fig. 12). Because epidote is a common accessory phase in some quartz-mica schists, minimum metamorphic temperatures are constrained by the reaction zoisite + paragonite + quartz + H_2O = lawsonite + albite; this reaction implies that syn-D_{1a} *HP/LT* metamorphism occurred above approximately 350°C.

2. Patrick (1995) analyzed two samples from the Koyukuk River pluton, a small body of granitic orthogneiss that crops out about 6 km to the north of the Clara Creek eclogite. Based on K-feldspar-plagioclase geothermometry and phengite-biotite-K-feldspar-quartz geobarometry, Patrick (1995) calculated metamorphic temperatures of 405 and 412°C and pressures of 10.9 and 9.2 kbar, respectively, for the two samples.

3. Rare white mica + epidote pseudomorphs have been described at several localities in rocks correlative to the host rocks of the Clara Creek eclogite (Till et al., 1988; Little et al., 1994) indicating that the Schist belt experienced an early phase of lawsonite-stable metamorphism. Replacement textures indicate that lawsonite may have broken down by the reaction lawsonite + albite = zoisite + paragonite + quartz + H_2O in response to increasing temperature at high metamorphic pressure (Fig. 12).

4. As discussed previously, rare relict crossite occurs in metagabbro from the southern outcrops of the Schist belt. Brown (1977) demonstrated that $Na(M_4)$ in amphibole increases with metamorphic pressure, and developed an empirical geobarometer based upon the $Na(M_4)$ content of amphibole in equilibrium with magnetite, chlorite, epidote, and albite. In metagabbro, crossite contains $Na(M_4)$ ranging from 1.51 to 1.87 per formula unit (normalized to 23 oxygens, Fig. 4b). These values are comparable to the $Na(M_4)$ content of amphiboles from the Shuksan and Sanbagawa high-pressure metamorphic terranes, and imply similar high metamorphic pressures in the Brooks Range Schist belt. In contrast, the $Na(M_4)$ contents of sodic-hornblende and barroisite that make up the bulk of the amphibole in metagabbro are significantly lower (0.43 to 1.21 per formula unit) and indicate that the main phase of amphibole growth occurred at somewhat lower metamorphic pressures than those that produced crossite.

Retrograde metamorphism

Eclogite. In eclogite, the secondary assemblage of zoned amphibole (actinolite to hornblende, rimward), epidote, chlorite, garnet, albite, and sphene indicates hydrous retrograde metamorphism of eclogite under lower amphibolite- to greenschist-facies conditions. Temperatures determined by garnet-hornblende geothermometry (T = 478°C ± 25°C) are consistent with the stability ranges of other minerals that comprise the retrograde assemblage. The abundance of epidote and the absence of pumpellyite indicates temperatures of retrograde metamorphism in excess of

about 350°C (curve 5, Fig. 12). Because albite is the sole plagioclase present in altered eclogite, temperatures apparently remained below the "oligoclase-in isograd." The transition from albite to oligoclase in metabasites occurs via series of sliding reactions that generally involve the breakdown of epidote + albite with increasing temperature (Laird, 1980; Apted and Liou, 1983); Winkler (1979) semiquantitatively places the albite to oligoclase transition at about 550°C (curve 6, Fig. 12). These relations indicate that temperatures derived by garnet-hornblende geothermometry are a reasonable estimate of metamorphic temperatures during retrograde metamorphism. The growth of actinolite and hornblende during retrograde metamorphism and the lack of sodic-amphibole indicates that metamorphic pressures were relatively low, and outside stability field of glaucophane.

Host rocks. In quartz-mica schist, the appearance of syn-D_{1b} albite, and the lack of glaucophane in D_{1b} assemblages suggests that the D_{1a} assemblage (quartz + phengite + paragonite + chlorite ± chloritoid ± glaucophane) was broken down to form the assemblage quartz + phengite + paragonite + chlorite ± albite ± chloritoid during D_{1b}. This probably occurred by a series reactions involving the formation of paragonite + chlorite at the expense of chloritoid and glaucophane, and the breakdown of glaucophane + paragonite to form albite + chlorite (Brown and Forbes, 1986; Guiraud et al., 1990). These reactions are largely pressure dependent, and indicate a significant but indeterminate decrease in metamorphic pressure between D_{1a} and D_{1b}. The replacement of crossite by barroisite in metagabbro records a similar change in metamorphic conditions from relatively high to moderate pressures.

Because marble, metachert, and mafic schist show evidence of complete dynamic recrystallization during D_{1b}, estimated temperatures from calcite-dolomite and garnet-biotite geothermometry are interpreted as reflecting syn- or possibly post-D_{1b} metamorphic conditions. These data indicate that metamorphism during D_{1b} occurred at temperatures between about 380 and 480°C.

CONCLUSIONS

Analysis of phase relations in the Clara Creek eclogite and in the enclosing rocks of the Schist belt indicates that both share a metamorphic history characterized by blueschist-facies metamorphism, followed by retrograde metamorphism within the lower amphibolite to greenschist facies. Structural and geochronological studies indicate that blueschist-facies metamorphism in the Schist belt was accompanied by penetrative deformation, and probably occurred in the Middle to Late Jurassic. Retrograde metamorphism and associated deformation occurred in the Early Cretaceous (Armstrong et al., 1986; Till and Patrick, 1991; Christiansen and Snee, 1994; Gottschalk and Snee, this volume, Chapter 13). The prograde assemblage in eclogite (garnet + sodic-augite + glaucophane + actinolitic-hornblende + epidote + rutile + apatite) is interpreted to have formed isofacially with phengite + paragonite + chlorite ± glaucophane ± chloritoid in the enclosing quartz-mica schists at temperatures between 350 and 550°C and at pressures above the lower stability limit of glaucophane (greater than approximately 7 kbar). These results are consistent with those of Patrick (1995) who estimated *HP/LT* metamorphic conditions in two samples from the Koyukuk River pluton at *T* = 405 and 412°C and *P* = 10.9 and 9.2 kbar, respectively. The occurrence of epidote as part of the *HP/LT* assemblage in eclogite and in the surrounding rocks indicates that blueschist-facies metamorphism occurred mainly under epidote-stable conditions. Isolated occurrences of pseudomorphs after lawsonite indicates that the Schist belt passed through the lawsonite stability field en route to peak metamorphic conditions.

Phase relations and mineral geothermometry indicate that hydrous lower amphibolite- to greenschist-facies alteration of eclogite and syndeformational recrystallization of the enclosing schistose rocks took place at temperatures between approximately 380 to 480°C. Metamorphic pressures during retrograde metamorphism are not precisely known, but were probably below the lower stability limit of glaucophane. In the Clara Creek eclogite, the abundance of veins containing the secondary assemblage indicates that retrograde metamorphism was accommodated, in large part, by the influx of metamorphic fluids into the eclogite body, perhaps in response to penetrative deformation and recrystallization of the surrounding schists.

Comparison of the major- and trace-element chemistry of eclogite and metabasites in the surrounding schists indicates that the occurrence of the paragenesis garnet + Na-pyroxene + rutile is compositionally controlled. Eclogite was derived from a silica-poor, Fe-Ti-rich mafic protolith, in contrast to mafic schists and metagabbros that have normal basaltic compositions.

ACKNOWLEDGMENTS

This study was supported by grants to J. S. Oldow, H. G. Avé Lallemant, and A. W. Bally from the U.S. Department of Energy (DE-ASO5-83ER13124), the National Science Foundation (EAR-8517384), and the Alaska Industrial Associates Program (AIAP) at Rice University and the University of Alaska, Fairbanks. The AIAP received contributions from the following companies: Arco, Amoco, Chevron, Standard Oil of Ohio, Mobil, and Gulf. Critical reviews by B. E. Patrick, E. D. Ghent, V. B. Sisson, H. G. Avé Lallemant, and J. S. Oldow significantly improved the quality of this manuscript. Thanks to C. L. Lum and A. E. Moran for assistance with Instrumental Neutron Activation Analysis, and to C. Chaika for loss on ignition analyses. B. E. Patrick kindly provided preliminary results of his thermobarometric study of granitic orthogneisses in the Brooks Range. Thanks also to D. Doering for his instruction in the use of Southern Methodist University's JEOL 733 Superprobe.

APPENDIX 1: ELECTRON MICROPROBE ANALYSIS AND CALCULATION OF ENDMEMBERS

Mineral compositions presented in this study were determined using a JEOL 733 Superprobe, and an ETEC Autoprobe electron microprobe analyzer, both equipped with wavelength dispersive PET, LIF, and TAP crystal spectrometers, and outfitted with Tracor-Northern automation. Analyses were carried out at an accelerating potential of 15 kv and a sample current of 15 nanoamps for micas, and 20 nanoamps for all other phases. Data collection and reduction were performed using the Tracor-Northern TASK IV software with matrix correction based on the method of Bence and Albee (1968) and Albee and Ray (1970). Natural mineral standards were used for all analyses, and the instrument was standardized frequently to maximize accuracy and precision.

Garnet and clinopyroxene endmembers were calculated stoichiometrically in the following order. *Garnet:* all Fe^{+2} as almandine ($Fe_3Al_2Si_3O_{12}$), all Fe^{+3} as andradite ($Ca_3Fe_2Si_3O_{12}$), remaining Ca as grossularite ($Ca_3Al_2Si_3O_{12}$), all Mg as pyrope ($Mg_3Al_2Si_3O_{12}$), and all Mn as spessartine ($Mn_3Al_2Si_3O_{12}$); Fe^{+2} and Fe^{+3} contents were estimated as follows: $Fe^{+3} = 2 - Al - Ti$; $Fe^{+2} = Fe_{Total} - Fe^{+2}$. *Clinopyroxene:* all Fe^{+3} as acmite ($Na\ Fe^{+3}Si_2O_6$), remaining Na as jadeite ($NaAlSi_2O_6$), remaining Al as Ca-tschermaks's molecule ($CaAl_2SiO_6$), Fe^{+2} as hedenbergite ($CaFeSi_2O_6$), remaining Ca as diopside ($CaMgSi_2O_6$), remaining Mg as enstatite ($Mg_2Si_2O_6$); Fe^{+2} and Fe^{+3} contents were estimated as follows: $Fe^{+3} = 12 -$ total cation charge assuming all Fe = Fe^{+2}; $Fe^{+2} = Fe_{Total} - Fe^{+3}$. For amphiboles, Fe^{+2} and Fe^{+3} contents were estimated using the algorithm of Spear and Kimball (1984). For chloritoid, $Fe^{+3} = 4 - Al^{IV}$; $Fe^{+2} = Fe_{Total} - Fe^{+3}$.

APPENDIX 2: LIST OF ABBREVIATIONS

ab = albite
ac = acmite
act = actinolite
ah = actinolitic-hornblende
am = amphibole (undifferentiated)
ap = apatite
alm = almandine
an = anorthite
and = andradite
ar = aragonite
bi = biotite
CaTs = calcium-tschermak's molecule
cc = calcite
chl = chlorite
cpx = clinopyroxene
ctd = chloritoid
cz = clinozoisite
di = diopside
en = enstatite
ep = epidote
gl = glaucophane
gr = grossularite
gt = garnet
hbl = hornblende
hd = hedenbergite
ilm = ilmenite
jd = jadeite
ky = kyanite
lw = lawsonite
op = opaque phase
ph = phengite
pg = paragonite

pm = pumpellyite
py = pyrope
qtz = quartz
rt = rutile
sph = sphene
stp = stilpnomelane
tour = tourmaline
tr = tremolite
wt mica = white mica (undifferentiated)
zo = zoisite

APPENDIX 3: WHOLE-ROCK CHEMICAL ANALYSIS

Rock samples were cut into a series of slabs and broken into pea-sized chips using a hydraulic jack rock splitter. Rock chips were washed in a 1% HCl solution for 30 minutes, followed by repeated rinses with deionized water in an ultrasonic cleaner. Chips were dried overnight at 110°C then powdered in a hardened steel shatterbox; silica sand was powdered between each sample run and the shatterbox was thoroughly cleaned to eliminate memory effects. Powders were dried at 110°C and cooled in a dessicator to remove adsorbed water.

Major and trace elements analyses were run on a Jarrell-Ash induction-coupled plasma spectrograph (ICP). Elements analyzed by ICP and their peak wavelengths in nannometers (nm) are as follows: Si (251.61 and 288.16), Ti (334.94), Al (396.22), Fe (238.20), Mn (257.61), Mg (280.27), Ca (393.43), Na (589.59), K (766.49), P (214.95), Sr (407.77), Ba (455.40), Zr (339.21 and 343.83), Y (371.03), Sc (361.39), V (290.88), Ni (231.60), and Cr (267.72). Background corrections were applied by counting on an interference-free region in the vicinity of each peak. U.S. Geological Survey standards were run before and after each run to check for analytical accuracy and standards were inserted every fifth sample to correct for instrumental drift. Correction procedures assume linear drift between each drift standard.

Trace elemental abundances in some samples were also measured by instrumental neutron activation analysis (INAA) at the Planetary Sciences Division of the NASA Johnson Space Center and Oregon State University. 20- to 50-mg samples were sealed in quartz tubes and irradiated for 28 hours with a neutron flux of 10^{13} n cm^{-2} sec^{-1}. Count sets were started at 7, 10, and 40 days after removal from the reactor. Sample spectra were normalized to U.S. Geological Survey standards run with unknowns, using the ORTEC Gamma software package.

REFERENCES CITED

Albee, A. L., and Ray, L., 1970, Correction factors for electron probe microanalysis of silicates, oxides, carbonates, phosphates, and sulphates: Analytical Chemistry, v. 42, p. 1408–1414.

Anovitz, L. M., and Essene, E. J., 1987, Phase equilibria in the system CaCO$_3$-MgCO$_3$-FeCO$_3$: Journal of Petrology, v. 28, p. 389–414.

Apted, M. J., and Liou, J. G., 1983, Phase relations among greenschist, epidote-amphibolite, and amphibolite in a basaltic system: American Journal of Science, v. 283-A, p. 328–354.

Armstrong, R. L., Harakal, J. E., Forbes, R. B., Evans, B. W., and Thurston, S. P., 1986, Rb-Sr and K-Ar study of metamorphic rocks of the Seward Peninsula and southern Brooks Range, Alaska, *in* Evans, B. W., and Brown, E. H., eds., Eclogites and blueschists: Geological Society of America Memoir 164, p. 185–203.

Bence, A. E., and Albee, A. L., 1968, Empirical correlation factors for the electron micro-analysis of silicates and oxides: Journal of Geology, v. 76, p. 382–403.

Biino, G., and Pognante, U., 1990, Paleozoic continental-type gabbros in the Gran Paradiso nappe (Western Alps, Italy): Early-Alpine eclogitization and geochemistry: Lithos, v. 24, p. 3–19.

Black, P. M., 1973, Mineralogy of New Caledonian metamorphic rocks II: Amphiboles from the Ouégoa District: Contributions to Mineralogy and Petrology, v. 39, p. 55–64.

Black, P. M., Brothers, R. N., and Yokoyama, K., 1988, Mineral parageneses in eclogite facies meta-acidites in northern New Caledonia, *in* Smith, D. C., ed., Eclogite and eclogite facies rocks: Amsterdam, Elsevier, p. 271–289.

Boettcher, A. L., and Wyllie, P. J., 1968, The calcite-aragonite transition measured in the system CaO-CO$_2$-H$_2$O: Journal of Geology, v. 76, p. 314–330.

Brown, E. H., 1977, The crossite content of Ca-amphibole as a guide to pressure of metamorphism: Journal of Petrology, v. 18, p. 53–72.

Brown, E. H., and Forbes, R. B, 1986, Phase petrology of eclogitic rocks in the Fairbanks district, Alaska, *in* Evans, B. W., and Brown, E. H., eds., Eclogites and blueschists: Geological Society of America Memoir 164, p. 155–167.

Carden, J. R., 1978, The comparative geology of blueschists and greenschists in the Brooks Range and Kodiak-Seldovia Schist belts [Ph.D. thesis]: Fairbanks, Alaska, University of Alaska, 242 p.

Christiansen, P. P., and Snee, L. W., 1994, Structure, metamorphism, and geochronology of the Cosmos hills and Ruby Ridge, Brooks Range Schist belt, Alaska: Tectonics, v. 13, p. 193–214.

Coleman, R. G., and Lee, D. E., 1963, Glaucophane-bearing metamorphic rock types of the Cazadero area, California: Journal of Petrology, v. 4, p. 260–301.

Dillon, J. T., Pessel, G. H., Chen, J. H., and Veach, N. C., 1980, Middle Paleozoic magmatism and orogenesis in the Brooks Range: Geology, v. 8, p. 338–343.

Dillon, J. T., Brosgé, W. P., and Dutro, J. T., Jr., 1986, Generalized geologic map of the Wiseman Quadrangle: U.S. Geological Survey Open File Report 86-219, scale 1:250,000.

Dungan, M. A., Vance, J. A., and Blanchard, D. P., 1983, Geochemistry of the Shuksan greenschists and blueschists, North Cascades, Washington: variably fractionated and altered metabasalts of oceanic affinity: Contributions to Mineralogy and Petrology, v. 82, p. 131–146.

Dupuy, C., and Dostal, J., 1984, Trace element geochemistry of some continental tholeiites: Earth and Planetary Science Letters, v. 67, p. 61–69.

Dusel-Bacon, C., Brosgé, W. P., Till, A. B., Doyle, E. O., Mayfield, C. F., Reiser, H. N., and Miller, T. P., 1989, Distribution, facies, age, and proposed tectonic associations of regionally metamorphosed rocks in northern Alaska: U.S. Geological Survey Professional Paper 1497-A, 44 p.

Ebihara, M., Wolf, R., and Anders, E., 1982, Are CI chondrites chemically fractionated? A trace element study: Geochimica et Cosmochimica Acta, v. 46, p. 1849–1862.

Ellis, D. J., and Green, D. H., 1979, An experimental study of the effect of Ca upon garnet-clinopyroxene Fe-Mg exchange equilibria: Contributions to Mineralogy and Petrology, v. 71, p. 13–22.

Essene E. J., and Fyfe, W. S., 1967, Omphacite in Californian metamorphic rocks: Contributions to Mineralogy and Petrology, v. 15, p. 1–23.

Evans, B. W., 1990, Phase relations of epidote-blueschists: Lithos, v. 25, p. 3–23.

Evans, B. W., and Patrick, B. E., 1987, Phengite (3T) in high-pressure metamorphosed granitic orthogneiss, Seward Peninsula, Alaska: Canadian Mineralogist, v. 25, p. 141–158.

Evans, B. W., Patrick, B. E., and Irving, A. J., 1987, Compositional control of blueschist/greenschist and genesis of Seward Peninsula metabasites: Geological Society of America Abstracts with Programs, v. 19, p. 375.

Forbes, R. B., Evans, B. W., and Thurston, S. P., 1984, Regional progressive high-pressure metamorphism, Seward Peninsula, Alaska: Journal of Metamorphic Geology, v. 66, p. 113–117.

Ganguly, J., and Saxena, S. K., 1984, Mixing properties of aluminosilicate garnets: constraints from natural and experimental data, and application to geothermometry: American Mineralogist, v. 69, p. 88–97.

Ghent, E. D., Stout, M. Z., Black, P. M., and Brothers, R. N., 1987, Chloritoid-bearing rocks associated with blueschists and eclogites, northern New Caledonia: Journal of Metamorphic Geology, v. 5, p. 239–254.

Gilbert, W. G., Wiltse, M. A., Carden, J. R., Forbes, R. B., and Hackett, S. W., 1977, Geology of the Ruby Ridge, southwestern Brooks Range, Alaska: Alaska Division of Geological and Geophysical Surveys Report 58, 16 p.

Gottschalk, R. R., 1987, Structural and petrologic evolution of the south-central Brooks Range, Alaska [Ph.D. thesis]: Houston, Texas, Rice University, 263 p.

Gottschalk, R. R., 1990, Structural evolution of the Schist belt, south-central Brooks Range fold and thrust belt, Alaska: Journal of Structural Geology, v. 12, p. 453–469.

Graham, C. M., and Powell, R., 1984, A garnet-hornblende geothermometer: Calibration, testing and application to the Pelona schist, southern California: Journal of Metamorphic Geology, v. 2, p. 13–21.

Green, T. H., and Adam, J., 1991, Assessment of the garnet-clinopyroxene Fe-Mg exchange thermometer using new experimental data: Journal of Metamorphic Geology, v. 9, p. 341–347.

Guiraud, M., Holland, T., and Powell, R., 1990, Calculated mineral equilibria in the greenschist-blueschist-eclogite facies in Na_2O-FeO-MgO-Al_2O_3-SiO_2-H_2O: methods results and geological applications: Contributions to Mineralogy and Petrology, v. 104, p. 85–98.

Heinrich, W., and Althaus, E., 1980, Die obere stabilitatagrenze von lawsonit plus albit bsw. jadeit: Forschritte die Mineralogie, v. 58, p. 49–50.

Hellman, P. L., Smith, R. E., and Henderson, P., 1979, The mobility of rare earth elements: evidence and implications from selected terranes affected by burial metamorphism: Contributions to Mineralogy and Petrology, v. 71, p. 23–44.

Himmelberg, G. R., and Papike, J. J., 1969, Coexisting amphiboles from blueschist facies metamorphic rocks: Journal of Petrology, v. 10, p. 102–114.

Hitzman, M. W., Proffett, J. M., Schmidt, J. M., and Smith, T. E., 1986, Geology and mineralization of the Ambler district, northwestern Alaska: Economic Geology, v. 91, p. 1592–1618

Hodges, K. V., and Spear, F. S., 1982, Geothermometry, geobarometry, and the Al_2O_3 triple point at Mt. Moosilauke, New Hampshire: American Mineralogist, v. 67, p. 1118–1134.

Hoinkes, G., 1986, Effect of grossular content in garnet on the partitioning of Fe anf Mg between garnet and biotite: Contributions to Mineralogy and Petrology, v. 92, p. 393–399.

Holland, T. J. B., 1979, Experimental determination of the reaction paragonite = jadeite + kyanite + H_2O, and internally consistent thermodynamic data for part of the system Na_2O-Al_2O_3-SiO_2-H_2O, with application to eclogites and blueschists: Contributions to Mineralogy and Petrology, v. 68, p. 293–301.

Hughes, C. J., 1982, Igneous petrology: Developments in Petrology 7: Amsterdam, Elsevier, 551 p.

Klemd, R., 1989, *P-T* evolution and fluid inclusion characteristics of retrograded eclogites, Münchberg gneiss complex, Germany: Contributions to Mineralogy and Petrology, v. 102, p. 221–229.

Koons, P. O., 1984, Implications to garnet-clinopyroxene geothermometry of non-ideal solid solution in jadeitic pyroxenes: Contributions to Mineralogy and petrology, v. 88, P. 340–347

Koons, P. O., Rubie, D. C., and Frueh-Green, G., 1987, The effects of disequilibrium and deformation on the mineralogical evolution of quartz diorite in the eclogite facies: Journal of Petrology, v. 28, p. 679–700.

Krogh, E. J., 1988, The garnet-clinopyroxene Fe-Mg geothermometer—a reinterpretation of existing experimental data: Contributions to Mineralogy and Petrology, v. 99, p. 44–48.

Laird, J., 1980, Phase equilibria in mafic schist from Vermont: Journal of Petrology, v. 21, p. 1–37.

Laird, J., and Albee, A. L., 1981, High-pressure metamorphism in mafic schist from northern Vermont: American Journal of Science, v. 281, p. 97–126.

Leake, B. E., 1978, Nomenclature of amphiboles: Canadian Mineralogist, v. 16, p. 501–520.

Le Roux, A. P., Dick, H. J. B., Erlank, A. J., Reid, A. M., Frey, F. A., and Hart, S. R., 1983, Geochemistry, mineralogy and petrogenesis of lavas erupted along the Southwest Indian Ridge between the Bouvet triple junction and 11 degrees East: Journal of Petrology, v. 24, p. 267–318.

Liou, J. G., Maruyama, S., and Cho, M., 1985, Phase equilibria and mineral parageneses of metabasites in low-grade metamorphism: Mineralogical Magazine, v. 49, p. 321–333.

Little, T. A., Miller, E. L., Lee, J., and Law, R. D., 1994, Extensional origin of ductile fabrics in the Schist belt, central Brooks Range, Alaska—I. Geologic and structural studies: Journal of Structural Geology, v. 16, p. 899–918.

Maresch, W. W., and Abraham, J., 1981, Petrography, mineralogy and metamorphic evolution of an eclogite from the Island of Margarita: Journal of Petrology, v. 22, p. 337–362.

Maruyama, S., Cho, M., and Liou, J. G., 1986, Experimental investigations of blueschist-greenschist transition equilibria: Pressure dependence of Al_2O_3 contents in sodic amphiboles—A new geobarometer, *in* Evans, B. W., and Brown, E. H., eds., Eclogites and blueschists: Geological Society of America Memoir 164, p. 1–16.

Moore, T. E., 1986, Petrology and tectonic implications of the blueschist-bearing Puerto Nuevo melange complex, Vizcaino Peninsula, Baja California Sur, Mexico, *in* Evans, B. W., and Brown, E. H., eds., Eclogites and blueschists: Geological Society of America Memoir 164, p. 43–58.

Moore, T. E., Wallace, W. K., Bird, K. J., Karl, S. M., Mull, C. G., and Dillon, J. T., 1994, Geology of northern Alaska, *in* Plafker, G., and Berg, H. C., eds., The geology of Alaska: Boulder, Colorado, Geological Society of America, The Geology of North America, v. G-1, p. 49–140.

Nelsen, C. J., 1979, The geology and blueschist petrology of the western Ambler Schist belt, southwestern Brooks Range, Alaska [M.S. thesis]: Albuquerque, New Mexico, University of New Mexico, 123 p.

Nelson, S. W., and Grybeck, D., 1982, Metamorphic rocks of the Survey Pass Quadrangle, U.S. Geological Survey Miscellaneous Field Studies Map MF-1176-C, scale 1:250,000.

Newton, R. C., 1986, Metamorphic temperatures and pressures of Group B and Group C eclogites, *in* Evans, B. W., and Brown, E. H., eds., Eclogites and blueschists: Geological Society of America Memoir 164, p. 17–30.

Owen, C., 1989, Magmatic differentiation and alteration in isofacial greenschist and blueschist, Shuksan Suite, Washington: statistical analysis of major element variation: Journal of Petrology, v. 30, p. 739–761.

Patrick, B. E., 1988, Synmetamorphic evolution of the Seward Peninsula blueschist terrane: Journal of Structural Geology, v. 10, p. 555–566.

Patrick, B. E., 1995, High pressure–low temperature metamorphism of granitic orthogneisses in the Brooks Range, northern Alaska: Journal of Metamorphic Geology, v. 13, p. 111–124.

Patrick, B. E., and Evans, B. W., 1989, Metamorphic evolution of the Seward Peninsula blueschist terrane: Journal of Petrology, v. 30, p. 531–555.

Patrick, B. E., and Lieberman, J. E., 1988, Thermal overprint of the Seward Peninsula: the Lepontine in Alaska: Geology, v. 16, p. 1100–1103.

Pattison, D. R. M., and Newton, R. C., 1989, Reversed experimental calibration of the garnet-clinopyroxene Fe-Mg exchange geothermometer: Contributions to Mineralogy and Petrology, v. 101, p. 87–103.

Pearce, J. A., 1982, Trace element characteristics of lavas from destructive plate boundaries, *in* Thorpe, R. S., ed., Orogenic andesites: Chichester, J. Wiley and Sons, p. 525–547.

Pearce, J. A., and Cann, J. R., 1973, Tectonic setting of basic volcanic rocks determined using trace element analyses: Earth and Planetary Science Letters, v. 19, p. 290–300.

Pearce, J. A., and Norry, M. J., 1979, Petrogenetic implications of Ti, Zr, Y and Nb variations in volcanic rocks: Contributions to Mineralogy and Petrology, v. 69, p. 33–47.

Perchuk, L. L., and Lavrent' eva, I. V., 1983, Experimental investigation of exchange equilibria in the system cordierite-garnet-biotite, *in* Saxena, S. K., ed., Kinetics and equilibrium in mineral reactions: Berlin, Heidelberg, New York, Springer-Verlag, p. 199–239.

Powell, R., 1985, Regression diagnostics and robust regression in geothermometer/geobarometer calibration: the garnet-clinopyroxene geothermometer revisited: Journal of Metamorphic Geology, v. 3, p. 231–243.

Powell, R., Condliffe, D. M., and Condliffe, E., 1984, Calcite-dolomite geothermometry in the system $CaCO_3$-$MgCO_3$-$FeCO_3$, an experimental study: Journal of Metamorphic Geology, v. 2, p. 33–41.

Rice, J. M., 1977, Contact metamorphism of impure dolomitic limestone in the Boulder aureole, Montana: Contributions to Mineralogy and Petrology, v. 59, p. 237–259.

Robinson, P., Spear, F. S., Schumacher, J. C., Laird, J., Klein, C., Evans, B. W., and Doolan, B. L., 1982, Phase relations of metamorphic amphiboles: Natural occurrence and theory, *in* Veblen D. R., and Ribbe, P. H., eds., Amphiboles: Petrology and experimental phase relations: Mineralogical Society of America Reviews in Mineralogy, v. 9B, p. 1–227.

Sainsbury, C. L., Coleman, R. G., and Kachadoorian, R., 1970, Blueschist and related greenschist facies rocks of the Seward Peninsula, Alaska: U.S. Geological Survey Professional Paper 700-B, p. B33–B42.

Schliestedt, M., 1986, Eclogite-blueschist relationships as evidenced by mineral equilibria in the high-pressure metabasic rocks of Sifnos (Cycladic Islands), Greece: Journal of Petrology, v. 27, p. 1437–1459.

Sorensen, S. S., 1986, Petrologic and geochemical comparison of the blueschist and greenschist units of the Catalina Schist terrane, southern California, *in* Evans, B. W., and Brown, E. H., eds., Eclogites and blueschists: Geological Society of America Memoir 164, p. 59–75.

Spear, F. S., and Kimball, K. L., 1984, RECAMP—A Fortran IV program for estimating Fe^{3+} contents of amphiboles: Computers and Geology, v. 10, p. 317–325.

Spry, A., 1969, Metamorphic textures: New York, Pergamon Press, 350 p.

Sun, S.-S., Nesbitt, R. W., and Sharaskin, A. Y., 1979, Geochemical characteristics of mid-ocean ridge basalts: Earth and Planetary Science Letters, v. 44, p. 119–138.

Thompson, R. N., Dickin, A. P., Gibson, I. L., and Morrison, M. A., 1982, Elemental fingerprints of isotopic contamination of Hebridean Paleocene mantle-derived magmas by Archean sial: Contributions to Mineralogy and Petrology, v. 79, p. 159–168.

Thurston, S. P., 1985, Structure, petrology, and metamorphic history of the Nome Group blueschist terrane, Salmon Lake area, Seward Peninsula, Alaska: Geological Society of America Bulletin, v. 96, p. 600–617.

Till, A. B., 1988, Evidence for two Mesozoic blueschist belts in the hinterland of the western Brooks Range fold and thrust belt: Geological Society of America Bulletin Abstracts with Programs, v. 20, p. A112.

Till, A. B., and Moore, T. E., 1991, Tectonic relations of the Schist belt, southern Brooks Range, Alaska: Eos (Transactions, American Geophysical Union), v. 72, p. 295.

Till, A. B., and Patrick, B. E., 1991, Ar-Ar evidence for a 110–105 Ma amphibolite facies overprint on blueschist in the south-central Brooks Range, Alaska: Geological Society of America Bulletin Abstracts with Programs, v. 23, p. A436.

Till, A. B., Schmidt, J. M., and Nelson, S. W., 1988, Thrust involvement of metamorphic rocks, southwestern Brooks Range, Alaska: Geology, v. 16, p. 930–933.

Turner, D. L., Forbes, R. B., and Dillon, J. T., 1979, K-Ar geochronology of the southwestern Brooks Range, Alaska: Canadian Journal of Earth Science, v. 16, p. 1789–1804.

Vierek, L. G., Flower, M. F. J., Hertogen, J., Schmincke, H.-U., and Jenner, G. A., 1989, The genesis and significance of N-MORB sub-types: Contributions to Mineralogy and Petrology, v. 102, p. 112–126.

Wayte, G. J., Worden, R. H., Rubie, D. C., and Droop, G. T. R., 1989, A TEM study of disequilibrium plagioclase breakdown at high pressure: Contributions to Mineralogy and Petrology, v. 101, p. 426–437.

Winkler, H. G. F., 1979, Petrogenesis of metamorphic rocks: Berlin, Heidelberg, New York, Springer-Verlag, 220 p.

Zayatz, M. R., 1987, Petrography of the Baird Mountains schistose lithologies, northwestern Alaska: U.S. Geological Survey Circular 998, p. 49–52.

MANUSCRIPT ACCEPTED BY THE SOCIETY SEPTEMBER 23, 1997

Geological Society of America
Special Paper 324
1998

Constraints on the cooling history of the central Brooks Range, Alaska, from fission-track and $^{40}Ar/^{39}Ar$ analyses

Ann E. Blythe* and John M. Bird
Department of Geological Sciences, Cornell University, Ithaca, New York 14853
Gomaa I. Omar
Department of Geology, University of Pennsylvania, Philadelphia, Pennsylvania 19104

ABSTRACT

Forty-five apatite and zircon fission-track and seven $^{40}Ar/^{39}Ar$ white mica analyses from a transect along the Dalton Highway of the central Brooks Range, Alaska, are used to constrain the cooling and exhumation history. In general, average cooling rates estimated for the time period from 130 to 95 Ma are consistent with the pattern expected across an active orogenic belt: The average rates are highest in the hinterland (the schist belt) and progressively lower toward the foreland (the northern fold-and-thrust belt). Deformation, however, had clearly begun prior to 130 Ma: white mica $^{40}Ar/^{39}Ar$ ages of 130–120 Ma from the schist belt document cooling through 350°C after peak temperatures of 450°C were reached. The zircon fission-track ages from the northern fold-and-thrust belt ranged from 224 to 130 Ma, and although some of these ages may be mixed ages representing more than one cooling event, we interpret a cluster of ~185 Ma ages as representing a distinct cooling event associated with the onset of Brooks Range deformation. A white mica age from the phyllite belt yielded a plateau age of ~113 Ma, which we interpret as the time of motion on normal faults juxtaposing the schist belt and phyllite belt. Apatite fission-track data from the southern fold-and-thrust belt are consistent with a hiatus in deformation by ~95 Ma and indicate that samples now at the surface were within ~2–3 km of the surface at that time. Apatite fission-track analyses from the northern fold-and-thrust belt and schist belt document a regionwide rapid cooling event (50°C in ~5 m.y.) at ~60 Ma. Although the amount of exhumation associated with the 60-Ma cooling event was small in terms of the entire orogenic history, it is significant because it suggests deformation while the active plate margin was >1,000 km to the south. Zircon fission-track ages suggest that the Doonerak window began to form at ~60 Ma also. Exhumation rates are calculated from cooling data obtained in this study by using a range of geothermal gradients; the geothermal gradient in the southern Brooks Range probably evolved from ~15°C/km early in its deformational history to ~30°C/km, the present-day geothermal gradient at Prudhoe Bay, following the end of active deformation.

*Present address: Department of Earth Sciences, University of Southern
California, Los Angeles, California 90089-0740.

Blythe, A. E., Bird, J. M., and Omar, G. I., 1998, Constraints on the cooling history of the central Brooks Range, Alaska, from fission-track and $^{40}Ar/^{39}Ar$ analyses, *in* Oldow, J. S., and Avé Lallemant, H. G., eds., Architecture of the Central Brooks Range Fold and Thrust Belt, Arctic Alaska: Boulder, Colorado, Geological Society of America Special Paper 324.

INTRODUCTION

The Brooks Range (Fig. 1, in Handschy, Chapter 1, this volume), Alaska, is a major collisional belt within the North American Cordillera that has undergone a complicated tectonic and structural evolution. The Brooks Range and its foreland basin, the Colville basin, formed primarily as the result of north-vergent (in present coordinates) contraction caused by collision with a southern island arc during Jurassic and Early Cretaceous time (e.g., Mull, 1982; Box, 1985). Large-scale extension may have substantially thinned the southern Brooks Range and accreted island arc, creating the Yukon-Koyukuk basin, during mid-Cretaceous time (e.g., Miller and Hudson, 1991). Regional contraction was occurring as late as early Tertiary time in the central Colville basin (Mull, 1982; Mull, 1985), and earthquake data indicate that deformation is still active in the northeastern Brooks Range (Estabrook et al., 1988).

Thermochronometric data, such as fission-track and $^{40}Ar/^{39}Ar$ analyses, date the time at which a rock cooled through specific temperatures and are commonly used to reconstruct regional cooling histories (e.g., Copeland and Harrison, 1990; Green, 1989). When combined with information on geothermal gradients, surface elevations, and surface temperatures, thermochronometric data can be used to estimate the timing and rates of exhumation. We have obtained 45 apatite and zircon fission-track and 7 $^{40}Ar/^{39}Ar$ white mica analyses from samples collected along a south–north transect (along the Dalton Highway) from the Yukon-Koyukuk basin to the southern margin of the Colville basin; the regional geology of this area was described in Chapter 1 of this volume. These data on the cooling and exhumation history of the central Brooks Range are used to better constrain tectonic events affecting northern Alaska.

ANALYTICAL METHODS

Fission-track method

A fission track is a damaged region in a crystal or glass formed by the natural fission of ^{238}U or the induced fission of ^{235}U. During fission, two charged particles that are approximately one-half of the size of the original nucleus are formed. Coulomb repulsion causes the particles to be ejected in opposite directions, creating a "track," or linear zone of defects within the crystal or glass. Etching with acids makes the tracks visible with an optical microscope.

At low temperatures, fission tracks are relatively stable; at high temperatures the fission tracks shorten and disappear; this process is referred to as "annealing." The rate of annealing and range of temperatures over which it occurs is dependent primarily on the rate of cooling of the host rock. At typical geologic cooling rates (1 to 100°C/m.y.), fission-track annealing in apatite occurs between ~60 and 110°C (e.g., Gleadow and Duddy, 1981; Gleadow et al., 1983). The orientation of tracks with respect to the crystallographic C-axis of apatite affects annealing rates; tracks perpendicular to the C-axis of the apatite

crystal shorten fastest, and those parallel, slowest (e.g., Green and Durrani, 1977; Donelick, 1991). Annealing rates also vary according to the Cl/(Cl+F) content of the apatite (Green et al., 1986; Crowley et al., 1990).

The range of temperatures over which fission-track annealing in apatite occurs is referred to as the partial annealing zone, or PAZ (Gleadow and Fitzgerald, 1987; Wagner et al., 1989). Induced fission tracks in apatite have an initial mean length of ~16 μm. In F-rich apatites, tracks are relatively stable and shorten at extremely low rates below 60°C; above 110°C, tracks are shortened and totally annealed soon (in geologic time) after they form (Green et al., 1986). The likelihood of a track to cross a planar surface within a mineral (where tracks are observed) is proportional to the length of the track (Laslett et al., 1982). Therefore, the fission-track "age" of a sample with shorter tracks is proportionally younger than a sample with longer tracks (Green, 1988). Samples that experience complicated or slow cooling within the PAZ may have a mixture of short and long tracks with fission-track ages that are not necessarily geologically meaningful. Track-length distributions in apatite, therefore, are used to evaluate the cooling history of the sample within the PAZ and the meaning of the fission-track age (e.g., Gleadow et al., 1986).

The zircon fission-track PAZ is poorly defined. Based on a recent study of a thermal halo around a pluton, Tagami and Shimada (1996) suggested that the zircon PAZ spanned a temperature range from 230 to 330°C. However, due to uncertainties in this range, and because past studies have suggested that the zircon closure temperature (Dodson, 1973) ranges from ~175 to 250°C (e.g., Zaun and Wagner, 1985), we will use a zircon closure temperature of ~250 ± 50°C.

$^{40}Ar/^{39}Ar$ method

The $^{40}Ar/^{39}Ar$ method (summary in McDougall and Harrison, 1988) is a variation of the conventional K-Ar ($^{40}K/^{39}Ar$) dating method. In the $^{40}Ar/^{39}Ar$ method, the sample is irradiated to produce ^{39}Ar from ^{39}K. Flux monitors (samples with known ages) are included in the irradiation package; these are used to generate a correction factor (J) for the age equation that accounts for the ratio of ^{40}K to ^{39}K (assumed to be a constant) and variables such as the neutron flux. A mass spectrometer is then used to analyze the $^{40}Ar/^{39}Ar$ released from a sample during an incremental "step" heating experiment, and an "age" is calculated for each step. The age represents the time at which the mineral became "closed" to argon diffusion.

Samples in this study consisted of white micas (a mixture of phengite and paragonite) from blueschist facies rocks. The closure temperature of phengite is not known, and a muscovite closure temperature of 350 ± 50°C is assumed (Purdy and Jäger, 1976). White mica $^{40}Ar/^{39}Ar$ analyses from blueschists (as discussed by Baldwin and Harrison, 1989) are often difficult to interpret because of incomplete recrystallization and low K concentrations in the white micas (paragonite has no K), and because the blueschists often have undergone prolonged slow cooling.

ANALYTICAL TECHNIQUES

The samples used in this study were collected during two field seasons in northern Alaska during the summers of 1987 and 1989. Table 1 lists the locations, lithologies, and type(s) of analyses for the 36 individual samples. Samples were crushed, ground, sieved, and washed, and minerals were separated from the ground rock using standard magnetic and heavy liquid techniques.

**TABLE 1. LOCALITIES, LITHOLOGIES, AND TYPES
OF ANALYSES DONE ON
CENTRAL BROOKS RANGE SAMPLES**

Sample	Lat. (°N)	Long. (°W)	Elev. (m)	Lithology	Unit	Analyses*
Fold and Thrust Belt						
6a-87	68.49,	149.41	1,370	Sandstone	Kfm	A
6b-87	68.49,	149.42	1,210	Conglomerate	Kfm	A
6c-87	68.49,	149.42	1,060	Conglomerate	Kfm	A
43-87	68.25,	149.41	950	Quartzite	MDks	Z
44-87	68.28,	149.37	865	Conglomerate	Dks	A, Z
45-87	68.31,	149.35	880	Conglomerate	Dks	A, Z
47-87	68.13,	149.43	1,265	Sparse conglomerate	Dhf	A, Z
48-87	68.17,	149.47	1,165	Graywacke	Dhf	Z
49-87	68.35,	149.33	900	Sandstone	Dke	A
50-87	68.41,	149.31	885	Graywacke	MDks	A, Z
51a-87	68.13,	149.47	1,620	Conglomerate	MDk	A, Z
51b-87	68.13,	149.48	1,470	Conglomerate	MDk	A
51c-87	68.13,	149.52	1,325	Conglomerate	MDk	A
51d-87	68.12,	149.55	1,220	Conglomerate	MDk	A
1-89	68.48,	149.41	1,000	Graywacke	Kfm	A
2-89	68.37,	149.30	980	Grit sandstone	Dke	A
3-89	68.34,	149.33	1,065	Grit sandstone	Dke	Z
4-89	68.30,	149.35	1,075	Grit sandstone	Dke	A, Z
5-89	68.27,	149.36	1,385	Grit sandstone	MDks	A, Z
6-89	68.28,	149.38	1,165	Sandstone	MDks	A, Z
7-89	68.21,	149.42	1,050	Sandstone	Dhf	A, Z
Doonerak Window and Central Belt						
8-87	67.91,	150.00	1,160	Fol. conglomerate	Mk	Z
10-87	67.96,	149.79	705	Conglomerate	Dhf	Z
12-89	67.75,	149.72	475	Schist	Obpm	Z
17-89	67.78,	149.67	1,430	Gabbro	Dg	A
18-89	67.86,	150.10	1,310	Arkose	Mkt	Z
21-89	67.97,	149.72	725	Volcanic schist	Df	A, Z
Schist Belt						
2a-87	67.40,	150.09	415	Schist	PzPqs	A, WM
25b-87	67.22,	150.23	335	Schist	PzPqs	A, WM
.26a-87	67.23,	150.20	350	Metagabbro	Dg	A
28-87	67.28,	150.48	370	Schist	PzPqs	A, WM
29-87	67.34,	150.13	325	Schist	PzPqs	A, WM
30-87	67.43,	150.07	395	Mafic schist	PzPqs	A, WM
31-87	67.36,	150.11	400	Schist	PzPqs	A, WM
52-87	67.18,	150.27	525	Sandstone	Dms	A
54-87	67.36,	150.16	370	Gneiss	Dgr	WM

*A = apatite fission track; Z = zircon fission track; WM = white mica $^{40}Ar/^{39}Ar$.

Fission-track analyses

A sufficient number of apatites and zircons for fission-track analyses were obtained from 28 and 17 samples, respectively. Apatites were mounted in epoxy and zircons in FEP Teflon. Sample surfaces were ground using 400 and 600 grit and polished using 6- and 1-μm diamond paste. Apatite mounts were etched in 7% HNO_3 for 20 to 25 seconds, zircon mounts in a eutectic mixture of 11.5 gm KOH and 8 gm NaOH at 220°C for 4 to 9.5 hours. An "external detector" (Naeser, 1976, 1979) of low U (<5ppb) Brazil Ruby muscovite was attached to the surface of each sample.

Samples were irradiated to induce fission in ^{235}U (recorded in the external detector) in 8 different packages (6 apatite and 2 zircon). Each package consisted of several unknown mounts, sandwiched between two NBS glasses (SRM-962; Carpenter and Reimer, 1974), with a standard, Fish Canyon Tuff (27.8 ± 0.2 Ma; Hurford and Hammerschmidt, 1985), in the middle. Samples collected during 1987 were irradiated in the thermal neutron facility (RT-4) of the National Bureau of Standards Research Reactor in Gaithersburg, Maryland. Samples collected during 1989 were irradiated in the Cornell University Triga 500-kilowatt Mark II reactor; samples were placed in the rotary specimen rack, a region of the reactor that at full power has an estimated thermal neutron flux of 3.5×10^{12} N/cm^2, and a fast neutron flux of 2.5×10^{11} N/cm^2, with a thermal to fast ratio of 14:1.

Following irradiation, the mica detector from each sample was etched in 48% HF for 30 minutes. The dosimeters indicated flux differences at each end of the irradiation packages; a linear interpolation was used to estimate the flux for each sample between the two dosimeters. Analyses were done using a 100× dry objective and a total magnification of 1,250×. Tracks were counted only in crystals with well-etched, clearly visible tracks, and sharp polishing scratches. For each sample, an attempt was made to count tracks in 10 to 20 crystals; however, in some samples fewer suitable crystals were available.

The standard fission-track equation was used to calculate ages (Hurford and Green, 1982). Zeta factors (Hurford and Green, 1983) for A. Blythe of 318 and 335 were determined for apatite and zircon, respectively, by using NBS glass SRM-962 as the dosimetry standard. Individual grain ages for a sample were assumed to have a Poissonian distribution, and the statistical probability of the distribution was estimated with the Chi-square (χ^2) test (Galbraith, 1981). A "pooled" age was calculated from the sums of the spontaneous and induced tracks from all grains; this age is weighted in relation to the number of tracks counted per grain. A pooled age is generally considered reliable if it passed the χ^2 test at >5% probability (Galbraith, 1981); all of the pooled ages in this study passed the χ^2 test at >23% probability, and most at >50% probability. The "error" of a sample age was calculated as a function of the total number of tracks counted (Green, 1981).

For track-length analyses in apatites, only horizontal, "confined" tracks were counted. Whenever possible, 100 track lengths were measured, however, in several samples fewer track lengths were measured because of a scarcity of apatite in the sample or low concentrations of uranium in the apatite.

$^{40}Ar/^{39}Ar$ analyses

Seven white mica separates were analyzed in the laboratory of T. M. Harrison at the State University of New York (S.U.N.Y.), in Albany. The specifications of the mass spectrometer and the analytical techniques used were described by Harrison and Fitzgerald (1986). Separates were wrapped in Sn foil and packed in an evacuated, flat-bottomed quartz vial; samples of biotite Fe-mica (with a known age of 307.3 Ma) were included as flux monitors. The package was irradiated with fast neutrons in the H-5 position in the Ford Reactor at the University of Michigan.

Individual samples were then heated in a double-vacuum, resistance-heated furnace (with temperatures accurate to better than ± 5°C) in order to extract Ar. An ion pump (with high H_2 throughput) and SAES getter pumps were used to purify the gas. Analyses were done with an automated Nuclide 4.5-60-RSS mass spectrometer, with a sensitivity of ~2×10^{-15} mol Ar/mV.

Ages were calculated using the decay constants and isotope abundances of Steiger and Jäger (1977). Interfering reactions of K and Ca were corrected with values of $(^{40}Ar/^{39}Ar)K = 0.029$, $(^{36}Ar/^{37}Ar)Ca = 0.000222$, and $(^{39}Ar/^{37}Ar)Ca = 0.000825$. One-sigma errors, including 0.5% error in the J-factor, are reported.

RESULTS AND INTERPRETATION OF DATA

Fission-track results are summarized in Table 2 (individual fission-track and $^{40}Ar/^{39}Ar$ analyses are available from A. Blythe).

Fold-and-thrust belt

Twelve zircon fission-track ages were obtained from the Brooks Range fold-and-thrust belt (FTB), immediately to the south of the Brooks Range front (Fig. 1). The zircon fission-track ages range from 224.6 ± 14.0 Ma to 130.3 ± 5.6 Ma. All of these ages are significantly younger than their Mississippian-Devonian stratigraphic ages, suggesting that they are fully reset. Conodont Alteration Index data indicate that rocks in the FTB were exposed to maximum paleotemperatures of >300°C (Harris et al., 1987); these data are also consistent with full resetting of zircon fission-track ages, however, without zircon track-length data these data cannot be fully interpreted (see Yamada et al., 1995). In general, zircon ages are younger in the Atigun Pass region (the continental divide), and older to the north, where there is an apparent relationship between elevation and zircon age.

Apatite fission-track ages from 18 samples (Fig. 1) range from 94.7 ± 4.5 Ma to 44.3 ± 3.3 Ma. Ages from the Fortress Mountain Formation at Atigun syncline (the Brooks Range "front") decrease with decreasing elevation (87.9 ± 5.7, 79.4 ± 4.9, 75.1 ± 4.5, and 44.6 ± 4.1 Ma at 1,370, 1,210, 1,060, and 1,000 m, respectively). All of the ages are younger than the stratigraphic age (113 to

95 Ma) of the Fortress Mountain Formation. Although the number of track lengths measured was low, the oldest and youngest samples have relatively short mean track lengths (12.3 and 12.5 μm), and the middle two samples have longer mean track lengths (13.8 and 13.9 μm). P. O'Sullivan (personal communication, 1991) has obtained apatite fission-track analyses of samples from different localities at Atigun syncline. His data, which include more track-length measurements, indicate that partial (but not complete) annealing of samples occurred. He interprets the oldest grain ages to be from samples with a high-Cl content that were never completely annealed.

Five samples to the south of Atigun syncline yielded apatite ages ranging from 54–65 Ma, with a mean age of 60 Ma. Track-length distributions, with means of 12.8 to 14.0 μm, are consistent with cooling through the PAZ (from 110 to 60°C) in 5 to 10 m.y. The pattern further south is more complicated. At Trevor Creek (Fig. 1), samples were collected from different elevations (1,385, 1,165, and 865 m). The two apatite ages from the highest elevations are the same within error (57.1 ± 2.4 and 60.1 ± 3.0 Ma), and both samples had relatively long mean track lengths (13.7 and 14.5 μm). The sample from the lowest elevation at Trevor Creek had a younger age (44.3 ± 3.3 Ma) and a shorter mean track length (12.1 μm), indicating a longer residence within the PAZ.

Apatite ages are older to the south of Trevor Creek. An age of 67.6 ± 4.3 Ma and a relatively short mean track length of 11.9 μm were obtained from sample 7-89. Sample 47-87, collected at a higher elevation north of Atigun Pass, yielded an age of 94.7 ± 4.5 Ma and a mean track length of 13.4 μm. The data indicate sample 47-87 cooled relatively rapidly through the PAZ, and that sample 7-89 remained within the PAZ (above 60°C) for a longer period of time.

Samples collected at Atigun Pass from different elevations (1,620, 1,470, 1,325, and 1,220 m) yielded ages of 93.1 ± 6.6, 89.0 ± 7.0, 82.5 ± 4.3, and 63.8 ± 5.2 Ma, respectively, and mean track lengths of 13.5, 13.1, 12.7, and 12.6 μm. The upper three sample ages are the same within 1σ error. These data indicate relatively rapid cooling of the uppermost sample through the PAZ at 93 Ma, and progressively slower cooling of the lower samples.

Central belt and Doonerak window

Fission-track ages from the central belt (CB) and Doonerak window (DW) are much younger than FTB ages (Fig. 2). No samples were collected from "basement" rocks exposed in the DW; all of the ages were obtained from samples interpreted to belong to the "upper plate" of the DW (Endicott allochthon). The zircon fission-track ages of these samples were 65.6 ± 3.8, 70.1 ± 7.1, 112.4 ± 7.6, and 102.6 ± 5.4 Ma. A sufficient number of apatites were obtained from only one of these samples, 21-89, which yielded an age of 29.4 ± 1.6 Ma and mean track length of 13.4 μm, indicating relatively rapid cooling through the PAZ. These ages, even though from "upper plate" rocks, probably were reset during formation of the DW basement-involved duplex. O'Sullivan et al. (this volume) obtained similar apatite ages from samples collected from basement rocks within the DW.

Zircon and apatite ages of 70.3 ± 3.5 Ma and 48.3 ± 8.7 Ma were obtained from two samples from within the CB, 15 km to the south of the DW. The zircon age is similar to the younger zircon ages from the DW region. No track-length measurements were obtained from the apatite sample because of the low number of tracks.

Schist, phyllite, and metagreywacke belts

^{40}Ar/^{39}Ar analyses were obtained from white-mica separates from six schist belt (SB) samples (Fig. 3). The spectra are complicated, in part because of low K concentrations in these samples. The sample with the highest concentration of K was 30-87; the age of each step increased from 120 to 130 Ma during the heating experiment. This spectrum is similar to those obtained from phengites in other studies (e.g., Wijbrans and McDougall, 1986; Baldwin and Harrison, 1989), which are generally interpreted to be caused by diffusional loss of radiogenic ^{40}Ar during slow cooling in subduction zones. Baldwin and Harrison (1989) have also suggested that the phengite age gradient could be caused by differences in diffusion rates of different-sized fractions of white mica.

Although samples 31-87, 54-87, 29-87, and 28-87 did not have well-developed plateaus, the variation in age between individual steps was not extreme and ages of 120–125 Ma could be interpreted. A more complex spectrum was obtained from sample 2a-87. Although 40% of the gas from this sample yielded an age of ~150 Ma, excess argon appeared to be a problem: early steps had relatively old ages, and late steps had younger ages. A significantly younger plateau age of 113 Ma was obtained from the only sample (25b-87) analyzed from the phyllite belt.

Apatite fission-track ages from six SB samples ranged from 54.6 ± 9.5 to 67.4 ± 11.4 Ma. The mean age of these samples is 63 Ma, with all of the ages intersecting the mean within a standard deviation of 1σ. Track-length measurements were done on only one sample, 2a-87, because of the low numbers of tracks in SB samples. A mean track length of 13.6 μm indicated relatively rapid cooling. Samples 25b-87 and 52-87, from the phyllite and metagreywacke belts south of the SB, yielded apatite fission-track ages of 57.6 ± 8.7 and 60.3 ± 3.3 Ma. Within 1σ, these ages are the same as those from the SB. Both of these samples also had long mean track lengths (14.0 and 14.7 μm), indicating relatively rapid cooling. Track-length analyses from the schist, phyllite, and metasedimentary belt samples are consistent with relatively rapid cooling through the PAZ (in ~5 m.y.) at ~60 Ma. These data are similar to results obtained from FTB samples.

COOLING AND EXHUMATION RATES

Thermochronometric data constrain the time at which a mineral cooled through a specific temperature or range of tempera-

tures, and therefore, cooling rates can be calculated from the thermochronometric ages of different minerals from the same rock. In order to extract exhumation histories, assumptions must be made about geothermal gradients and surface elevations (see discussion in Parrish, 1983). Recent studies indicate that geothermal gradients vary spatially and temporally throughout an orogenic belt (e.g., Hubbard et al., 1991). Data on the paleogeothermal gradient are rarely available and therefore, a linear geothermal gradient is used to interpret thermochronometric data (e.g., Parrish, 1983; Baldwin and Harrison, 1989). In this study, exhumation rates were calculated with geothermal gradients ranging from 30 to 15°C/km for the schist and central belt data and 30 to 20°C/km for the fold-and-thrust belt data. The uppermost value, 30°C/km, was chosen because it is the present-day geothermal gradient (~30°C/km) at Prudhoe Bay (Lachenbruch et al., 1982). For the SB and CB, the lowermost value, 15°C/km, was chosen because these belts include metamorphic rocks exposed to high pressure–low temperature conditions. Thus, following the end of subduction-related contraction, the geothermal gradient probably evolved from ~10–12°C/km to more normal values (e.g., England and Thompson, 1984); it has been proposed that Seward Peninsula blueschist facies rocks (which are often correlated with the Brooks Range SB rocks) underwent decompression as the geothermal gradient increased from 12.5 to 30°C/km over ~55 m.y. (Patrick and Evans, 1988). A lower value of 15°C/km was chosen for the SB and CB because the oldest cooling age (130 Ma) used in the exhumation calculation is probably significantly younger than the time (>160 Ma?) at which the geothermal gradient was at its lowest value. A slightly higher value of 20°C/km was used for the minimum geothermal gradient in the fold-and-thrust belt because models for the thermal evolution of a fold-and-thrust belt are consistent with a slightly lowered geothermal gradient near the toe of an orogenic wedge (e.g., Barr and Dahlen, 1989).

A mean annual surface temperature of 0°C was used for the present-day surface temperature. We chose this value as an average over the last 60 Ma, although clearly the mean annual surface temperature has varied through time: The present-day mean annual temperature on the North Slope is –10 to –12°C, but recent studies have shown that the mean annual surface temperature in the Arctic was substantially warmer (0–10°C) until ~5 Ma (Herman and Hopkins, 1980).

Estimated cooling rates are summarized in Table 3, and cooling and exhumation rates, which are discussed below, are illustrated in Figure 4. Errors for the cooling rates were estimated by using the single largest source of error in any one calculation. These consisted of (1) the ±50°C error on the white mica closure temperature for the white mica ^{40}Ar/^{39}Ar age to apatite fission-track age cooling rate, (2) the ±50°C error on the zircon fission-track closure temperature for the zircon fission-track to apatite fission-track age cooling rate, (3) a range of times of 10 to 2.5 m.y. for rapid cooling through the apatite PAZ, and (4) a range of times of ±5 m.y. for rapid cooling through the apatite PAZ to a surface temperature of 0°C. For the estimated exhumation rates, the greatest source of error was the

A. E. Blythe and others

TABLE 2. FISSION TRACK ANALYTICAL RESULTS*

Sample	Mineral	Number Grains	Standard Track Density (x10^{6}cm^{-2})	Fossil Track Density (x10^{6}cm^{-2})	Induced Track Density (x10^{6}cm^{-2})	Chi Square Prob. (%)	Fission Track Age (Ma)	Mean Track Length (μm)	Standard Track Dev. (μm)
Fold-and-Thurst Belt									
6a-87	Apatite	23	1.8460† (4615)	0.3214 (314	1.0676 (1043)	97	87.9± 5.7	12.3±0.7 (10)	2.3
6b-87	Apatite	18	1.7684† (4421)	0.2907 (341)	1.0247 1202)	23	79.4± 4.9	13.8±0.3 (24)	1.5
6c-87	Apatite	19	1.8159† (4540)	0.3207 (348)	1.2277 (1332)	76	75.1± 4.5	13.9±0.4 (7)	1.1
1-89	Apatite	13	1.9422† (10198)	0.1348 (134)	0.9316 (926)	97	44.6± 4.1	12.5±0.5 (20)	2.1
43-87	Zircon	8	0.1798† (944)	15.3893 (2016)	3.3588 (440)	81	138.5± 7.3		
44-87	Apatite	15	1.6955† (4239)	0.2435 (214)	1.4790 (1300)	>99	44.3± 3.3	12.1±0.4 (9)	2.4
	Zircon	10	0.1761† (925)	13.3016 (838)	2.5396 (160)	>99	154.8±13.4		
45-87	Apatite	20	1.7783† (4446)	0.3878 (553)	2.0252 (2888)	>99	54.0± 2.5	13.0±0.2 (100)	1.8
	Zircon	8	0.1779† (934)	20.3067 (1523)	3.7733 (283)	>99	160.5±10.4		
47-87	Apatite	12	1.7901† (4475)	0.9851 (595)	2.9437 (1778)	98	94.7± 4.5	13.4±0.2 (47)	1.5
	Zircon	15	0.1815 (953)	14.7865 (2839)	3.4583 (664)	99	130.3± 5.6		
48-87	Zircon	17	0.1768† (928)	12.8138 (1858)	2.5448 (369)	>99	149.5± 8.5		
49-87	Apatite	13	1.9565† (4891)	0.6057 (467)	3.0791 (2374)	>99	61.0± 3.1	12.8±0.2 (57)	1.6
50-87	Apatite	15	1.9130† (4783)	0.4855 (302)	2.5563 (1590)	>99	57.6± 3.6	13.7±0.2 (67)	1.6
	Zircon	3	0.1853 (973)	20.3714 (713)	4.7714 (167)	>99	132.3±11.4		
51a-87	Apatite	7	1.7797† (4449)	0.9371 (268)	2.8322 (810)	>99	93.1± 6.6	13.5±0.3 (27)	1.6
	Zircon	5	0.1823† (957)	15.9487 (622)	3.4872 (136)	<99	139.6±13.2		
51b-87	Apatite	7	1.9117† (4779)	0.7909 (208)	2.6844 (706)	>99	89.0± 7.0	13.1±0.2 (38)	1.5
51c-87	Apatite	11	1.9273† (4818)	0.6765 (458)	2.5007 (1693)	89	82.5± 4.3	12.7±0.3 (55)	2.0
51d-87	Apatite	12	1.7755† (4439)	0.4741 (183)	2.0907 (807)	97	63.8± 5.2	12.6±0.5 (12)	1.6
2-89	Apatite	8	1.4248 (7480)	1.1717 (273)	4.0730 (949)	82	65.0± 4.5	13.8±0.5 (9)	1.4
3-89	Zircon	9	0.3116 (1636)	12.1429 (1020)	3.5119 (295)	74	180.1±11.9		
4-89	Apatite	8	1.4537 (7632)	0.8906 (285)	3.2938 (1054)	95	62.3± 4.2	14.0±0.2 (14)	0.8
	Zircon	10	0.3131 (1644)	10.8365 (1127)	3.0288 (315)	99	187.6±12.0		
5-89	Apatite	18	1.4825 (7783)	0.6911 (734)	2.8484 (3025)	99	57.1± 2.4	13.7±0.2 (46)	1.5
	Zircon	10	0.3131 (1644)	12.8113 (1358)	2.9906 (317)	99	224.6±14.0		
6-89	Apatite	11	1.5091† (3773)	1.2067 (502)	4.8029 (1998)	36	60.1± 3.0	14.5±0.2 (24)	1.1
	Zircon	11	0.3156 (1657)	13.8387 (1716)	3.9919 (495)	51	182.9± 9.3		

TABLE 2. FISSION TRACK ANALYTICAL RESULTS* (continued - page 2)

Sample	Mineral	Number Grains	Standard Track Density (x10⁶cm⁻²)	Fossil Track Density (x10⁶cm⁻²)	Induced Track Density (x10⁶cm⁻²)	Chi Square Prob. (%)	Fission Track Age (Ma)	Mean Track Length (μm)	Standard Track Dev. (μm)
Fold-and-Thurst Belt (continued)									
7-89	Apatite	11	1.5805† (39.51)	1.0329 (314)	3.8257 (1163)	90	67.6 ± 4.3	11.9 ± 0.2 (20)	1.3
	Zircon	10	0.3175 (1667)	12.2133 (916)	3.4267 (257)	98	189.2 ±13.4		
Doonerak Window and Central Belt									
8-87	Zircon	12	0.1859 (976)	5.5325 (935)	2.6272 (444)	89	65.6 ± 3.8		
10-87	Zircon	7	0.1756 (922)	7.7442 (333)	3.2558 (140)	>99	70.1 ± 7.1		
12-89	Zircon	10	0.3175 (1667)	8.3273 (916)	6.2909 (692)	74	70.3 ± 3.5		
17-89	Apatite	8	1.8519 (9724)	0.0373 (36)	0.2269 (219)	>99	48.3 ± 8.7		
18-89	Zircon	5	0.3198 (1679)	11.2459 (686)	5.3607 (327)	54	112.4 ± 7.6		
21-89	Apatite	20	1.6271 (8542)	0.3476 (383)	3.0381 (3348)	72	29.4 ± 1.6	13.4 ± 0.3 (32)	1.4
	Zircon	10	0.3198 (1679)	11.7955 (1038)	6.1591 (542)	55	102.6 ± 5.4		
Schist and Phyllite Belts									
2a-87	Apatite	8	1.9844 (4961)	0.1245 (59)	0.6189 (294)	>99	63.0 ± 9.0	13.6 ± 0.3 (28)	1.7
25b-87	Apatite	6	2.0844 (5211)	0.3006 (52)	1.7225 (298)	98	57.6 ± 8.7	14.0 ± 0.6 (12)	2.2
26a-87	Apatite	13	2.0400 (51.00)	0.0628 (90)	0.3250 (466)	>99	62.4 ± 7.2		
28-87	Apatite	10	1.7636 (4409)	0.0510 (42)	0.2136 (176)	>99	66.6 ±11.4		
29-87	Apatite	6	1.7064 (4266)	0.0863 (44)	0.3451 (176)	>99	67.4 ±11.4		
31-87	Apatite	10	1.7064 (4266)	0.0851 (88)	0.3646 (377)	92	63.0 ± 7.5		
52-87	Apatite	13	1.762 (4405)	0.9191 (409)	4.4090 (1962)	86	60.3 ± 3.3	14.7 ± 0.2 (19)	0.8
54-87	Apatite	3	1.5680 (3920)	0.2548 (40)	1.1592 (182)	74	54.6 ± 9.5		

*Numbers shown in parentheses show number of tracks counted. Standard and induced track densities measured on mica external detectors (g = 0.5), and fossil track densities on internal mineral surfaces. Ages calculated using zeta = 318 (apatite) and 335 (zircon) for dosimeter glass SRM 962a (e.g., Hurford and Green, 1983).
†Weighted mean standard track density for samples with multiple mounts, each with different standard track densities.

range of geothermal gradients, and the uncertainty of how the gradients evolved over time.

Fold-and-thrust belt

Figure 5 illustrates variations in the cooling and exhumation patterns between the northern (all samples north of and including sample 7-89 on Fig. 1) and the southern FTB. In the northern FTB, the linear relationship between elevation and zircon age suggests very slow exhumation between 185 and 130 Ma. However, it is unlikely that the spatial relationship between samples that is now observed existed at 130 Ma (as the result of thrust-faulting after 130 Ma), and therefore we suggest that the apparent slope is coincidental. The oldest zircon age (225 ± 14 Ma) is from the highest elevation at Trevor Creek. We suspect that this age is only partially reset and therefore cannot be interpreted as repre-

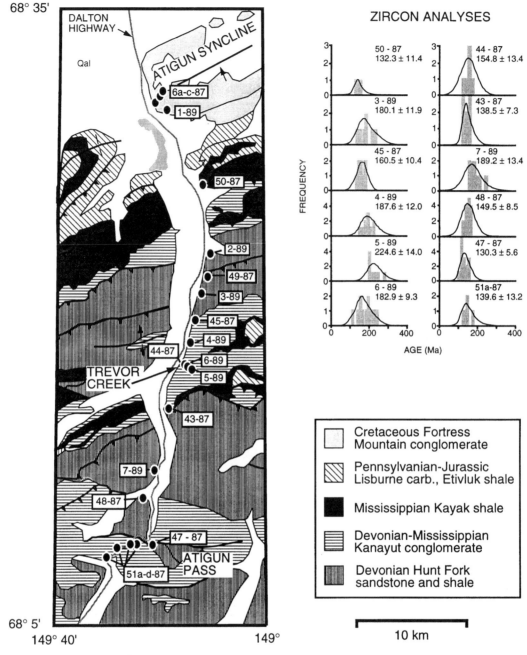

Figure 1 (on this and facing page). Simplified geologic map of the fold-and-thrust belt, with sample local-ities (geology after Mull and Adams, 1989). Histograms of individual zircon and apatite fission-track grain ages for each sample are plotted with the probability density curve shown as a black line. Histograms of apatite track length analyses are also shown. N = total number of track lengths measured; track length totals on histograms have been normalized to 100. This format is also used on Figures 2 and 3.

senting a distinct cooling event because samples collected at lower elevations at Trevor Creek are substantially younger. How-ever, the clusters of zircon fission-track ages at ~185 and 140 Ma suggest that there might have been two pulses of cooling and exhumation at these times in the northern FTB. The 185-Ma age and the youngest zircon age of ~130 Ma (sample 50-87 is 132 ± 11 Ma) are used as the upper and lower ages for calculating cool-

ing rates for the northern FTB. In the southern FTB (the Atigun Pass region), the three zircon ages range from ~130 to 150 Ma; these ages are used for calculating cooling rates.

The apatite ages used in cooling rate calculations were deter-mined by the pattern of track length distributions: Only ages with long mean track lengths (>13.5 μm) were considered as repre-senting a distinct cooling event. In Figure 6, sample elevation

APATITE ANALYSES

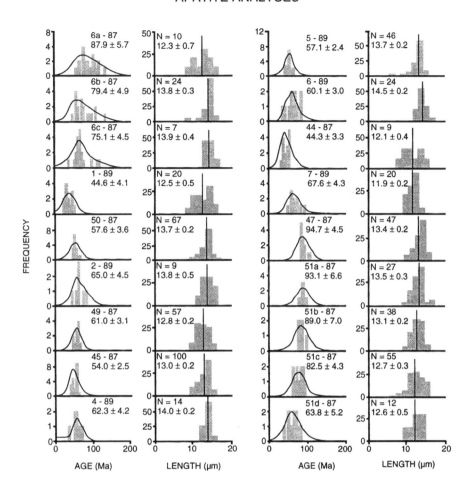

versus age and mean track length versus age plots indicate that two pulses of relatively rapid cooling (from ~110 to 60°C in ~5 m.y.) occurred, one at ~100 Ma in the southern FTB, and the other at 60 Ma in the northern FTB.

For the northern FTB, cooling rates of 0 (assuming no cooling from 185 to 130 Ma) to 2.0°C/m.y. (130 to 60 Ma), 10°C/m.y. (60 to 55 Ma), and 1.1°C/m.y. (55 Ma to the present) are estimated. At geothermal gradients of 20 to 30°C/km-depth, these cooling rates correspond to exhumation rates of 0 to 0.1, 0.8 to 1.2, and 0.04 to 0.06 km/m.y., respectively. For the southern FTB, cooling rates of 0 to 3.7°C/m.y. are determined for 150 to 100 Ma, and rates of 10 and 0.7°C/m.y. for 100 to 95 Ma and 95 Ma to the present, respectively; these rates correspond to exhumation rates of 0.10 to 0.19, 0.8 to 1.2, and 0.02 to 0.03 km/m.y.

Doonerak window and Central belt

No information on the cooling history of the DW region prior to 112 Ma (the oldest zircon age) was obtained. The zircon fission-track ages ranged from 112 to 66 Ma; two ages are ~105 Ma and three ages are ~67 Ma (within 1σ); suggesting two pulses of rapid cooling at those times. However, no systematic

pattern is evident in age versus elevation plots (Fig. 7). The random pattern of ages with respect to elevation may be the result of vertical offset along faults in the DW region (not shown on generalized geologic map in Fig. 2) after 65 Ma.

Exhumation rates were estimated using the range of zircon ages and the single DW apatite analysis, which is consistent with cooling through the PAZ over a 5-m.y. time interval (from 30 to 25 Ma). The estimated cooling rates for 112 to 30 Ma, 30 to 25 Ma, and 25 Ma to the present, are 0 to 3.8, 10, and 2.4°C/m.y. Exhumation rates are estimated from these cooling rates by assuming geothermal gradients of 15 to 30°C/km-depth; exhumation rates are 0 to 0.25, 0.8 to 1.6, and 0.08 to 0.16 km/m.y., respectively.

Cooling and exhumation rates are not calculated for the central belt, because only single zircon and apatite ages were obtained. The zircon age (70 Ma) is similar to DW zircon ages, although the apatite age (48 Ma) appears to be transitional between DW and SB apatite ages.

Schist belt

SB white mica [40]Ar/[39]Ar ages of 130 to 120 Ma are interpreted to represent the time of cooling through ~350°. However,

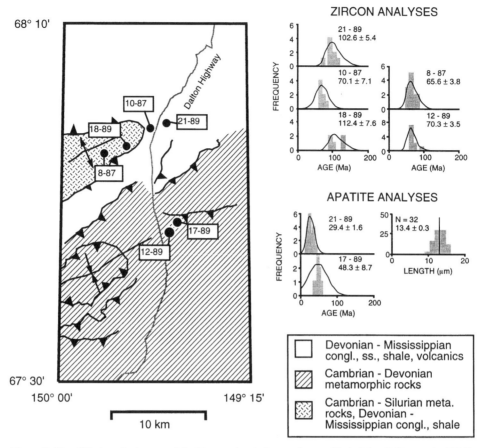

Figure 2. Simplified geologic map of the Doonerak window and central belt (after Dillon et al., 1986), with sample localities and apatite and zircon fission-track analyses. See Figure 1 for explanation of histograms.

maximum temperatures and pressures of ~450°C and 12 kbar for the SB (Patrick, 1995; Gottschalk, this volume) are consistent with the onset of deformation and cooling long before 130 Ma. SB apatite fission-track data are similar to the analyses from the northern FTB and indicate relatively rapid cooling through the apatite PAZ from ~60 to 55 Ma. The estimated average cooling rate for 120 to 60 Ma was 4.0°C/m.y. Rates estimated for 60 to 55 Ma and for 55 Ma to the present were 10 and 1.1°C/m.y., respectively, the same as those estimated for the FTB. At geothermal gradients of 15 to 30°C/km-depth, the cooling rate for 130 to 65 Ma corresponds to exhumation rates of 0.13 to 0.27 km/m.y. Rates, however, may have been much higher (>1.0 km/m.y.) initially (from 130 to 100 Ma) and slower (0.13 km/m.y.) later on (from 100 to 65 Ma), as the result of isothermal decompression in the SB; this scenario is shown as a dark line (the suggested exhumation path) in the schist belt and central belt exhumation boxes on Figure 4. SB exhumation rates calculated for 60 to 55 Ma, and 55 Ma to the present are 0.8 to 1.6, and 0.04 to 0.07 km/m.y.

The phyllite belt white-mica $^{40}Ar/^{39}Ar$ age of 113 Ma is significantly younger than SB ages of 130 to 120 Ma; younger amphibole and white mica plateau ages of 115 and 118 Ma were also obtained by Gottschalk et al. (this volume) from southern SB samples. We suggest that these younger ages were not the result of regional cooling, but instead, were reset by fluid migration and recrystallization of micas near the fault zone during extensional faulting between the phyllite belt and SB. Apatite fission-track analyses from the phyllite and metagreywacke belts are similar to those from the SB, indicating similar cooling and exhumation patterns from 60 Ma to the present.

DISCUSSION

Zircon and apatite fission-track and $^{40}Ar/^{39}Ar$ white-mica analyses indicate that cooling rates varied spatially and temporally throughout the central Brooks Range. Cooling rates (summarized in Table 3) derived from the thermochronometric data are used to constrain central Brooks Range tectonic events as follows:

The Brooks Range exhumation had probably begun by

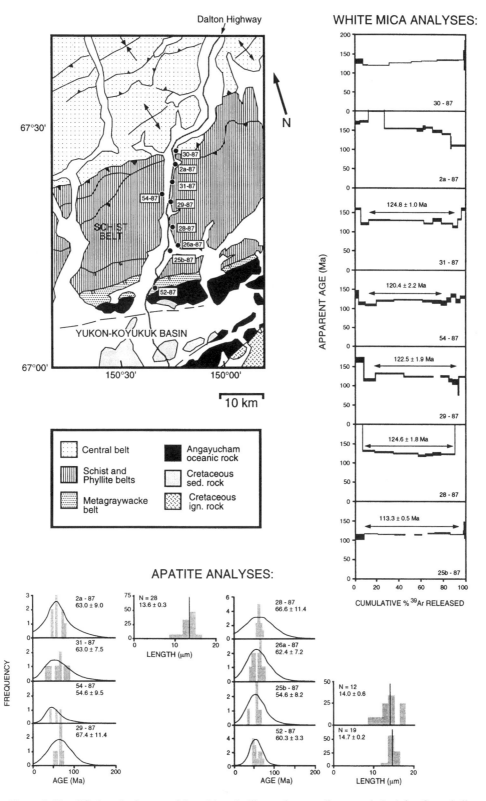

Figure 3. Simplified geologic map of the schist, phyllite, and metasedimentary belts (after Gottschalk, 1987) with sample localities and ^{40}Ar/^{39}Ar white-mica and apatite fission-track analyses. See Figure 1 for explanation of histograms.

TABLE 3. SUMMARY OF ESTIMATED COOLING RATES

Region	Time (m.y.)	Cooling rate (°C/m.y.)	Method used to calculate rate
Northern fold-and-thrust belt	185-60	0.0 to 2.0 ± 0.7	Z/A
	60-55	10 +10/-5	A
	55-0	1.1 ± 0.2	A/surface
Southern fold-and-thrust belt	130-95	0 to 3.7 ± 0.7	Z/A
	95-90	10 +10/-5	A
	90-0	0.7 ± 0.1	A/surface
Doonerak Window	112-30	0 to 3.8 ± 1.5	Z/A
	30-25	10 +10/-5	A
	25-0	2.4 ± 0.6	A/surface
Schist Belt	130-60	4.0 ± 0.8	WM/A
	60-55	10 +10/-5	A
	55-0	1.1 ± 0.2	A/surface

*Z/A = zircon fission-track ages and closure temperature (250 ± 50°C) and bottom of the apatite PAZ (110°), A = apatite fission-track ages and track length data for cooling from 110 to 70°C, WM/A = white mica 40Ar/39Ar ages and closure temperature (350 ± 50°C) and bottom of the apatite PAZ, A/surface = top of the apatite PAZ (60°C) and mean surface temperature (0°C—see text for discussion).

~185 Ma, but may have begun earlier. Zircon fission-track data from the northern FTB are consistent with very slow cooling from 185 to 130 Ma. Exhumation had also begun in the southern FTB and SB by 130 Ma. We estimate that ~50°C and 100°C of cooling had occurred by 130 Ma in the FTB and SB, respectively, by subtracting the zircon fission-track and ^{40}Ar/^{39}Ar white mica closure temperatures from the maximum paleotemperatures of >300°C in the southern FTB (Harris et al., 1987) and >450°C in the SB (Patrick, 1995; Gottschalk, this volume, Chapter 9). Exhumation prior to 130 Ma is consistent with the 150-Ma age of the earliest Brooks Range–derived sediments (Curtis et al., 1982) and with data suggesting that emplacement of the western Brooks Range ophiolites onto the continental margin occurred at 146 to 134 Ma (Wirth et al., 1993).

From 130 to 100 Ma, average exhumation rates were highest in the SB and lower to the north in the southern and northern FTB. During this time, the geothermal gradient in the SB was probably increasing and the exhumation rate may have been much higher than estimated (~1 km/m.y.). This scenario is consistent with sedimentation patterns: the Colville Basin sedimentation rates were greatest from ~112 to 95 Ma (Mull, 1982); at the same time, thick sequences of Brooks Range–derived sediments were deposited in the Yukon-Koyukuk basin (Mull, 1985).

Miller and Hudson (1991) have proposed that large-scale extension occurred in the southern Brooks Range from 130 to 90 Ma. Although the distribution of ^{40}Ar/^{39}Ar white-mica ages does not provide evidence for extension within the SB, phyllite belt data suggests that extension may have occurred on the southern margin of the SB at ~113 Ma.

From ~100 to 95 Ma, a pulse of more rapid cooling occurred in the southern FTB; from 95 to 60 Ma, cooling rates were low throughout the FTB. Exhumation rates in the DW can not be determined for these times; the zircon fission-track ages might represent either (1) two pulses of cooling and exhumation of unknown rate at ~112 and ~66 Ma, or (2) extremely slow cooling from 112 to 66 Ma. For both of these cases, however, the exhumation rate in the DW from ~90 to 65 Ma would be extremely low. Exhumation rates in the SB from 95 to 60 Ma would also have been lower if, as argued above, rates were higher from 130 to 100 Ma. Lower rates throughout the central Brooks Range from ~95 to 60 Ma are consistent with the cessation of tectonic activity: The opening of the Canada Basin apparently ended at ~100 Ma (e.g., Halgedahl and Jarrard, 1987), and subduction along the southern margin of the Brooks Range had ended by ~112 Ma (e.g., Box and Patton, 1989).

From ~60 to 55 Ma, a pulse of more rapid cooling occurred throughout the central Brooks Range. Apatite fission-track data from the northeastern Brooks Range and Colville Basin (e.g., O'Sullivan and Decker, 1990; O'Sullivan et al., 1993) and the western Brooks Range (Wirth et al., 1993) suggest that this event was widespread. From ~30 to 25 Ma, more rapid exhumation also occurred in the Doonerak window region (see also O'Sullivan et al., this volume). We suggest that both of these events may be related to major episodes of magmatism in the Alaska Range–Aleutian Arc batholiths from 74 to 55 Ma and 42 to 21 Ma (Reed and Lanphere, 1973; Hudson, 1979) that have been attributed to plate reorientations at 74 and 43 Ma (Wallace and Engebretson, 1984).

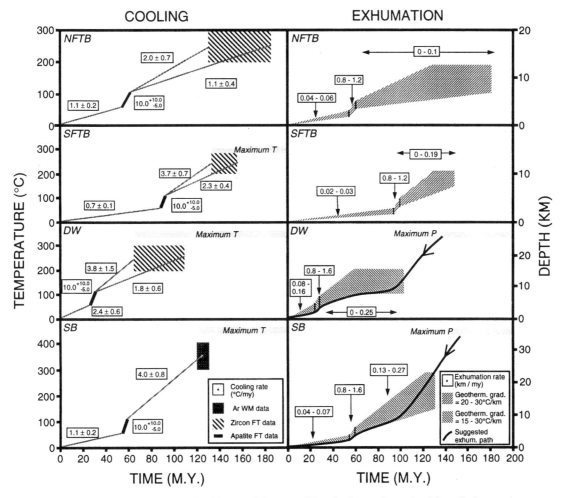

Figure 4. Cooling and exhumation history of the central Brooks Range determined from fission-track and ⁴⁰Ar/³⁹Ar analyses (discussed in text), for the northern fold-and-thrust belt (NFTB), southern fold-and-thrust belt (SFTB), Doonerak window (DW), and schist belt (SB). The boxes contain cooling or exhumation rate estimates for the time intervals indicated. See text for further discussion.

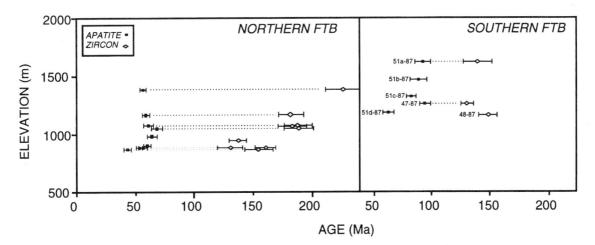

Figure 5. Zircon and apatite fission-track ages versus sample elevation for northern and southern fold-and-thrust belt samples. The dashed line indicates analyses from the same samples.

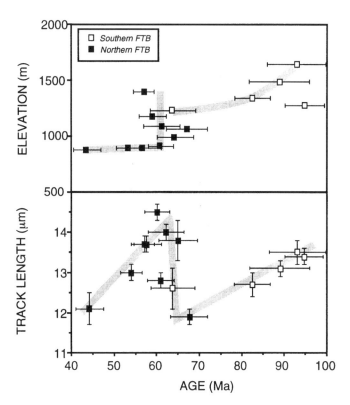

Figure 6. Fold-and-thrust belt apatite fission-track ages plotted against sample elevation (upper plot) and mean track length (lower plot).

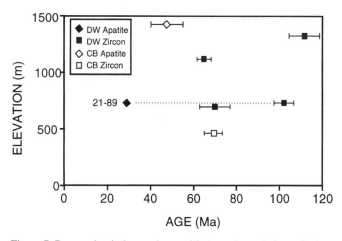

Figure 7. Doonerak window and central belt apatite and zircon fission-track ages versus sample elevation. The dashed line indicates analyses from the same sample.

ACKNOWLEDGMENTS

This work was supported by the National Aeronautics and Space Administration (NASA; NAGW-1287), and A. Blythe was funded by a NASA Graduate Student Researchers Program Fellowship. Partial field support was provided by the Geological Society of America, the American Association of Petroleum Geologists, Sigma Xi, and the Cornell Chapter of Sigma Xi. Helicopter time was provided by the U.S. Geological Survey. J. Zender and K. Wirth assisted with sample collection, and G. Mull, A. Till, E. Julian, M. Seidenstecker, and R. Gottschalk advised us on sample localities. $^{40}Ar/^{39}Ar$ analyses were done in the laboratory of T. M. Harrison at the State University of New York, Albany, with the assistance of M. Heizler and P. Copeland. P. O'Sullivan and J. Murphy greatly improved an early version of this manuscript; R. Gottschalk and R. Donnelick provided thoughtful and thorough reviews.

REFERENCES CITED

Baldwin, S. L., and Harrison, T. M., 1989, Geochronology of blueschists from west-central Baja California and the timing of uplift in subduction complexes: Journal of Geology, v. 97, p. 149–163.

Barr, T. D., and Dahlen, F. A., 1989, Brittle frictional mountain building 2. Thermal structure and heat budget: Journal of Geophysical Research, v. 94, no. B4, p. 3923–3947.

Box, S. E., 1985, Early Cretaceous orogenic belt in northwestern Alaska: Internal organization, lateral extent, and tectonic interpretation, in Howell, D. G., ed., Tectonostratigraphic terranes of the Circum-Pacific region: Houston, Texas, Circum-Pacific Council for Energy and Mineral Resources, p. 137–145.

Box, S. E., and Patton, W. W., Jr., 1989, Igneous history of the Koyukuk terrane, western Alaska: Constraints on the origin, evolution, and ultimate collision of an accreted island arc terrane: Journal of Geophysical Research, v. 94, no. B11, p. 15821–15841.

Carpenter, B. S., and Reimer, G. M., 1974, Standard reference materials: calibrated glass standards for fission track use: National Bureau of Standards Special Publication 260-49, 16 p.

Copeland, P., and Harrison, M. T., 1990, Episodic rapid uplift in the Himalaya revealed by $^{40}Ar/^{39}Ar$ analysis of detrital K-feldspar and muscovite, Bengal fan: Geology, v. 18, p. 354–357.

Crowley, K. D., Cameron, M., and McPherson, B. J., 1990, Annealing of etchable fission-track damage in F-, OH-, Cl-, and Sr-apatite: 1. Systematics and preliminary interpretations [abs.]: Nuclear Tracks and Radiation Measurements, v. 17, p. 409–410.

Curtis, S. M., Ellersieck, I., Mayfield, C. F., and Tailleur, I. L., 1982, Reconnaissance geologic map of southwestern Misheguk Mountain Quadrangle, Alaska: U.S. Geological Survey Open-File Report OF 82-611, scale 1:63,360.

Dillon, J. T., Brosgé, W. P., and Dutro, J. T., Jr., 1986, Generalized geologic map of the Wiseman Quadrangle, Alaska: U.S. Geological Survey Open File Report OF 86-219, scale 1:250,000.

Dodson, M. H., 1973, Closure temperature in cooling geochronological and petrological systems: Contributions to Mineralogy and Petrology, v. 40, p. 259–274.

Donelick, R. A., 1991, Crystallographic orientation dependence of mean etchable fission track length in apatite: an empirical model and experimental observations: American Mineralogist, v. 76, p. 83–91.

England, P. C., and Thompson, A. B., 1984, Pressure-temperature-time paths of regional metamorphism I. Heat transfer during the evolution of regions of thickened continental crust: Journal of Petrology, v. 25, p. 894–928.

Estabrook, C. H., Stone, D. B., and Davies, J. N., 1988, Seismotectonics of northern Alaska: Journal of Geophysical Research, v. 93, no. B10, p. 12026–12040.

Galbraith, R. F., 1981, On statistical methods of fission track counts: Mathematical Geology, v. 13, p. 471–478.

Gleadow, A. J. W., and Duddy, I. R., 1981, A natural long-term track annealing experiment for apatite: Nuclear Tracks, v. 5, p. 169–174.

Gleadow, A. J. W., and Fitzgerald, P. G., 1987, Uplift history and structure of the Trans-Antarctic Mountains: New evidence from fission track dating of basement apatites in the Dry Valleys area, southern Victoria Land: Earth and Planetary Science Letters, v. 82, p. 1–14.

Gleadow, A. J. W., Duddy, I. R., and Lovering, J. F., 1983, Fission track analysis: A new tool for the evaluation of thermal histories and hydrocarbon potential: Australian Petroleum Exploration Association Journal, v. 23, p. 93–102.

Gleadow, A. J. W., Duddy, I. R., Green, P. F., and Lovering, J. F., 1986, Confined fission track lengths in apatite: a diagnostic tool for thermal history analysis: Contributions to Mineralogy and Petrology, v. 94, p. 405–415.

Gottschalk, R. R., 1987, Structural and petrologic evolution of the south-central Brooks Range near Wiseman, Alaska [Ph.D. thesis]: Houston, Texas, Rice University, 263 p.

Green, P. F., 1981, A new look at statistics in fission track dating: Nuclear Tracks, v. 5, p. 77–86.

Green, P. F., 1988, The relationship between track shortening and fission track age reduction in apatite: combined influences of inherent instability, annealing anisotropy, length bias and system calibration: Earth and Planetary Science Letters, v. 89, p. 335–352.

Green, P. F., 1989, Thermal and tectonic history of the East Midlands shelf (onshore UK) and surrounding regions assessed by apatite fission track analysis: Journal of the Geological Society, London, v. 146, p. 755–773.

Green, P. F., and Durrani, S. A., 1977, Annealing studies of tracks in crystals: Nuclear Track Detection, v. 1, p. 33–39.

Green, P. F., Duddy, I. R., Gleadow, A. J. W., Tingate, P. T., and Laslett, G. M., 1986, Thermal annealing of fission tracks in apatite: 1—A qualitative description: Isotope Geoscience, v. 59, p. 237–253.

Halgedahl, S., and Jarrard, R., 1987, Paleomagnetism of the Kuparuk River Formation from oriented drill core; Evidence for rotation of the North Slope block, in Tailleur, I. L., and Weimer, P., eds., Alaskan North Slope geology: Society of Economic Paleontologists and Mineralogists, Pacific Section, Special Publication 50, p. 581–620.

Harris, A. G., Lane, H. R., Tailleur, I. L., and Ellersieck, I., 1987, Conodont thermal maturation patterns in Paleozoic and Triassic rocks, northern Alaska—Geologic and exploration implications, in Talleur, I., and Weimer, P., eds., Alaska North Slope geology: Society of Economic Paleontologists and Mineralogists, Pacific Section, Publication 50, p. 181–191.

Harrison, T. M., and Fitzgerald, J. D., 1986, Exsolution in hornblende and its consequences for $^{40}Ar/^{39}Ar$ age spectra and closure temperature: Geochimica et Cosmochimica Acta, v. 50, p. 247–253.

Herman, Y., and Hopkins, D. M., 1980, Arctic oceanic climate in late Cenozoic time: Science, v. 209, p. 557–562.

Hubbard, M., Royden, L., and Hodges, K., 1991, Constraints on unroofing rates in the High Himalaya, eastern Nepal: Tectonics, v. 10, no. 2, p. 287–298.

Hudson, T., 1979, Calc-alkaline plutonism along the Pacific rim of southern Alaska: U.S. Geological Survey Open File Report 79-953, 31 p.

Hurford, A. J., and Green, P. F., 1982, A user's guide to fission track dating calibration: Earth and Planetary Science Letters, v. 59, p. 343–354.

Hurford, A. J., and Green, P. F., 1983, The zeta age calibration of fission track dating: Isotope Geoscience, v. 1, p. 285–317.

Hurford, A. J., and Hammerschmidt, K., 1985, $^{40}Ar/^{39}Ar$ and K/Ar dating of the Bishop and Fish Canyon Tuffs: calibration ages for fission-track dating standards: Chemical Geology, v. 58, p. 23–32.

Lachenbruch, A. H., Sass, J. H., Marshall, B. V., and Moses, T. H., Jr., 1982, Permafrost, heat flow, and the geothermal regime at Pruhoe Bay, Alaska: Journal of Geophysical Research, v. 87, no. B11, p. 9301–93106.

Laslett, G. M., Kendall, W. S., Gleadow, A. J. W., and Duddy, I. R., 1982, Bias in measurement of fission track length distributions: Nuclear Tracks, v. 6, p. 79–85.

McDougall, I., and Harrison, T. M., 1988, Geochronology and thermochronology by the $^{40}Ar/^{39}Ar$ method: New York, Oxford University Press, Oxford Monographs on Geology and Geophysics No. 9, 212 p.

Miller, E. L., and Hudson, T. L., 1991, Mid-Cretaceous extensional fragmentation of a Jurassic–Early Cretaceous compressional orogen, Alaska: Tectonics, v. 10, no. 4, p. 781–796.

Mull, C. G., 1982, Tectonic evolution and structural style of the Brooks Range, Alaska: An illustrated summary, in Powers, R. B., ed., Geologic studies of the Cordilleran Thrust belt: Denver, Colorado, Rocky Mountain Association of Geologists, p. 1–45.

Mull, C. G., 1985, Cretaceous tectonics, depositional cycles of the Nanushuk Group, Brooks Range and Arctic Slope, Alaska, in Huffman, A. C., ed., Geology of the Nanushuk Group and related rocks, North Slope, Alaska: U.S. Geological Survey Bulletin, v. 1614, p. 7–36.

Mull, C. G., and Adams, K. E., 1989, Dalton Highway, Yukon River to Prudhoe Bay, Alaska: Bedrock geology of the eastern Koyukuk Basin, central Brooks Range and Arctic Slope: Alaska Division of Geological and Geophysical Surveys, Guidebook 7, 2 volumes, 327 p.

Naeser, C. W., 1976, Fission track dating: U.S. Geological Survey Open-File Report 76-190, 65 p.

Naeser, C. W., 1979, Fission track dating and geologic annealing of fission tracks, in Jäger, E., and Hunziker, J. C., eds., Lectures in isotope geology: New York, Springer-Verlag, p. 154–169.

O'Sullivan, P. B., and Decker, J., 1990, Apatite fission track evidence for Paleocene and Oligocene uplift events in the northeastern Brooks Range, Alaska: Nuclear Tracks and Radiation measurements, v. 17, n. 3, p. 367–371.

O'Sullivan, P. B., Green, P. F., Bergman, S. C., Decker, J., Duddy, I. R., Gleadow, A. J. W., and Turner, D. L., 1993, Multiple phases of Tertiary uplift and erosion in the Arctic National Wildlife Refuge, Alaska, revealed by apatite fission track analysis: American Association of Petroleum Geologists Bulletin, v. 77, n. 3, p. 359–385.

Parrish, R. R., 1983, Cenozoic thermal evolution and tectonics of the Coast Mountains of British Columbia 1. Fission track dating, apparent uplift rates, and patterns of uplift: Tectonics, v. 2, p. 601–631.

Patrick, B. E., 1995, High pressure–low temperature metamorphism of granitic orthogneisses in the Brooks Range, northern Alaska: Journal of Metamorphic Geology, v. 13, p. 111–124.

Patrick, B. E., and Evans, B. W., 1988, Metamorphic evolution of the Seward Peninsula blueschist terrane: Journal of Petrology, v. 30, p. 531–555.

Purdy, J. W., and Jäger, E., 1976, K-Ar ages on rock-forming minerals from the Central Alps: Padova, Italy, University of Padova Institute of Geology and Mineralogy Memoir 30, 31 p.

Reed, B. L., and Lanphere, M. A., 1973, The Alaskan-Aleutian Range batholith: geochronology, chemistry, and relation to circum-Pacific plutonism: Geological Society of America Bulletin, v. 84, p. 2583–2610.

Steiger, R. H., and Jäger, E., 1977, Subcommission on geochronology: Convention on the use of decay constants in geo- and cosmochronology: Earth and Planetary Science Letters, v. 36, p. 359–362.

Tagami, T., and Shimada, C., 1996, Natural long-term annealing of the zircon fission-track system around a granitic pluton: Journal of Geophysical Research, v. 101, p. 8245–8255.

Wagner, G. A., Gleadow, A. J. W., and Fitzgerald, P. G., 1989, The significance of the partial annealing zone in apatite fission-track analysis: Projected track length measurements and uplift chronology of the Transantarctic Mountains: Isotope Geoscience, v. 79, p. 295–305.

Wallace, W. K., and Engebretson, D. C., 1984, Relationships between plate motions and Late Cretaceous to Paleogene magmatism in southwestern Alaska: Tectonics, v. 3, p. 295–315.

Wijbrans, J. R., and McDougall, I., 1986, $^{40}Ar/^{39}Ar$ dating of white micas from an Alpine high pressure metamorphic belt on Naxos (Greece): the resetting of the argon isotopic system: Contributions to Mineralogy and Petrology, v. 93, p. 187–194.

Wirth, K. R., Bird, J. M., Blythe, A. E., Harding, D. J., and Heizler, M. T., 1993, Age and tectonic evolution of western Brooks Range ophiolites, Alaska: results from $^{40}Ar/^{39}Ar$ thermochronometry: Tectonics, v. 12, p. 410–432.

Yamada, R., Tagami, T., and Nishimura, S., 1995, Confined fission track length measurement of zircon: Assessment of factors affecting the paleotemperature estimate: Chemical Geology, v. 119, p. 293–306.

Zaun, P. E., and Wagner, G. A., 1985, Fission track stability in zircons under geological conditions: Nuclear Tracks, v. 10, p. 303–307.

MANUSCRIPT ACCEPTED BY THE SOCIETY SEPTEMBER 23, 1997

Geological Society of America
Special Paper 324
1998

Tertiary uplift of the Mt. Doonerak antiform, central Brooks Range, Alaska: Apatite fission-track evidence from the Trans-Alaska crustal transect

Paul B. O'Sullivan
School of Earth Sciences, La Trobe University, Bundoora, Victoria 3083, Australia
Thomas E. Moore
U.S. Geological Survey, 345 Middlefield Road, Menlo Park, California 94025
John M. Murphy
Department of Geology/Geophysics, University of Wyoming, Laramie, Wyoming 82071

ABSTRACT

The Mt. Doonerak antiform is a northeast-trending, doubly plunging antiform located along the axial part of the central Brooks Range. This antiform is a crustal-scale duplex estimated to have a vertical displacement of ~15 km. The antiform folds the Amawk thrust, which separates relatively less displaced lower plate rocks in a window in the core of the antiform from allochthonous upper plate rocks of the Endicott Mountains allochthon. Because regional geological relations indicate that displacement on the Amawk thrust occurred between early Neocomian and early Albian time, uplift of the antiform is post-early Neocomian in age.

Zircon fission-track data from the Mt. Doonerak antiform suggest ~8–12 km of vertical denudation has occurred within the antiform region since ~70–65 Ma, whereas apatite fission-track data indicate the antiform has experienced a minimum of ~4–6 km of denudation since late Oligocene time. Following rapid denudation at ~24 ± 3 Ma, the rocks have experienced continued denudation to present surface conditions at a slower rate.

We conclude from the relative relations and timing that the Mt. Doonerak duplex was constructed in part during the late Oligocene by reactivation of an older duplex formed during the latest Cretaceous to Paleocene. Deformation and uplift of Oligocene age for the axial part of the Brooks Range orogen is anomalously young, but it is the same age as the youngest episode of north-vergent contractional uplift in the northeastern Brooks Range. Because the Mt. Doonerak antiform displays structural characteristics similar to those of antiforms in the northeastern Brooks Range and because both regions experienced simultaneous rapid denudation, we suggest that the Mt. Doonerak antiform formed in response to an episode of contractional deformation that affected both areas in the late Oligocene.

O'Sullivan, P. B., Moore, T. E., and Murphy, J. M., 1998, Tertiary uplift of the Mt. Doonerak antiform, central Brooks Range, Alaska: Apatite fission-track evidence from the Trans-Alaska crustal transect, *in* Oldow, J. S., and Avé Lallemant, H. G., eds., Architecture of the Central Brooks Range Fold and Thrust Belt, Arctic Alaska: Boulder, Colorado, Geological Society of America Special Paper 324.

INTRODUCTION

The Mt. Doonerak antiform is located along the transition between thrusted and imbricated sedimentary rocks of the northern Brooks Range and penetratively and semipenetratively deformed metasedimentary rocks of the southern Brooks Range (Fig. 1). Rocks in the core of the antiform consist of deformed lower Paleozoic argillite and metavolcanic rocks and an unconformably overlying distinctive succession of Lower Mississippian to Upper Triassic strata that are lithologically similar to the coeval part of the Ellesmerian sequence in the subsurface of the North Slope foreland basin (Fig. 2). The rocks adjacent to the antiform, both to the north and to the south, display significantly different stratigraphy and are allochthonous relative to those in the core of the antiform (Fig. 2). The allochthonous rocks are generally thought to have been emplaced northward over the core rocks of the antiform during early Brookian orogenic deformation in the Early Cretaceous (e.g., Dutro et al., 1976; Mull, 1982; Mull et al., 1987a; Oldow et al., 1987; Grantz et al., 1991; Moore et al., 1994).

The stratigraphic relations in the core of the Mt. Doonerak antiform have been the focus of many geologic studies because they provide evidence for the allochthonous character of the sedimentary rocks of the northern Brooks Range (Dutro et al., 1976; Mull, 1982; Mull et al., 1987a, 1989). More recent investigations (Oldow et al., 1987; Phelps et al., 1987; Seidensticker et al., 1987; Julian and Oldow, this volume, Chapter 5) have mapped and analyzed structures within the core of the Mt. Doonerak antiform and concluded that the Mt. Doonerak antiform constitutes an exhumed large-scale duplex. A duplex origin for the antiform was shown on the cross sections of Oldow et al. (1987), Grantz et al. (1991), and Moore et al. (1994) and confirmed by seismic reflection data (Fuis et al., 1995).

On the basis of regional stratigraphic relations, the thrust faults in the Mt. Doonerak antiform that juxtapose contrasting Mississippian to Triassic sequences are interpreted as large-displacement thrust faults (e.g., Dutro et al., 1976; Mull, 1982; Mull et al., 1987a, b, 1989; Oldow et al., 1987). The large-displacement thrust faults are thought to be Neocomian in age (Mull et al., 1987a), although a younger age for these faults is possible (Oldow et al., 1987). Most workers agree that the large-displacement thrust faults are deformed by the Mt. Doonerak antiform (e.g., Mull et al., 1987a; Oldow et al., 1987; Phelps et al., 1987; Seidensticker et al., 1987; Moore et al., 1994), indicating a post-Neocomian age for the antiform. The precise time of formation of the Mt. Doonerak antiform is uncertain, however, because of the absence of Cretaceous and younger geologic units that would provide direct geologic evidence of the age of the structure.

Regional stratigraphic and structural relations seemed to suggest that deformation in the core of the Brooks Range ceased by earliest Tertiary time (Mull, 1982; Mayfield et al., 1988; Mull et al., 1987a). Mull et al. (1989) suggested that the uplift and folding of the Mt. Doonerak antiform occurred during the Albian to Late Cretaceous. Oldow et al. (1987) con-cluded from a balanced cross section that the thrust faults within the duplex at Mt. Doonerak were active during the last major tectonic phase of the Brookian orogeny, in the Late Cretaceous or possibly in the Cenozoic.

This chapter reports the results of an apatite fission track study of samples from the Mt. Doonerak antiform, which indicate that the antiform experienced rapid cooling from temperatures >110 to <40°C during late Oligocene time at ~24 ± 3 Ma (time scale of Harland et al., 1990). These data suggest that the Mt. Doonerak antiform experienced an episode of rapid denudation as recently at the late Oligocene. Denudation of this age is considerably younger than Albian and latest Cretaceous to Paleocene denudation ages for the adjacent parts of the central Brooks Range (O'Sullivan et al., 1993a, 1997; Murphy et al., 1994; Blythe et al., this volume, Chapter 10) but are correlative with ages in a younger foldbelt that forms a distinct salient in the northeastern Brooks Range (O'Sullivan, 1994; O'Sullivan et al., 1993b; Fig. 1).

STRUCTURAL AND STRATIGRAPHIC SETTING

Regional geology

Earliest Brooks Range deformation is characterized by large-displacement thrusting of middle Paleozoic and younger sedimentary rocks and emplacement of ophiolite at high structural levels during the Late Jurassic and Neocomian (Moore et al., 1994, and references therein). Relatively younger Brooks Range deformation is indicated by setting of mid-Cretaceous K-Ar cooling ages in metamorphic rocks in the southern Brooks Range and folding of large-displacement thrust faults. In response to rapid kilometer-scale denudation of the orogen during younger phases of the deformation, as much as 8 km of Albian and Upper Cretaceous sedimentary strata were deposited in the North Slope foreland basin to the north (Bird and Molenaar, 1992), and in the Koyukuk basin to the south (Patton and Box, 1989; Fig. 1). During the Tertiary, a younger north-vergent fold-thrust belt developed north of the main axis of the Brooks Range in the northeastern Brooks Range (Wallace and Hanks, 1990; Fig. 1). The younger fold-thrust belt has deformed, uplifted, and exposed the northern part and foreland of the older orogen. Structures in the northeastern Brooks Range are characterized by relatively small displacement thrust faults that were active during latest Cretaceous and Cenozoic time (Wallace and Hanks, 1990; O'Sullivan et al., 1993b; Moore et al., 1994).

The Mt. Doonerak antiform is a large double-plunging antiform that can be traced for ~110 km southwest from the Dalton Highway in the central Brooks Range (Fig. 2). The antiform lies along the southwestward projection of the northwest-facing mountain front of the northeastern Brooks Range (Fig. 1). Exposed structural relief of the Mt. Doonerak antiform is ~2 km, but structural relief in excess of ~10 km has been estimated based on regional mapping (Mull et al., 1987b; Oldow et al., 1987; Grantz et al., 1991) and about 15 km by seismic reflection data (Fuis et al., 1995; Fig. 3).

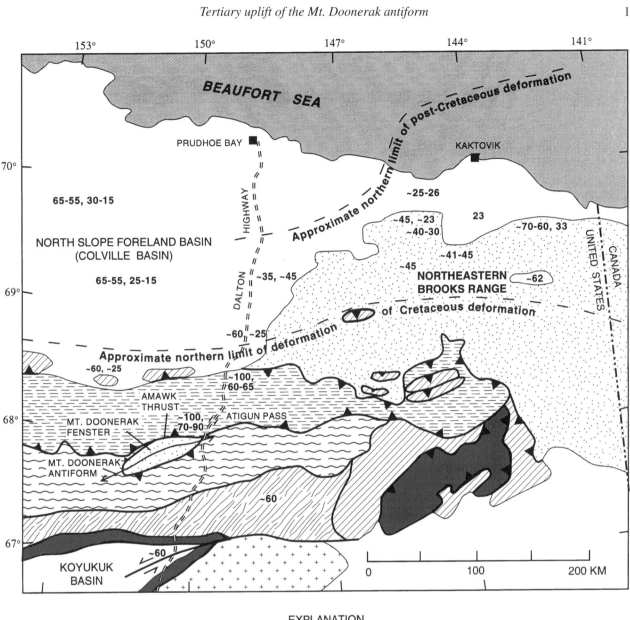

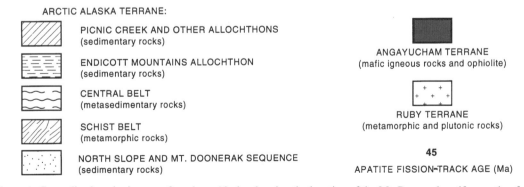

Figure 1. Generalized geologic map of northern Alaska showing the location of the Mt. Doonerak antiform, other features mentioned in the text, and apatite fission-track ages from the eastern Brooks Range and North Slope (O'Sullivan, 1996, unpublished data; O'Sullivan et al., 1993a, 1997). Fission-track data indicate episodes of kilometer-scale denudation occurred at ~100 Ma, ~60 Ma, and ~25 Ma and were widespread. Ages between ~70 and 90 Ma in the Atigun Pass area are partially reset from ~100 Ma by kilometer-scale denudation at ~60 Ma (Murphy et al., 1994; O'Sullivan et al., 1993a, 1997). The ~70- to 90-Ma ages are located at the same elevation as Oligocene ages reported from the Mt. Doonerak area in this chapter.

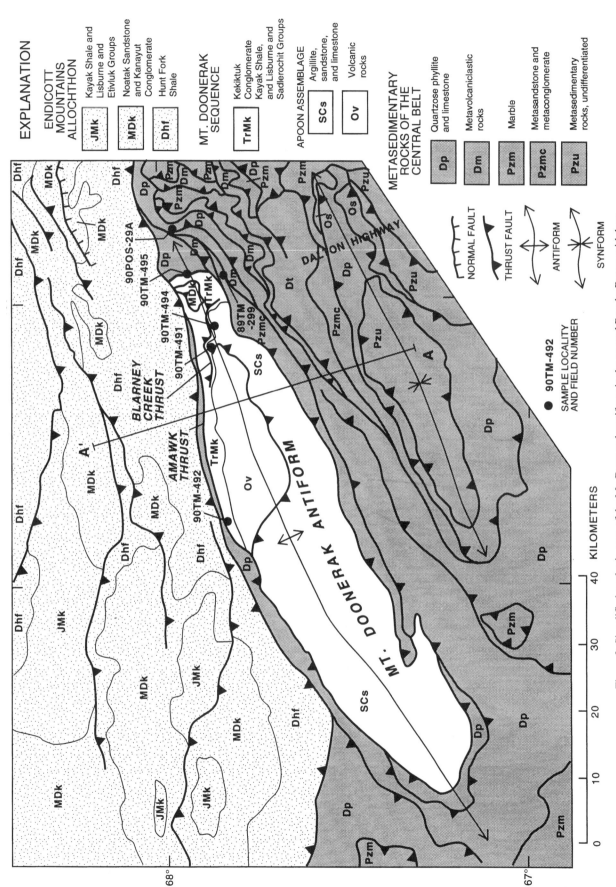

Figure 2. Simplified geologic map of the Mt. Doonerak antiform in the central Brooks Range with locations of apatite fission-track samples. Also shown is location of the geologic cross section A–A' shown in Figure 3. See Figure 1 for definition of symbols.

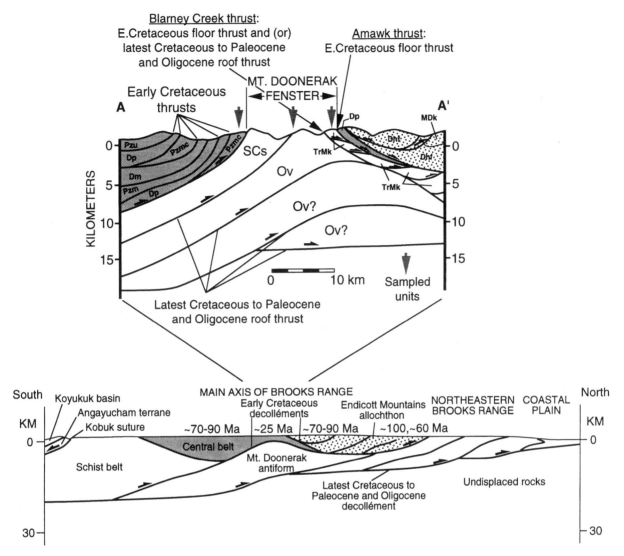

Figure 3. Schematic cross section across the Mt. Doonerak antiform showing regional relations (modified from Oldow et al., 1987). Proposed timing of displacement on faults in Mt. Doonerak antiform area from fission-track ages (top) and conceptual model relating latest Cretaceous to Paleocene and Oligocene shortening in the Mt. Doonerak area to shortening in the northeastern Brooks Range (bottom). Fission-track ages and structural relations indicate detachment and emplacement of Endicott Mountains and other allochthons at relatively higher structural levels in the Early Cretaceous, whereas detachment in the latest Cretaceous to Paleocene and Oligocene occurred at deeper structural levels. Fission-track ages are shown in bold type. Ages of ~25 Ma are from this study. Other ages (O'Sullivan et al., 1993a) are projected into the line of section from a sampled section located along the Dalton Highway. See Figure 2 for rock-unit symbol usage in this figure.

Three major structural and stratigraphic packages are recognized in the Mt. Doonerak antiform (Figs. 2 and 3). These packages are (1) the Endicott Mountains allochthon (Upper Devonian to Neocomian), which is exposed on the northern flank of the antiform; (2) metasedimentary rocks (Devonian and older) of the southern Brooks Range, which compose the southern flank of the antiform; and (3) the Mt. Doonerak sequence (lower Paleozoic to Triassic), which is exposed in the core of the antiform in a structural window termed the Mt. Doonerak fenster (Mull, 1982). The Endicott Mountains allochthon structurally overlies the Mt.

Doonerak sequence on the Amawk thrust of Mull (1982; Figs. 2 and 3). The Mt. Doonerak fenster is significant because of the lithologic similarity of the Mt. Doonerak sequence to the stratigraphic succession in the North Slope subsurface, and its contrast with the adjacent stratigraphic succession of the Endicott Mountains allochthon. The correlation of the Mt. Doonerak and North Slope sequences implies that the Endicott Mountains allochthon was thrust northward on the Amawk thrust over the Mt. Doonerak sequence a minimum of between ~88 km (Mull et al., 1987a, 1989) and ~145 km (Howell et al., 1992).

Structural geology of the Mt. Doonerak antiform

Near the eastern termination of the antiform, recent detailed structural analysis indicates that the oldest fabric elements in the lower Paleozoic rocks are younger than the sub-Mississippian unconformity in the Mt. Doonerak sequence (Seidensticker et al., 1987). This observation indicates that (1) pre-Mississippian deformation, if present, is nonpenetrative in character, and (2) penetrative fabric elements present in the antiform are post-Mississippian, and presumably Brookian (Mesozoic and (or) Cenozoic) in age. Phelps et al. (1987) reported that most map-scale thrust faults in the eastern end of the antiform are northwest vergent and are accompanied by a widespread penetrative slaty cleavage that is axial-planar to isoclinal folds. The slaty cleavage is present in the metasedimentary rocks of the southern flank of the antiform, the Mt. Doonerak sequence, and near the base of the Endicott Mountains allochthon, although the cleavage is more strongly expressed toward the south across the antiform. The slaty cleavage is folded by chevron-style kink folds; this folding may have occurred during the final stages of thrusting. The thrust faults have detachment levels in the Hunt Fork Shale at the base of the Endicott Mountains allochthon, within the Sadlerochit Group, Kayak Shale, and Apoon assemblage of the Mt. Doonerak sequence, and beneath the metasedimentary rocks of the central belt (Figs. 2 and 3). Particular attention has been drawn to the complex detachment fault in the Kayak Shale, the Blarney Creek thrust, which separates south-dipping structures in the southern flank of the antiform from north-dipping structures on the northern flank of the antiform (Seidensticker et al., 1987; Figs. 2 and 3). The Blarney Creek thrust is interpreted to result first from duplexing of the Kayak Shale, Lisburne Group, and Sadlerochit Group above a floor thrust in the Kayak Shale and a roof thrust in the Sadlerochit Group. Relatively younger deformation resulted in development of duplexes in the structurally lower part of the Apoon assemblage and Kekiktuk Conglomerate on a floor thrust in the Apoon and a roof thrust in the Kayak Shale at the Blarney Creek thrust (Seidensticker et al., 1987).

High-angle faults cut the rocks of the Mt. Doonerak antiform and are mainly east–west trending faults along the northern flank of the antiform (Dillon et al., 1986; Mull et al., 1987a; Phelps et al., 1987). On the basis of detailed structural analysis, Phelps et al. (1987) suggested that the majority of high-angle faults are left-lateral strike-slip faults that resulted from northeast-directed contractional deformation. Because these faults postdate all other structures in the area and differ in orientation from those associated with large-scale Brookian thrust faults, they presumably reflect the youngest episode of deformation.

Estimated timing of deformation

Regional stratigraphic relations and isotopic data indicate that the main contractional phase of the Brooks Range occurred during the Late Jurassic and Early Cretaceous as a result of southward subduction of a south-facing (present coordinates) passive

margin (the Arctic Alaska terrane) beneath an intra-oceanic arc (Box, 1985; Moore et al., 1994). Mull (1982) and Mull et al. (1987b) reported thrust imbricates that contain rocks as young as Valanginian are unconformably overlain by gently deformed Albian sedimentary rocks (Fortress Mountain Formation) in the Colville basin along the Brooks Range mountain front. These relations suggest that large-scale thrusting and emplacement of the Endicott Mountains allochthon occurred in late Neocomian time. Since the thrust faults that bound the allochthon are folded by the Mt. Doonerak antiform, the antiform was formed in post-Neocomian time. The absence of younger geologic units or dated structures in the area of the Mt. Doonerak antiform, however, prevents more precise dating of the time of formation of the antiform from stratigraphic and structural relations.

The main contractional phase of Brooks Range deformation was followed by younger deformation in the mid-Cretaceous and in the Late Cretaceous and Tertiary (e.g., Mull, 1982; Moore et al., 1994; Bird and Molenaar, 1992). Denudation of the southern Brooks Range resulted in setting of K-Ar white-mica cooling ages in metamorphic rocks at 130–86 Ma, erosion, and deposition of thick and extensive sedimentary strata in foreland and hinterland basins during the Albian and Cenomanian (Turner et al., 1979; Mull, 1982; Till, 1992). Whether this denudation was the product of continued contractional deformation (e.g., Oldow et al., 1987) or large-scale extensional deformation (Miller and Hudson, 1991) is uncertain. Evidence for younger contractional deformation, however, is readily apparent in the folding of foreland basin sedimentary rocks that are at least as young as late Tertiary in the northeastern Brooks Range foothills (Kelley and Foland, 1987).

Fission-track ages from the Endicott Mountains allochthon north of the Mt. Doonerak antiform suggest that the axial part and northern flank of the central Brooks Range experienced several major episodes of cooling (Murphy et al., 1994; O'Sullivan et al., 1993a, 1997). The earliest episode occurred at ~100 ± 5 Ma (Albian) and was approximately concurrent with regional setting of K-Ar cooling ages in the metamorphic rocks of the southern Brooks Range and filling of the Colville and Koyukuk basins. Fission-track data from metagranitic rocks of the southern Brooks Range (Murphy et al., 1994) and from the Colville basin (O'Sullivan, 1996) suggest that regional denudation extending southward into the southern Brooks Range and northward into the Colville basin occurred at ~60 ± 4 Ma (Paleocene). These data indicate that, while the main contractional phase of Brookian deformation occurred in the Late Jurassic and Neocomian, additional episodes of regional cooling and uplift of the orogen occurred in the Albian and latest Cretaceous and Paleocene.

The apatite fission-track ages from the northeastern Brooks Range are younger than those of the axial part of the Brooks Range and generally decrease in age northward (O'Sullivan, 1994; O'Sullivan et al., 1993b). Apatite ages from the southern part of the northeastern Brooks Range suggest that denudation occurred at ~60 ± 4 Ma and decreased northward to ~43 ± 3 Ma and ~34 ± 3 Ma. In the northernmost part of the uplift, fission-track cooling ages are ~25 ± 3 Ma. O'Sullivan et al. (1993b) suggested that the northward younging of fission-track ages in the northeastern Brooks Range records progressive denudation due to thrusting within a northward-advancing thrust front.

EXPERIMENTAL DETAILS, RESULTS, AND INTERPRETATION

Samples were collected for apatite fission track analysis from the eastern end of the Mt. Doonerak antiform during the 1990 field season of the Trans-Alaska Crustal Transect (TACT) project of the U.S. Geological Survey. A pilot study consisting of six samples collected at a range of elevations was designed to investigate the uplift history of the Mt. Doonerak antiform. Two samples were collected from the Lower Mississippian Kekiktuk Conglomerate, one from the Ordovician(?) intermediate volcanic rocks of the Apoon assemblage, one from the Upper Triassic Karen Creek Sandstone, and two from metaclastic units of the central belt. The lithology, unit, and geographic coordinates for all samples are shown in Table 1; sample locations are shown in Figure 2.

Apatite grains were separated from the samples by conventional heavy liquid and magnetic techniques. The methodology used to process the apatite mounts, determine apatite ages, and measure confined track lengths has been described in detail by Green (1986). The apatite data have been interpreted using the understanding of apatite annealing described by Green et al. (1989). Thermal history interpretations are based on a quantitative treatment of annealing achieved by forward computer modeling (Green et al., 1989). The uncertainty in predicted track length values is ~0.25 to 0.5 µm, which is considered equivalent to an uncertainty in the absolute paleotemperature estimates of about ±10°C (Green et al., 1989).

Analytical results of the samples from the Mt. Doonerak antiform are presented in Table 2. Basic counting data, including individual crystal ages, are available in O'Sullivan et al. (1993a). Confined track-length distributions and single-grain age distributions for representative samples are shown in Figure 4. Apatite fission-track ages for the six samples from the Paleozoic and Mesozoic volcanic and sedimentary rocks exposed in the Mt. Doonerak region are concordant and range from 22.0 ± 4.0 to 33.9 ± 16.0 Ma, with a weighted mean value of 24.1 ± 2.8 Ma. The fission-track ages are significantly less than the Ordovician to Triassic stratigraphic ages of the samples. Each sample has single-grain age distributions with strong peaks at ~20–30 Ma (Fig. 4), and in all but one sample the single-grain ages are consistent with a single-age population.

The confined track-length distributions for most samples from the core of the antiform contain well-defined peaks at ~14 µm with narrow distributions (standard deviations <1.3 µm; Table 2). Narrow track-length distributions for most samples imply that most tracks have been formed at low temperatures. The mean track-length distribution for one sample, collected along strike away from the core of the fenster (90POS29A), is much lower than for the other samples, suggesting that its thermal history is more complicated. The complicated track-length distribution for this sample may be explained by its residence for an extended period of time in the temperature range ~110–60°C.

Since the apatite fission-track ages of the Mt. Doonerak region rocks are significantly younger than the stratigraphic ages and were formed at low temperatures, the apatites from these samples must have been totally annealed prior to rapid cooling. Forward modeling of the track-length distributions of samples from the Mt. Doonerak antiform, following the procedures outlined by Green et al. (1989), suggests that cooling below ~110°C commenced in the late Oligocene at the time given by the fission-track ages (~24 ± 3 Ma) and proceeded rapidly at a minimum rate of ~14–23°C/m.y. (from >110 to ≤40°C in 3 to 5 m.y.). Forward modeling of the track-length distribution for the sample from outside the fenster along the structural trend suggests the sample first cooled to between ~100–70°C at some time between ~70–50 Ma and then underwent rapid cooling during the late Oligocene.

COOLING HISTORY

The thermal history inferred for the Mt. Doonerak antiform from the apatite fission-track data is shown in Figure 5. In addition to the apatite fission track data reported here, Figure 5 utilizes the zircon fission ages of 65.6 ± 3.8 Ma, 70.1 ± 7.1 Ma, 112.4 ± 7.6 Ma, and 102.6 ± 5.4 Ma from rocks on the flank of

TABLE 1. DETAILS OF OUTCROP SAMPLES

Sample	Latitude (N)	Longitude (W)	Elevation (ft/m)	Rock Type	Unit	Stratigraphic Subdivision
Doonerak fenster						
89TM-299A	67°54'50"	149°58'02"	4,000/1,200	Metasandstone	Undiff. Paleozoic metamorphic rocks	Devonian(?)
90TM-491A	67°56'04"	150°14'35"	2,940/882	Volcanic dike	Apoon assemblage	Ordovician
90TM-492A	67°55'38"	150°42'40"	1,900/570	Sandstone	Karen Creek Formation	Triassic
90TM-494A	67°55'59"	150°07'22"	4,840/1,452	Sandstone	Kekiktuk Conglomerate	Mississippian
90TM-495A	67°57'46"	149°54'34"	2,560/768	Sandstone	Kekiktuk Conglomerate	Mississippian
90POS-29A	67°57'59"	149°47'03"	2,200/671	Metasandstone	Undiff. Paleozoic metamorphic rocks	Devonian

TABLE 2. APATITE FISSION-TRACK ANALYTICAL RESULTS

Sample	Grains Counted	Standard Track Density* ($\times 10^6 cm^{-2}$)	Fossil Track Density† ($\times 10^5 cm^{-2}$)	Induced Track Density* ($\times 10^6 cm^{-2}$)	Chi Square Probability (%)	Age§ (Ma ± 2σ)	Mean Track Length (μm ± 1σ)	N (No. of tracks measured)	Standard Deviation
Within Doonerak fenster									
89TM299A	25	1.354 (3,047)	1.402 (142)	1.522 (1,541)	36.5	22.0 ± 0.4	13.63 ± 0.12	101	1.25
90TM491A	12	1.354 (3,047)	3.992 (97)	2.704 (657)	0.06	33.9 ± 16.0**	14.15 ± 0.18	24‡	0.63
90TM492A	25	1.354 (3,047)	2.227 (99)	2.215 (985)	81.8	24.0 ± 5.2	13.88 ± 0.28	47‡	1.32
90TM494A	16	1.354 (3,047)	1.948 (88)	1.474 (666)	11.1	31.5 ± 7.2	13.14 ± 0.26	31	1.46
90TM495A	5	1.354 (3,047)	2.735 (16)	2.051 (120)	70.5	31.8 ± 17.0	14.65 ± 0.17	32‡	0.50
Along Structural Trend									
90POS29A	25	1.142 (5,146)	2.203 (193)	1.961 (1,643)	77.1	23.6 ± 2.8	11.58 ± 0.25	83	2.26

Parentheses show number of tracks counted.
*Standard and induced track densities measured on mica external detectors (g = 0.5).
†Fossil track densities measured on internal mineral surfaces.
§Ages calculated using a personal zeta z = 352.7 for dosimeter glass SRM612.
**Central age, used where pooled data fail χ^2 test at 5 percent.
‡Track-length data from grains on multiple mounts.

the Mt. Doonerak antiform reported by Blythe et al. (this volume, Chapter 10). The two samples with zircon ages greater than ~100 Ma were collected at higher stratigraphic levels than the samples with the two younger ages. The older ages record an episode of cooling within the Brooks Range during the Early to mid-Cretaceous, an event that correlates with K-Ar cooling ages in metamorphic rocks to the south and with apatite fission-track data in the Atigun Pass area (O'Sullivan et al., 1993a, 1997; Blythe et al., this volume, Chapter 10; Fig. 1). These ages probably indicate that kilometer-scale of denudation of the Brooks Range was widespread during the mid-Cretaceous, but it is uncertain if the Mt. Doonerak antiform was active at that time.

Because all fission-track ages for samples within the core of the anticline are latest Cretaceous or younger, there is no direct control on the time-temperature history of the Mt. Doonerak antiform prior to ~70–65 Ma (latest Cretaceous to early Paleocene). However, the zircon fission track data of Blythe et al. (this volume, Chapter 10) and apatite fission-track data of O'Sullian et al. (1993a) indicate that the rocks on the flank of the Mt. Doonerak antiform experienced cooling in the mid-Cretaceous, but the cooling did not bring the rocks in the core of the antiform to temperatures low enough to allow fission tracks to be preserved. The younger zircon fission-track ages indicate that the core rocks of the Mt. Doonerak antiform cooled below ~240°C during the latest Cretaceous to Paleocene. The apatite fission-track data indicate that the antiform experienced an second episode of rapid cooling during the late Oligocene at ~24 ± 3 Ma. Although

the data are sparse, the apparent absence of fission track ages between ~100 and 70 Ma and between ~55 and ~30 Ma suggests that the cooling was episodic. Following the second episode of rapid cooling, the sampled section probably continued to cool at a slower rate to present ambient surface temperatures.

DISCUSSION

Two types of thermotectonic history are plausible given the cooling history portrayed in Figure 5. Temperatures above those necessary to completely anneal fission tracks in apatite in the Tertiary may have been produced by long-term burial due to continued deposition and (or) tectonic loading under conditions of normal geothermal gradients (~20–30°C/km). In response to kilometer-scale Tertiary denudation, the Mt. Doonerak antiform may have undergone rapid cooling through the closure temperature for apatite. Alternatively, rocks of the antiform may have been heated sufficiently to anneal apatite by a short-lived heat pulse associated with a Tertiary igneous event. In this case, some denudation would still be necessary to expose the rocks at the surface, but on a smaller scale.

If the second type of thermal history is to explain the apatite data for the Mt. Doonerak antiform, igneous rocks of Tertiary age should be expected in the region since a heating event of approximately early to late Oligocene age would be required. Igneous rocks of Tertiary age are unknown from the Mt. Doonerak antiform (Dillon et al., 1986; Brosgé and Reiser, 1964), and have

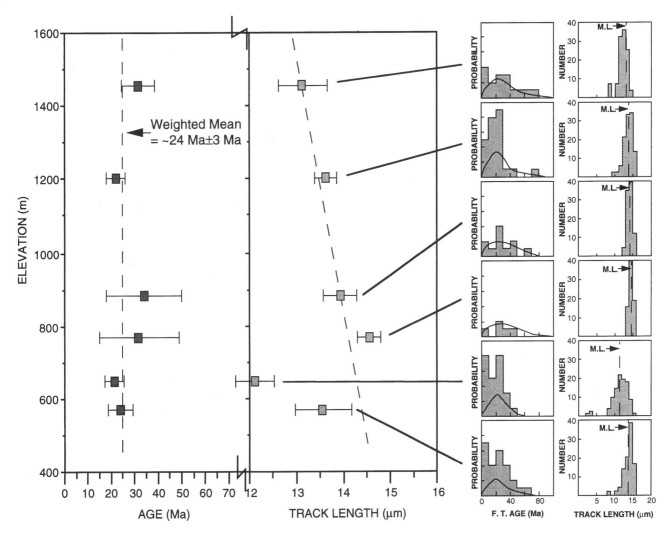

Figure 4. Composite vertical profile from the Mt. Doonerak antiform summarizing apatite fission-track data. Fission-track ages show no signs of change with changing elevation, while mean track lengths seem to decrease with decreasing elevation. Also shown to the right of the profiles are track-length and individual grain-age histograms for the samples. The track-length histograms have been normalized to 100 tracks. The curve on the individual grain-age histograms is a smoothed probability function obtained from the addition of Gaussian distributions representing each individual grain and its corresponding error. The age and length scales are identical for all samples.

been reported from only one restricted location in northern Alaska, in the easternmost Brooks Range (Barker, 1982; Brosgé et al., 1976), over 150 km east of the antiform. Granitic rocks are present in the metamorphic core of the Brooks Range, but these yield U-Pb crystallization ages of ~390 Ma (Aleinikoff et al., 1993) and white-mica K-Ar cooling ages of ~90 to 130 Ma (Turner et al., 1979). Since metasedimentary rocks throughout the southern Brooks Range also yield mid-Cretaceous K-Ar mica ages, the K-Ar ages are generally interpreted to represent cooling below 350°C due to regional uplift of the Brooks Range orogen following the termination of large-displacement thrusting instead of a mid-Cretaceous or younger thermal event (Turner et al., 1979; Dillon et al., 1987; Oldow et al., 1987; Mayfield et al., 1988; Moore et al., 1994). Thus, there is no direct evidence of a

Tertiary igneous event in the central Brooks Range, and we therefore conclude that the fission-track data were unaffected by heating at high structural levels in the Tertiary.

On the other hand, tectonic burial of the rocks of the Mt. Doonerak antiform by the Endicott Mountains allochthon and metasedimentary rocks of the central belt during the Neocomian is well documented by regional stratigraphic and structural relations (Oldow et al., 1987; Moore et al., 1994). Assuming a minimum thickness for the Endicott Mountains allochthon of about 5 km (Fuis et al., 1995), a minimum thickness of ~5 km must have been removed from the Mt. Doonerak area in order to expose the structurally lower Mt. Doonerak sequence. If thicknesses are added to account for (1) the central belt rocks that are inferred to lie between the Mt. Doonerak sequence and Endicott

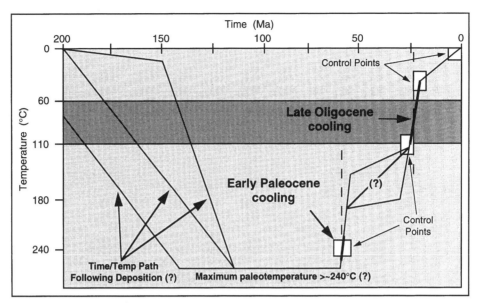

Figure 5. Schematic proposed time-temperature history for the Mt. Doonerak antiform inferred from apatite fission-track data presented in this report, as well as zircon fission-track data from Blythe et al. (this volume, Chapter 10). The zircon data indicate the region must have been exposed to maximum paleotemperatures greater than 240°C prior to cooling to less than ~240°C during the latest Cretaceous to Paleocene. The apatite data indicate the region must have experienced major cooling from greater than ~110 to ~40°C during the late Oligocene. Late Oligocene cooling was probably related to kilometer-scale denudation. Control points are from modelling the apatite and zircon fission-track data.

Mountains allochthon (Fig. 3), and (2) other allochthonous sequences that may have eroded from structural levels above the Endicott Mountains allochthon (e.g., Mayfield et al., 1988), the structural cover for the Mt. Doonerak sequence probably was far greater. Maximum paleotemperature data from Johnsson et al. (1992) suggest that the structural cover must have been more than ~10 km in the Mt. Doonerak area (Howell et al., 1992). The seismic reflection-refraction models (Fuis et al., 1995) show about 15 km of uplift in the Mt. Doonerak region. Thus, we estimate a minimum of 15 km of overburden must have been removed since emplacement of the Endicott Mountains allochthon in order to expose rocks presently at the surface. The now-absent tectonic load and the presence of a major contractional structure (that is, the Mt. Doonerak antiform), which both postdates the tectonic load and displays the sufficient vertical displacement to explain the apatite data, indicates that the thermal history of the Mt. Doonerak antiform can best be explained by long-term burial followed by kilometer-scale denudation.

If the cooling ages may be attributed to kilometer-scale denudation, then the fission-track data discussed here indicate that one or more episodes of rapid denudation occurred within the Mt. Doonerak region during the Tertiary. The fission-track data suggest rapid cooling (e.g., kilometer-scale denudation) of the rocks from temperatures greater than ~110°C to temperatures less than 40°C occurred during the late Oligocene (Fig. 5). Furthermore, it is possible that the latest Cretaceous to Paleocene zircon cooling ages (Blythe et al., this volume, Chapter 10)

record an episode of rapid cooling that cooled the rocks below ~240°C during the latest Cretaceous and Paleocene. Rocks only 1–3 km higher in the structural section, however, yield mid-Cretaceous zircon fission-track ages and, therefore, were not buried deep enough to be above ~240°C at any time since the mid-Cretaceous. These relations indicate that the rocks of the Mt. Doonerak antiform must have laid near the annealing zone for zircon during the latest Cretaceous and Paleocene.

We propose that most of the kilometer-scale denudation of the Mt. Doonerak antiform recorded by the fission-track data occurred principally during discrete episodes of rapid cooling in the latest Cretaceous/Paleocene and Oligocene. Howell et al. (1992) previously concluded that all uplift and erosion during the Tertiary occurred in response to regional isostatic rebound related to deflexing of the lithosphere. We believe, however, that simple isostatic rebound cannot explain the distribution of apatite ages along the Dalton Highway presented in O'Sullivan et al. (1993a, 1997) and Blythe et al. (this volume, Chapter 10; Figs. 1 and 3). Samples collected to the north of the Mt. Doonerak antiform at higher structural levels but at lower elevations yield apatite fission-track ages between ~80 and 90 Ma, much older than the late Oligocene ages from within the antiform. This relation is only possible if the antiform has experienced substantially more denudation than the region to the north and now exposes deeper structural levels. Since rocks located within the core of the antiform, at the highest elevations, were rapidly cooled below ~110°C in the late Oligocene, while those at lower elevations to the north cooled to temperatures below ~110°C during the Early to mid-Cretaceous, we conclude that denudation resulted from construction of at least part of the structural relief of the antiform in the late Oligocene.

AMOUNT AND RATE OF TERTIARY UPLIFT AND EROSION

To determine the amount of denudation related to cooling events during the Tertiary in the Mt. Doonerak area, assumptions must be made concerning the paleogeothermal gradient during that time. Unfortunately, the present-day geothermal gradient for the region has not been determined. Geothermal gradients for wells within the deformed northern foothills belt of the Brooks Range mountain front, however, are typically between ~20 and 30°C (e.g., Lisburne #1, Kemik #2, and Lupine #1; Lachenbruch et al., 1987). Due to the lack of control within the Mt. Doonerak region, this range of temperature is arbitrarily used in the calculations that follow.

Likewise, there is little control on the paleosurface temperature in the Mt. Doonerak region, so we estimated this parameter from regional observations as follows. The present estimated mean annual temperatures across the North Slope foreland basin increase from ~−12°C near the northern coastline to ~−4°C in the foothills belt (Lachenbruch et al., 1987). Prior to ~5 Ma when the Bering land bridge isolated the Arctic and northern Alaska began to cool significantly (Herman and Hopkins, 1980), the mean annual temperature during the Cretaceous to Pliocene ranged from ~0 to 10°C based on interpretations of plant fossils (e.g., Smiley, 1967; Spicer and Parrish, 1986). Since this interval corresponds to the time of uplift of the Mt. Doonerak antiform, a paleo-surface temperature of ~5° is assumed for our calculations.

Using these assumptions, the estimated amounts and rates of Tertiary uplift and erosion are presented in Figure 6. To satisfy the 15 km of minimum total uplift estimated from balanced cross sections and seismic data, a minimum of ~3 km of uplift and erosion may have occurred before ~70–65 Ma. At ~70–65 Ma, the region cooled below ~240°C (located at a depth of ~8–12 km depending on paleogeothermal gradient), as determined from the zircon data from Blythe et al. (this volume, Chapter 10). Between ~70–65 and 25 Ma, when the region cooled to >110°C (depth of ~3.7–5.5 km), the average rate of denudation was ~0.11–0.16 km/m.y. Between ~25 and 20 Ma when the region cooled to ~40°C, the rate of denudation was between ~0.46 and 0.70 km/m.y. Following the rapid phase of denudation in the Oligocene, the region has experienced slow denudation at a rate of ~0.06–0.09 km/m.y.

TECTONIC IMPLICATIONS

Potassium argon cooling ages of metamorphic rocks in the southern Brooks Range, deposition of huge thicknesses of mid-Cretaceous clastic strata from the Brooks Range orogen into surrounding basins, and the ~100 Ma fission-track data from the northern Brooks Range indicate that regional uplift of the Brooks Range orogen began in the Neocomian and continued into the mid-Cretaceous. Recent apatite fission-track ages from the Brooks Range show that a distinctly younger kilometer-scale denudation event occurred during the early Paleocene at ~60 ± 4 Ma and affected the entire mountain belt (Murphy et al., 1994; O'Sullivan et al., 1997) and extended well north of the Brooks Range mountain front into the North Slope foreland basin (O'Sullivan, 1996). The regional nature of the latest Cretaceous to Paleocene event suggests that it may have largely created the present-day Brooks Range and folded the older fore-

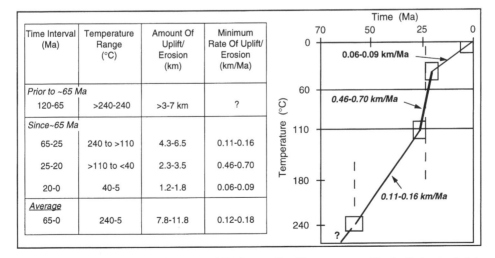

Figure 6. Relationship between the proposed Tertiary cooling history suggested by the fission-track data and the proposed rates of uplift and erosion for each time interval. A geothermal gradient between ~20 and 30°C/km is assumed.

land basin strata. Still younger fission-track ages from the northeastern Brooks Range suggest that it was formed by episodic northward migration of a thrust front during the Tertiary. This deformation resulted in northward migration of foredeep deposition and deformation of Tertiary strata in northeastern Alaska (O'Sullivan et al., 1993b).

Zircon and apatite fission-track data from the Mt. Doonerak antiform indicate that the Mt. Doonerak region experienced the regional cooling events during the mid-Cretaceous and latest Cretaceous to Paleocene. These cooling events postdate the emplacement of the Endicott Mountains allochthon over the Mt. Doonerak region on the Amawk thrust in the Early Cretaceous. In addition, fission-track data from the Mt. Doonerak region indicate that the region experienced a cooling event during the Oligocene. Apatite ages from nearby parts of the Brooks Range along the Dalton Highway (Fig. 1) are substantially older (O'Sullivan et al., 1993a, 1997; Blythe et al., this volume, Chapter 10) and indicate that Oligocene ages are restricted to the Mt. Doonerak area (Fig. 3). This geographic distribution of ages appears to require that the major affects of the Oligocene uplift were of local, rather than regional, character in the central Brooks Range. Since the structural relations of the Mt. Doonerak antiform indicate that it was formed by crustal-scale duplexing of the overthrusted foreland of the Brooks Range orogen after emplacement of the Endicott Mountains allochthon, final uplift by structural duplexing in the core of the antiform was not completed until late Oligocene time.

Because the Mt. Doonerak antiform displays perhaps ~15 km of vertical displacement, and only about 4 km of vertical displacement are required to expose completely annealed apatite at the surface, 10 km or more of vertical displacement on the antiform must have occurred prior to the late Oligocene. Therefore, uplift of the structure may have begun at an earlier time. The zircon fission-track data support this interpretation, since uplift of the entire 10–15 km of crustal section during the Oligocene should have also locked in Oligocene zircon fission-track ages in the Mt. Doonerak antiform. We therefore conclude that the Mt. Doonerak antiform region was also denuded in post-Albian time, probably in the latest Cretaceous to Paleocene as part of a regional crustal shortening event and then was reactivated in the Oligocene (Fig. 3).

Duplexes have been recognized at two structural levels in the Mt. Doonerak antiform: above and below the Blarney Creek thrust (Seidensticker et al., 1987). Shortening in the upper Paleozoic rocks of the Mt. Doonerak sequence (stratigraphic thickness of about 1 km) above the Blarney Creek thrust has resulted in structural thickening of probably 1–2 km (Fig. 3). The stratigraphic thickness of mainly lower Paleozoic rocks below the Blarney Creek thrust is unknown but is substantial, possibly in excess of 5 km. Duplexes developed in a stratigraphic section of such a thickness could result in many kilometers of uplift. Considering the 4+ km of denudation for both the latest Cretaceous to Paleocene and Oligocene events, we suggest thickening above the Blarney Creek thrust alone is insufficient in magnitude to

explain the denudation documented by the fission-track data. Shortening beneath the Blarney Creek thrust along detachments at deep structural levels, however, would be sufficient to explain both the latest Cretaceous to Paleocene and Oligocene uplift events as well as the physical expression of the antiform. We therefore conclude that both the latest Cretaceous to Paleocene and Oligocene uplift events occurred principally because of duplexing of lower Paleozoic rocks (Apoon assemblage) beneath a roof thrust at the Blarney Creek thrust (Fig. 3). However, it is unclear from the present data whether shortening in the section above the Blarney Creek thrust also occurred during the latest Cretaceous and (or) Oligocene, or whether it resulted from older, presumably Early Cretaceous, shortening.

If the Mt. Doonerak antiform is a crustal-scale duplex that was formed initially in latest Cretaceous to Paleocene time and did not attain its full development as a major antiform until late Oligocene time, what was its role in the Brooks Range orogen? Its position in the axial part of the Brooks Range orogen appears anomalous, since major uplift of that region occurred in the mid-Cretaceous and Paleocene. The Mt. Doonerak antiform, however, is also located along the southwestward projection of the northeast-trending mountain front of the northeastern Brooks Range (Fig. 1), and thus is positioned in the hinterland region of the younger foldbelt in the northeastern salient of the Brooks Range. Apatite fission-track data from the northeastern Brooks Range indicate that kilometer-scale denudation and cooling events occurred at ~60, ~45, ~35, and ~25 Ma (O'Sullivan, 1994; O'Sullivan et al., 1993b). The northeastern Brooks Range is characterized by large, northward-convex anticlinoria that are northeast trending along the western margin of the uplift. The anticlinoria can be traced for as much as 100 km; the distance from the crest of the anticlines to the troughs of adjacent synclines typically is about 15 km (Wallace and Hanks, 1990). The anticlinoria fold the Ellesmerian sequence, the foreland for the main axis of the Brooks Range, and comprise regional north-vergent duplexes with floor thrusts deep in the stratigraphically and structurally complex pre-Mississippian rocks (Wallace and Hanks, 1990). Like the Mt. Doonerak area, the anticlinoria in the northeastern Brooks Range display subsidiary detachment folds in its covering Mississippian and younger rocks and have detachment faults in the Kayak Shale (Wallace, 1993).

A comparison of the trend, scale, deformational style, and depth to detachment of the Mt. Doonerak antiform shows that it is similar to antiforms in the northeastern Brooks Range. These similarities and the location of the Mt. Doonerak antiform along the southern projection of the margin of the uplifted region of the northeastern Brooks Range suggest that the Mt. Doonerak antiform may be genetically related to the northeastern Brooks Range. Our fission-track results indicate that denudation of the Mt. Doonerak antiform occurred most recently at ~24 ± 3 Ma, which correlates well with the youngest uplift ages for the northeastern Brooks Range. We therefore conclude that the Mt. Doonerak antiform formed in response to the same contractional deformation that formed the northeastern Brooks Range. The young age of the Mt. Doonerak antiform and its position south of the area of Oligocene apatite ages in the northeastern Brooks Range (O'Sullivan et al., 1993b; Fig. 1) suggest that the final phase of duplex formation in the Mt. Doonerak antiform was caused by thrusting far behind the Oligocene deformation front. Older (latest Cretaceous to Paleocene) denudation of the Mt. Doonerak antiform, indicated by the zircon fission-track ages, suggests that the Oligocene deformation reactivated an older duplex developed mainly in the Paleocene. These relations provide evidence that the detachment faults for the Oligocene structures in the northeastern Brooks Range extended at midcrustal levels at least as far south as the southern Brooks Range.

Despite recent work that has elucidated the age, structural style, and displacement history of the northeastern Brooks Range (e.g., Wallace and Hanks, 1990; O'Sullivan et al., 1993b; Wallace, 1993; O'Sullivan, 1994), the fundamental tectonic cause of this orogen is not well understood. Most workers have suggested that the northeastern Brooks Range is a geographically restricted effect of contractional strain transmitted across Alaska by Pacific–North American plate interactions in southern Alaska (Oldow et al., 1987; Grantz et al., 1991). Apatite fission-track analysis of wells in the North Slope, however, suggest that an episode of rapid cooling at some time between 30 and 15 Ma may have affected most of the Colville basin (O'Sullivan, 1996). Late Oligocene uplift has also been documented by fission-track data from the Mt. Arrigetch region of the southern Brooks Range, about 200 km west-southwest of the Mt. Doonerak antiform (Murphy et al., 1994). These data suggest that kilometer-scale denudation during the late Oligocene may not have been restricted to the area of the northeastern Brooks Range, but instead extended across most of northern Alaska. If late Oligocene denudation was indeed widespread, we suggest that the underlying tectonic cause of the Mt. Doonerak antiform and the northeastern Brooks Range should be reevaluated.

SUMMARY AND CONCLUSIONS

Fission-track data from the central Brooks Range near the eastern end of the Mt. Doonerak antiform suggest that the region (1) cooled below ~240°C during the latest Cretaceous to Paleocene, and (2) experienced a phase of rapid cooling due to kilometer-scale denudation during the Oligocene at ~24 ± 3 Ma. The Mt. Doonerak region experienced an estimated average rate of denudation during the Paleocene to Oligocene (~70–65–25 Ma) of ~0.11–0.16 km/m.y. (assuming geothermal gradients of ~20–30°C/km). The rate of denudation during the Oligocene event was on the order of ~0.46–0.70 km/m.y. Since the Oligocene, the region has experienced an average rate of ~0.06–0.09 km/m.y. The minimum total denudation required by the fission-track data for Tertiary time is about 8 km.

Geologic relations show that the Mt. Doonerak antiform is a crustal-scale duplex that was formed after emplacement of the Endicott Mountains allochthon in Neocomian time. We conclude from the relative age data, the fission-track data, and the amount of vertical displacement required for its construction that the thrust faults underlying the antiform were initiated in latest Cretaceous or Paleocene time and were active into late Oligocene time. Despite its presence along the main deformational axis of the Jurassic and Early Cretaceous Brooks Range orogen, the similarity in age, scale, structural style, and depth to detachment of the Mt. Doonerak antiform to the Tertiary structures in the northeastern Brooks Range indicate that the Mt. Doonerak antiform was formed as part of the younger northeastern Brooks Range. The position in the hinterland of the north-vergent northeastern Brooks Range indicate that the Mt. Doonerak antiform was formed by thrusting well behind the Oligocene thrust front of that orogen. Additionally, these data show that the northeastern Brooks Range structures are in part superimposed on structures of the axial part of the Brooks Range orogen.

Because the fission-track data we have presented represent only a small pilot study, we believe that additional fission-track data are required to confirm our initial conclusions. Additional data should (1) clarify the regional extent of rocks uplifted during the Oligocene in the Mt. Doonerak region, and (2) better detail the higher temperature (≥240°C) thermal evolution of the structure.

ACKNOWLEDGMENTS

Acknowledgment is made to the Donors of The Petroleum Research Fund, administered by the American Chemical Society, for major support of this research (#24101-AC2). Field support was provided by the Trans-Alaska Crustal Transect (TACT) project of the U.S. Geological Survey. Grants provided by Exxon Exploration and Production Company and Arco Oil and Gas Company contributed to sample processing. O'Sullivan and Murphy were supported by La Trobe University Post-Graduate Scholarships. Sample irradiations performed at the Australian Atomic Energy Commission HIFAR Reactor were paid from an AINSE grant to the La Trobe University Fission Track Research Group. We thank Gary Fuis, John Oldow, and Wes Wallace for sharing unpublished data and ideas that significantly improved our understanding of the structural geology of the Mt. Doonerak antiform and the northeastern Brooks Range. Reviews by Ann Blythe, David Howell, Mark Johnsson, and Frances Julian greatly improved the manuscript.

REFERENCES CITED

Aleinikoff, J. N., Moore, T. E., Walter, M., and Nokleberg, W. J., 1993, U-Pb ages of zircon, monazite, and sphene from Devonian metagranites and meta-felsites, central Brooks Range, Alaska, *in* Dusel-Bacon, C., and Till, A. B., eds., Geologic studies in Alaska by the U.S. Geological Survey, 1992: U.S. Geological Survey Bulletin 2068, p. 59–70.

Barker, J. C., 1982, Reconnaissance of rare-metal occurrences associated with the Old Crow batholith, eastern Alaska–northwestern Canada: Short notes on Alaska Geology—1981: Alaska Division of Geological and Geophysical Surveys Geologic Report 73, p. 43–49.

Bird, K. J., and Molenaar, C. M., 1992, The North Slope foreland basin, Alaska, *in* Macqueen, R., and Leckie, D. A., eds., Foreland basins and foldbelts: American Association of Petroleum Geologists Memoir 55, p. 363–393.

Box, S. E., 1985, Early Cretaceous orogenic belt in northeastern Alaska: Internal organization, lateral extent, and tectonic interpretation, *in* Howell, D. G., ed., Tectonostratigraphic terranes of the circum-Pacific region: Houston, Texas, Circum-Pacific Council for Energy and Mineral Resources Earth Science Series, v. 1, p. 137–145.

Brosgé, W. P., and Reiser, H. N., 1964, Geologic map and section of the Chandalar Quadrangle, Alaska: U.S. Geological Survey Miscellaneous Geologic Investigations Map I-375, scale 1:250,000, 1 sheet.

Brosgé, W. P., Reiser, H. N., Dutro, J. T., Jr., and Detterman, R. L., 1976, Reconnaissance geologic map of the Table Mountain Quadrangle, Alaska: U.S. Geological Survey Open-File Report 76-546, scale 1:200,000.

Dillon, J. T., Brosgé, W. P., and Dutro, J. T., Jr., 1986, Generalized geologic map of the Wiseman Quadrangle, Alaska: U.S. Geological Survey Open-File Report OF 86-219, scale 1:250,000.

Dillon, J. T., Tilton, G. R., Decker, J., and Kelly, M. J., 1987, Resource implications of magmatic and metamorphic ages for Devonian igneous rocks in the Brooks Range, *in* Tailleur, I., and Weimer, P., eds., Alaskan North Slope geology: Society of Economic Paleontologists and Mineralogists, Pacific Section, Publication 50, p. 713–723.

Dutro, J. T., Jr., Brosgé, W. P., Lanphere, M. A., and Reiser, H. N., 1976, Geologic significance of Doonerak structural high, central Brooks Range, Alaska: American Association of Petroleum Geologists Bulletin, v. 60, p. 952–961.

Fuis, G. S., Levander, A. R., Lutter, W. J., Wissinger, E. S., Moore, T. E., and Christensen, N. I., 1995, Seismic images of the Brooks Range, Arctic Alaska, reveal crustal-scale duplexing: Geology, v. 23, p. 65–68.

Grantz, A., Moore, T. E., and Roeske, S. M., 1991, Continent-ocean transect A-3: Gulf of Alaska to Arctic Ocean: Boulder, Colorado, Geological Society of America, scale 1:500,000, 3 sheets.

Green, P. F, 1986, On the thermo-tectonic evolution of northern England: evidence from fission track analysis: Geology, v. 5, p. 493–506.

Green, P. F., Duddy, I. R., Laslett, G. M., Hegarty, K. A., Gleadow, A. J. W., and Lovering, J. F., 1989, Thermal annealing of fission tracks in apatite 4. Qualitative modelling techniques and extensions to geological timescales: Chemical Geology (Isotope Geoscience Section), v. 79, p. 155–182.

Harland, W. B., Armstrong, R. L., Cox, A. V., Craig, L. E., Smith, A. G., and Smith, D. G., 1990, A geologic time scale 1989: Cambridge, Cambridge University Press, 263 p.

Herman, Y., and Hopkins, D. M., 1980, Arctic oceanic climate in late Cenozoic time: Science, v. 209, p. 557–562.

Howell, D. G., Bird, K. J., Huafu, L., and Johnsson, M. J., 1992, Tectonics and petroleum potential of the Brooks Range fold and thrust belt—A progress report, *in* Bradley, D. C., and Ford, A. B., eds., Geologic studies in Alaska by the U.S. Geological Survey during 1990: U.S. Geological Survey Bulletin 1999, p. 112–126.

Johnsson, M. J., Pawlewicz, J., Harris, A. G., and Valin, Z. C., 1992, Vitrinite reflectance and conodont color alteration index data from Alaska: U.S. Geological Survey Open-File Report 92-409, 3 computer disks, 1 sheet.

Kelley, J. S., and Foland, R. L., 1987, Structural style and framework geology of the coastal plain and adjacent Brooks Range, *in* Bird, K. J., and Magoon, L. B., eds., Petroleum geology of the northern part of the Arctic National Wildlife Refuge, northeastern Alaska: U.S. Geological Survey Bulletin 1778, p. 255–270.

Lachenbruch, A. H., Sass, J. H., Lawver, L. A., Brewer, M. C., Marshall, B. V., Munroe, R. J., Kennelly, J. P., Jr., Galanis, S. P., Jr., and Moses, T. H., Jr., 1987, Temperature and depth of permafrost on the Alaskan North Slope, *in* Tailleur, I., and Weimer, P., eds., Alaskan North Slope geology: Society of Economic Paleontologists and Mineralogists, Pacific Section, Publication 50, p. 545–558.

Mayfield, C. F., Tailleur, I. L., and Ellersieck, I. E, 1988, Stratigraphy, structure, and palinspastic synthesis of the western Brooks Range, northwestern Alaska, *in* Gryc, G., ed., Geology and exploration of the National Petroleum Reserve in Alaska, 1974 to 1982: U.S. Geological Survey Professional Paper 1399, p. 143–186.

Miller, E. L., and Hudson, T. L., 1991, Mid-Cretaceous extensional fragmentation of a Jurassic–Early Cretaceous compressional orogen, Alaska: Tectonics, v. 10, p. 781–796.

Moore, T. E., Wallace, W. K., Bird, K. J., Karl, S. M., Mull, C. G., and Dillon, J. T., 1994, Geology of northern Alaska, *in* Plafker, G., and Berg, H. C., eds., The geology of Alaska: Boulder, Colorado, Geological Society of America, The Geology of North America, v. G-1, p. 59–140.

Mull, C. G., 1982, Tectonic evolution and structural style of the Brooks Range and Arctic Slope, Alaska, *in* Powers, R. B., ed., Geologic studies of the Cordilleran Thrust belt: Denver, Colorado, Rocky Mountain Association of Geologists, p. 1–45.

Mull, C. G., Adams, K. E., and Dillon, J. T., 1987a, Stratigraphy and structure of the Doonerak fenster and Endicott Mountains allochthon, central Brooks Range, Alaska, *in* Tailleur, I., and Weimer, P., eds., Alaskan North Slope geology: Society of Economic Paleontologists and Mineralogists, Pacific Section, Publication 50, p. 663–679.

Mull, C. G., Roeder, D. H., Tailleur, I. L., Pessel, G. H., Grantz, A., and May, S. D., 1987b, Geological sections and maps across Brooks Range and Arctic Slope to Beaufort Sea: Geological Society of America Map and Chart series, MC-28S, scale 1:500,000, 1 sheet.

Mull, C. G., Adams, K. E., and Dillon, J. T., 1989, Stratigraphy and structure of the Doonerak fenster and Endicott Mountains allochthon, central Brooks Range, Alaska, *in* Mull, C. G., and Adams, K. E., eds., Dalton Highway, Yukon River to Prudhoe Bay, Alaska, bedrock geology of the eastern Koyukuk basin, central Brooks Range, and eastcentral Arctic Slope: Alaska Geological and Geophysical Surveys Guidebook 7, p. 203–218.

Murphy, J. M., O'Sullivan, P. B., and Gleadow, A. J. W., 1994, Apatite fission track evidence of episodic Early Cretaceous to late Tertiary cooling and uplift, central Brooks Range, Alaska, *in* Thurston, D., and Fujita, K., eds., 1992 Proceedings International Conference on Arctic Margins: U.S. Minerals Management Service Outer Continental Shelf Study 94-0040, p. 257–262.

Oldow, J. S., Seidensticker, C. M., Phelps, J. C., Julian, F. E., Gottschalk, R. R., Boler, K. W., Handschy, J. W., and Avé Lallemant, H. G., 1987, Balanced cross sections through the central Brooks Range and North Slope, Arctic Alaska: American Association of Petroleum Geologists, Special Publication, 19 p., 8 pl.

O'Sullivan, P. B., 1994, Timing of Tertiary episodes of uplift and erosion, northeastern Brooks Range, Alaska, *in* Thurston, D., and Fujita, K., eds., 1992 Proceedings International Conference on Arctic Margins: U.S. Minerals Management Service Outer Continental Shelf Study 94-0040, p. 269–274.

O'Sullivan, P. B., 1996, Late Mesozoic and Cenozoic thermal-tectonic evolution of the North Slope foreland basin, Alaska, *in* Johnsson, M. J., and Howell, D. G., eds., Thermal evolution of sedimentary basins in Alaska: U.S. Geological Survey Bulletin 2142, p. 45–79.

O'Sullivan, P. B., Murphy, J. M., Moore, T. E., and Howell, D. G., 1993a, Results of 110 apatite fission track analyses from the Brooks Range and North Slope of northern Alaska, completed in cooperation with the Trans-Alaska Crustal Transect (TACT): U.S. Geological Survey Open-File Report 93-545, 104 p.

O'Sullivan, P. B., Green, P. F., Bergman, S. C., Decker, J., Duddy, I. R., Gleadow, A. J. W., and Turner, D. L., 1993b, Multiple phases of Tertiary uplift and erosion in the Arctic National Wildlife Refuge, Alaska, revealed by apatite fission track analysis: American Association of Petroleum Geologists Bulletin, v. 77, no. 3, p. 359–385.

O'Sullivan, P. B., Murphy, J. M., and Blythe, A. E., 1997, Late Mesozoic and Cenozoic thermotectonic evolution of the central Brooks Range and adjacent North Slope foreland basin, Alaska: Including fission track results form the Trans-Alaska Crustal Transect (TACT): Journal of Geophysical Research, v. 102, no. B9, p. 20321–20845.

Patton, W. W., and Box, S. E., 1989, Tectonic setting of the Yukon-Koyukuk basin and its borderlands, western Alaska: Journal of Geophysical Research, v. 94, p. 15807–15820.

Phelps, J. C., Avé Lallemant, H. G., and Seidensticker, C. M., 1987, Late-stage high-angle faulting, eastern Doonerak window, central Brooks Range, Alaska, *in* Tailleur, I., and Weimer, P., eds., Alaskan North Slope geology: Society of Economic Paleontologists and Mineralogists, Pacific Section, Publication 50, p. 685–690.

Seidensticker, C. M., Julian, F. E., and Phelps, J. C., 1987, Lateral continuity of the Blarney Creek Thrust, Doonerak window, central Brooks Range, Alaska, *in* Tailleur, I., and Weimer, P., eds., Alaskan North Slope geology: Society of Economic Paleontologists and Mineralogists, Pacific Section, Publication 50, p. 681–683.

Smiley, C. J., 1967, Cretaceous floras from Kuk River area, Alaska: stratigraphic and climatic interpretations: Geological Society of America Bulletin, v. 77, p. 1–14.

Spicer, R. A., and Parrish, J. T., 1986, Paleobotanical evidence for cool north polar climates in middle Cretaceous (Albian-Cenomanian) time: Geology, v. 14, p. 703–706.

Till, A. B., 1992, Detrital blueschist facies metamorphic mineral assemblages in Early Cretaceous sediments of the foreland basin of the Brooks Range, Alaska, and the implications for orogenic evolution: Tectonics, v. 11, p. 1207–1223.

Turner, D. L., Forbes, R. B., and Dillon, J. T., 1979, K-Ar geochronology of the southwestern Brooks Range, Alaska: Canadian Journal of Earth Sciences, v. 16, p. 1789–1804.

Wallace, W. K., 1993, Detachment folds and a passive-roof duplex: examples from the northeastern Brooks Range, Alaska, *in* Solie, D. N., and Tannian, F., eds., Short notes on Alaskan geology 1993: Alaska Division of Geological and Geophysical Surveys Professional Report 113, p. 81–99.

Wallace, W. K., and Hanks, C. L., 1990, Structural provinces of the northeastern Brooks Range, Arctic National Wildlife Refuge, Alaska: American Association of Petroleum Geologists Bulletin, v. 74, p. 1100–1118.

MANUSCRIPT ACCEPTED BY THE SOCIETY SEPTEMBER 23, 1997

Geological Society of America
Special Paper 324
1998

Geology and Mesozoic structural history
of the south-central Brooks Range, Alaska

Richard R. Gottschalk
Exxon Production Research Company, P.O. Box 2189, Houston, Texas 77252
John S. Oldow
Department of Geology and Geological Engineering, University of Idaho, Moscow, Idaho 83844-3022
H. G. Avé Lallemant
Department of Geology and Geophysics, Rice University, Houston, Texas 77005-1892

ABSTRACT

Rocks exposed in the south-central Brooks Range are distributed in four east-west–trending, fault-bounded lithologic belts. These include, from south to north, the Devonian to Lower Jurassic pillow basalt, diabase, gabbro, and radiolarian chert of the Angayucham terrane; the Devonian metasedimentary rocks of the Metagreywacke belt; the Devonian (?) metasedimentary rocks of the Phyllite belt; and the Precambrian to Middle Paleozoic metasedimentary rocks of the Schist belt.

We recognized two major synmetamorphic deformational events in the south-central Brooks Range. The oldest of these (early Brookian phase, D_{1a}) took place in the Middle to Late Jurassic, probably in response to impingement of an island arc on the Alaskan continental margin. Convergence between the arc and continent resulted in penetrative deformation of continental margin sediments (now exposed in the Schist belt) under high pressure–low temperature metamorphic conditions and resulted in the imbrication of the Angayucham terrane at shallower structural levels; penetrative deformation of the Metagreywacke belt at low metamorphic grades, and isoclinal folding of the Phyllite belt under pumpellyite-actinolite facies is proposed to have occurred at this time. The second major deformational event (Main Brookian phase, D_{1b}) occurred in the Early Cretaceous and resulted in penetrative deformation of the Schist belt under lower amphibolite- to greenschist-facies conditions. Petrofabric and microstructural analysis indicates that D_{1b} folds in the Schist belt formed in a strain regime dominated by top-to-the-north shear, consistent with formation during the main phase of contraction in the Brooks Range fold and thrust belt; the Metagreywacke belt and Phyllite belt are affected by semipenetrative to penetrative deformational fabrics that also may be Early Cretaceous in age. D_{1b} structures and fabrics in the Schist belt are cross-cut by the Minnie Creek thrust, a south-dipping thrust fault whose structural relations indicate that it is an "out-of-sequence" fault of latest Early to Late Cretaceous age. North-directed displacement of the Schist belt along the Minnie Creek thrust may have resulted in the formation of the Wiseman-Emma Dome arch and subsidiary post-D_{1b} contractional folds and thrust faults.

The south-central Brooks Range was affected by north-south extension in the latest Early to Late Cretaceous that locally postdated contractional deformation, but was broadly contemporaneous with thrust faulting elsewhere in the Brooks Range. Exten-

Gottschalk, R. R., Oldow, J. S., and Avé Lallemant, H. G., 1998, Geology and Mesozoic structural history of the south-central Brooks Range, Alaska, *in* Oldow, J. S., and Avé Lallemant, H. G., eds., Architecture of the Central Brooks Range Fold and Thrust Belt, Arctic Alaska: Boulder, Colorado, Geological Society of America Special Paper 324.

sion resulted in (1) juxtaposition along normal faults of rocks deformed and metamorphosed at differing structural levels within the contractional orogen; (2) formation of ductile extensional structures in the Schist belt and Phyllite belt; and (3) brittle extensional deformation of the Angayucham terrane and Metagreywacke belt.

Extension in the south-central Brooks Range was followed by the formation of south-vergent contractional folds, brittle faulting related to right-lateral offset on the Kobuk-Malemute fault, and a period of east-west contraction that resulted in gentle north-south trending flexures; the age of these postextensional deformational events is poorly constrained, but geologic relations indicate that they are probably Late Cretaceous to Tertiary in age.

INTRODUCTION

The southern Brooks Range exposes a large area of polydeformed igneous, metamorphic, and sedimentary rocks that make up the hinterland of the Brooks Range fold and thrust belt (Fig. 1). Excellent exposures of these rocks in the south-central Brooks Range provide an unique opportunity to study crustal processes in a collisional orogen. In this chapter, we present the results of our detailed mapping and structural analysis along the Middle Fork Koyukuk River in the vicinity of Wiseman, Alaska (Fig. 2); detailed work was augmented with reconnaissance mapping and structural analysis over much of the area shown in Figure 1b. In this chapter, we describe the general geology of the south-central Brooks Range and outline the Mesozoic structural evolution of the area. In addition, we present a model for the tectonic history of the south-central Brooks Range, with emphasis on relationships between arc-continent collision, regional metamorphism, and penetrative deformation of Alaskan continental margin sedimentary rocks. We also examine the role of late- to postmetamorphic deformation, including Late Cretaceous extensional deformation and Late Cretaceous–Early Tertiary strike-slip faulting in modifying the contractional architecture of the southern Brooks Range.

GENERAL GEOLOGY

The south-central Brooks Range is composed of a series of east-west–trending lithologic belts that are separated by major structural boundaries (Fig. 1). From south to north, these are (1) the Upper Devonian to Lower Jurassic seamount–ocean plateau rocks of the Angayucham terrane (Jones et al., 1988; Barker et al., 1988; Gottschalk, 1987); (2) the Middle Devonian Metagreywacke belt (Gottschalk, 1987; Murphy and Patton, 1988); (3) the Devonian (?) basinal sedimentary rocks of the Phyllite belt; (4) the Middle Devonian and older (?) epicontinental basinal sedimentary rocks of the Schist belt, metamorphosed to epidote-blueschist-facies (Gottschalk, 1990, and this volume, Chapter 9); and (5) the predominantly greenschist-facies Cambrian to Middle Devonian and possibly Precambrian platformal and basinal sedimentary rocks of the Skajit allochthon, (Dillon et al., 1987; Boler, 1989; Dusel-Bacon et al., 1989; Oldow et al., this volume, Chapters 7 and 8; Gottschalk and Snee, this

volume, Chapter 13). The Brooks Range is bounded on the south by the Yukon-Koyukuk basin where Jurassic to Lower Cretaceous rocks of island-arc affinity are exposed (Patton, 1973; Box and Patton, 1989; Patton and Box, 1989). The Yukon-Koyukuk island arc and the Angayucham terrane are both unconformably overlain by Albian and younger marine sedimentary rocks composed of detritus shed from the Brooks Range and the Ruby geanticline (Patton, 1973; Dillon and Smiley, 1984; Murphy et al., 1989; Nilsen, 1989).

Angayucham terrane

A regionally extensive, east-west trending belt of structurally dismembered pillow basalt, basalt flows, diabase, gabbro, radiolarian chert, and limestone crops out along the southern flank of the central Brooks Range (Brosgé and Reiser, 1962, 1964, 1971; Patton et al., 1977; Bird, 1977; Dillon et al., 1986; Gottschalk, 1987; Barker et al., 1988; Jones et al., 1988; Fig. 1b). These rocks are generally interpreted as part of an extensive ocean plateau–seamount terrane that was obducted onto the southern margin of Alaska in the Early to Late Jurassic (Patton et al., 1977; Roeder and Mull, 1978; Murchey and Harris, 1985; Barker et al., 1988; Pallister et al., 1989; Wirth et al., 1993; Gottschalk and Snee, this volume, Chapter 13). Oceanic rocks of plateau-seamount affinity are exposed in the Narvak panel of Patton and Box (1989) and the Copter Peak allochthon of Mayfield et al. (1988). In the western Brooks Range, the Copter Peak allochthon is structurally overlain by the Brooks Range ophiolite, which consists of 187- to 169-Ma ultramafic rocks and cumulate gabbros that are interpreted to have formed in a suprasubduction zone setting (Patton et al., 1977; Nelson et al., 1979; Mayfield et al., 1982; Nelson and Nelson, 1982; Ellersieck et al., 1982; Harris, 1989, 1992, this volume, Chapter 17; Wirth et al., 1993; Moore et al., 1993). Studies of dynamothermally metamorphosed rocks at the base of the Brooks Range ophiolite indicate that it was tectonically emplaced above the Copter Peak allochthon in the Middle Jurassic (171 to 163 Ma) shortly after crystallization (Zimmerman and Frank, 1982; Wirth and Bird, 1992; Wirth et al., 1993; Harris, this volume, Chapter 17). The rocks of the Brooks Range ophiolite, the Copter Peak allochthon, and the Narvak panel are collectively referred to as the *Angayucham terrane* on

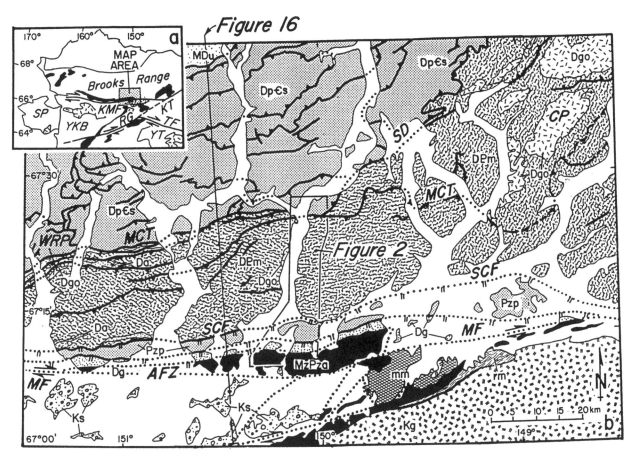

Figure 1. a, Selected tectonic elements of central and northern Alaska, with location of study area; black = ophiolitic rocks; hatchured pattern = Jurassic and Cretaceous intrusive and volcanic rocks, undivided. KMF, Kobuk-Malemute fault; KT, Kaltag fault; RG, Ruby geanticline; SP, Seward Peninsula; TF, Tintina fault; YKB, Yukon-Koyukuk basin; YT, Yukon-Tanana terrane. b, Generalized geology of the south-central Brooks Range, modified after Brosgé and Reiser (1964, 1971), Dillon et al. (1986), Oldow et al. (1987), and Jones et al. (1988). Ks, Lower Cretaceous sedimentary rocks of the Yukon-Koyukuk basin; Kg, Lower Cretaceous Hodzana granite; MDu, Devonian and Mississippian Endicott Group clastic rocks and Devonian Beaucoup Formation, undivided; DpCs, Precambrian–Middle Devonian rocks of the Skajit allochthon; DPm, Proterozoic to Devonian metamorphic rocks of the Schist belt; Da, Ambler volcanogenic sequence; Dgo, Devonian granitic orthogneiss; Dg, Devonian metagreywacke (Metagreywacke belt); Pzp, Paleozoic(?) phyllite (Phyllite belt). MzPza, Paleozoic and Mesozoic basalt, diabase, gabbro, chert, and limestone of the Angayucham terrane; mm, metamorphic rocks of the Mosquito terrane of Jones et al. (1988), undivided; rm, metamorphic rocks of the Ruby geanticline, undivided; MCT, Minnie Creek thrust; SD, Skajit decollement; SCF, Slate Creek fault; AFZ, Angayucham fault zone; MF, Malemute fault; WRP, Wild River pluton; CP, Chandalar pluton. Thin lines are geologic contacts. Heavy lines with barbs are thrust faults, with teeth on upper plate. Heavy lines with hatchures are normal faults; hatchures on upper plate. Heavy lines without ornamentation are steeply dipping faults.

the tectonostratigraphic terrane map of Silberling et al. (1994); however, because of the island-arc affinities of the Brooks Range ophiolite, Harris (1992, and this volume, Chapter 17) and Patton (1993) have argued that the Brooks Range ophiolite is genetically unrelated to the rocks of seamount/ocean plateau origin that lie structurally beneath it.

Our field work in the Angayucham terrane was carried out principally in the Cathedral Mountain area on the eastern flank of the Middle Fork Koyukuk River (Fig. 3). These rocks have been assigned to the Narvak panel of the Angayucham terrane by Patton and Box (1989). Basalt is the dominant rock type, occurring as pillow basalt and locally vesiculated flows that range from

a few centimeters to several meters in thickness. Basalt flows dip steeply to the south, but are locally vertical or north dipping and overturned. The ultramafic rocks of the Brooks Range ophiolite are not present in the study area, but are regionally extensive (Patton et al., 1994), and may have been present prior to uplift and erosion in the Cathedral Mountain area.

Bedded chert, locally with intercalated argillite, occurs in lenticular bodies within the basaltic volcanic pile. Individual chert lenses range from 20 cm to tens of meters in thickness, and in general, are oriented parallel to the flow foliation in the surrounding basalt; however, the presence of numerous folds in the chert lenses, and the lack of similar structures in the adjacent

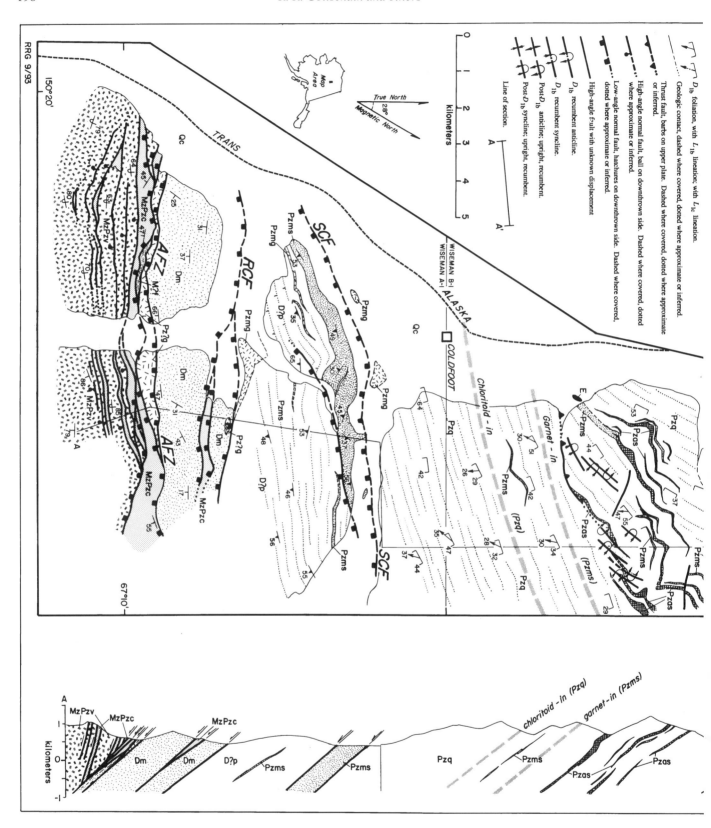

Figure 2. Geologic transect map of the south-central Brooks Range along the Middle Fork Koyukuk River, simplified and modified from Gottschalk (1987). AFZ, Angayucham fault zone; MCT, Minnie Creek thrust; RCF, Rosie Creek fault; SCF, Slate Creek fault.

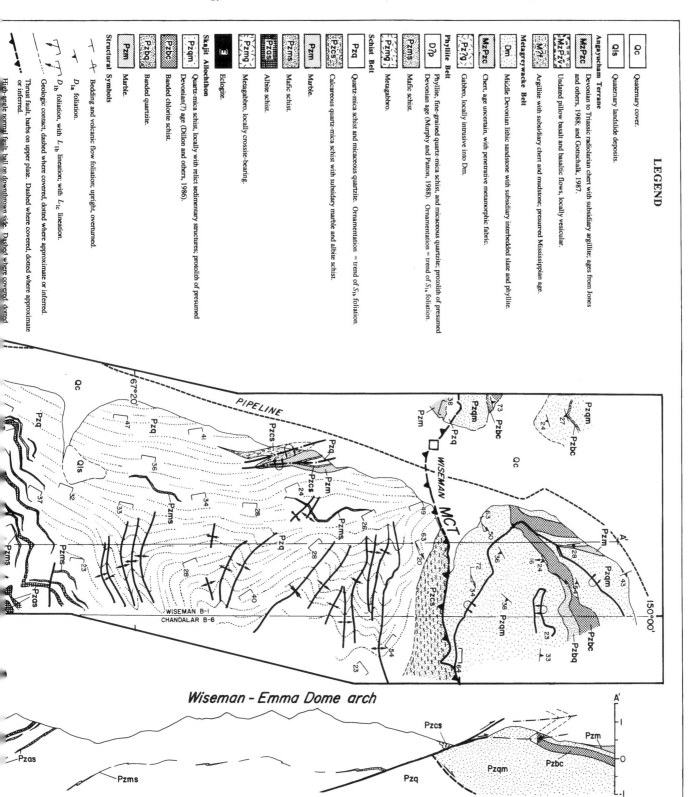

LEGEND

Qc Quaternary cover.

Qls Quaternary landslide deposits.

Angayucham Terrane

MzPzc Devonian to Triassic radiolarian chert with subsidiary argillite; ages from Jones and others, 1988; and Gottschalk, 1987.

MzPzv Undated pillow basalt and basaltic flows, locally vesicular.

Mx? Argillite with subsidiary chert and mudstone; presumed Mississippian age.

Metagreywacke Belt

Dm Middle Devonian lithic sandstone with subsidiary interbedded slate and phyllite.

MzPzc Chert, age uncertain, with penetrative metamorphic fabric.

Pzq Gabbro, locally intrusive into Dm.

Phyllite Belt

D?p Phyllite, fine-grained quartz-mica schist, and micaceous quartzite; protolith of presumed Devonian age (Murphy and Patton, 1988). Ornamentation = trend of S_{1a} foliation.

Schist Belt

Pzmg Metagabbro.

Pzq Quartz-mica schist and micaceous quartzite. Ornamentation = trend of S_{1b} foliation.

Pzcs Calcareous quartz-mica schist with subsidiary marble and albite schist.

Pzm Marble.

Pzms Mafic schist.

Pzas Albite schist.

Pzmg Metagabbro, locally crossite-bearing.

 Eclogite.

Skajit Allochthon

Pzqm Quartz-mica schist, locally with relict sedimentary structures; protolith of presumed Devonian(?) age (Dillon and others, 1986).

Pzbc Banded chlorite schist.

Pzbq Banded quartzite.

Pzm Marble.

Structural Symbols

 Bedding and volcanic flow foliation; upright, overturned.

 D_{1a} foliation.

 D_{1b} foliation, with L_{1b} lineation; with L_{1c} lineation.

 Geologic contact, dashed where covered, dotted where approximate or inferred.

 Thrust fault, barbs on upper plate. Dashed where covered, dotted where approximate or inferred.

 High-angle normal fault, ball on downthrown side. Dashed where covered, dotted

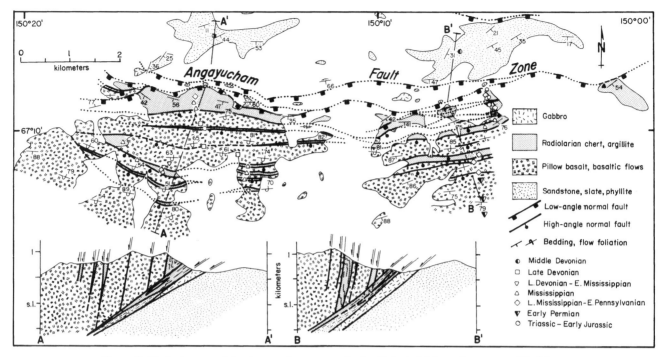

Figure 3. Geologic map of the Angayucham terrane in the Cathedral Mountain area, modified from Gottschalk (1987). Paleontologic data from Jones et al. (1988) and Gottschalk (1987).

basalts indicates that contacts are dominantly tectonic, rather than sedimentary. Fault-bound chert slivers are oriented roughly east-west, and are continuous along strike for as much as 3 km; a steep dip is inferred for the bounding faults based on their sublinear map pattern over topographically rugged terrain (Fig. 3). Jones et al. (1988) and Gottschalk (1987) have shown that cherts ranging in age from Late Devonian to Early Jurassic are spatially nonsystematically distributed (Fig. 3), indicating that cherts and basalts have been complexly interleaved during one or more post–Early Jurassic faulting events. Basalts that surround fault-bound chert slivers are undated.

Metamorphism of the Angayucham terrane is very low grade, and, in basalt, primary igneous minerals and textures are well preserved. Chlorite with oxidized patches is the chief product of metamorphism; it occurs along with magnetite as the replacement of what was probably glassy matrix, and partially replaces primary augite and olivine. Chlorite also occurs with calcite in cross-cutting veins, and as secondary fill in vesicles. Calcite is often present in irregular clots within the basaltic matrix and occurs together with pumpellyite as dustings on igneous plagioclase grains. All other metamorphic minerals, which include quartz, albite, prehnite, pumpellyite, and sphene are restricted to cross-cutting veins. These relations indicate that the Angayucham terrane was metamorphosed at temperatures between 200 and 300°C and pressures below about 2.5 kbar (Frey et al., 1991). It is uncertain whether metamorphism was tectonically induced or related to sea-floor alteration.

Metagreywacke belt

Along the Middle Fork Koyukuk River, the Metagreywacke belt is poorly exposed and consists mainly of weakly metamorphosed, polydeformed clastic sedimentary rocks (Dillon et al., 1981a, b; Gottschalk, 1987; Murphy and Patton, 1988). These rocks are assigned to the Narvak panel of the Angayucham terrane by Patton et al. (1994), and are part of the Rosie Creek allochthon of Oldow et al. (1987). Faults that separate the Metagreywacke belt from the adjacent Phyllite belt and from the pillow basalts and flows exposed at Cathedral Mountain were originally proposed to be south-dipping thrust faults (Brosgé and Reiser, 1964, 1971; Dillon et al., 1986), but more recently have been argued to be south-dipping normal faults (Gottschalk and Oldow, 1988; Gottschalk, 1990; Little et al., 1994). Elsewhere in the central Brooks Range, the boundary between the Phyllite and Metagreywacke belts may be sedimentary rather than tectonic (Dillon et al., 1981a, b; Little et al., 1994). The Metagreywacke belt is part of a regionally extensive belt of "greywacke" that has been mapped along strike in the Brooks Range by Brosgé and Reiser (1964, 1971), Dillon et al. (1986), and Jones et al. (1988), and also appears along the northeastern flank of the Ruby geanticline as far south as the Ray Mountains (Murphy and Patton, 1988; Jones et al., 1988). The "greywacke" belt yields largely Devonian pollen and spores (Brosgé and Reiser, 1962; Gottschalk, 1987; Murphy and Patton, 1988), and locally contains tectonically emplaced (?) blocks of Triassic chert and Mississippian limestone (Jones et al., 1988; Murphy and Patton, 1988).

The Metagreywacke belt consists of a moderately south-dipping sequence of medium to fine-grained sandstone, with subordinate interbeds of siltstone, mudstone, and claystone, now metamorphosed to phyllite or slate. Sedimentary rocks are locally intruded by gabbro and diabase with whole-rock and trace element compositions that are similar to gabbro from the Angayucham terrane and mafic schists of the Phyllite and Schist belt, but are compositionally distinct from basalts of the Angayucham terrane (Barker et al., 1988; Gottschalk, 1987; Gottschalk, this volume, Chapter 9; Fig. 4). A diabase plug analyzed by Barker et al. (1988) from the Metagreywacke belt yielded a chondrite-normalized REE profile with a marked light rare earth enrichment; the Barker et al. data suggest that gabbro and diabase intrusions may have crystallized from mantle-derived N-MORB that has undergone significant interaction with continental crust (Thompson et al., 1982; Dupuy and Dostal, 1984). Sedimentary rocks and mafic intrusives both show evidence of metamorphism at low grades, indicating that intrusion occurred prior to metamorphism and deformation of the Metagreywacke belt.

In the Middle Fork Koyukuk River area, the Metagreywacke belt is composed chiefly of a poorly sorted, medium- to fine-grained sandstone, consisting of angular to subangular quartz and chert, with detrital zircon, tourmaline, and magnetite in minor amounts. The angularity of quartz and chert grains in sandstones indicates that deposition of the metagreywacke occurred proximal to a source region that, according to Murphy and Patton (1988, p. 106), was "dominated by granitic plutonic and unfoliated quartzose metamorphic rock, and by unfossiliferous unfoliated chert." The source terrane for the metagreywacke is not known, but may be Proterozoic crystalline basement similar to that described by Till (1989), Karl and Aleinikoff (1990), and Nelson et al. (1993).

The most common sedimentary structures in sandstones are graded bedding and flute casts that occur at interfaces with underlying phyllite or slate horizons. Flute casts indicate sedimentary transport from a northerly source. Tangential cross-beds, hummocky cross-stratification, and centimeter-scale oscillation ripple marks are also locally present. Plant fragments were found at several locations, and four samples yielded palynofloral assemblages of a Siegenian-Emsian (Lower Devonian) age; samples yielded identifiable specimens of *Retusotriletes, Calamospora, Hystrichosporites, Aneurospora, Verrucosisporites, Dictyotriletes,* and *Rhabdosporites*, among others (Gottschalk, 1987).

The depositional environment and tectonic setting of the metagreywacke unit is controversial. Regional mapping by Murphy and Patton (1988) has shown that the Metagreywacke belt consists largely of turbidites, argued by them to have been deposited in a deep-marine or possibly continental-slope environment. Based upon the presumed deep-marine affinities of the Metagreywacke belt, Murphy and Patton (1988) and Patton and Box (1989) have included it as part of the Angayucham terrane. In contrast, Moore et al. (1994) have argued that the metagreywacke unit is more likely part of the Alaskan continental margin assemblage. The following geologic relations in the Middle Fork Koyukuk River area support a continental margin setting for deposition of the metgreywacke belt. (1) Oscillation ripple-marks and hummocky cross stratification indicate local deposition of the metagreywacke unit above wave base in a lower shoreface rather than a deep-marine basin setting. (2) The possible derivation of gabbro and diabase from contaminated basaltic magma suggests that the basement upon which the metagreywacke unit was deposited may have been continental rather than oceanic. (3) Petrographic studies cited above and carried out by us suggest that the metagreywacke unit may have been derived from and deposited in close proximity to a continental source region.

The metagreywacke is separated from the volcanic rocks of the Angayucham terrane by the Angayucham fault zone. This structure is equivalent to the "Angayucham thrust" of Dillon et al. (1986), the "Angayucham fault" of Little et al. (1994), and has locally been referred to as the "Cathedral Mountain fault zone" by Gottschalk (1987, 1990) and Gottschalk and Oldow (1988). In contrast to other faults in the study area, the Angayucham fault zone is relatively well exposed and was studied in detail where it is exposed along the Middle Fork Koyukuk River; here the Angayucham fault zone is characterized by a series of east-west–trending, anastomosing faults that bound and juxtapose slivers of basalt, chert, gabbro, basaltic igneous breccia, metagreywacke, and fine-grained clastic rocks with interbedded chert (Fig. 3). Locally, the fault zone is a tectonic melange within which fragments of various lithologies are suspended in a highly tectonized argillaceous matrix. A shallow to moderately south-dipping attitude for the Angayucham fault zone is inferred from numerous south-dipping mesoscopic faults within the fault zone (Fig. 5a; Avé Lallemant et al., this volume, Chapter 15). Bedded chert and argillite within the Angayucham fault zone are deformed in folds with generally south to southwest dipping axial planes, and fold axes that are distributed in a great circle girdle (Fig. 5b). Because folds within the Angayucham fault zone are dismembered by numerous small-scale faults, fold asymmetries are difficult to ascertain.

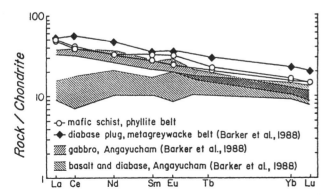

Figure 4. Comparative rare-earth element (REE) geochemistry of metabasites and mafic igneous rocks from the southern Brooks Range. Phyllite belt: Gottschalk, unpublished data. All other analyses: Barker et al. (1988).

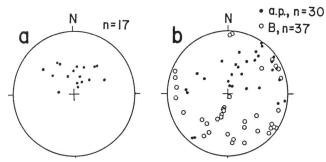

Figure 5. Lower hemisphere, equal-area projections of: a, poles to minor faults; and b, fold elements from the Angayucham fault zone. B = fold axes; a.p. = poles to axial planes.

The contact between the Metagreywacke belt and the Phyllite belt juxtaposes rocks of clear sedimentary character (the Metagreywacke belt) with the pumpellyite-actinolite-facies metamorphic tectonites of the Phyllite belt. In outcrop, metagreywacke is similar in appearance to the rocks assigned to the lowest grade and southernmost portion of the Phyllite belt, but in thin section they may be differentiated by the lack of relict detrital chert grains and the presence of abundant metamorphic quartz in the metasedimentary rocks of the Phyllite belt. Also, rocks of the Phyllite belt contain abundant, pervasive centimeter-scale folds, in contrast to the less intensely deformed Metagreywacke belt. Textural and structural differences between the Metagreywacke and Phyllite belts suggest that the contact is a fault that juxtaposes the two lithotectonic belts (the Rosie Creek fault, RCF, Fig. 2). A tectonic contact also is indicated by the presence of isolated, possibly fault-bound outcrops of metagabbro and metachert in the contact area with foliations discordant to the regional trend. Juxtaposition of rocks with differing metamorphic grades and strain histories indicates that offset on the inferred structure is postmetamorphic, and postdates synmetamorphic deformation in both belts. Due to the scarcity of outcrops in the Middle Fork Koyukuk River area, we cannot determine the dip or the sense of offset on the Rosie Creek fault.

Phyllite belt

The Phyllite belt is a narrow belt of pumpellyite-actinolite-facies phyllite, fine-grained schist, and metabasite that crops out along southern flank of the Schist belt (Brosgé and Reiser, 1964, 1971; Dillon et al., 1986; Gottschalk, 1987). These rocks are assigned to the Slate Creek subterrane of the Arctic Alaska terrane by Patton et al. (1994), and are included in the Rosie Creek allochthon of Oldow et al. (1987). The Phyllite belt is separated from the Schist belt on the north by the Slate Creek fault (SCF, Fig. 1b and 2), a south-dipping fault or shear zone with normal displacement (Gottschalk, 1990; Little et al., 1994; Law et al., 1994). The Phyllite belt in the central Brooks Range is lithologically similar to the Schist belt, but lacks granitic orthogneiss, marble, and calc-schist, and shows no evidence of blueschist-

facies metamorphism. The protolith of the Phyllite belt is undated, but is presumed to be Devonian in age (Brosgé and Reiser, 1964, 1971; Dillon et al., 1986; Murphy and Patton, 1988).

In the Middle Fork Koyukuk River area, the Phyllite belt is composed mainly of interfoliated phyllite, fine-grained pelitic and semipelitic quartz-mica schist, mafic schist, and metagabbro. The rocks of the Phyllite belt are exposed in a moderately south-dipping section, with the metamorphic grade and degree of synmetamorphic strain increasing downsection to the north (Gottschalk, 1987). Quartz-mica phyllites and pelitic and semipelitic quartz-mica schists are mineralogically similar, and differ mainly in grain size and quartz content. They are composed chiefly of quartz, white mica, chlorite, and subsidiary albite, with accessory graphite, apatite, tourmaline, and ilmenite. Quartz-rich and mica-rich layers are interfoliated on a scale of centimeters to meters, suggesting that much of the Phyllite belt was derived from a protolith of interbedded organic-rich shales and siliciclastic sediments. In the southernmost and lowest grade exposures of the Phyllite belt, relict sedimentary grains of quartz and plagioclase (now albite), and detrital zircon, all occur within a matrix of fine-grained metamorphic quartz; no sedimentary structures are preserved. In higher grade phyllites and fine-grained schists exposed in the northern Phyllite belt, relict sedimentary grains are obliterated, and foliation-forming minerals have been partially overgrown and recrystallized during a later deformation that refolds the foliation. The structural geology of the Phyllite belt is discussed in detail below.

Metagabbro from the Phyllite belt occurs as conformable intrafolial layers within the metasedimentary pile. The concordant relationship of metagabbro bodies to their enclosing sedimentary rocks may have resulted from intrusion as sills into the sedimentary succession, or from transposition of formerly discordant intrusive bodies during deformation. If contact metamorphic aureoles existed, they have since been obliterated by syndeformational metamorphism. Metagabbro is composed primarily of albite, chlorite, epidote, actinolite, and pumpellyite, with minor sphene, calcite, stilpnomelane, and quartz, and accessory pyrite and apatite. Relict crystals of igneous augite are occasionally present, and are fringed with epitaxial overgrowths of actinolite. Igneous plagioclase is altered to a fine-grained intergrowth of albite and pumpellyite. In most samples, ophitic to subophitic textures are well preserved.

Mafic schist also occurs as intrafolial layers within the metasedimentary pile, and is readily distinguishable from metagabbro by its lack of relict igneous textures. Mafic schist is composed of metamorphic albite, chlorite, epidote, actinolite, and sphene with minor calcite, quartz, white mica, stilpnomelane, apatite, and pyrite. No relict igneous minerals are preserved. Mafic schists are basaltic in composition, with trace element data indicative of a MORB-like protolith (Gottschalk, 1987). Chondrite-normalized REE profiles of two mafic schists show a marked light rare earth enrichment, with $(La/Sm)_N = 1.56$ and 1.77 (Fig. 4); these data are consistent with derivation of mafic schists from mantle-derived N-MORB that has undergone significant interaction with continental crust (Thompson et al., 1982; Dupuy and Dostal, 1984). We therefore infer an epicontinental depositional setting for the sedimentary protolith of the Phyllite belt.

Schist belt

The Schist belt (the Coldfoot subterrane of Silberling et al., 1994) is a regionally extensive, lithologically diverse belt of sedimentary and igneous rocks that were metamorphosed and penetratively deformed under blueschist-facies and greenschist- to amphibolite-facies conditions (Gilbert et al., 1977; Carden, 1978; Turner et al., 1979; Nelsen, 1979; Nelson and Grybeck, 1982; Hitzman et al., 1986; Dillon et al., 1986; Zayatz, 1987; Gottschalk, 1987, 1990, this volume, Chapter 9; Till, 1988; Till et al., 1988; Dusel-Bacon et al., 1989; Till and Patrick, 1991; Moore et al., 1994; Little et al., 1994; Patrick, 1995).

The Schist belt has been divided by Moore et al. (1994) into four principal lithologic assemblages. The structurally lowest unit in the Schist belt is composed of Proterozoic and lower Paleozoic metasedimentary rocks (Hitzman et al., 1982, 1986; Karl et al., 1989; Moore et al., 1994) that locally are intruded by Proterozoic granitic rocks (Karl et al., 1989). The recent Nd- and Sr-isotopic study of Nelson et al. (1993) indicates that Proterozoic metaigneous rocks in the Schist belt and elsewhere in the southern Brooks Range may have been derived, at least in part, from continental crust approximately 2.0 Ga old; these data imply that the Proterozoic to lower Paleozoic assemblage and younger rocks of the Schist belt were deposited on continental basement approximately 2.0 Ga in age. The Proterozoic and lower Paleozoic assemblage is structurally overlain, locally in tectonic contact, by the *quartz-mica schist unit,* a thick succession consisting predominantly of pelitic and semipelitic quartz-mica schists derived mainly from latest Devonian and older sedimentary rocks (cf. Moore et al., 1994, and references cited therein). The *Ambler volcanogenic sequence* of Hitzman et al. (1982; Da, Fig. 1b) is a distinctive lithologic succession within the quartz-mica schist unit consisting of felsic and mafic metavolcanic rocks, marble, chloritic schist, and graphitic schist; the Ambler volcanogenic sequence yields poorly preserved Late Devonian to Mississippian corals (Hitzman et al., 1986; Smith et al., 1978), and felsic metavolcanic rocks yield Late Devonian to Early Mississippian U-Pb and Pb-Pb ages (Dillon et al., 1980). In the western Brooks Range, the quartz-mica schist unit is conformably overlain by the *Bornite carbonate sequence* of Hitzman et al. (1986), a complex assemblage of marble and dolostone yielding mainly Middle Devonian to Early Mississippian age fossils.

The rocks that we have examined in detail along the Middle Fork Koyukuk River are assigned by Moore et al. (1994) to the quartz-mica schist unit of the Coldfoot subterrane. Here, the quartz-mica schist unit is exposed in a 7-kilometer-thick section composed mainly of metasedimentary rocks, including graphitic quartz-mica schist, marble, calc-schist, albite schist, metaconglomerate, and metachert; these rocks were probably derived from a sedimentary succession composed mainly of interbedded organic-rich shale, siliciclastic sediments, limestone, and marl. The metamorphic grade of the Schist belt increases down section and to the north, as indicated by the appearance of chloritoid in quartz-mica schists, and garnet in metabasites (Fig. 2).

Metaigneous rocks are an important minor component of the quartz-mica schist unit in the central Brooks Range, and include mafic schist, metagabbro, eclogite, and granitic orthogneiss. Rocks with recognizable volcanic protoliths (metarhyolites and meta-andesites, locally with preserved pyroclastic textures) have also been reported (Dillon et al., 1981a, b; Little et al., 1994). A metarhyolite layer from the Ambler volcanogenic sequence has been dated by the U-Pb zircon method, and yields a Devonian age (Dillon et al., 1980). Granitic rocks are intrusive into the metasedimentary sequence (Dillon et al., 1986; Newberry et al., 1986), and yield U-Pb zircon ages ranging from 402 to 357 Ma in age (Dillon et al., 1980; Dillon, 1989). Mafic schist and metagabbro are probably derived from basaltic flows and gabbro intrusions into the quartz-mica schist unit. An epicontinental setting for the protolith of the Schist belt is supported by the trace element signatures in metabasites; these data indicate that metabasites were derived from a MORB-like basaltic protolith that underwent significant contamination by continental crust (Gottschalk, this volume, Chapter 9). Based on the bimodal character of middle Paleozoic igneous rocks in the Schist belt of the western Brooks Range, Hitzman et al. (1986) have argued that igneous activity in the Schist belt records an event of Devonian intracontinental extension that eventually led to the formation of the Alaskan passive continental margin, and the opening of the Angayucham ocean basin. More recently, Nelson et al. (1993) have argued that Devonian granitic rocks have isotopic characteristics compatible with formation in a continental margin volcanic-arc setting.

The Schist belt is separated from the Skajit allochthon by two regionally extensive faults. (1) The oldest of these is the Skajit decollement (SD, Fig. 1b), the gently? north-dipping basal fault of the Skajit allochthon. Oldow et al. (1987) have argued that the Skajit decollement is a thrust fault along which the shelfal and basinal sedimentary rocks of the Skajit allochthon were carried northward over the "parautochthonous" basinal sedimentary rocks of the Schist belt. The displacement history of the Skajit decollement may have been complex, as Till and Moore (1991) contend that the last displacement was south directed. (2) The Skajit decollement, as well as subsidiary thrust faults within the Skajit allochthon, are apparently truncated by the Minnie Creek thrust (MCT, Fig. 1b; Oldow et al., 1987; Gottschalk, 1990). Cross-cutting relationships indicate that the Minnie Creek thrust was an "out-of-sequence" or envelopment thrust that breached the Skajit decollement, and placed rocks of the Schist belt northward over the metasedimentary rocks of the Skajit allochthon (Oldow et al., 1987). As defined by Bally et al. (1966), envelopment thrusts are thrust faults that originate from a deep detachment, and cross-cut and offset older faults that originate from shallower detachment levels.

The Yukon-Koyukuk basin and the Malemute fault

The Brooks Range is bounded to the south by the lowlands of the Yukon-Koyukuk basin (YKB, Fig. 1a). A thick sequence of andesitic volcanics, volcaniclastics, and hypabyssal intrusives ranging in age from 137 to 118 Ma, is exposed in the western Yukon-Koyukuk basin (Patton, 1973; Box et al., 1985; Nilsen, 1989). These rocks unconformably overlie a series of tonalite and trondjemite plutons that range in age from 173 to 154 Ma (Patton et al., 1994). Compositionally, both suites are characteristic of subduction-related volcanics erupted in an oceanic setting and are thought to represent an island-arc that was active from the Middle Jurassic through Early Cretaceous time (Box et al., 1985; Box and Patton, 1989; Patton and Box, 1989). Geologic evidence outlined in Patton et al. (1994) suggests that Middle Jurassic to Early Cretaceous igneous rocks intruded and were extruded onto a basement of mafic and ultramafic rocks similar to, and probably correlative to, the Angayucham terrane of the Brooks Range. Lower Cretaceous island-arc volcanic rocks as well as the ocean plateau–seamount rocks of the Angayucham terrane that rim the Yukon-Koyukuk basin are overlain by Albian and younger marine and nonmarine sedimentary rocks derived from the Angayucham terrane and metasedimentary rocks in the southern Brooks Range (Patton, 1973; Dillon and Smiley, 1984; Box et al., 1985; Murphy et al., 1989; Nilsen, 1989). These sediments provide a record of uplift and erosion in the Brooks Range, and subsidence in the Yukon-Koyukuk basin in the Early Cretaceous. Subsidence and marine sedimentation is coeval with late-syn or post-subduction alkalic plutonism and volcanism in the western Yukon-Koyukuk basin and on the Seward Peninsula (Box et al., 1985; Arth et al., 1989; Miller, 1989; Box and Patton, 1989; Patton and Box, 1989; Miller and Hudson, 1991).

The boundary between the Brooks Range and the Yukon-Koyukuk basin is also the locus of the right-lateral Kobuk-Malemute strike-slip fault (Fig. 1a). Patton and Miller (1966) argue that the Kobuk-Malemute fault was active from the Late Cretaceous through the early Tertiary, but recent seismological activity suggests that it is locally active at present (Gedney and Marshall, 1981).

STRUCTURAL ANALYSIS

In this section, we present a comparative geometric and kinematic analysis of structures in the lithologic belts of the south-central Brooks Range. All of the belts have protracted polyphase deformational/metamorphic histories that differ significantly from belt to belt. Whereas the Angayucham terrane and Metagreywacke belt have accommodated strain by a combination of folding and brittle faulting, the Phyllite belt and Schist belt have deformed mainly by the formation of polyphase ductile folds. In spite of differences in lithology, metamorphic grade, and strain history, structures in all of the lithologic belts display a remarkable consistency in orientation and superposition relationships, permitting us to correlate deformational phases over the study region. Metamorphic textural and microstructural relations (Zwart, 1962; Spry, 1969) were utilized to determine the relative timing of mineral growth and folding phases; these criteria constrain the relative timing of structural and metamorphic events in each belt.

Because deformation in fold and thrust belts is spatially and temporally transgressive, it is unlikely that all structures assigned to a single deformational phase formed simultaneously. Instead, we suggest that deformation is progressive and tied to the propagation of contraction-related structures northward (in present-day coordinates) and structurally down section. this chapter concludes with a tectonic model that attempts to explain the polyphase deformational/metamorphic history of the south-central Brooks Range in the context of progressive Brookian contractional deformation and the effects of later extensional and strike-slip deformation.

D_{1a}, D_{1b}, and D_{1c} structures

Structures and fabrics designated D_{1a}, D_{1b}, and D_{1c} are those that predate structures of demonstrable extensional origin (D_{1d}). Although the nature of pre-D_{1d} deformation is complex in detail, we can make several generalizations: (1) The number of pre-D_{1d} fold phases increases down section and to the north. One pre-D_{1d} fold phase (designated D_{1a}) is present in the Angayucham terrane, and two pre-D_{1d} fold phases occur in the Metagreywacke, Phyllite, and Schist belts (designated D_{1a} and D_{1b}); a third pre-D_{1d} fold phase (D_{1c}) is locally present in the northernmost outcrops of the Schist belt and in the southern Skajit allochthon. (2) In all of the lithologic belts, peak metamorphic conditions existed during D_{1a}. The metamorphic grade associated D_{1a} appears to increases gradationally down section within each belt, and increases abruptly to the north across the Rosie Creek and Slate Creek faults. (3) The intensity of deformation and the grade of metamorphism associated with D_{1b} also increases stepwise to the north across the Rosie Creek and Slate Creek faults, ranging from a weakly penetrative slaty cleavage in the Metagreywacke belt, to synmetamorphic isoclinal and sheath folds in the Schist belt.

In the following section we present detailed descriptions of the geometry and orientation of pre-D_{1d} structures and fabrics in each lithologic belt.

Angayucham terrane

D_{1a} structures in the Angayucham terrane are difficult to decipher due to the intensity of the extensional and strike-slip overprint in our study area. Pre-extensional structures consist of thrust faults, isoclinal folds in fault-bound chert slivers, and a locally developed spaced cleavage in basalts (Gottschalk, 1987; Avé Lallemant et al., this volume, Chapter 15). D_{1a} folds are isoclines, chevron folds, and flattened chevron folds, with steeply north-dipping axial planes, and fold axes that form a girdle distribution with a down-dip maximum (Fig. 6n).

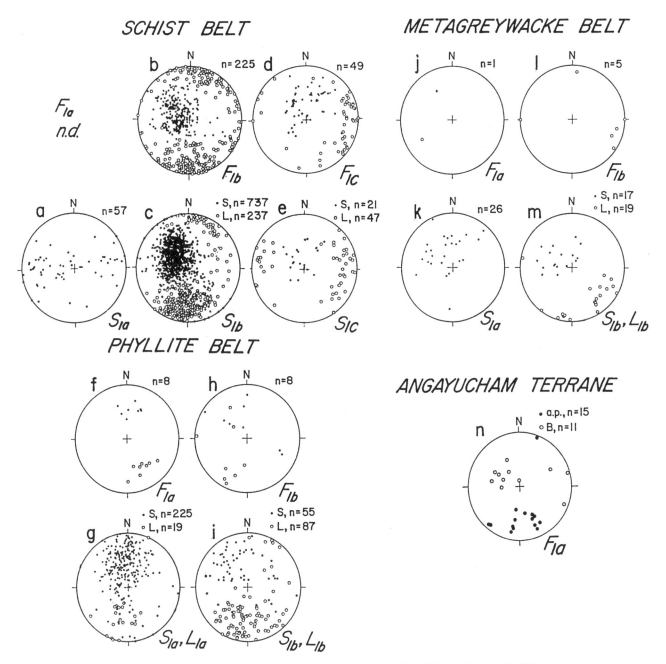

Figure 6. Lower hemisphere, equal-area projections of pre-D_{1d} fold and fabric elements. F = folds; open circles = fold axes; filled circles = poles to axial planes; S = poles to penetrative surfaces; L = lineations; B = fold axes; a.p. = poles to axial planes; n.d. = no data.

Metagreywacke belt

D_{1a}. The oldest structures in the Metagreywacke belt (D_{1a}) are isoclinal folds associated with a penetrative axial-planar foliation in phyllitic horizons, and a weak, spaced cleavage in sandy layers (S_{1a}). Although common in thin section, only one outcrop-scale D_{1a} fold was measured, with a moderately south-dipping axial plane and southwest-plunging fold axis (Fig. 6j). The foliation is axial planar to D_{1a} folds, and dips to the south, subparallel to bedding (Fig. 6k); it is composed of oriented grains of white mica, chlorite, and biotite or stilpnomelane that grow along anastomosing dissolution planes. Metamorphic quartz is also rarely present as tiny segregations that parallel the foliation. Calcite is locally present as a metamorphic phase, and plagioclase occurs as end-member albite suggesting that detrital plagioclase grains were albitized during metamorphism. D_{1a} in the Metagreywacke belt took place under low-grade metamorphic conditions, probably similar to those recorded in the overlying Angayucham terrane.

D_{1b}. The second fold phase in the Metagreywacke belt is manifested mainly as a spaced cleavage (S_{1b}) that occurs as weakly developed parting surfaces within the sandstones and as a penetrative crenulation cleavage in the phyllite and slate horizons (Fig. 7d). The cleavage dips to the south and southeast (Fig. 6m), and intersections of S_{1b} with bedding and S_{1a} plunge gently to the southeast and south (L_{1b}). Crenulation lineations parallel to intersection lineations are developed on S_{1a} foliation surfaces and are oriented parallel to axes of meter-scale D_{1b} folds (Fig. 6l and m). Because D_{1b} folds in the Metagreywacke belt are poorly exposed, their sense of asymmetry could not be determined.

Phyllite belt

D_{1a}. The oldest folds in the Phyllite belt (D_{1a}) are isoclinal and sheath folds (F_{1a}) with a pervasive axial-planar metamorphic foliation that comprises the main fabric in the Phyllite belt. In the transect area S_{1a} dips mainly to the south, but due to refolding during several later folding events, a plot of poles to S_{1a} shows a

diffuse girdle about an east-west–trending, subhorizontal axis (Fig. 6g). Intersections between lithologic layering (S_0) and foliation produce a lineation (L_{1a}) that plunges gently to the south (Figs. 6g and 8), parallel to isoclinal fold axes and sheath folds axes (Fig. 6f).

In phyllite and fine-grained quartz-mica schist, S_{1a} is defined by the preferred orientation of white mica and chlorite grains, and by weakly developed metamorphic quartz segregations. In mafic schist, S_{1a} is defined by the alignment of chlorite, actinolite, and sphene, which occur within a matrix of larger albite, epidote, and calcite porphyroblasts; chlorite, actinolite, and sphene occur as tiny, aligned inclusions within porphyroblasts, indicating significant post-D_{1a} growth of albite, epidote, and calcite. Metamorphic minerals that comprise the D_{1a} foliation in metagabbro include chlorite, epidote, sphene, calcite, stilpnomelane, and quartz; relict igneous augite grains are fringed by epitaxial overgrowths of metamorphic actinolite, and igneous plagioclase is replaced by a fine-grained intergrowth of albite + pumpellyite. Mineral assemblages in rocks from the Phyllite belt indicate that D_{1a} structures

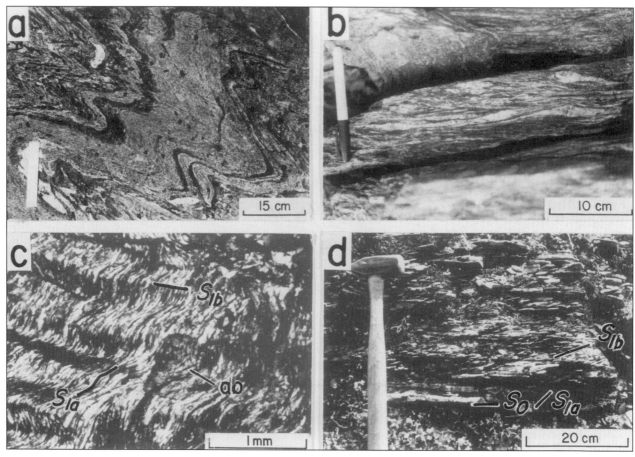

Figure 7. D_{1b} structures and fabrics. a, Close to tight D_{1b} folds in the Schist belt, with penetrative axial planar crenulation cleavage S_{1b}. b, Differentiated crenulation cleavage (S_{1b}) created by transposition of foliation (S_{1a}) in quartz-mica schist from Schist belt. Rootless hinges and boudinaged limbs occur where syn-D_{1a} quartz-segregations are folded. c, Photomicrograph of D_{1b} crenulation cleavage in the Phyllite belt; ab = albite porphyroblast. d, Bedding (S_0)-cleavage (S_{1a}) relations in the Metagreywacke belt. Photograph shows interbedded phyllite, siltstone, and sandstone with bedding-parallel foliation (S_{1a}) and cross-cutting penetrative crenulation cleavage (S_{1b}).

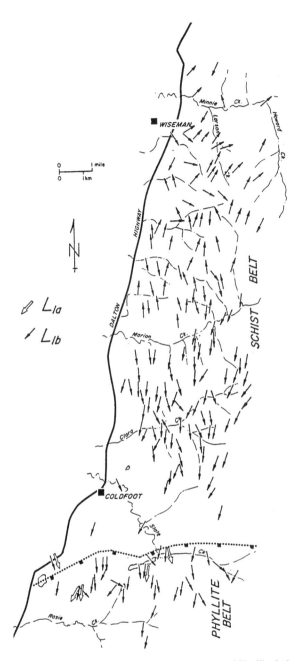

Figure 8. Trend of lineations in the Schist belt and Phyllite belt along the Middle Fork Koyukuk River.

and fabrics formed under pumpellyite-actinolite-facies metamorphic conditions (T =200–350°C, P = 3–7 kbar, from Liou et al., 1985). Deformation microstructures produced in quartz during D_{1a} are indicative of deformation by dislocation climb and recovery processes; temperatures are thus inferred to have been in excess of approximately 300°C during D_{1a} (Tullis and Yund, 1980; Lloyd and Knipe, 1992).

D_{1b}. The second fold phase in the Phyllite belt is manifested as a weakly differentiated crenulation cleavage (S_{1b}) that is axial planar to D_{1b} folds (Fig. 7c). S_{1b} dips generally to the south (Fig. 6i), usually at slightly steeper angles than the older foliation (S_{1a}). Intersections between S_{1a} and S_{1b} produce north-south–trending lineations (L_{1b}, Fig. 6i). D_{1b} folds have south-easterly dipping axial planes, and north-south–trending fold axes parallel that are parallel to L_{1b} intersection and crenulation lineation (Fig. 6h).

Microlithon boundaries created during D_{1b} folding served as the locus for growth of white mica and chlorite neoblasts in phyllite and quartz-mica schist, and actinolite, chlorite, sphene, and calcite in mafic schist. Dissolution seams formed preferentially at microlithon boundaries, creating anastomosing arrays along which S_{1a} minerals were dissolved and where insoluble residues remain. Quartz grains from syn-D_{1a} veins and segregations contain subgrains elongated parallel to S_{1b}, surrounded by tiny, strain-free quartz neoblasts (core-mantle structure of White, 1976). Microstructures of this type indicate significant dynamic recrystallization of quartz during D_{1b} in the Phyllite belt. Microstructural and textural relations indicate that temperatures were sufficiently high to produce greenschist-facies mineral assemblages, and to allow accommodation of D_{1b} strain by dynamic recrystallization of quartz and solution creep (greater than approximately 300°C).

In the Phyllite belt, D_{1b} folding is followed by a period of late- to postkinematic mineral growth. In phyllite and fine-grained quartz-mica schist, postkinematic mineral growth is characterized by albite grains that helicitically enclose D_{1b} structures (Fig. 7c); in mafic schist, epidote, albite, and calcite porphyroblasts helicitically overgrow D_{1b} crenulations. These relations indicate that temperatures remained sufficiently high to produce metamorphic mineral phases as D_{1b} deformation waned.

Schist belt

D_{1a}. The oldest folds in the Schist belt (D_{1a}) are poorly preserved, centimeter-scale isoclinal folds and sheath folds with a penetrative axial planar foliation (S_{1a}). The orientation of D_{1a} folds could not be measured directly, but interference patterns generated by superposed D_{1b} folds indicate that D_{1a} and D_{1b} folds are coaxial (Fig. 9). Because D_{1b} fold axes trend predominantly north-south in the Schist belt, the axes of D_{1a} isoclines and sheath folds are inferred to also trend north-south.

S_{1a} is readily distinguishable in mica-rich rocks where it is defined by metamorphically differentiated quartz- and mica-rich layers and by the preferred orientation of mica and chlorite grains. Syn-D_{1a} mineral assemblages in rocks from the Schist belt were determined from textural and microstructural relations, and are as follows: (1) in quartz-mica schist: quartz + phengite + paragonite + chlorite ± chloritoid ± pseudomorphs after glaucophane; (2) in calc schist: calcite + dolomite + quartz + white mica + chlorite; (3) in albite schist: chlorite + white mica + quartz + epidote + sphene. The assemblages listed above are consistent with formation under epidote-blueschist-facies metamorphic conditions and are interpreted have formed under the same pressure/temperature conditions that produced the Clara Creek

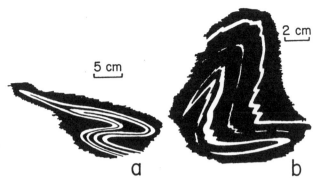

Figure 9. Type-3 interference patterns (Ramsay, 1967) produced by refolding of: a, a D_{1a} isoclinal fold, and b, a D_{1a} sheath fold from the Schist belt. Similar D_{1a} fold morphologies also exist in the Phyllite belt.

eclogite (Gottschalk, 1990, and this volume, Chapter 9). Petrologic studies by Patrick (1995) indicate that blueschist-facies metamorphism in the Schist belt occurred at temperatures of 375–425°C and pressures of 9–12 kbar.

D_{1b}. The principal effect of D_{1b} in the Schist belt was to refold S_{1a} into close to isoclinal, noncylindrical folds that range from microscopic crenulae to meter- and perhaps kilometer-scale folds. D_{1b} folds are associated with an axial planar differentiated crenulation cleavage that is the dominant planar fabric in the Schist belt. Axial planes of D_{1b} folds dip mainly to the southeast and east, but poles to S_{1b} are distributed in a north-northwest–south-southeast–trending girdle due to post-D_{1b} formation of the east-west–trending Wiseman-Emma Dome arch (Figs. 2, 6b, 6c). D_{1b} fold axes trend predominantly north-south, but some trend more easterly, forming a great circle about the point maximum of poles to axial planes (Fig. 6b). The noncylindrical nature of D_{1b} folds suggests that they may be sheathlike in geometry.

Rocks from the Schist belt often display a well-developed L-S tectonite fabric (Turner and Weiss, 1963). Lineations (L_{1b}) are S_{1a}/S_{1b} intersections that occur either as color bands on S_{1b} surfaces, or as prominent quartz rods where syn-D_{1a} quartz veins are folded. Like D_{1b} fold axes, intersection lineations concentrate in a north-south orientation (Figs. 6c and 8).

Textural and microstructural relations outlined by Gottschalk (1987, 1990) indicate synkinematic growth of the following minerals during D_{1b}: in quartz-mica schist: quartz + phengite + paragonite + chlorite ± albite ± chloritoid; in calc-schist: quartz + white mica + calcite + dolomite ± albite ± epidote; and in albite schist: quartz + white mica + chlorite + albite. Minerals indicative of high pressure/low temperature (*HP/LT*) metamorphism such as glaucophane were not observed in textural configurations that unequivocally indicate syn-D_{1b} growth. These relations therefore indicate that D_{1b} deformation took place under greenschist- to amphibolite-facies metamorphic conditions. Gottschalk (this volume, Chapter 9) has argued that D_{1b} deformation took place at temperatures between 380 to 480°C and at pressures below the stability field

of glaucophane. Other workers in the Schist belt have reported syn-D_{1b} glaucophane, indicating that epidote-blueschist-facies conditions locally existed during D_{1b} time (A. B. Till, personal communication, 1993).

In other rock types, D_{1a} and D_{1b} do not form discrete penetrative fabrics. Marble horizons are generally gray in color, but often contain boudinaged and isoclinally folded layers of white calcite suggesting a refolded older fabric (S_{1a}). In thin section, however, white calcite layers and the enclosing matrix have been recrystallized into a single metamorphic foliation comprised of a polygonal-granoblastic aggregate of slightly elongate calcite grains. These relations indicate that the principal fabric in marbles may have resulted from dynamic recrystallization during D_{1b}. Metachert is composed of slightly elongate quartz grains in polygonal-granoblastic aggregates with subsidiary stilpnomelane, biotite, and garnet forming a single foliation. We also interpret this fabric to result from dynamic recrystallization during D_{1b}. Mafic schist occurs in outcrop as weakly banded, relatively undeformed layers within the complexly deformed metasedimentary rocks. In thin section, mafic schists are composed of oriented grains of actinolite, chlorite, and sphene, surrounded and enclosed by porphyroblasts of albite, epidote, calcite, and garnet. The relationship of the single foliation in mafic schist to D_{1a} or D_{1b} is not known. However, the presence of greenschist-facies assemblages in mafic schists suggests that the penetrative fabric in mafic schist resulted mainly from syn-D_{1b} mineral growth.

D_{1a} and D_{1b} fabrics in the Schist belt are cross-cut by metamorphic minerals with random orientations, indicating late- to postkinematic recrystallization and mineral growth (Gottschalk, 1990). Post-D_{1b} minerals include white mica, chlorite, and albite in quartz-mica and albite schists; and epidote, albite, calcite, chlorite, biotite, and stilpnomelane in mafic schists. These relations indicate that greenschist-facies metamorphic conditions persisted as D_{1b} deformation waned and eventually ceased.

D_{1c}. The principal result of D_{1c} deformation was the formation of the Wiseman-Emma Dome arch, a northeast- to east-northeast–trending structural culmination (Dillon et al., 1986) with a steeply dipping north limb, and a moderately dipping south limb (Fig. 2). A number of subsidiary D_{1c} structures are present; these include northwest- to east-northwest–trending open folds with wavelengths of tens to hundreds of meters that occur mainly in the crestal region of the Wiseman-Emma Dome arch.

Penetrative D_{1c} deformation is restricted to the vicinity of the Minnie Creek thrust (i.e., in the Schist belt on north-dipping flank of the Wiseman-Emma Dome arch and the southernmost exposures of the Skajit allochthon); here D_{1c} folds are moderate to close north-vergent folds with gently south dipping to subhorizontal axial planes and east-west–trending, subhorizontal fold axes (Fig. 6d). D_{1c} folds range in size from microscopic crenulae to the kilometer-scale recumbent synform-antiform pair developed in the footwall of the Minnie Creek thrust (Fig. 2). Near the Minnie Creek thrust, D_{1c} folds are associated with subhorizontal thrust faults along which minor top-to-the-north displacement has occurred (Fig. 10). Conjugate D_{1c} fold sets indicate that, dur-

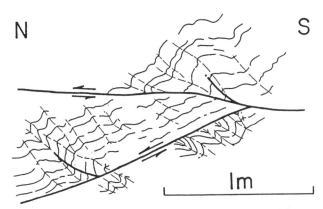

Figure 10. Line drawing from photograph of D_{1c} folds and thrusts in the Schist belt, in the immediate vicinity of the Minnie Creek thrust. N = north; S = south.

ing D_{1c}, the shortening axis lay in a north-south, subhorizontal orientation. A penetrative, gently south dipping axial planar cleavage (S_{1c}), and a prominent east-west–trending crenulation/ intersection lineation (L_{1c}) occur in association with D_{1c} folds (Fig. 6d and e). On a microscopic scale, S_{1c} is a weakly differentiated crenulation cleavage; microlithon boundaries serve as the locus for limited growth of white mica and chlorite, as well as the dissolution of pre-D_{1c} minerals. Deformation microstructures produced in quartz during D_{1c} are indicative of deformation by dislocation climb and recovery processes (Gottschalk, 1990); therefore, we infer that D_{1c} occurred at temperatures greater than about 300°C.

D_{1d} structures

D_{1d} structures in the southern Brooks Range are extensional features that range from centimeter- to meter-scale normal faults and extensional shear zones to the regionally extensive, east-west–trending faults that separate the lithologic belts. The Schist

belt, Phyllite belt, and southern Skajit allochthon are cross-cut by numerous ductile extensional shear zones, and an extensional crenulation cleavage (Platt and Vissers, 1980) is locally pervasive (Fig. 11). Extensional shear zones are generally east-west–trending and dip to the north or south, terminating into the older D_{1b} foliation (Fig. 12a). Shear zones are locally abundant, and may reach 10 m or more in length. Microstructural and textural relations indicate that extensional shear zones are areas of localized strain softening in which dynamic recrystallization of quartz and dissolution mass-transfer processes were the chief deformation mechanisms (Gottschalk, 1990). In the Schist belt, displacement on extensional shear zones was accompanied by growth of chlorite and white mica neoblasts.

In the Metagreywacke belt and Angayucham terrane, D_{1d} deformation resulted in the formation of numerous small-scale brittle extensional faults. These features and their kinematic significance are discussed in Avé Lallemant et al. (this volume, Chapter 15).

In the Angayucham terrane, extension also resulted in the formation of numerous south-vergent folds that refold thrust-related isoclinal D_{1a} folds. D_{1d} folds in the Angayucham terrane occur mainly in fault-bound chert slivers where they occur as open folds with south-vergent asymmetries and interlimb angles of 90° or less. Axial planes dip to the south and southeast and fold axes trend in an east-west, subhorizontal orientation (Fig. 12c). The pervasiveness of south-vergent folds in chert slivers strongly suggests that the faults that bound chert slivers are normal faults, or thrust faults that have been reactivated as normal faults.

D_2 structures

D_2 folds ranging from few centimeters to hundreds of meters in wavelength occur in all of the lithologic belts, but are most prominently developed in the Schist belt and Phyllite belt. In the Schist belt, D_2 structures are gentle to moderate folds (Fig. 13a) with moderately to steeply north dipping axial planes, and east-west–trending, subhorizontal fold axes (Fig.

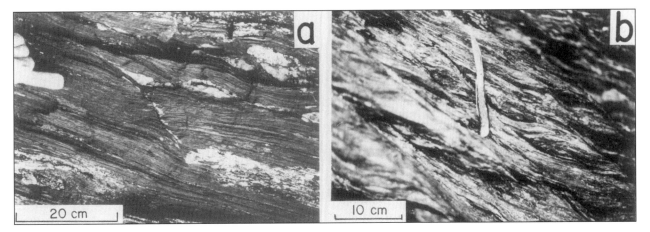

Figure 11. D_{1d} (extensional) structures. a, Ductile extensional fault from the southern Skajit allochthon. b, Extensional shear band cleavage from the Schist belt; folded surface is S_{1b}.

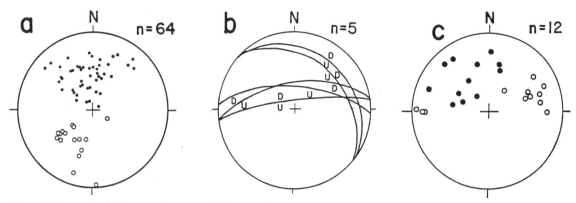

Figure 12. Lower hemisphere, equal-area projections: a, poles to extensional shear zones in the Schist belt, open circles = north-dipping, with down to north offset; filled circles = south-dipping with down to south offset; b, mesoscopic normal faults, southern Skajit allochthon; c, south-vergent folds in fault-bounded chert slivers, Angayucham terrane. Filled circles = poles to axial planes; open circles = fold axes).

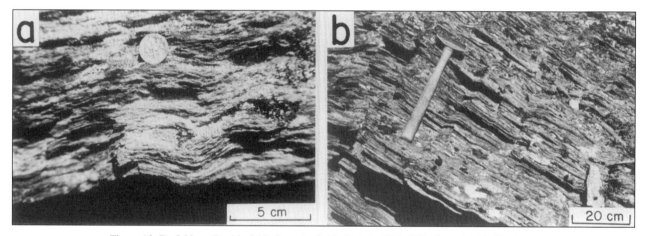

Figure 13. D_2 folds. a, Buckle folds from the Schist belt; b, Kink folds from the Phyllite belt.

14a). Crenulation lineations on the D_{1b} surface are common and are parallel to D_2 fold axes (Fig. 14a and b). D_2 folds generally are south vergent and locally are associated with a north-dipping thrust fault along which an unknown but probably minor amount of south-directed displacement has occurred. In the Phyllite belt, D_2 folds are south-vergent folds ranging from a few centimeters to meters in wavelength. These folds are associated with numerous subsidiary kink folds (Fig. 13b) with north-dipping axial planes and east-west–trending, subhorizontal fold axes (Fig. 14c). L_2 crenulation lineations parallel to D_2 fold axes are common (Fig. 14d). Conjugate kink folds in the Phyllite belt indicate that D_2 folds are contractional in origin, with the compression axis lying in a north-south, subhorizontal orientation. The relative age of D_2, D_{1c}, and D_{1d} folds could not be established due to a lack of cross-cutting or superposition relationships in the study area.

D_3 structures

The youngest folds in the southern Brooks Range are a series of broad, gentle flexures with north-south–trending fold axes and near-vertical, north-south–striking axial planes. These folds are typically several hundred meters in wavelength and affect all of the lithologic belts.

Strike-slip faults

In our study area, mesoscopic strike-slip faults subsidiary to the Kobuk-Malemute fault are intensely developed in outcrops of the Angayucham terrane along the West Fork of the Chandalar River. Fault planes strike east-west, are subvertical in orientation, and are striated with subhorizontal slickenside lineations; where measurable, the sense of displacement is right lateral (Avé Lallemant et al., this volume, Chapter 15). Right-lateral strike-

SCHIST BELT

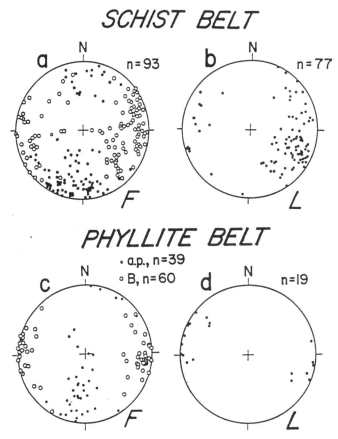

PHYLLITE BELT

Figure 14. Lower hemisphere, equal-area projections of fold and fabric elements for D_2 structures. F = folds; filled circles = poles to axial planes (a.p.); open circles = fold axes (B); L = lineations.

slip faults clearly postdate normal faults in the Angayucham terrane; the relationship of strike-slip faults to south-vergent D_2 folds and north-south–striking D_3 folds could not be demonstrated from cross-cutting or superposition relationships.

KINEMATICS AND TIMING OF DEFORMATION

D_{1a}

Because D_{1a} structures and fabrics are generally obscured by intense post-D_{1a} deformation, our kinematic interpretation of D_{1a} deformation is somewhat speculative. However, the common occurrence of D_{1a} sheath folds in the Schist belt and Phyllite belt indicates that during D_{1a}, both belts were deformed in a noncoaxial strain regime and were subjected to large-magnitude shear strains (cf. Quinquis et al., 1978; Cobbold and Quinquis, 1980). Moreover, Quinquis et al. (1978) and Cobbold and Quinquis (1980) have shown that the axes of sheath folds and isoclinal folds are oriented approximately parallel to the orientation of the shear couple. Because the axes of D_{1a} sheath folds and isoclines in the Phyllite belt trend mainly south-southeast, we infer a north-northwest–south-southeast orientation for the tectonic transport

direction (orientation of shear couple); in the Metagreywacke belt, a single mesoscopic D_{1a} sheath fold with a north-northwest–trending fold axis suggests a similar transport direction. Because the oldest folds in the Metagreywacke, Phyllite, and Schist belts are similar in geometry and orientation, and because they imply a consistent tectonic transport direction, we infer a similar origin for D_{1a} folds in all three belts despite differences in metamorphic grade from belt to belt. Due to the intensity of post-D_{1a} deformation, we cannot uniquely determine the sense of shear in any of the three lithologic belts; therefore, our data are consistent with formation of D_{1a} folds in a strain regime dominated either by top-to-the-north *or* top-to-the-south shear.

In the Angayucham terrane, the origin of D_{1a} structures is obscured by the pervasive occurrence of post-D_{1a} extensional and strike-slip faults and related folds. However, it is clear that D_{1a} was characterized by partitioning of relatively ductile deformational strain into chert horizons while the surrounding basalts deformed in a more brittle fashion. Because faults of unequivocal pre-extensional origin are difficult to document, and because these faults were apparently reactivated during extension (cf. Avé Lallemant et al., this volume, Chapter 15), much of the information on the kinematic origin of D_{1a} structures is derived from D_{1a} folds in chert horizons. Strain in chert horizons during D_{1a} was sufficient to produce isoclinal folds, and if strain was noncoaxial D_{1a} may have caused the passive rotation of fold axes into the tectonic transport direction (cf. Bryant and Reed, 1969; Sanderson, 1973; Escher and Watterson, 1974). The steep down-dip plunge of the of the majority of isoclinal fold axes therefore suggests a near-vertical transport direction (Fig. 6n). However, the steep dip of basalt flows and chert beds indicates that the Angayucham terrane has been steeply tilted about an east-west axis, possibly during post-D_{1a} extension (Gottschalk, 1987; Gottschalk and Oldow, 1988). These relations suggest that D_{1a} fold axes restore to a north-northwest–south-southeast subhorizontal orientation, implying a north-northwest–south-southeast transport direction for D_{1a} in the Angayucham terrane.

The age of D_{1a} deformation in the south-central Brooks Range is constrained only in the Schist belt, based on the following: (1) Gottschalk and Snee (this volume, Chapter 13) have analyzed white micas from the Schist belt that yield convex-upward Ar-release spectra indicating a minimum age for *HP/LT* metamorphism of 142 Ma. This age is in good agreement with studies of Till and Patrick (1991) and Christiansen and Snee (1994), indicating that *HP/LT* metamorphism in the Schist belt occurred prior to 149 Ma and 171 Ma, respectively. (2) The Devonian Koyukuk River pluton is overprinted by blueschist-facies metamorphic assemblages (Patrick, 1995), establishing a maximum age for *HP/LT* metamorphism. Because the southern Alaskan continental margin was a zone of relative tectonic quiescence through much of the late Paleozoic and early Mesozoic (cf. Moore et al., 1994), it is most likely that *HP/LT* metamorphism is related to early Brookian orogenesis and may be Middle to Late Jurassic in age (Armstrong et al., 1986; Christiansen and Snee, 1994). Based on structural correlations discussed above, we infer that D_{1a} structures in the Angayucham terrane, Metagreywacke belt, and Phyllite belt are also Middle to Late Jurassic in age.

D_{1b}

Structural relations in the Schist belt indicate that D_{1b} structures and fabrics formed in a strain regime dominated by simple shear. Shear-sense indicators are common and include snowball-type albite and garnet porphyroblasts, rotated white mica grains with sigmoidal graphitic inclusion trails, sigmoidal white mica grains ("mica fish"), and asymmetric strain shadows around garnet, albite, and epidote porphyroblasts. Our investigation revealed that of the 11 samples that yielded an unambiguous sense of shear, eight indicate top-to-the-north shear, and three indicate top-to-the-south shear. The method and results of microstructural studies are summarized in Gottschalk (1987, 1990).

Quartz c-axis petrofabric analysis was carried out on 18 samples from the eastern part of the study area (Fig. 15). All of the samples examined were dynamically recrystallized during D_{1b}; thus quartz petrofabrics most likely reflect local deformational strains associated with D_{1b}. Samples from the Schist belt yield diffuse to strong type-I cross-girdle fabrics (Lister, 1977) with a marked degree of asymmetry with respect to the foliation plane. Fabrics of this type are common in quartz-rich tectonites subjected to high shear strains, and numerous studies have shown that the inclination of the girdle with respect to the foliation plane reflects the sense of shear (e.g., Price, 1985; Schmid and Casey, 1986). Of the 12 samples from the Schist belt, nine yielded lattice preferred orientations (LPOs) indicative of top-to-the-north shear, two (Fig. 15i and m) yielded LPOs indicative of top-to-the-south shear, and one yielded an LPO with no marked sense of asymmetry (Fig. 15g). Weak LPOs indicative of top-to-the-south shear were also found in samples from the Phyllite belt (Fig. 15n), and the northernmost Ruby geanticline (Fig. 15r). Samples from the southern Skajit allochthon (Fig. 15a–d) yield essentially random quartz c-axis distributions. Petrofabric analysis also indicates that D_{1b} fold axes and lineations correspond to the pole-free area of c-axis diagrams, indicating that D_{1b} fold axes and lineations in the Schist belt lie approximately parallel to the *X*-axis of finite strain ($X \geq Y \geq Z$; Tullis et al., 1973; Tullis, 1977; Mainprice et al., 1986).

Shear-sense indicators and quartz c-axis petrofabrics indicate that D_{1b} folds and fabrics in the Schist belt formed in a strain regime dominated by top-to-the-north simple shear. The large-magnitude shear strains indicated by the predominance of close to isoclinal folds, the concentration of D_{1b} fold axes and L_{1b} lineations in a north-south orientation, and the parallelism of D_{1b} fold axes to the *X*-axis of finite strain indicates that fold axes have been passively rotated into an orientation that is approximately parallel to the tectonic transport direction (Gottschalk, 1990). Discrete faults or shear zones associated with D_{1b} folds are rare, indicating that deformation was accommodated mainly by distributed shear. In the Metagreywacke and Phyllite belts, D_{1b} deformation is markedly weaker than in the Schist belt, and kine-

matic indicators are rare. However, the similarity in orientation and geometry of D_{1b} structures in all belts, including the concentration of D_{1b} fold axes and L_{1b} intersection lineations in a north-south orientation, suggests that D_{1b} structures and fabrics in the Metagreywacke and Phyllite belts also may have originated in a strain-regime dominated by top-to-the-north shear. Because top-to-the-north shear is consistent with the northerly vergence of the Brooks Range fold and thrust belt, D_{1b} structures and fabrics have been interpreted as the result of contractional tectonic processes (Gottschalk, 1990).

The age of D_{1b} structures and fabrics in the Schist belt of the south-central Brooks Range is constrained by a number of $^{40}Ar/^{39}Ar$ plateau and isochron dates on white mica and amphibole that range from 135 to 110 Ma (Blythe et al., this volume, Chapter 10; Gottschalk and Snee, this volume, Chapter 13). Gottschalk and Snee (this volume, Chapter 13) interpret Early Cretaceous dates as a direct record of the age of D_{1b} deformation; they argue that D_{1b} deformation resulted in the degassing of syn-D_{1a} white micas at temperatures below their closure temperature, thereby obliterating evidence of Middle Jurassic *HP/LT* metamorphism and resulting in a preponderance of Early Cretaceous plateau dates. Elsewhere in the Schist belt, *HP/LT* metamorphic fabrics are deformed by contractional folds that formed under amphibolite-facies conditions between 110 to 105 Ma (Till and Patrick, 1991); similarly, Christiansen and Snee (1994) have shown that *HP/LT* metamorphism is postdated by a penetrative deformational event that occurred under greenschist-facies conditions between 130 to 125 Ma. If the events described by Till and Patrick (1991) and Christiansen and Snee (1994) correlate to our D_{1b}, then the Schist belt underwent regional contractional deformation under greenschist- to amphibolite-facies metamorphic conditions in the Early Cretaceous to earliest Late Cretaceous.

The age of D_{1b} in the Phyllite belt and Metagreywacke belt is poorly constrained. From the Phyllite belt, Blythe et al. (this volume, Chapter 10) report a white mica plateau date of 113 Ma, and Gottschalk and Snee (this volume, Chapter 13) report a white mica isochron date of 124 Ma. The exact significance of these dates is not known, but Gottschalk and Snee (this volume, Chapter 13) suggest that they may be similar in origin to white mica dates in the Schist belt and record the age of D_{1b} in the Phyllite belt. The upper limit on the age of ductile structures in the Metagreywacke belt and Phyllite belt is constrained by the presence of metagreywacke and phyllite clasts in Aptian, Albian, and Cenomanian (?) sediments of the Yukon-Koyukuk basin (Dillon and Smiley, 1984; Murphy et al., 1989).

D_{1c}

Our structural data indicate that D_{1c} folds and possibly the Wiseman-Emma Dome arch formed as a result of post-D_{1b} thrust faulting. The proximity of north-vergent D_{1c} folds to the Minnie Creek thrust and their relationship to small-scale thrust faults with north-directed displacement indicates that D_{1c} structures are con-

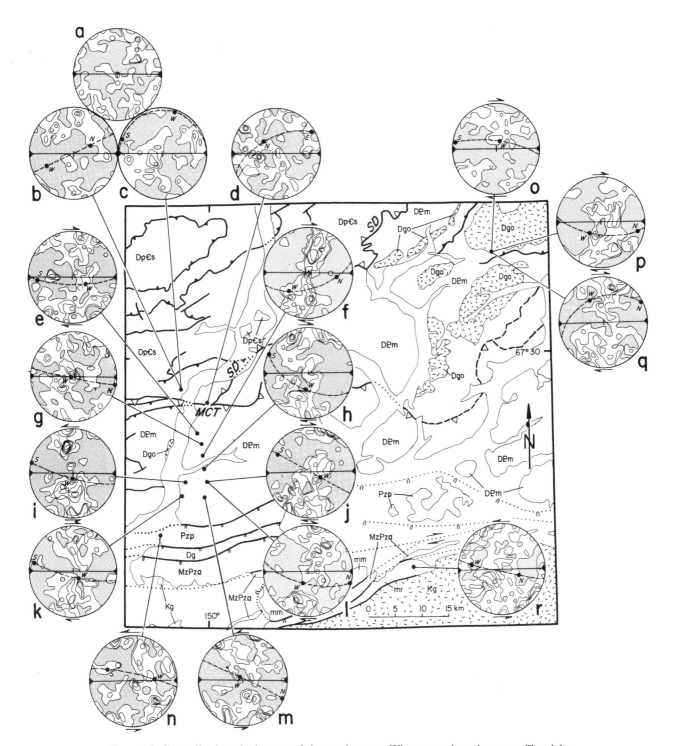

Figure 15. Generalized geologic map of the southeastern Wiseman and southwestern Chandalar Quadrangles with quartz c-axis petrofabric data and sample locations. Geologic units and abbreviations same as Figure 1b. Data are shown on lower hemisphere, equal-area projections. Contour interval: 1% per 1% area. Grey regions, less than 1% pcr 1% area. Solid line = foliation; triangle = lineation; filled circles = geographic coordinates, S = south, N = north, W = west; dashed line = horizontal. Arrows indicate sense of shear. Plots e–m from Gottschalk (1990).

tractional in origin, and formed as the Schist belt was displaced northward along the Minnie Creek thrust; moreover, the shallow south dips of F_{1c} axial planes imply that, along the Middle Fork Koyukuk River, the Minnie Creek thrust is subhorizontal to shallowly south dipping. These relations suggest that the Wiseman-Emma Dome arch may be hanging-wall anticline that formed in response to displacement of the Schist belt over a flattening bend in the Minnie Creek thrust; note, however, that the geometry of the arch may have been modified during extensional deformation (Gottschalk, 1990). The age of offset on the Minnie Creek thrust is not known, but postdates Early Cretaceous D_{1b} deformation and metamorphism; the upper age limit of D_{1c} deformation is constrained by apatite fission-track data that indicates the Schist belt cooled through approximately 120°C between about 65 to 55 Ma (Blythe et al., this volume, Chapter 10).

D_{1d} (extensional deformation)

The widespread occurrence of brittle and ductile extensional structures that cross-cut contractional folds and fabrics indicates that the south-central Brooks Range was affected by a period of extensional deformation that locally postdated contraction. Analysis of fault populations in the Angayucham terrane indicates that extension was north-south directed (Avé Lallemant et al., this volume, Chapter 15), and Gottschalk (1987, 1990), Little et al. (1994), and Law et al. (1994) have argued that east-west–trending ductile shear zones in the Schist belt, Phyllite belt, and southern Skajit allochthon formed in response to north-south–directed extension. Because of the similarity in their implied orientation of extension, we believe that brittle and ductile extensional structures are manifestations of a single deformational event, with ductile structures forming at structural deeper levels within the extending orogen than brittle structures.

We consider mesoscopic extensional structures to be subsidiary to the major east-west faults that bound lithologic belts in the south-central Brooks Range. A post-peak-metamorphic extensional history for the Angayucham fault, Rosie Creek fault, and Slate Creek fault has been implied by their juxtaposition of rocks with significantly different metamorphic grades, and their omission of more that 8 km of section between the Angayucham terrane and the Schist belt (Gottschalk and Oldow, 1988). The extensional nature of the Slate Creek fault has recently been shown by Little et al. (1994), who studied its along-strike equivalent (their Florence Creek fault) and conclude that it is a ductile shear zone with down-to-the-south displacement. It is not known whether the Angayucham, Rosie Creek, and Slate Creek faults have experienced a purely extensional history, or whether they are thrust faults that were reactivated during extension.

The origin of metamorphic fabrics in the southern Brooks Range has been much debated in the recent literature. The principal question is whether metamorphic fabrics are contractional in origin and formed at deep structural levels within the Brooks Range fold and thrust belt, or whether they resulted from postcontractional ductile extension (cf. Gottschalk, 1990; Little et al.,

1994; Law et al., 1994). Central to this issue is the difficulty in distinguishing metamorphic fabrics formed in contractional versus extensional environments when strain is accommodated mainly by simple shear. Little et al. (1994) and Law et al. (1994) report an abundance of shear-sense indicators in the Schist belt that show top-to-the-south shear; because the sense of shear is opposite to that to the north-vergent Brooks Range fold and thrust belt, and because down-to-the-south extensional displacement is inferred for major east-west–trending faults in the south-central Brooks Range, Little et al. (1994) have argued that the D_{1b} fabrics are extensional rather than contractional in origin. We suggest, however, that the top-to-the-south shear that we, Little et al. (1994), and Law et al. (1994), report in the Schist and Phyllite belts postdates D_{1b}, and originated during D_{1d} ductile extensional deformation. Extensional shear zones with down-to-the-south offset are numerous and occur throughout south-dipping portions of the Schist belt and Phyllite belt. Although the amount of offset on individual shear zones is limited (Gottschalk, 1990), shear zones cumulatively accommodate significant top-to-the-south shear strain. Extensional shear zones terminate into S_{1b}, indicating that shear strain was transferred to older D_{1b} surfaces. Therefore top-to-the-south shear resulting from D_{1d} extension may have been accommodated by shear along post-D_{1b} extensional shear zones, and by reactivation of S_{1b}. We argue that sufficient strain may have been accommodated along S_{1b} to produce microstructures and quartz c-axis petrofabrics indicative of top-to-the-south shear deformation.

The age of extensional deformation in the south-central Brooks Range is not precisely known. Ductile extensional shear zones postdate Early Cretaceous D_{1b} structures, and formed prior to cooling of the Schist belt below 120°C (dated by apatite fission-track method at 65 to 55 Ma). These relations indicate that extension in the south-central Brooks Range is similar to that in the western Brooks Range where white micas from a ductile extensional shear zone yield dates of 103 to 95 Ma (Christiansen and Snee, 1994), and rocks as young as Upper Cretaceous are affected by extensional structures (Carlson, 1985; Box, 1987).

D_2

South-vergent folding is regional in extent and has affected rocks throughout the southern Brooks Range south of the Doonerak window (Avé Lallemant and Oldow, this volume, Chapter 14). The association of D_2 folds with thrust faults and the geometry of D_2 conjugate kink folds indicates that D_2 structures formed during an event of late-stage, north-south-directed contraction. The relationship of D_2 folds to D_{1c} contractional folds or D_{1d} extensional structures could not be demonstrated from cross-cutting or superposition relationships; therefore it is possible that some south-vergent folds assigned to D_2 formed contemporaneously with north-vergent D_{1c} folds and are related to emplacement of the Schist belt along the Minnie Creek thrust (as argued by Gottschalk, 1987). Because D_2 folds fold D_{1b} fabrics, they are post-latest Early Cretaceous in age.

D_3

The Angayucham terrane, Metagreywacke belt, Phyllite belt, and Schist belt are all affected by a series of gentle folds and broad-scale flexures whose geometries are indicative of east-west–directed shortening. North-south–trending folds occur throughout northern and interior Alaska, and affect rocks as young as Upper Cretaceous in age (Patton and Tailleur, 1977; Grantz et al., 1981; Karl and Long, 1990). The age of east-west contraction in northern Alaska is not known, but may be as young as early Tertiary (Grantz et al., 1981).

TECTONIC MODEL

The south-central Brooks Range fold and thrust belt is composed of Precambrian to Middle Paleozoic rocks that were structurally juxtaposed during, and possibly prior to, Jurassic and younger contraction. In the following section we present a model for the Mesozoic and younger tectonic history of the south-central Brooks Range. This model is based on regional geologic relations in the southern Brooks Range, and is constrained by our structural analysis and by recent thermochronologic studies.

Pre-Jurassic paleogeography

There has been considerable debate over the relative predeformational positions of allochthonous rocks in the southern Brooks Range; although it is generally agreed that the Angayucham terrane and Metagreywacke belt originated south of the Schist belt (in present-day coordinates), the relative position of the Schist belt and Skajit allochthon (or Central Belt) is disputed (e.g., compare Oldow et al., 1987; and Moore et al., 1994). The controversy revolves around interpretation of regional structural relations between the Schist belt and Skajit allochthon and their implications for palinspastic restoration. Several published cross sections indicate that the Skajit allochthon lies structurally below the Schist belt, separated by a south-dipping thrust fault (e.g., Mull et al., 1987; Fig. 16a); this structural relationship implies that the Schist belt originated from a position south of the Skajit allochthon. Other interpretations (e.g., Oldow et al., 1987; Moore et al., 1994) depict several key relationships that imply the Skajit allochthon may restore, in part, to a position *south* of the Schist belt (Fig. 16b): (1) The Skajit allochthon is internally dismembered, lies structurally *above* the Schist belt, and is separated from the Schist belt by a folded decollement surface (the Skajit decollement, SD, Fig. 1b). (2) Detailed structural analysis by Oldow et al. (this volume, Chapter 8), indicates that thrust sheets within the Skajit allochthon were thrust northward, implying an overall northerly tectonic transport for the Skajit allochthon. (3) The Skajit allochthon is terminated on the south by the Minnie Creek thrust, a relatively late feature that truncates thrust faults in the Skajit allochthon as well as truncating the Skajit decollement. (4) The Schist belt is depicted as internally imbricated and folded in duplex fashion beneath the Skajit decolle-

ment; duplexing apparently resulted in folding of the Skajit decollement and overlying imbricates. These relations suggest a structural history in which the Skajit allochthon was thrust northward over the Schist belt, followed by duplexing of the Schist belt beneath the Skajit allochthon; the Skajit allochthon and the Schist belt duplex were then truncated by the "out-of-sequence" Minnie Creek thrust. Because we believe that the cross sections of Oldow et al. (1987) and Moore et al. (1994) are most consistent with structural relations in the south-central Brooks Range, we interpret the Skajit allochthon to have resided, at least in part, on the southernmost portion of the Alaskan margin, north of the Angayucham ocean basin. In our model, the Skajit assemblage and its basement were separated from the lower Paleozoic rocks of the Doonerak window by a basin in which the protolith for the Schist belt was deposited (Fig. 17a). Because the southern Brooks Range has been affected by large-magnitude normal and strike-slip faults, the original thrust architecture of the root-zone of Skajit allochthon has been severely modified. Therefore it is not clear whether the rocks of the Phyllite belt and Metagreywacke belt have affinities to the Schist belt, or whether they represent the trailing edge of the Skajit allochthon. We prefer the interpretation that the Metagreywacke belt, Phyllite belt, and Schist belt originated as part of the same basinal assemblage because the three belts share comparable structural histories, and because the Metagreywacke belt, Phyllite belt, and Schist belt were derived from protoliths of similar age and lithology.

Structural relations and stratigraphic relations, along with the predominance of Devonian fossil ages from rocks of the south-central Brooks Range (Hitzman et al., 1986; Gottschalk, 1987; Murphy and Patton, 1988; Dillon et al., 1987; Boler, 1989; Oldow et al., this volume, Chapter 7), suggest that the sedimentary basins on the southern margin of the Alaska formed during a Devonian–Early Mississippian rifting event, as proposed by (Hitzman et al., 1986). Middle Paleozoic extension ultimately led to the formation of a south-facing passive continental margin, and the formation of the Angayucham ocean basin (Mayfield et al., 1988; Hitzman et al., 1986). The Angayucham ocean was the site of extensive seamount/ocean plateau basaltic volcanism from the Mississippian (?) into the early Mesozoic (Barker et al., 1988; Box and Patton, 1989; Patton and Box, 1989; Pallister et al., 1989).

Initiation of island-arc activity and subduction (Early Jurassic)

A common element to many recent tectonic models (e.g., Box, 1985; Box and Patton, 1989; Moore et al., 1994) is the assertion that *HP/LT* metamorphism in the Schist belt occurred in response to collision of the Yukon-Koyukuk island arc with the Arctic Alaskan continental margin in the Late Jurassic to Early Cretaceous. Recent geochronologic studies (Armstrong et al., 1986; Till and Patrick, 1991; Christiansen and Snee, 1994; Gottschalk and Snee, this volume, Chapter 13) have shown, however, that *HP/LT* metamorphism is significantly older than much of the magmatism in the Yukon-Koyukuk arc, and may even predate the oldest dated plutons in the Yukon-Koyukuk basin (com-

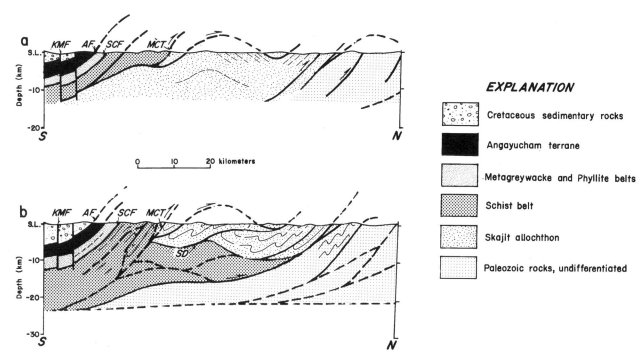

Figure 16. Cross sections showing alternative structural interpretations in the south-central Brooks Range. Location of section line shown in Figure 1a. a, from Mull et al. (1987); b, from Moore et al. (1994). Geologic units and abbreviations same as Figure 1b.

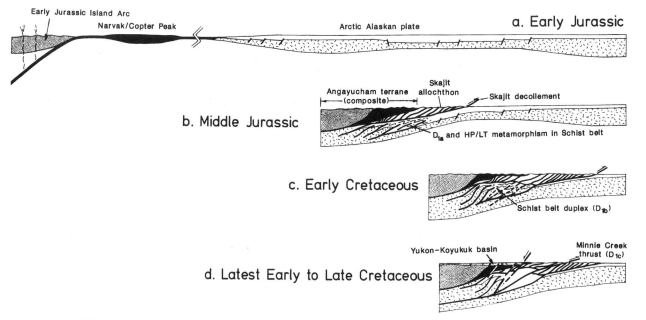

Figure 17. Tectonic model for the Jurassic to Early Cretaceous evolution of the south-central Brooks Range. See text for explanation.

pare >171 Ma for *HP/LT* metamorphism in the western Brooks Range, and 173 Ma as the age of the oldest dated pluton in the Yukon-Koyukuk basin). It is therefore an open question as to what actually did collide with the Arctic Alaskan continental margin to produce Middle Jurassic–aged *HP/LT* metamorphic assemblages in the Schist belt. A possible solution to this problem is provided by rocks of island-arc affinity exposed in the Brooks Range ophiolite of the western Brooks Range that range in age from 187 to 169 Ma (Harris, 1989; Wirth et al., 1993; Moore et al., 1993). These relations indicate that an island arc may have existed south of the Arctic Alaskan continental margin (in present-day coordinates) as far back as the Early Jurassic (cf. Harris, this volume, Chapter 17), and, as speculated by Miller and Hudson (1991), this Early Jurassic island arc may be the object that collided with Arctic Alaskan continental margin.

The polarity of subduction beneath the Early Jurassic arc is not known. It is probable that at least some degree of south-directed subduction accommodated convergence between the Early Jurassic arc and the Arctic Alaskan continental margin (cf. Harris, this volume, Chapter 17). It is not clear, however, whether south-directed subduction was the principal cause of Jurassic arc magmatism (as proposed by Box, 1985), or whether magmatism was related mainly to north-directed subduction (in present-day coordinates) along the southern margin of the Early Jurassic island arc (e.g., Oldow et al., 1989; Miller and Hudson, 1991). In the models of Oldow et al. (1989) and Miller and Hudson (1991), the island arc is part of a regionally extensive fringing arc system that is outboard of North America in the Early Jurassic, and includes the rocks of the Talkeetna/Bonanza arc volcanics exposed in southern Alaska.

Although the relative positions of the Early Jurassic arc system and the Arctic Alaskan continental margin are speculative, the limited time between ophiolite genesis (187 to 169 Ma) and emplacement above the Copter Peak/Narvak assemblage (171–163 Ma; Harris, this volume, Chapter 17) suggests that the Early Jurassic island arc formed in close proximity the Copter Peak/Narvak ocean-plateau/seamount assemblage.

Early Brookian contraction (Middle to Late Jurassic)

Although the timing of events in the Jurassic is poorly resolved, several events occurred in the Middle to Late Jurassic that comprise the Early Brookian phase of contraction in the south-central Brooks Range. (1) In the Middle Jurassic, the rocks of the Early Jurassic island arc (exposed in the Brooks Range ophiolite) were tectonically emplaced above the ocean plateau/ seamount assemblage of the Copter Peak allochthon/Narvak panel of the Angayucham terrane. Although the rocks of the Brooks Range ophiolite are not present in the south-central Brooks Range, they are regionally extensive (cf. Patton et al., 1994), and may have been present prior to uplift and erosion of the Angayucham terrane. (2) *HP/LT* metamorphism in the Schist belt occurred prior to 142 Ma in the south-central Brooks Range (Gottschalk and Snee, this volume, Chapter 13), and may be 171 Ma or older if

HP/LT metamorphism in the central Brooks Range was contemporaneous with *HP/LT* metamorphism in the western Brooks Range (Christiansen and Snee, 1994). In the south-central Brooks Range, the age of collision is constrained to Middle Jurassic or younger by the occurrence Lower Jurassic–aged radiolarian cherts, indicating that open marine conditions existed in the Angayucham ocean as late as the Early Jurassic.

The age relations outlined above indicate that Early Brookian contraction consisted of (1) amalgamation of the Early Jurassic island arc and the Devonian to Early Jurassic ocean plateau/seamount assemblages to form the composite Angayucham terrane, and (2) the collision of the Angayucham terrane with the thinned southern margin of the Arctic Alaskan continental block in the Middle to Late Jurassic (Fig. 17b). Contraction-related D_{1a} structures in the Angayucham terrane of the Cathedral Mountain area presumably formed when the oceanic rocks were obducted northward over continental margin sedimentary rocks. Following the emplacement of the Angayucham terrane, the locus of decollement transferred from the base of the obducted oceanic allochthon to a detachment level within the continental margin sedimentary wedge (Fig. 17b). If our reconstruction of the Arctic Alaskan continental margin is correct, deformation following the obduction of the Angayucham terrane involved Precambrian to Middle Paleozoic and younger basinal and platformal rocks now exposed in the the Skajit allochthon, and coeval basinal rocks now exposed in the Schist belt. Basinal rocks of the Schist belt were deformed at deep structural levels within the contractional orogen, resulting in the formation of D_{1a} folds and fabrics and *HP/LT* metamorphic assemblages.

Geologic relations in the southern Brooks Range indicate that there may be a significant time lag between collision-related *HP/LT* metamorphism in the Schist belt and the main phase of thrusting in the Brooks Range fold and thrust belt (see also Till, 1992). The onset of Brookian thrusting is indicated by the deposition of the Upper Jurassic–Neocomian deep-marine clastic turbidites of the Okpikruak Formation (Mayfield et al., 1988), the oldest sedimentary unit containing detritus derived from the nascent fold and thrust belt. If the >171 Ma age of *HP/LT* metamorphism and associated deformation proposed by Christiansen and Snee (1994) is correct, the initiation of contractional deformation in the central Brooks Range may have preceded the main phase of thrusting and foreland basin formation by as much as 40 m.y. This time lag corresponds to a period (Middle to Late Jurassic) when little apparent structural activity occurred in the south-central Brooks Range. The cause for this period of relative quiescence is not known, but may have occurred because relative convergence between the Early Jurassic island-arc and the Arctic Alaskan plate slowed significantly or ceased. It is also noteworthy that initial arc-continent collision occurred along the southern flank of the Arctic Alaskan block, which lay 500 km or more south of the North Slope prior to telescoping of the continental margin sequence (Mayfield et al., 1988; Oldow et al., 1987). If convergence between arc and continent *was* continuous through the Middle Jurassic and Late Jurassic, the early phases of contractional deformation may have involved mainly thinned, relatively weak continental crust; thus the imbricate stack may not have prograded onto crust with sufficient flexural rigidity to generate a significant foreland basin until the latest Jurassic time.

Yukon-Koyukuk arc (Middle Jurassic to Early Cretaceous)

Island-arc activity in the Yukon-Koyukuk basin began in the Middle Jurassic and continued intermittently through the Early Cretaceous. The relationship of these rocks to the Early Jurassic island-arc assemblage is not clear. However, if the ultramafic rocks that underlie the Yukon-Koyukuk basin correlate with the ultramafic rocks of island-arc affinity exposed in the Brooks Range ophiolite, the Yukon-Koyukuk arc may be built upon a substrate that includes the Early Jurassic arc complex. These relations suggest that Yukon-Koyukuk arc magmatism may have occurred in response to north-directed subduction (in present-day coordinates) beneath the accreted Early Jurassic arc as proposed by Miller and Hudson (1991).

Main Brookian contractional phase (Early to early Late? Cretaceous)

In the Early Cretaceous the south-central Brooks Range was affected by a second and locally a third penetrative deformational event. These structures formed during the main phase of Brookian orogenesis (Late Jurassic to latest Early Cretaceous), during which much of the shortening in the Brooks Range fold and thrust belt occurred (Hubbard et al., 1987; Mayfield et al., 1988; Moore et al., 1994). The *HP/LT* metamorphics of the Schist belt in the central Brooks Range were penetratively deformed under greenschist-facies conditions between 135 and 110 Ma, possibly as a result of internal dismemberment along ductile shear zones. Although we did not identify syn-D_{1b} shear zones in areas of the Schist belt that we studied in detail, tectonic contacts between lithologic units of the Schist belt are described in Moore et al. (1994), and may be Early Cretaceous–aged structures. Oldow et al. (1987) and Moore et al. (1994) infer a duplex structure for the Schist belt, implying that shear zones in the Schist belt lift off the basal detachment of the imbricate stack, and terminate into the overlying Skajit decollement (Fig. 17c).

Note that the shortening accommodated by ductile duplexing in the Schist belt was transferred at the northern terminus of the duplex to the Skajit decollement; resultant imbrication in the Skajit allochthon therefore may be partially contemporaneous with D_{1b} folding in the Schist belt. Note that imbrication of epicontinental sedimentary rocks requires the "disposal" of significant volumes of continental crust. "Disposal" of this crust may have been accommodated by crustal-scale imbrication and production of a root zone in front of the Yukon-Koyukuk arc, and partial subduction of thinned continental crust southward beneath the arc.

The youngest structures that formed during the main phase of Brookian contraction are the Minnie Creek thrust and asso-

ciated D_{1c} folds (possibly including the Wiseman-Emma Dome arch; Fig. 17d). The Minnie Creek thrust may be similar in origin to "out-of-sequence" thrusts in other contractional orogens, which commonly develop as a means of maintaining critical taper of the prograding imbricate stack (Dahlen et al., 1983). Geologic relations indicate that offset on the Minnie Creek thrust occurred in the latest Early to Late Cretaceous, and may be partially contemporaneous with extensional deformation along the southern flank of the Schist belt (Gottschalk and Oldow, 1988).

Extensional phase (latest Early to Late Cretaceous)

Although much of the penetrative deformation of rocks in southern Brooks Range formed as a result of north-vergent contractional deformation, the present-day distribution of allochthons is largely the result of Late Cretaceous extension. Figure 18a schematically shows the pre-extensional configuration of the southern Brooks Range and the locus of future normal faults. Down-to-the-south offset along these faults juxtaposed rocks deformed and metamorphosed at widely differing structural levels within the contractional orogen (Fig. 18b), causing the formation of brittle extensional structures in the Angayucham terrane and Metagreywacke belt, and ductile extensional structures in the Phyllite belt and Schist belt. Gottschalk and Oldow (1988) and Miller and Hudson (1991) have proposed that mid-Cretaceous subsidence and sedimentation in the northern Yukon-Koyukuk basin may have been extensionally controlled.

Although evidence for Cretaceous extensional deformation is widespread in the southern Brooks Range and elsewhere in central Alaska (Carlson, 1985; Box, 1987; Gottschalk and Oldow, 1988; Gottschalk, 1990; Miller and Hudson, 1991; Pavlis et al., 1993; Roeske et al., 1991; Little et al., 1994; Law et al., 1994; Avé Lallemant et al., this volume, Chapter 15), the driving force for extension is poorly understood and has yet to be resolved. Models for the origin and tectonic setting of extension fall into two broad categories, (1) those in which extensional in central Alaska is simultaneous with thrusting in the foreland of the Brooks Range fold and thrust belt, and (2) those in which extension interrupts contractional deformation. Following the model of Platt (1986), Gottschalk and Oldow (1988) proposed that extension in the southern Brooks Range was driven by thrust-related crustal thickening, with normal faulting occurring at high structural levels within the imbricate stack in order to maintain critical taper. More recently, Pavlis (1989) and Pavlis et al. (1993) have offered several scenarios based upon a Carpathian/Hellenic analog in which extension in central Alaska is of the "back-arc" type, accommodated by contraction at a "free-surface" of oceanic or thinned continental crust; as proposed by Pavlis (1989) and Pavlis et al. (1993), extension affects the entire thickness of the crust, and contraction in the foreland of the Brooks Range is driven by extensional collapse in the hinterland (cf. Dewey, 1988). A second category of tectonic model has been proposed by Miller and Hudson (1991), who argue that mid-Cretaceous extension affects all of northern Alaska, effec-

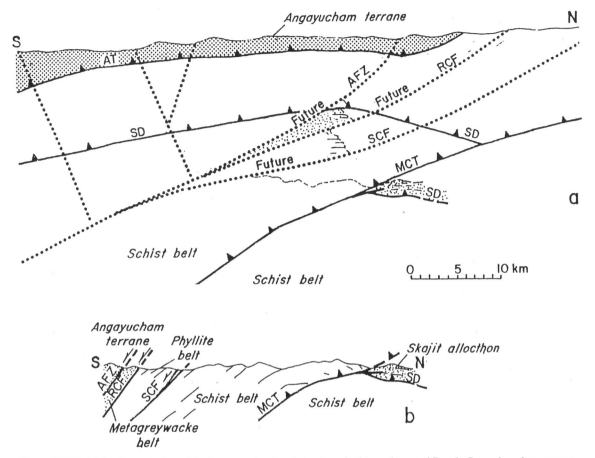

Figure 18. Model for the evolution of Cretaceous extensional structures in the south-central Brooks Range based on geometric relations along the Middle Fork Koyukuk River. a, Schematic pre-extensional configuration. Dashed lines represent approximate location of present-day erosional surface. Heavy lines with barbs are thrust faults formed during contractional phase. Dotted lines show the locus of future extensional faults and shear zones. b, Present-day configuration. AFZ = Angayucham fault zone; RCF = Rosie Creek fault; SCF = Slate Creek fault; MCT = Minnie Creek thrust; SD = Skajit decollement; AT = Angayucham thrust.

tively interrupting contractional processes in the Brooks Range and the North Slope. A comparison of these models is presented in Till et al. (1993).

Strike-slip faulting (post-Albian)

Detailed analysis of mesoscopic faults in the Angayucham terrane and Metagreywacke belt indicates that dextral strike-slip offset on the Kobuk-Malemute fault postdates extension in the south-central Brooks Range (Avé Lallemant et al., this volume, Chapter 15). Offset of the Late Cretaceous Hodzana pluton (Brosgé and Reiser, 1964; Blum et al., 1987) and distinctive staurolite-bearing schists indicates approximately 80 km offset along the Malemute fault (Avé Lallemant et al., this volume, Chapter 15). Strike-slip faults in northern Alaska may have formed in response to differential movement between the Siberian and Arctic Alaskan portions of the North American plate in the Late Cretaceous (Patton and Tailleur, 1977; Harbert et al., 1987),

or in response to oblique convergence along the Alaskan segment of the North American continental margin (cf. Oldow et al., 1989).

Late Brookian contraction (Paleocene to late Oligocene)

Avé Lallemant and Oldow (this volume, Chapter 14) have proposed that regionally developed south-vergent D_2 folds result from antithetic shear associated with the formation of the Doonerak window duplex; they suggest that the duplex may have acted as an impediment to slip along the basal decollement of the fold and thrust belt, resulting in shortening south of the Doonerak window by the formation of south-vergent folds. Apatite fission-track studies by O'Sullivan and Moore (this volume, Chapter 11) and Blythe et al. (this volume, Chapter 10) indicate that the Doonerak window is a late orogenic feature that formed between 65–24 Ma. Therefore, Avé Lallemant and Oldow (this volume, Chapter 14) argue that D_2 folds in the southern Brooks Range formed in the Paleocene to latest Oligocene.

East-west contraction

North-south–trending contractional folds (D_3) in the south-central Brooks Range appear to be part of a regionally developed set of folds that occur throughout northern Alaska and possibly in the Yukon-Koyukuk basin (Patton, 1973; Grantz et al., 1981; Karl and Long, 1987, 1990). Patton and Tailleur (1977) attributed east-west contraction and activity on strike-slip faults (including the Kobuk-Malemute fault) to oroclinal bending of the Brooks Range in the Late Cretaceous and early Tertiary. Grantz et al. (1981) and Karl and Long (1987) have argued that east-west contraction is a later event that has been superimposed on the Brooks Range fold and thrust belt. Although we cannot distinguish between these two models, it is unlikely that D_3 folds in the south-central Brooks Range are kinematically related to strike-slip faults. Experimental and field studies on strike-slip structures indicate that contractional folds form at angles of 45° or less to the principal displacement zone (e.g., Wilcox et al., 1973; Sylvester, 1988). Thus if D_3 structures were related to the east-west–striking, right-lateral Kobuk-Malemute fault, they should be oriented northeast–southwest to east-northeast–west-southwest rather than the observed north-south orientation.

SUMMARY AND CONCLUSIONS

Regional geologic relations in the south-central Brooks Range attest to large magnitude shortening, resulting in the juxtaposition of ocean plateau–seamount rocks (the Angayucham terrane) with sedimentary rocks of the Alaskan continental margin (the Schist belt, Phyllite belt, and Metagreywacke belt). Mesozoic contractional deformation occurred in two major pulses. (1) Early Brookian deformation (D_{1a}) occurred in the Middle to Late Jurassic with the collision of an island arc with the Arctic Alaskan continental margin. Collision resulted in imbrication of the Angayucham terrane as it was obducted onto the continental margin in advance of the converging island arc, and caused the penetrative deformation of Precambrian and younger continental margin rocks, locally under lawsonite- to epidote-blueschist–facies metamorphic conditions; *HP/LT* metamorphic rocks are presently exposed in the Schist belt of the south-central Brooks Range. Deformation of the Metagreywacke belt at low metamorphic grades and isoclinal folding of the Phyllite belt under pumpellyite-actinolite metamorphic conditions may also have occurred at this time. (2) In the Early Cretaceous, the south-central Brooks Range was affected by the main phase of Brookian deformation (D_{1b}). The Schist belt was penetratively deformed and pervasively folded under lower amphibolite- to greenschist-facies metamorphic conditions in response to renewed contractional deformation of the *HP/LT* metamorphic welt. D_{1b} also resulted in the refolding and local penetrative deformation in the Metagreywacke and Phyllite belts, and possibly caused additional imbrication of the Angayucham terrane.

D_{1b} structures and fabrics are cross-cut by the Minnie Creek thrust, a south-dipping thrust fault whose structural relations indicate that it is an "out-of-sequence" fault of latest Early to Late Cretaceous age. North-directed displacement of the Schist belt along the Minnie Creek thrust may have resulted in the formation of the Wiseman-Emma Dome arch and subsidiary post-D_{1b} contractional folds and thrust faults.

The distribution of allochthons in the south-central Brooks Range was strongly modified by north-south directed extension in the latest Early to Late Cretaceous. Although extension largely postdated contraction in the southern Brooks Range, it was broadly contemporaneous with thrust faulting in the frontal portion of the fold and thrust belt. Extension resulted in the juxtaposition across normal faults or extensional shear zones of rocks deformed and metamorphosed at widely differing structural levels within the contractional orogen. Rocks between the bounding structures were internally deformed during extension, with the Schist belt and Phyllite belt cross-cut by numerous ductile shear zones, and the Angayucham terrane and Metagreywacke belt dissected by numerous small-scale brittle normal faults.

Extension in the south-central Brooks Range was followed by (1) the formation of south-vergent contractional folds, (2) faulting related to right-lateral offset on the Kobuk-Malemute fault, and (3) an event of east-west contraction that resulted in gentle north-south–trending flexures. The absolute and relative ages of these postextensional deformational events are poorly constrained, but regional geologic relations indicates that they are probably Late Cretaceous to Tertiary in age. South-vergent folds may have been caused by late Brookian back-thrusting in the region south of the Doonerak window, and north-south–trending flexures may be related to east-west–directed contraction in the Lisburne Hills of the westernmost Brooks Range. Activity on the Kobuk-Malemute fault may have resulted from differential movement between the Siberian and Arctic Alaskan portions of the North American plate in the Late Cretaceous, or oblique convergence along the Alaskan segment of the North American continental margin.

ACKNOWLEDGMENTS

This study was supported by grants to J. S. Oldow, H. G. Avé Lallemant, and A. W. Bally from the Department of Energy (DE-ASO5-83ER13124), the National Science Foundation (EAR-8517384), and the Alaska Industrial Associates Program (AIAP) at Rice University and the University of Alaska, Fairbanks. The AIAP received contributions from the following companies: Arco, Amoco, Chevron, Standard Oil of Ohio, Mobil, and Gulf. We thank Satya Gargi and Richard Loftin for their assistance with quartz c-axis petrofabric analysis. This work benefited greatly from many stimulating discussions with our colleagues including John Anderson, Anne Blythe, Kent Boler, Steve Box, Dick Casey, Peter Coney, John Dillon, Bill Dinklage, Jim Handschy, Ron Harris, Davey Jones, Betsy Julian, Rick Law, Bill Leeman, Elizabeth Miller, Tom Moore, John Murphy, Brian Patrick, Terry Pavlis, Jim Phelps, Mike Seidensticker, Alison Till, and Karl Wirth. Able and invaluable assistance in the field was provided by John Giltner, Chin-See Ming, and Richard Torres. We are grateful to Sue Karl and Brian Patrick who provided thoughtful and thorough reviews of this manuscript. This paper is dedicated to the memory of John Giltner, whose efforts and cheerful nature in face of adversity helped make this work possible. Thanks, John.

REFERENCES CITED

Armstrong, R. L., Harakal, J. E., Forbes, R. B., Evans, B. W., and Thurston, S. P., 1986, Rb-Sr and K-Ar study of metamorphic rocks of the Seward Peninsula and southern Brooks Range, Alaska, *in* Evans, B. W., and Brown, E. H., eds., Eclogites and blueschists: Geological Society of America Memoir 164, p. 185–203.

Arth, J. G., Criss, R. E., Zmuda, C. C., Foley, N. K., Patton, W. W., Jr., and Miller, T. P., 1989, Remarkable isotopic and trace element trends through sodic Cretaceous plutons of the Yukon-Koyukuk basin, Alaska, and the nature of the lithosphere beneath the Yukon-Koyukuk terrane: Journal of Geophysical Research, v. 94, p. 15957–15968.

Bally, A. W., Gordy, P. L., and Stewart, G. A., 1966, Structure, seismic data, and orogenic evolution of the southern Canadian Rockies: Canadian Society of Petroleum Geologists Bulletin, v. 14, p. 337–381.

Barker, F., Jones, D. L., Budahn, J. R., and Coney, P. J., 1988, Ocean plateau-seamount origin of basaltic rocks, Angayucham terrane, central Alaska: Journal of Geology, v. 96, p. 368–374.

Bird, K. J., 1977, Late Paleozoic carbonates from the south-central Brooks Range: U.S. Geological Survey Circular 751-B, p. B19–B20.

Blum, J. D., Blum, A. E., Davis, T. E., and Dillon, J. T., 1987, Petrology of cogenetic silica-saturated and oversaturated plutonic rocks in the Ruby geanticline of north-central Alaska: Canadian Journal of Earth Science, v. 24, p. 159–169.

Boler, K. B., 1989, Stratigraphy, structure, and tectonics of the central Brooks Range near Dietrich Camp, Alaska [M.A. thesis]: Houston, Texas, Rice University, 163 p.

Box, S. E., 1985, Early Cretaceous orogenic belt in northwestern Alaska: internal organization, lateral extent, and tectonic interpretation, *in* Howell, D. G., ed., Tectonostratigraphic terranes of the Circum-Pacific region: Houston, Texas, Circum-Pacific Council for Energy and Mineral Resources, p. 137–145.

Box, S. E., 1987, Late Cretaceous or younger southwest-directed extensional faulting, Cosmos Hills, Brooks Range, Alaska: Geological Society of America Abstracts with Programs, v. 19, p. 361.

Box, S. E., and Patton, W. W., Jr., 1989, Igneous history of the Koyukuk terrane, western Alaska: constraints on the origin, evolution, and ultimate collision of an accreted island arc terrane: Journal of Geophysical Research, v. 94, p. 15843–15867.

Box, S. E., Patton, W. W., Jr., and Carlson, C., 1985, Early Cretaceous evolution of the northeast Yukon-Koyukuk basin, west-central Alaska: U.S. Geological Survey Circular 967, p. 21–24.

Brosgé, W. P., and Reiser, H. N., 1962, Preliminary geologic map of the Christian Quadrangle, Alaska: U.S. Geological Survey Open File Map 62-229, scale 1:250,000.

Brosgé, W. P., and Reiser, H. N., 1964, Geologic map and section of the Chandalar Quadrangle, Alaska: U.S. Geological Survey Miscellaneous Geologic Investigations Map I-375, scale 1:250,000.

Brosgé, W. P., and Reiser, H. N., 1971, Preliminary bedrock geologic map: Wiseman and eastern Survey Pass Quadrangles, Alaska: U.S. Geological Survey Open File Map 479, scale 1:250,000.

Bryant, B., and Reed, J. C., 1969, Significance of lineation and minor folds near major thrust faults in the southern Appalachians and the British and Norwegian Caledonides: Geological Magazine, v. 106, p. 412–429.

Carden, J. R., 1978, The comparative geology of blueschists and greenschists in the brooks Range and Kodiak-Seldovia Schist belts [Ph.D. thesis]: Fairbanks, Alaska, University of Alaska, 242 p.

Carlson, C., 1985, Large-scale south-dipping low-angle normal faults in the south-central Brooks Range, Alaska: Eos (Transactions, American Geophysical Union), v. 66, p. 1074.

Christiansen, P. P., and Snee, L. W., 1994, Structure, metamorphism, and geochronology of the Cosmos Hills and Ruby Ridge, Brooks Range Schist belt, Alaska: Tectonics, v. 13, p. 193–214.

Cobbold, P. R., and Quinquis, H., 1980, Development of sheath folds in shear regimes: Journal of Structural Geology, v. 2, p. 119–126.

Dahlen, F. A., Suppe, J., and Davis, D., 1983, Mechanics of fold thrust belts and accretionary wedges: Journal of Geophysical Research, v. 88, p. 1153–1172.

Dewey, J. F., 1988, Extensional collapse of orogens: Tectonics, v. 7, p. 1123–1139.

Dillon, J. T., 1989, Structure and stratigraphy of the southern Brooks Range and northern Koyukuk basin near the Dalton Highway, *in* Mull, C. G., and Adams, K. E., eds., Dalton Highway, Yukon River to Prudhoe Bay, Alaska: Alaska Division of Geological and Geophysical Surveys Guidebook 7, v. 2, p. 157–187.

Dillon, J. T., and Smiley, C. J., 1984, Clasts from the Early Cretaceous Brooks Range orogen in Albian and Cenomanian molasse deposits of the Yukon-Koyukuk basin, Alaska: Geological Society of America Abstracts with Programs, v. 16, p. 279.

Dillon, J. T., Pessel, G. H., Chen, J. H., and Veach, N. C., 1980, Middle Paleozoic magmatism and orogenesis in the Brooks Range, Alaska: Geology, v. 8, p. 338–343.

Dillon, J. T., Hamilton, W. B., and Lueck, L., 1981a, Geologic map of the Wiseman A-3 Quadrangle, Alaska: Alaska Division of Geological and Geophysical Surveys Map AOF-119, scale 1:63,360.

Dillon, J. T., Pessel, G. H., Lueck, L., and Hamilton, W. B., 1981b, Geologic map of the Wiseman A-4 quadrangle, Alaska: Alaska Division of Geological and Geophysical Surveys Map AOF-124, scale 1:63,360.

Dillon, J. T., Brosgé, W. P., and Dutro, J. T., Jr., 1986, Generalized geologic map of the Wiseman Quadrangle, Alaska: U.S. Geological Survey Open File Report OF 86-219, scale 1:250,000.

Dillon, J. T., Harris, A. G., and Dutro, J. T., Jr., 1987, Preliminary description of lower Paleozoic fossil-bearing strata in the Snowden Mountain area of the south-central Brooks Range, Alaska, *in* Taileur, I., and Weimer, P., eds., Alaskan North Slope geology: Society of Economic Paleontologists and Mineralogists, Pacific Section, Publication 50, p. 337–345.

Dupuy, C., and Dostal, J., 1984, Trace element geochemistry of some continental tholeiites: Earth and Planetary Science Letters, v. 67, p. 61–69.

Dusel-Bacon, C., Brosgé, W. P., Till, A. B., Doyle, E. O., Mayfield, C. F., Reiser, H. N., and Miller, T. P., 1989, Distribution, facies, age, and proposed tectonic associations of regionally metamorphosed rocks in northern Alaska: U.S. Geological Survey Professional Paper 1497-A, 44 p.

Ellersieck, I., Curtis, S. M., Mayfield, C. F., and Tailleur, I. L., 1982, Reconnaissance geologic map of the south-central Misheguk Mountain Quadrangle, Alaska: U.S. Geological Survey Open File Report OF 82-612, scale 1:63,360.

Escher, A., and Watterson, J., 1974, Stretching fabrics, folds, and crustal shortening: Tectonophysics v. 22, p. 223–231.

Frey, M., De Capitini, C., and Liou, J. G., 1991, A new petrogenetic grid for low-grade metabasites: Journal of Metamorphic Geology, v. 9, p. 497–509.

Gedney, L., and Marshall, D. L., 1981, A rare earthquake sequence in the Kobuk Trench, northwestern Alaska: Bulletin of the Seismological Society of America, v. 71, p. 1587–1592.

Gilbert, W. G., Wiltse, M. A., Carden, J. R., Forbes, R. B., and Hackett, S. W., 1977, Geology of the Ruby Ridge, southwestern Brooks Range, Alaska: Alaska Division of Geological and Geophysical Surveys Report 58, 16 p.

Gottschalk, R. R., 1987, Structural and petrologic evolution of the south-central Brooks Range near Wiseman, Alaska [Ph.D. thesis]: Houston, Texas, Rice University, 263 p.

Gottschalk, R. R., 1990, Structural evolution of the Schist belt, south-central Brooks Range fold and thrust belt, Alaska: Journal of Structural Geology, v. 12, p. 453–469.

Gottschalk, R. R., and Oldow, J. S., 1988, Low-angle normal faults in the south-central Brooks Range fold and thrust belt, Alaska: Geology , v. 16, p. 395–399.

Grantz, A., Eittriem, S., and Whitney, O., 1981, Geology and physiography of the continental margin north of Alaska and implications for the origin of the Canada basin, *in* Nairn, A. E. N., Churkin, M., Jr., and Stehli, F. G., eds., The ocean basins and margins, v. 5: New York, Plenum, p. 439–492.

Harbert, W., Frei, L. S., Cox, A., and Engebretson, D. C., 1987, Relative plate motions between Eurasia and North America in the Bering Sea region: Tectonophysics, v. 134, p. 239–261.

Harris, R. A., 1989, Processes of allochthon emplacement with special reference to the Brooks Range ophiolite and Timor, Indonesia [Ph.D. thesis]: London, United Kingdom, University of London, 490 p.

Harris, R. A., 1992, Peri-collisional extension and the formation of Oman-type ophiolites in the Brooks Range and Banda arc, *in* Parsons, L. M., Murton, B. J., and Browning, P., eds., Ophiolites and their modern oceanic analogs: Geological Society of London Special Publication No. 60, p. 310–325.

Hitzman, M. W., Smith, T. E., and Proffett, J. M., 1982, Bedrock geology of the Ambler district, southwestern Brooks Range, Alaska: Alaska Division of Geological and Geophysical Surveys Geologic Report 75, scale 1:125,000.

Hitzman, M. W., Proffett, J. M., Schmidt, J. M., and Smith, T. E., 1986, Geology and mineralization of the Ambler district, northwestern Alaska: Economic Geology, v. 91, p. 1592–1618.

Hubbard, R. J., Edrich, S. P., and Rattey, R. P., 1987, Geologic evolution and hydrocarbon habitat of the Arctic Alaskan microplate: Marine and Petroleum Geology, v. 4, p. 2–34.

Jones, D. L., Coney, P. J., Harms, T. A., and Dillon, J. T., 1988, Interpretive geologic map and supporting radiolarian data from the Angayucham terrane, Coldfoot area, southern Brooks Range, Alaska: U.S. Geological Survey Miscellaneous Field Studies Map MF-1993, scale 1:63,360.

Karl, S. M., and Aleinikoff, J. N., 1990, Proterozoic U-Pb zircon age of granite in the Kallarichuk Hills, western Brooks Range, Alaska: evidence for Precambrian basement in the Schist belt: U.S. Geological Survey Bulletin 1946, p. 95–100.

Karl, S. M., and Long, C. L., 1987, Evidence for tectonic truncation of east-west trending structures in the central Baird Mountains Quadrangle, western Brooks Range, Alaska: Geological Society of America Abstracts with Programs, v. 19, p. 392.

Karl, S. M., and Long, C. L., 1990, Folded Brookian thrust faults: Implications of three geologic/geophysical transects in the western Brooks Range, Alaska: Journal of Geophysical Research, v. 95, p. 8581–8592.

Karl, S. M., Dumoulin, J. A., Ellersieck, I., Harris, A. G., and Schmidt, J. M., 1989, Preliminary geologic map of the Baird Mountains quadrangle, western Brooks Range, Alaska: U.S. Geological Survey Open-File Report 88-42, 65 p., scale 1:250,000.

Law, R. D., Miller, E. L., Little, T. A., and Lee, J., 1994, Extensional origin of ductile fabrics in the Schist belt, central Brooks Range, Alaska—II. Microstructural and petrofabric evidence: Journal of Structural Geology, v. 16, p. 919–940.

Liou, J. G., Maruyama, S., and Cho, M., 1985, Phase equilibria and mineral parageneses of metabasites in low-grade metamorphism: Mineralogy Magazine, v. 49, p. 321–333.

Lister, G. S., 1977, Crossed-girdle c-axis fabrics in quartzites plastically deformed by plane strain and progressive simple shear: Tectonophysics, v. 39, p. 51–54.

Little, T. A., Miller, E. L., Lee, J., and Law, R. D., 1994, Extensional origin of ductile fabrics in the Schist belt, central Brooks Range, Alaska—I. Geologic and structural studies: Journal of Structural Geology, v. 16, p. 899–918.

Lloyd, G. E., and Knipe, R. J., 1992, Deformation mechanisms accommodating faulting of quartzite under upper crustal conditions: Journal of Structural Geology, v. 14, p. 127–143.

Mainprice, D., Bouchez, J.-L., Blumenfeld, P., and Tubià, J. M., 1986, Dominant c-slip in naturally deforming quartz: implication for dramatic plastic softening at high temperatures: Geology, v. 14, p. 819–822.

Mayfield, C. F., Curtis, S. M., Ellersieck, I., and Tailleur, I. L., 1982, Reconnaissance geological map of the southeastern part of the Misheguk Mountain Quadrangle, Alaska: U.S. Geological Survey Open File Report OF

82-613, scale 1:63,360.

Mayfield, C. F., Tailleur, I. L., and Ellersieck, I., 1988, Stratigraphy, structure and palinspastic synthesis of the western Brooks Range, northwestern Alaska, *in* Gryc, G., ed., Geology and exploration of the National Petroleum Reserve in Alaska, 1974 to 1982: U.S. Geological Survey Professional Paper 1399, p. 143–186.

Miller, E. L., and Hudson, T. L., 1991, Mid-Cretaceous extensional fragmentation of a Jurassic–Early Cretaceous compressional orogen, Alaska: Tectonics, v. 10, p. 781–796.

Miller, T. P., 1989, Contrasting plutonic rocks suites of the Yukon-Koyukuk basin and the Ruby geanticline, Alaska: Journal of Geophysical Research, v. 94, p. 15969–15987.

Moore, T. E., Aleinikoff, J. N., and Walter, M., 1993, Middle Jurassic U-Pb age for the Siniktanneyak Mountain ophiolite, Brooks Range, Alaska: Geological Society of America Abstracts With Programs, v. 25, p. 124.

Moore, T. E., Wallace, W. K., Bird, K. J., Karl, S. M., Mull, C. G., and Dillon, J. T., 1994, Geology of northern Alaska, *in* Plafker, G., and Berg, H. C., eds., The geology of Alaska: Boulder, Colorado, Geological Society of America, The Geology of North America, v. G-1, p. 49–140.

Mull, C. G., Roeder, D. H., Tailleur, I. L., Pessel, G. H., Grantz, A., and May, S. D., 1987, Geologic sections and maps across Brooks Range and Arctic Slope to Beaufort Sea, Alaska: Geological Society of America Map MC-28S, scale 1:500,000.

Murchey, B., and Harris, A. G., 1985, Devonian to Jurassic sedimentary rocks in the Angayucham Mountains of Alaska: Possible seamount or oceanic plateau deposits: Eos (Transactions, American Geophysical Union), v. 66, p. 1102.

Murphy, J. M., and Patton, W. W., Jr., 1988: Geologic setting and petrography of the phyllite and metagreywacke thrust panel, north-central Alaska: U.S. Geological Survey Circular 1016, p. 104–108.

Murphy, J. M., Moore, T. E., Patton, W. W., Jr., and Saward, S. E., 1989, Stratigraphy of Cretaceous conglomerates, NE Koyukuk basin, Alaska: Unroofing of the southeastern Brooks Range: Geological Society of America Abstracts with Programs, v. 21, p. 120.

Nelson, B. K., Nelson, S. W., and Till, A. B., 1993, Isotopic evidence for Proterozoic and Paleozoic crustal evolution in the Brooks Range, northern Alaska: Journal of Geology, v. 101, p. 435–450.

Nelsen, C. J., 1979, The geology and blueschist petrology of the western Ambler Schist belt, southwestern Brooks Range, Alaska [M.S. thesis]: Albuquerque, New Mexico, University of New Mexico, 123 p.

Nelson, S. W., and Grybeck, D., 1982, Metamorphic rocks of the Survey Pass Quadrangle: U.S. Geological Survey Miscellaneous Field Studies Map MF-1176-C, scale 1:250,000.

Nelson, S. W., and Nelson, W. H., 1982, Geology of the Siniktanneyak Mountain ophiolite, Howard Pass Quadrangle, Alaska: U.S. Geological Survey Miscellaneous Field Studies Map MF-1441, scale 1:63,360.

Nelson, S. W., Nokleberg, W. J., Miller-Hoare, M., and Mullen, M. W., 1979, Siniktanneyak Mountain ophiolite: U.S. Geological Survey Circular 804B, p. B14–B16.

Newberry, R. J., Dillon, J. T., and Adams, D. D., 1986, Skarns and skarn-like deposits of the Brooks Range, northern Alaska: Economic Geology, v. 81, p. 1728–1752.

Nilsen, T. H., 1989, Stratigraphy and sedimentology of the mid-Cretaceous deposits of the Yukon-Koyukuk basin, west-central Alaska: Journal of Geophysical Research, v. 94, p. 15925–15940.

Oldow, J. S., Seidensticker, C. M., Phelps, J. C., Julian, F. E., Gottschalk, R. R., Boler, K. W., Handschy, J. W., and Avé Lallemant, H. G., 1987, Balanced cross-sections through the central Brooks Range and North Slope, Arctic Alaska: American Association of Petroleum Geologists, Special Publication, 19 p. and 8 pl.

Oldow, J. S., Bally, A. W., Avé Lallemant, H. G., and Leeman, W. P., 1989, Phanerozoic evolution of the North American Cordillera: United States and Canada, *in* Bally, A. W., and Palmer, A. R., eds., The geology of North America, An overview: Boulder, Colorado, Geological Society of

America, The Geology of North America, v. A, p. 139–232.

Pallister, J. S., Budahn, J. R., and Murchey, B., 1989, Pillow basalts of the Brooks Range: Oceanic-plateau and island crust accreted to the Brooks Range: Journal of Geophysical Research, v. 94, p. 15901–15923.

Patrick, B. E., 1995, High pressure–low temperature metamorphism of granitic orthogneisses in the Brooks Range, northern Alaska: Journal of Metamorphic Geology, v. 13, p. 111–124.

Patton, W. W., Jr., 1973, Reconnaissance geology of the northern Yukon-Koyukuk province, Alaska: U.S. Geological Survey Professional Paper 774-A, p. A1–A17.

Patton, W. W., Jr., 1993, Ophiolitic terranes of northern and central Alaska and their correlatives in Canada and northeastern Russia: Geological Society of America Abstracts with Programs, v. 25, p. 132.

Patton, W. W., Jr., and Box, S. E., 1989, Tectonic setting of the Yukon-Koyukuk basin and its borderlands, western Alaska: Journal of Geophysical Research, v. 94, p. 15807–15820.

Patton, W. W., Jr., and Miller, T. P., 1966, Regional geologic map of the Hughes Quadrangle, Alaska: U.S. Geological Survey Miscellaneous Geologic Investigations Map I-459, scale 1:250,000.

Patton, W. W., Jr., and Tailleur, I. L., 1977, Evidence in the Bering Strait region for differential movement between North America and Eurasia: Geological Society of America Bulletin, v. 88, p. 1298–1304.

Patton, W. W., Jr., Tailleur, I. L., Brosgé, W. P., and Lanphere, M. A., 1977, Preliminary report on the ophiolites of northern and western Alaska: Oregon Department of Geology and Mineral Industries, v. 95, p. 51–57.

Patton, W. W., Jr., Box, S. E., Moll-Stalcup, E. J., and Miller, T. P., 1994, Geology of west-central Alaska, *in* Plafker, G, and Berg, H. C., eds., The geology of Alaska: Boulder, Colorado, Geological Society of America, The Geology of North America, v. G-1, p. 241–269.

Pavlis, T. L., 1989, Mid-Cretaceous orogenesis in the northern Cordillera: A Mediterranean analog of collision-related tectonics: Geology, v. 17, p. 947–950.

Pavlis, T. L., Sisson, V. B., Foster, H. L., Nokleberg, W. J., and Plafker, G, 1993, Mid-Cretaceous tectonics of the Yukon-Tanana terrane, Trans-Alaska Crustal Transect (TACT), east-central Alaska: Tectonics, v. 12, p. 103–122.

Platt, J. P., 1986, Dynamics of orogenic wedges and the uplift of high-pressure metamorphic rocks: Geological Society of America Bulletin, v. 97, p. 1037–1053.

Platt, J. P., and Vissers, R. L. M., 1980, Extensional structures in anisotropic rocks: Journal of Structural Geology, v. 2, p. 397–410.

Price, G. P., 1985, Preferred orientation in quartzites, *in* Wenk, H-R., ed., Preferred orientation in metals and rocks: An introduction to modern textural analysis: San Diego, California, Academic Press, p. 385–406.

Quinquis, H., Audreu, C., Brun, J. P., and Cobbold, P. R., 1978, Intense progressive shear in the Ile de Groix blueschists and compatibility with obduction or subduction: Nature, v. 273, p. 43–45.

Ramsay, J. G., 1967, Folding and fracturing of rocks: New York, McGraw-Hill, 568 p.

Roeder, D., and Mull, C. G., 1978, Tectonics of the Brooks Range ophiolites, Alaska: American Association of Petroleum Geologists Bulletin, v. 62, p. 1696–1713.

Roeske, S. M., Walter, M., and Aleinikoff, J. N., 1991, Cretaceous deformation of granitic rocks in the southwest Ruby terrane, central Alaska: Geological Society of America Abstracts with Programs, v. 23, p. 93.

Sanderson, D. J., 1973, The development of fold axes oblique to regional trend: Tectonophysics v. 16, p. 55–70.

Schmid, S. M., and Casey, M., 1986, Complete fabric analysis of some commonly observed quartz c-axis patterns: American Geophysical Union Monograph 36, p. 263–286.

Silberling, N. J., Jones, D. L., Monger, J. W. H., Coney, P. J., Berg, H. C., and Plafker, G, 1994, Lithotectonic terrane map of Alaska and adjacent parts of Canada, *in* Plafker, G., and Berg, H. C., eds., The geology of Alaska: Boulder, Colorado, Geological Society of America, The Geology of North America, v. G-1, scale 1:2,500,000.

Smith, T. E., Webster, G. D., Heatwole, D. A., Profett, J. M., Kelsey, G., and

Glavinovich, P. S., 1978, Evidence for mid-Paleozoic depositional age of volcanogenic base metal sulfide occurrences and enclosing strata, Ambler district, northwest Alaska: Geological Society of America, Abstracts with Programs, v. 10, p. 148.

Spry, A., 1969, Metamorphic textures: New York, Pergamon Press, 350 p.

Sylvester, A. G., 1988, Strike-slip faults: Geological Society of America Bulletin, v. 100, p. 1666–1703.

Thompson, R. N., Dickin, A. P., Gibson, I. L., and Morrison, M. A., 1982, Elemental fingerprints of isotopic contamination of Hebridean Paleocene mantle derived magmas by Archean sial: Contributions to Mineralogy and Petrology, v. 79, p. 159–168.

Till, A. B., 1988, Evidence for two Mesozoic blueschist belts in the hinterland of the western Brooks Range fold and thrust belt: Geological Society of America Abstracts with Programs, v. 20, p. A112.

Till, A. B., 1989, Proterozoic rocks of the western Brooks Range: U.S. Geological Survey Bulletin 1903, p. 20–25.

Till, A. B., 1992, Detrital blueschist-facies metamorphic mineral assemblages in Early Cretaceous sediments of the foreland basin of the Brooks Range, Alaska, and implications for orogenic evolution: Tectonics, v. 11, p. 1207–1223.

Till, A. B., and Moore, T. M., 1991, Tectonic relations of the Schist belt, southern Brooks Range, Alaska: Eos (Transactions, American Geophysical Union), v. 72, p. 295.

Till, A. B., and Patrick, B. E., 1991, $^{40}Ar/^{39}Ar$ isotopic evidence for a 110–105-Ma amphibolite-facies overprint on blueschist in the south-central Brooks Range, Alaska: Geological Society of America, Abstracts with Programs, v. 23, p. A436.

Till, A. B., Schmidt, J. M., and Nelson, S. W., 1988, Thrust involvement of metamorphic rocks, southwestern Brooks Range, Alaska: Geology, v. 16, p. 930–933.

Till, A. B., Box, S. E., Roeske, S. M., and Patton, W. W., Jr., 1993, Comment on "Mid-Cretaceous extensional fragmentation of a Jurassic-Early Cretaceous compressional orogen, Alaska" by E. L. Miller and T. L. Hudson: Tectonics, v. 12, p. 1076–1081.

Tullis, J., 1977, Preferred orientation of quartz produced by slip during plane strain: Tectonophysics, v. 39, p. 87–102.

Tullis, J., Christie, J. M., and Griggs, D. T., 1973, Microstructures and preferred orientation of experimentally deformed quartzites: Geological Society of America Bulletin, v. 84, p. 297–314.

Tullis, J., and Yund, R. A., 1980, Hydrolytic weakening of experimentally deformed Westerly granite and Hale albite rock: Journal of Structural Geology, v. 2, p. 439–452.

Turner, D. L., Forbes, R. B., and Dillon, J. T., 1979, K-Ar geochronology of the southwestern Brooks Range, Alaska: Canadian Journal of Earth Science, v. 16, p. 1789–1804.

Turner, F. J., and Weiss, L. E., 1963, Structural analysis of metamorphic tectonites: New York, McGraw-Hill, 545 p.

White, S. H., 1976, The effects of strain on microstructures, fabrics, and deformation mechanisms in quartz: Philosophical Transactions of the Royal Society of London, v. A283, p. 69–86.

Wilcox, R. E., Harding, T. P., and Seely, D. R., 1973, Basic wrench tectonics: American Association of Petroleum Geologists Bulletin, v. 57, p. 74–96.

Wirth, K. R., and Bird, J. M., 1992, Chronology of ophiolite crystallization, detachment, and emplacement: evidence from the Brooks Range, Alaska: Geology, v. 20, p. 75–78.

Wirth, K. R., Bird, J. M., Blythe, A. E., Harding, D. J., and Heizler, M. T., 1993, Age and tectonic significance of the western Brooks Range ophiolites: results from $^{40}Ar/^{39}Ar$ thermochronometry: Tectonics, v. 12, p. 410–432.

Zayatz, M. R., 1987, Petrography of the Baird Mountain schistose lithologies, northwestern Alaska: U.S. Geological Survey Circular 998, p. 49–52.

Zimmerman, J., and Frank, C. O., 1982, Possible obduction-related metamorphic rocks at the base of the ultramafic zone, Avan Hills Complex, De Long Mountains: U.S. Geological Survey Circular 844, p. 27–28.

Zwart, H. J., 1962, On the determination of polymetamorphic mineral associations and its application to the Bosost area (coastal Pyrenees): Geologische Rundschau 52, p. 38–65.

MANUSCRIPT ACCEPTED BY THE SOCIETY SEPTEMBER 23, 1997

Geological Society of America
Special Paper 324
1998

Tectonothermal evolution of metamorphic rocks in the south-central Brooks Range, Alaska: Constraints from $^{40}Ar/^{39}Ar$ geochronology

Richard R. Gottschalk
Exxon Production Research Company, P.O. Box 2189, Houston, Texas 77252
Lawrence W. Snee
Branch of Isotope Geology, U.S. Geological Survey, MS-963, Denver Federal Center, Denver, Colorado 80225

ABSTRACT

To constrain the age of deformational/metamorphic events in the south-central Brooks Range, we analyzed 16 samples of white mica, amphibole, and biotite using the $^{40}Ar/^{39}Ar$ incremental heating technique. Metamorphic rocks in the study area (between 151°W and 148°W) occur in three principal east-west–trending fault-bounded belts. These are, from south to north, the pumpellyite-actinolite–facies rocks of the Phyllite belt, the high-pressure/low-temperature (*HP/LT*) metamorphic rocks of the Schist belt, and the predominantly greenschist-facies metamorphic rocks of the Skajit allochthon. All three belts have been affected by two penetrative deformational/metamorphic events. The oldest of these (D_{1a}) resulted in tight to isoclinal folding and was accompanied by pumpellyite-actinolite–facies metamorphism in the Phyllite belt, blueschist-facies metamorphism in the Schist belt, and blueschist- to greenschist-facies metamorphism in the Skajit allochthon. Two white micas samples, one from the Schist belt, and one from the sodic-amphibole–bearing schists north of the Minnie Creek thrust (MCT) yielded convex-upward Ar-release spectra with maximum apparent ages of 142 and 129 Ma, respectively; we interpret maximum apparent ages as a minimum age for D_{1a} deformation and *HP/LT* metamorphism in the south-central Brooks Range.

A second synmetamorphic deformational event (D_{1b}) affected all but the northernmost rocks of the Skajit allochthon, resulting in pervasive dynamic recrystallization accompanied by growth of metamorphic minerals; in the Schist belt and in sodic-amphibole–bearing schists north of the MCT, D_{1b} occurred under lower amphibolite- to greenschist-facies metamorphic conditions, and in the Phyllite belt, D_{1b} occurred under lower greenschist(?)-facies conditions. Plateau, preferred, and isochron dates on white mica, amphibole, and biotite are Early Cretaceous in age, and range from 135 to 110 Ma. We argue that the predominance of Early Cretaceous dates result from the degassing of *HP/LT* metamorphic minerals during dynamic recrystallization associated with D_{1b} folding and the pervasive growth of syn-D_{1b} neoblasts. These ages are significantly older than Late Cretaceous to Tertiary(?) extensional structures in the southern Brooks Range, and support the interpretation that D_{1b} structures formed during Brookian contractional deformation.

Gottschalk, R. R., and Snee, L. W., 1998, Tectonothermal evolution of metamorphic rocks in the south-central Brooks Range, Alaska: Constraints from $^{40}Ar/^{39}Ar$ geochronology, *in* Oldow, J. S., and Avé Lallement, H. G., eds., Architecture of the Central Brooks Range Fold and Thrust Belt, Arctic Alaska: Boulder, Colorado, Geological Society of America Special Paper 324.

INTRODUCTION

The south-central Brooks Range exposes a series of east–west–trending, fault-bounded lithologic belts composed of metamorphic rocks derived from Alaskan continental margin sediments and basement; these are, from south to north, the pumpellyite-actinolite–facies rocks of the Phyllite belt (Pzp, Fig. 1b), the high-pressure/low-temperature *(HP/LT)* metamorphic rocks of the Schist belt (DPm), and the predominantly greenschist-facies rocks of the Skajit allochthon (of Oldow et al., 1987; DpCs; Dusel-Bacon et al., 1989; Dillon, 1989; Moore et al., 1994; Gottschalk, this volume, Chapter 9; Gottschalk et al., this volume, Chapter 12;

Oldow et al., this volume, Chapter 8). All three lithologic belts are characterized by polyphase folds and penetrative metamorphic fabrics that record a history of deformation and metamorphism at great depth within an island arc–continental collisional orogen, followed by uplift and exhumation accommodated by a combination of contractional and extensional processes (Box, 1985; Gottschalk and Oldow, 1988; Gottschalk, 1990; Till, 1992; Miller and Hudson, 1991; Little et al., 1994; Law et al., 1994; Moore et al., 1994; Gottschalk et al., this volume, Chapter 12; Avé Lallemant and other, this volume, Chapter 15).

Despite differences in the lithology, structural style, and metamorphic history of the Phyllite belt, Schist belt, and Skajit

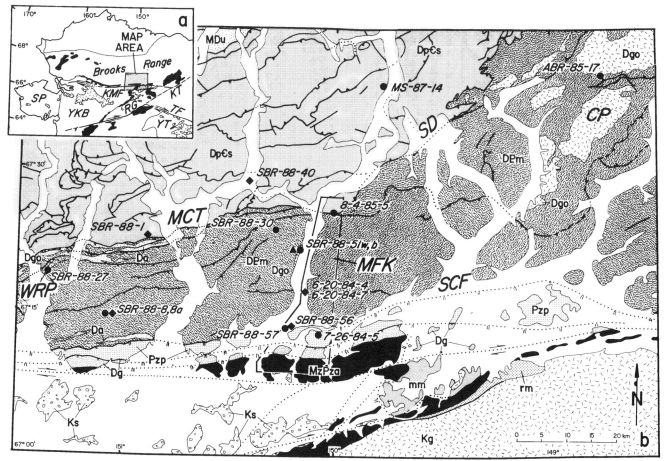

Figure 1. a, Selected tectonic elements of central and northern Alaska, with location of study area; black = ophiolitic rocks; hatchured pattern = Jurassic and Cretaceous intrusive and volcanic rocks, undivided; KMF, Kobuk-Malemute fault; KT, Kaltag fault; RG, Ruby Geanticline; SP, Seward Peninsula; TF, Tintina fault; YKB, Yukon-Koyukuk basin; YT, Yukon-Tanana terrane. b, Generalized geology of the south-central Brooks Range with $^{40}Ar/^{39}Ar$ sample locations modified after Brosgé and Reiser (1964, 1971), Dillon et al. (1986), Oldow et al. (1987), and Jones et al. (1988). Ks, Early Cretaceous sedimentary rocks of the Yukon-Koyukuk basin; Kg, Early Cretaceous granite; MDu, Devonian and Mississippian Endicott Group clastic rocks and Devonian Beaucoup Formation undivided; DpCs, Precambrian(?) to Devonian rocks of the Skajit allochthon; DPm, Proterozoic to Devonian metamorphic rocks of the Schist belt; Dgo, Devonian granitic orthogneiss; Dg, Devonian metagreywacke (Metagreywacke belt); Pzp, Paleozoic(?) phyllite (Phyllite belt); MzPza, Paleozoic and Mesozoic basalt, diabase, gabbro, chert, and limestone of the Angayucham terrane; mm, metamorphic rocks of the Mosquito terrane of Jones et al. (1988) undivided; rm, metamorphic rocks of the Ruby Geanticline, undivided; CP, Chandalar pluton; MCT, Minnie Creek thrust; MFK, Middle Fork Koyukuk River; SD, Skajit decollement; SCF, Slate Creek fault; WRP, Wild River pluton. Thin lines are geologic contacts. Heavy lines with barbs are thrust faults, with teeth on upper plate. Heavy lines with hatchures are low-angle normal faults, with hatchures on upper plate. Heavy lines without ornamentation are steeply dipping faults. Sample locations: Diamonds, amphibole; circles, white mica; triangle, biotite.

allochthon, two penetrative, synmetamorphic deformational events have been recognized that correlate from belt to belt (Gottschalk et al., this volume, Chapter 12). We refer to these deformational events, from oldest to youngest, as D_{1a} and D_{1b}. The D_{1a} and D_{1b} structures and fabrics in all three lithologic belts are transected by numerous ductile extensional shear zones that are subsidiary to major east-west–trending normal faults and shear zones that separate the belts. The nature of these deformational events, and their relationship to regional metamorphism in the south-central Brooks Range is outlined below, and discussed in detail in Gottschalk (1990), Oldow et al. (this volume, Chapter 8), and Gottschalk et al. (this volume, Chapter 12).

Although the polyphase nature of deformation in the south-central Brooks Range has been recognized for some time, the nature and timing of deformational events is a subject of debate. Discussion has focused mainly on the Schist belt, which has been the focus of most of the detailed structural analysis. Dillon (1989) argued that the oldest structural fabric in the Schist belt (equivalent to D_{1a}?) is pre-Devonian in age and was transposed into a second metamorphic fabric during Mesozoic contractional deformation (D_{1b}). Gottschalk (1990) proposed a model in which D_{1a} and D_{1b} structures formed progressively during top-to-the-north shear related to Jurassic-Cretaceous contraction with no significant time gap between the two events; in his model, D_{1a} folding occurred under blueschist-facies conditions, with D_{1b} occurring at somewhat lower pressures under lower amphibolite- to greenschist-facies conditions. In a third model, Till and Patrick (1991) argued that D_{1a} and D_{1b} folds are discrete deformational events, with D_{1a} forming under blueschist-facies conditions prior to 149 Ma, and D_{1b} structures forming under amphibolite-facies conditions in the Early Cretaceous; Till and Patrick (1991) also infer a contractional origin for both folding events. Little et al. (1994) infer a timing of events similar to that proposed by Till and Patrick (1991), but argue that D_{1b} structures formed as the result of ductile extension rather than contraction; whereas Gottschalk (1990) proposed that post-D_{1b} extensional structures in the south-central Brooks Range formed during a discrete event that postdated D_{1b}, Little et al. (1994) have argued that extensional shear zones and related features are late-kinematic features related to their proposed D_{1b} extension.

The main purpose of our study was to place quantitative constraints on the age of deformational/metamorphic events in the Phyllite belt, Schist belt, and Skajit allochthon, and to test the correlations of events that heretofore have been based mainly on similarities in structural style and superposition relationships. As our principal means of investigation, we carried out a reconnaissance geochronologic study of metamorphic rocks in the south-central Brooks Range between 151 and 148°W (Fig. 1). Our study focused mainly on the Schist belt of the south-central Brooks Range, but samples from the Skajit allochthon and Phyllite belt were also collected and analyzed. For this study we analyzed 16 samples of white mica, amphibole, and biotite using the $^{40}Ar/^{39}Ar$ incremental heating technique; sample locations are shown in Figure 1b. Geographic coordinates of sample locations,

sample mineralogies, and a summary of analytical results are presented in Table 1. Details of $^{40}Ar/^{39}Ar$ analyses are tabulated in Appendix 1. To constrain metamorphic temperatures during deformational events, we have analyzed a number of samples by electron microprobe for mineral geothermometry.

The $^{40}Ar/^{39}Ar$ incremental heating technique was chosen for this study because of its ability to sample and analyze different Ar reservoirs within a single sample. By utilizing detailed isochron analysis, the $^{40}Ar/^{39}Ar$ incremental heating technique provides the means to assess and mitigate problems related to polyphase mineral growth, partial degassing due to later thermal or deformational events, and incorporation of nonradiogenic ^{40}Ar. Moreover, the use of a single aliquot to measure ^{40}Ar and K (actually K-derived ^{39}Ar) provides greater accuracy and analytical precision than is currently available using conventional K-Ar dating techniques.

PREVIOUS GEOCHRONOLOGIC STUDIES

Few published K-Ar isotopic dates are available from our study area. Brosgé and Reiser (1964) and Grybeck et al. (1977) report isotopic dates on white mica and biotite from the Chandalar pluton and adjacent schistose rocks (Fig. 1b). K-Ar dates on three biotite separates range from 495 to 103 Ma, and a single white mica sample yielded an apparent age of 100 Ma.

In contrast to the poorly dated south-central Brooks Range, correlative metamorphic rocks in the western Brooks Range have been extensively studied using K-Ar and Rb-Sr dating techniques. Turner et al. (1979) carried out the first systematic isotopic dating in the Brooks Range and showed that glaucophanes from the Schist belt yielded late Precambrian K-Ar dates, while tremolite/actinolite separates yielded K-Ar dates ranging from 314 to 159 Ma. In addition, Turner et al. (1979) dated muscovite (some probably phengitic), biotite, and paragonite from the Schist belt, which yielded apparent ages ranging from 100 to 130 Ma, 94 to 129 Ma, and 239 to 191 Ma, respectively. Based on these results, Turner et al. (1979) argued that the Schist belt was metamorphosed under *HP/LT* conditions in the late Precambrian and was uplifted and cooled in the Early Cretaceous. Turner et al. (1979) recognized, however, that Precambrian ages were derived from glaucophane, a low-K amphibole, and warned that old apparent ages may have resulted from incorporation of excess ^{40}Ar. Since publication of the pioneering study of Turner et al. (1979), it has been shown that rocks yielding fossil assemblages and isotopic dates as young as the Devonian are affected by high-pressure metamorphism (Dillon et al., 1980; Hitzman et al., 1986; Dillon, 1989; Moore et al., 1994; Patrick, 1995).

With the clarification of the "Precambrian blueschist" problem in the Schist belt, work has continued to focus on constraining the age and cooling history of *HP/LT* metamorphic rocks in the southern Brooks Range. Based on K-Ar and Rb-Sr dating of samples from the western Brooks Range and Seward Peninsula, Armstrong et al. (1986) concluded that *HP/LT* metamorphism occurred prior to 160 Ma followed by Cretaceous uplift and cool-

TABLE 1. SUMMARY OF GEOCHRONOLOGIC DATA

Sample	Mineralogy*	Location		Mineral Analyzed*	Plateau Date (Ma)	Isochron Date (Ma)	Total-Gas Date (Ma)
		(N)	(W)				
PHYLLITE BELT							
Fine-grained quartz-mica schist							
7-26-84-5	Qtz + wm + chl + ab	67°12'45";	150°04'25"	Wm	None	124 ± 1	125.8 ± 05
SCHIST BELT							
Quartz-mica schist							
SBR88-8a	Qtz + wm + chl + ctd + (gl)	67°14'35";	151°03'00"	Wm	123.3 ± 0.7	124.2 ± 0.3	125.7 ± 5
SBR88-30	Qtz + wm + chl + ctd + (gl)	67°23'45";	150°16'25"	Wm	135 ± 2	145 ± 2	154 ± 2
SBR-88-57	Qtz + wm + chl + ab	67°13'30";	150°13'35"	Wm	118.3 ± 0.3†	121.0 ± 0.7	116.4 ± 0.3
Granitic orthogniess							
SBR88-27	Qtz + kfs + wm + plag + bi + oc	67°18'55";	151°20'45"	Wm	110.4 ± 0.3†	110.4 ± 0.2	110.1 ± 0.3
ABR85-17	Qtz + kfs + wm + plag + bi + ep + sph	67°40'06";	148°45'28"	Wm	120.7 ± 0.3	119 ± 1	119.2 ± 0.3
Feldspathic schist							
SBR88-51w	Ab + wm + bi + chl + qtz + carb + ep	67°21'35";	150°09'38"	Wm	None	131 ± 1§	None
SBR88-51w (rerun)	Ab + wm + bi + chl + qtz + carb + ep	67°21'35";	150°09'38"	Wm	None	132 ± 1§	135.2 ± 0.4
SBR88-51b	Ab + wm + bi + chl + qtz + carb + ep	67°21'35";	150°09'38"	Bi	128.8 ± 0.6	128.5 ± 0.2	128.6 ± 0.4
SBR88-51b (rerun)	Ab + wm + bi + chl + qtz + carb + ep	67°21'35";	150°09'38"	Bi	129.3 ± 0.4	129.7 ± 0.2	129.0 ± 0.4
Metabasite							
SBR88-8	Ab + chl + act + ep + carb	67°14'35";	151°03'00"	Act	None	160 ± 13	197 ± 4
Metagabbro							
SBR88-56	Ab + bar-hbl + chl + oc + stp + sph + mgt	67°13'35";	150°11'55"	Bar-hbl	115 ± 1	114 ± 1	147 ± 2
Eclogite (retrograded)							
6-20-84-4	Am + gt + chl + ep + ab + sph	67°17'02";	150°08'20"	Hbl	None	188 ± 6§	187 ± 6
6-20-84-7	Am + gt + chl + ep + ab + sph	67°17'02";	150°08'20"	Act	None	None	174.4 ± 0.6
SKAJIT ALLOCHTHON							
Metabasite							
SBR88-1	Ab + am + chl + qtz + ep + sph + bi	67°23'00";	150°53'05"	Act	None	425 ± 12†	379 ± 2
SBR88-40	Qtz + am + ab + ep + wm +sph + chl	67°28'51";	150°23'46	Act	None	None	893 ± 3
Calc-phyllite							
MS87-14	Wm + qtz + oc + chl	67°39'00";	149°46'50"	Wm	None	125.3 ± 0.7	123.3 ± 0.4
8-4-85-5	Qtz + wm + chl + ab + gt	67°24'57";	150°00'25"	Wm	None	127 ± 2†	125.5 ± 0.4

*ab = albite; am = amphibole, undifferentiated; act = actinolite; bar-hbl = barroisite-hornblende; bi = biotite; carb = carbonate, undifferentiated; cc = calcite; chl = chlorite; ctd = chloritoid; ep = epidote; (gl) = pseudomorphs of ab + chl ± qtz after glaucophane; gt = garnet; kfs = k-feldspar; mgt = magnetite; plag = plagioclase, undifferentiated; sph = sphene; stp = stiplnomemlane; qtz = quartz; wm = white mica, undifferentiated.
†Preferred date. See text for explanation.
§Isochron date determined from all temperature steps. See text for discussion.

ing; the 160-Ma age was based chiefly on dates from the Seward Peninsula and was extended to the Brooks Range on the assumption that the *HP/LT* rocks of Schist belt and the Seward Peninsula are correlative. A Middle Jurassic or older age for *HP/LT* metamorphism for the Schist belt of the western Brooks Range is indicated by a ⁴⁰Ar/³⁹Ar plateau date of 171 Ma on phengitic white mica obtained by Christiansen and Snee (1994); they interpret this date as a minimum age for *HP/LT* metamorphism.

Detailed ⁴⁰Ar/³⁹Ar geochronologic investigations by Till and Patrick (1991), Till (1992), Christiansen and Snee (1994), Till and Snee (1995), and Blythe et al. (this volume, Chapter 10), have shown that data do not support the simple tectonic model in which Middle Jurassic or older *HP/LT* metamorphism is followed by

uplift and cooling in the Early Cretaceous. A number of events are now recognized. (1) A clustering of white mica plateau and isochron dates of approximately 120 to 130 Ma clearly postdates *HP/LT* metamorphism in the Schist belt, and have been interpreted as recording either synmetamorphic deformation of the Schist belt under lower amphibolite- to greenschist-facies conditions (Christiansen and Snee, 1994), or as dating the uplift and cooling of the Schist belt (Blythe et al., this volume, Chapter 10). These dates are somewhat younger than dates of 110 to 105 Ma obtained by Till and Patrick (1991) on amphibolite-facies metamorphism that overprints *HP/LT* assemblages in the Schist belt to the immediate west of our study area, and was accompanied by the formation of ductile north-vergent folds (Till and Patrick,

1991). (2) In the western Brooks Range, Till and Snee (1995) have shown that continental rocks of the Central belt (roughly correlative to the Skajit allochthon of the central Brooks Range, or the Hammond terrane of Moore et al., 1994) experienced *HP/LT* metamorphism at about 120 Ma, much later than the Jurassic blueschist-facies metamorphism in the Schist belt. (3) Based upon a number of white mica plateau dates between 105 to 95 Ma on rocks affected by penetrative extensional deformation, Christiansen and Snee (1994) have argued for extensional uplift and denudation of the Schist belt in the late Albian to Cenomanian; extensional faults in the southern Brooks Range affect rocks as young as Late Cretaceous in age (Box, 1987).

GENERAL GEOLOGY

Phyllite belt

The Phyllite belt is composed predominantly of phyllite and fine-grained quartz-mica schist derived from turbidites of Devonian(?) age (Murphy and Patton, 1988). Metasedimentary rocks are intercalated with lenticular bodies of mafic schist and metagabbro derived from mafic extrusive(?) and intrusive igneous rocks (Gottschalk, 1987). The rocks of the Phyllite belt are affected by two phases of synmetamorphic penetrative deformation. The earliest structures are isoclinal folds and sheath folds with an axial planar foliation (S_{1a}) that formed under pumpellyite-actinolite–facies metamorphic conditions; phase relations indicate that temperatures during D_{1a} folding were approximately 200 to 350°C and pressures were between 3 and 7 kbar (Gottschalk, 1987). The second folding event resulted in crenulation of the D_{1a} foliation, and the formation of a locally penetrative cleavage (S_{1b}). Deformation microstructures indicate that strain during D_{1a} and D_{1b} was accommodated, in part, by diffusion-aided dislocation climb in quartz; thus the temperature during both deformational events was probably in excess of approximately 300°C (Gottschalk et al., this volume, Chapter 12).

Schist belt

Within the area examined in this study, the Schist belt is composed of quartz-mica schist, with interfoliated marble, mafic schist, feldspathic schist, metagabbro, calc-schist, metarhyolite, granitic orthogneiss, and eclogite. These rocks are derived from Precambrian to middle Paleozoic protoliths of sedimentary and volcanic rocks intruded by Precambrian- and Devonian-aged granitic plutons (Dillon et al., 1980; Hitzman et al., 1986; Dillon, 1989; Karl and Aleinikoff, 1990; Moore et al., 1994). The Schist belt experienced a pressure/temperature history that defines a clockwise loop in pressure/temperature space, passing through the lawsonite-blueschist facies en route to peak pressure conditions in the epidote-blueschist facies. During subsequent decompression, epidote-blueschist–facies minerals were overprinted and partially to completely replaced by lower amphibolite- to greenschist-facies assemblages (Gottschalk, 1990, and this volume, Chapter 9; Till et al., 1988; Dusel-Bacon et al., 1989; Till and Patrick, 1991; Little et al., 1994; Patrick, 1995). Structural analyses presented in

Gottschalk (1987), Little et al. (1994), and Gottschalk et al. (this volume, Chapter 12) indicates that D_{1a} isoclines and sheath folds in the Schist belt formed under epidote-blueschist–facies metamorphic conditions, while D_{1b} structures and fabrics formed during the lower amphibolite- to greenschist-facies metamorphic event. Locally, syn-D_{1b} deformation was sufficiently pervasive to obliterate the earlier D_{1a} metamorphic fabric. Both D_{1a} and D_{1b} metamorphic fabrics are overprinted by a late- to postkinematic phase of mineral growth under greenschist-facies conditions.

Quantitative estimates of metamorphic conditions within the study area are presented in Patrick (1995) and Gottschalk (this volume, Chapter 9) and are summarized in Table 2. Data from three sites pertinent to this study are tabulated. Along the Middle Fork Koyukuk River (area MFK, Fig. 1b) estimated conditions for syn-D_{1a} epidote-blueschist metamorphism range from 405 to 490°C and 9.2 to 10.9 kbar; estimated temperatures for the syn-D_{1b} lower amphibolite- to greenschist-facies overprint range from 350 to 550°C at pressures below the lower stability limit of glaucophane (less than approximately 7 kbar). Epidote-blueschist–facies metamorphic conditions in the Chandalar and Wild River plutons (Fig. 1) are estimated at T = 372–393°C, P = 7.0–9.8 kbar, and T = 377°C, P = 11.3 kbar, respectively; no quantitative data are available on post-peak metamorphic pressure conditions at these sites.

D_{1a} and D_{1b} fabrics are crosscut by numerous ductile extensional shear zones that occur throughout the Schist belt and Phyllite belt. The metamorphic grade under which post-D_{1b} extensional shear zones formed is not known. However, deformation microstructures in quartz within the shear zones are indicative of

TABLE 2. SUMMARY OF METAMORPHIC CONDITIONS IN THE SCHIST BELT

M₁ - High-Pressure/Low-Temperature Metamorphism

Middle Fork Koyukuk River		Wild River Pluton	
Kfs-plag*	T = 405°C, 412°C	Kfs-plag*	T = 377°C
Gt-cpx†	T = 490°C	PBKQ*	P = 11.3 kbar
PBKQ*	P = 10.9 kbar, 9.2 kbar		

Chandalar Pluton
Kfs-plag*   T = 393°C, 372°C
PBKQ*   P = 7.0 kbar, 9.8 kbar

M₂ - Retrograde Lower Amphibolite- to Greenschist-Facies Metamorphism

Middle Fork Koyukuk River
Cc-dol†   T = 462°C, 479°C, 483°C   P <lower stability limit of
Gt-bi†   T = 381°C   glaucophane†
Gt-hbl†   T = 478°C

*Data from Patrick, 1995.
†Data from Gottschalk, this volume, Chapter 9.
Kfs-plag = nominal temperatures calculated from two-feldspar geothermometry; gt-cpx = garnet-clinopyroxene geothermometry; PBKQ = phengite-biotite-k-feldspar-quartz geobarometry; cc-dol = calcite-dolomite geothermometry; gt-bi = garnet-biotite geothermometry; gt-hbl = garnet hornblende geothermometry.

dynamic recrystallization, suggesting that shear zones formed at temperatures in excess of about 300°C (Gottschalk et al., this volume, Chapter 12).

Metamorphic rocks north of the Minnie Creek thrust

In our study, we have analyzed several samples from the large area of metamorphic rocks that crop out north of the Minnie Creek thrust (MCT, Fig. 1b). These rocks have been assigned by Oldow et al. (1987) to the Skajit allochthon, locally characterized by clastic and carbonate metasedimentary rocks of Precambrian (?) through Middle Devonian age derived from platformal to basinal assemblages (Brosgé and Reiser, 1964, 1971; Dillon et al., 1986, 1987; Boler, 1989; Oldow et al., this volume, Chapter 7). Metamorphic rocks north of the MCT display a variable strain history, with D_{1a} structures and fabrics present throughout the Skajit allochthon, and D_{1b} structures and metamorphic fabrics present in the southern Skajit allochthon dying out gradually to the north; both deformational phases are interpreted by Oldow et al. (this volume, Chapter 8) as having formed during internal thrust imbrication of the Skajit allochthon. Along the southernmost margin of the Skajit allochthon, D_{1a} and D_{1b} fabrics are overprinted by a third-fold phase D_{1c} related to displacement on the MCT.

The nature and location of the contact between the Schist belt and the Skajit allochthon in the study area is disputed. Based upon the absence of pseudomorphs after glaucophane in schists that he examined from north of the MCT, Gottschalk (1987) proposed that the MCT was the northern tectonic boundary of the Schist belt. However, recent work has shown that relict sodic-amphibole occurs at least 5 km north of the MCT in rocks previously assigned by Oldow et al. (1987) and Gottschalk (1990) to the Skajit allochthon (A. B. Till, written communication, 1993). The extent to which sodic-amphiboles exist north of the MCT is not known, and the nature of the transition to the sodic-amphibole–free rocks of the Skajit allochthon has yet to be determined (i.e., tectonic contact versus gradational change in metamorphic grade). It is not clear whether sodic-amphibole–bearing rocks north of the MCT are part of the Schist belt (Till and Moore, 1991) or whether they represent a part of the Skajit allochthon that has undergone *HP/LT* metamorphism (Oldow et al., this volume, Chapter 7). In this paper, we separate sodic-amphibole–bearing rocks north of the MCT from areas where sodic-amphiboles have not been identified, and refer to sodic-amphibole–free rocks north of the MCT as the Skajit allochthon senso stricto.

Our reconnaissance study of metamorphic conditions north of the MCT yielded estimated metamorphic temperatures that varied significantly over the study area (Fig. 2). Biotite-garnet and calcite-dolomite geothermometry produced temperature estimates ranging from 227 ± 60°C to 486 ± 5°C (Tables 3 and 4). No pressure estimates were obtained. Samples 8-4-85-5, 8-5-85-3, and SBR88-70 lie within the belt of known sodic-amphibole–bearing rocks north of the MCT. Other samples were

collected from areas where there is no existing evidence for *HP/LT* metamorphism.

ANALYTICAL PROCEDURES AND SAMPLE PREPARATION

White mica, amphibole, and biotite mineral separates were prepared from crushed and sized powders using conventional heavy liquid and magnetic separation techniques and hand picked to greater than 99% purity. Weighed concentrates, together with K_2SO_4 and CaF_2 salts, and irradiation standard MMhb-1 hornblende were wrapped in Al foil, encapsulated in quartz vials, and irradiated in the central thimble position of the U.S. Geological Survey TRIGA reactor in Denver, Colorado, for 30 megawatt hours. Irradiation standards were spaced between every two samples to measure neutron-flux gradients within the irradiation assembly. Irradiated samples were incrementally heated to at least 1,350°C in a resistance-heated furnace; temperatures were monitored using a tungsten thermocouple, and kept constant to within ± 5°C using a Eurotherm temperature controller. Samples were maintained at each temperature increment for 20 minutes, and evolved gas was cleaned using a molecular-sieve desiccant and titanium, zirconium-aluminum, and vanadium-iron getters maintained at appropriate temperatures. The crucible was baked at 1,550°C for at least 30 minutes between samples, and procedural blanks were within ± 5% of atmospheric $^{40}Ar/^{39}Ar$.

Isotopic analyses were carried out using a Mass Analyzer Products 215 Rare Gas mass spectrometer, using a Faraday cup with a sensitivity of 9.736×10^{-13} moles/V and a detection limit of 5×10^{-17} moles Ar. Measurements were corrected for atmospheric contamination, mass fractionation, trap current, and radioactive decay of ^{37}Ar and ^{39}Ar. Abundances of interfering isotopes of Ar were corrected using production ratios measured on K_2SO_4 and CaF_2 salts. Corrections for Cl-derived ^{36}Ar were determined using the method of Roddick (1983). The production ratios used in corrections are given in Appendix 1.

J-values were calculated using the irradiation monitor MMhb-1 (520.4 ± 1.7 Ma; Samson and Alexander, 1987) and the reproducibility of split gas fractions from each monitor was used to calculate imprecisions in J. J values for unknowns were determined by interpolation between monitors. Apparent ages for each temperature step were calculated using the decay constants and isotopic abundances listed in Steiger and Jäger (1977); uncertainties in apparent ages were calculated according to the equations of Dalrymple et al. (1981). Plateau and total-gas dates were statistically calculated using the method described in Snee et al. (1988). A plateau age is determined if contiguous fractions representing at least 50% of the total $^{39}Ar_K$ released yielded statistically indistinguishable ages at the 1-sigma level. A preferred age is calculated from contiguous fractions that overlap within 3-sigma level. $^{40}Ar/^{36}Ar$ versus $^{39}Ar/^{36}Ar$ isochrons were computed using the methods described in York (1969); errors were determined at the 2-sigma level. $^{39}Ar_K/^{37}Ar_{Ca}$ ratios for each sample are plotted to show the relative distribution of K and Ca during incremental heating; an approximate K/Ca ratio can be

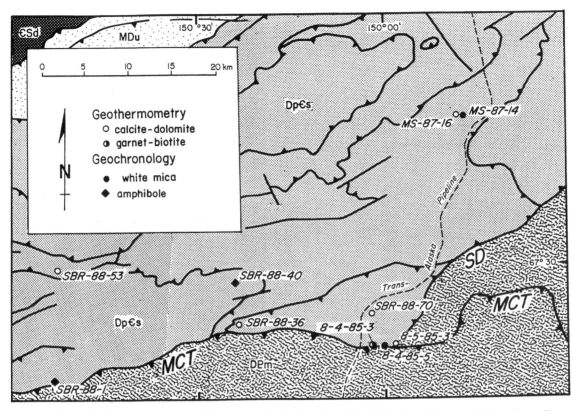

Figure 2. Generalized structural geology of the western Skajit allochthon (simplified from Oldow et al., this volume, Chapter 7) showing the location of $^{40}Ar/^{39}Ar$ samples, and samples analyzed for calcite-dolomite and garnet-biotite geothermometry. Єsd, Cambrian-Silurian rocks of the Doonerak window. Other geologic units same as Figure 1b.

obtained by multiplying $^{39}Ar_K/^{37}Ar_{Ca}$ by 0.5. Analytical results are tabulated in Appendix 1.

ANALYTICAL RESULTS

Schist belt

White mica from quartz-mica schists. Samples SBR88-8a, SBR88-30, and SBR88-57 are white micas separated from quartz-mica schist, the dominant rock type in the Schist belt of the south-central Brooks Range. Mineralogically, these rocks are composed of quartz + white mica + chlorite + chloritoid ± albite, and locally contain albite + chlorite ± quartz pseudomorphs after glaucophane (Gottschalk, 1987, 1990). Microstructural and textural analysis of the analyzed samples indicates that three generations of white micas are present (Fig. 3). White micas are the principal constituent of the oldest fabric in the Schist belt (S_{1a}), which is argued to have formed during *HP/LT* metamorphism (Gottschalk, 1990; Little et al., 1994). A second generation of white micas grew during D_{1b} folding, with syn-D_{1b} white micas overgrowing and replacing older white micas at microlithon boundaries. White micas also occur in textural configurations indicating late syn- to post-D_{1b} growth under greenschist-facies conditions. All of the quartz-mica schist samples analyzed are mixtures of the three generations of white mica.

Although we did not carry out a detailed microprobe analysis of each sample, the general nature and composition of white micas in quartz-mica schist from the Schist belt are known and have been discussed elsewhere (Gottschalk, 1987; and this volume, Chapter 9). Quartz-mica schist commonly contains two white micas, phengite and paragonite; phengite and paragonite may exist either as separate grains, or may be intergrown on a subgrain scale. Phengite from quartz-mica schist ranges from 3.25 to 3.48 Si per formula unit (normalized to 11 oxygens), with Mg/(Mg+Fe) between 0.61 to 0.77; Mg/(Mg+Fe) is independent of Si content, and is probably controlled by the bulk composition of the rock. Phengites are virtually devoid of Ca, and K/Ca ratios are in excess of 1,000. Paragonite exhibits little compositional variation and is nearly pure end-member paragonite with Mg/(Mg+Fe) ranging between 0.21 and 0.50, and K/Ca ranging between 4.3 and 12.4. The proportion of phengite to paragonite is highly variable from sample to sample; although phengite is the dominant white mica in most samples, paragonite is common, and in some samples was the sole white mica detected. Therefore the large sample to sample variation in $^{39}Ar_K/^{37}Ar_{Ca}$ is probably related to differences in the proportion of phengite to paragonite.

Sample SBR88-57 produced a relatively simple spectrum, with a preferred date of 118.3 ± 0.3 Ma on steps 4 to 9 (Fig. 4a); because the temperature steps that make up this plateau overlap

**TABLE 3. MICROPROBE ANALYSES OF GARNET AND BIOTITE FROM SAMPLE 8-4-85-3
AND RESULTS OF GARNET–BIOTITE GEOTHERMOMETRY***

	Garnet		Biotite	
SiO_2	37.58		36.62	$(Mb/Fe)_{gt} = 0.062$
TiO_2	0.11		1.45	$(Mg/Fe)_{bi} = 0.920$
Al_2O_3	20.88		16.80	
FeO[†]	27.03		20.04	$lnK_D= ln ([Mg/Fe]_{gt}/[Mg/Fe]_{bi})$
MnO	2.90		0.11	$lnK_D = -2.697$
MgO	0.95		10.34	
CaO	10.65		0.04	
Na_2O	b.d.		0.06	
K_2O	b.d.		9.30	
Σ	100.10%		94.74%	
Si	3.005	Si	2.805	For arbitary pressure P = 5 kbar,
		Al[iv]	1.195	T = 438°C, using geothermome-
Al	1.967			ter of Hodges and Spear, 1982.
Fe^{+3}[§]	0.029	Al[vi]	0.322	
Ti	0.004	Ti	0.083	
		Mg	1.180	
Fe^{+2}**	1.779	Fe	1.283	
Mn	0.195	Mn	0.006	Error based on reproducibility of
Mg	0.112	Σ	2.874	microprobe analyses: ± 15°C.
Ca	0.911			
Σ	2.997	Ca	0.002	
		Na	0.006	
O	6.000	K	0.908	
		Σ	0.916	
		O	11.000	

*Analytical methds summarized in Gottschalk (this volume, Chapter 9).
[†]Total Fe = FeO
[§]$Fe^{+3} = 2 - Al - Ti$
**$Fe^{+2} = 12$ - total cation charge assuming all Fe = Fe^{+2}

at the 3-sigma level, we consider this to be a preferred date. An isochron date calculated from steps 4 to 9 (121.0 ± 0.7 Ma, $(^{40}Ar/^{36}Ar)_i = 176 ± 33$) is slightly older than the preferred date (Fig. 4c). The $(^{40}Ar/^{36}Ar)_i$ value is significantly less than the present-day atmospheric value of 295.5, indicating that younger apparent ages in low-temperature fractions result from ^{40}Ar loss, probably occurring later than 98 Ma. The convex-upward shape of the $^{39}Ar_K/^{37}Ar_{Ca}$ spectrum indicates that sample SBR88-57 was a mixture of low-K and high-K white micas, probably phengite and paragonite (Fig 4b).

Sample SBR88-8a produced an Ar release spectrum with a maximum apparent age of 132 Ma on step 2 dropping to a plateau date of 123.3 ± 0.7 Ma on steps 4 through 8 (Fig. 4d). The isochron date of this sample using all temperature steps is slightly younger than the plateau date (121.5 ± 0.3 Ma), and yielded a $(^{40}Ar/^{36}Ar)_i$ value of 373 ± 2. These data suggest that older ages in steps 1, 2, 3, 9, 10, and 11 result from excess ^{40}Ar. An isochron on steps 4 to 8 produced an apparent age of 124.2 ± 0.3 Ma, identical to the plateau date, with a $(^{40}Ar/^{36}Ar)_i$ value of 255 ± 9 (Fig. 4f); these data indicate that temperature steps 4 to 8 contained no significant excess ^{40}Ar, and may have tapped an Ar reservoir that suffered minor ^{40}Ar loss. The convex-upward shape of the

$^{39}Ar_K/^{37}Ar_{Ca}$ spectrum and relatively low K/Ca values produced by this sample indicate that SBR88-8a is a mixture of white micas, including a significant proportion of paragonite (Fig. 4e).

Sample SBR88-30 produced a saddle-shaped spectrum, with a plateau date of 135 ± 2 Ma on steps 5 through 8 (Fig. 4g). Isochron analysis indicates that the sample is dominated by excess ^{40}Ar, probably residing mainly in the low- and high-temperature fractions that yield apparent ages older than the plateau age. An isochron on steps 5 to 8 produced an apparent age of 145 ± 2 Ma, with $(^{40}Ar/^{36}Ar)_i = 260 ± 7$ (Fig. 4i). While the isochron and plateau dates correlate poorly, the low $(^{40}Ar/^{36}Ar)_i$ value indicates that no significant nonradiogenic ^{40}Ar was released in steps 5 to 8; poor correlation may result from the close grouping of $^{40}Ar/^{36}Ar$ and $^{39}Ar/^{36}Ar$ values used to calculate the isochron. These relations suggest that the plateau date may be somewhat more reliable than the isochron date on the corresponding temperature steps. Low measured $^{39}Ar_K/^{37}Ar_{Ca}$ values (maximum of 79 in the 800°C temperature step) indicate that the sample analyzed is composed mainly of low K/Ca white mica, and is nearly pure paragonite.

White micas from granitic orthogneiss. We have analyzed white micas from two orthogneiss bodies in the Schist belt, both derived from granitic protoliths of Devonian age (Dillon et al.,

TABLE 4. MICROPROBE ANALYSES OF COEXISTING CALCITE AND DOLOMITE IN METAMORPHIC ROCKS NORTH OF THE MINNIE CREEK THRUST, AND RESULTS OF CALCITE-DOLOMITE GEOTHERMOMETRY*

Sample	SBR88-36	SBR88-53	SBR88-70	MS87-16	8-5-85-3
MgO	0.11	0.86	0.19	0.33	1.63
SrO	0.34	0.03	0.43	0.19	0.19
CaO	54.07	54.20	54.60	54.20	51.73
BaO	0.02	0.02	0.03	0.01	0.01
MnO	0.67	0.35	0.07	0.30	0.49
FeO	1.36	0.32	0.80	1.51	2.00
CO_3[†]	43.90	43.90	43.90	43.90	43.90
Σ	100.45%	99.67%	100.02%	100.44%	99.95%
X_{Mg}	0.003	0.021	0.005	0.008	0.040
X_{Fe}	0.018	0.005	0.011	0.021	0.028
(°C)					
T_1[§]	n.a.	340 ± 110	n.a.	160 ± 130	513 ± 10
T_2**	227 ± 60	399 ± 15	263 ± 15	301 ± 25	486 ± 5

*Analytical methods summarized in Gottschalk (this volume, Chapter 9).
[†]CO_3 determined stoichiometrically.
[†]T_1 = °C. Temperatures determined from Anovitz and Essene, 1987, geothermometer.
**T_2 = °C. Temperatures determined from Rice, 1977.
n.a. = thermometer not applicable.

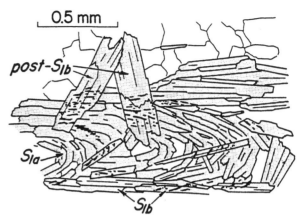

Figure 3. Line drawing from photomicrograph showing microstructural and textural relations in quartz-mica schist from the Schist belt. White micas grains that make up the early foliation (S_{1a}) grew under *HP/LT* metamorphic conditions. D_{1b} folding resulted in the close to isoclinal folding of S_{1a} and the growth of new white mica grains along microlithon boundaries; D_{1b} folding and dynamic recrystallization took place under lower amphibolite- to greenschist-facies conditions, creating an axial-planar differentiated crenulation cleavage, S_{1b}. Late- to postkinematic white mica grains crosscut S_{1a}, S_{1b}, and D_{1b} folds, and also grew under greenschist-facies conditions.

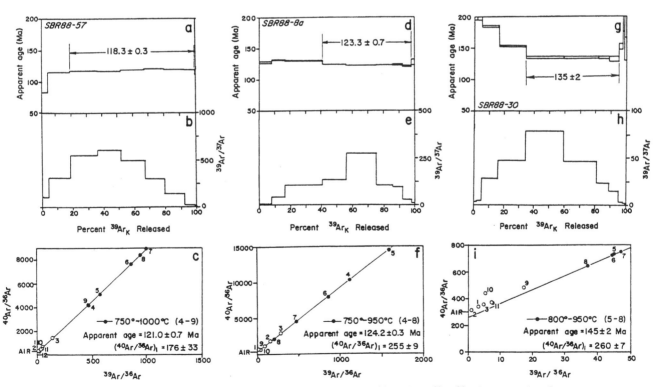

Figure 4. $^{40}Ar/^{39}Ar$ age spectra, $^{39}Ar/^{37}Ar$ spectra, and $^{40}Ar/^{36}Ar$ versus $^{39}Ar/^{36}Ar$ isochron plots for white micas from quartz-mica schist, Schist belt. AIR = atmospheric $^{40}Ar/^{36}Ar$ ratio (295.5).

1980; Dillon, 1989). Samples SBR88-27 and ABR85-17 were separated from deformed parts of the Wild River and Chandalar plutons, respectively (WRP and CP, Fig. 1b). Both samples were separated from blastomylonitic granitic orthogneisses made up mainly of quartz + K-feldspar + white mica + plagioclase + biotite + calcite. In contrast to the polyphase folds and fabrics that characterize the surrounding schistose rocks, granitic orthogneisses have a single, well-defined metamorphic foliation consisting mainly of metamorphic white micas, biotite, and dynamically recrystallized quartz. Although we have not studied the petrology of these granitic orthogneisses in detail, our samples are mineralogically and texturally similar to samples from the Wild River and Chandalar plutons described by Patrick (1995); the mineral chemistry of white micas described in his paper is therefore inferred to be comparable to those analyzed in this study. In both plutonic bodies, nearly all white micas are phengites of metamorphic origin; two compositionally distinct populations are present: In the Chandalar pluton, low-Si phengites average 3.19 Si per formula unit (per 11 oxygens), with Mg/(Mg+Fe) = 0.45; high-Si phengites average 3.30 per formula unit, with Mg/(Mg+Fe) = 0.49. In the Wild River pluton, low-Si phengites average 3.28 Si per formula unit, with Mg/(Mg+Fe) = 0.40; the high-Si phengite population averages 3.48 Si per formula unit, with Mg/(Mg+Fe) =

0.46. Thermobarometric analysis by Patrick (1995) indicates that high-Si phengites formed during *HP/LT* metamorphism of the Schist belt, and low-Si phengites formed at significantly lower pressures as a result of isothermal decompression. K/Ca values in phengites analyzed by Patrick (1995) range from ~500 to >1,000, with no discernible difference between the low-Si and high-Si populations. Because we were not able to physically separate low-Si from high-Si phengites, the samples that we analyzed in this study are probably a mixture of both phengites.

Samples SBR88-27 and ABR85-17 produced relatively simple Ar release spectra, with plateaus developed at 110.4 ± 0.3 Ma and 120.7 ± 0.3 Ma, respectively; isochron dates are identical to plateau dates within analytical error (Fig. 5). For samples SBR88-27 and ABR85-17, isochrons calculated using all temperature steps yielded $(^{40}Ar/^{36}Ar)_i$ values of 244 ± 10 and 198 ± 20, respectively, indicating that young apparent ages in the low-temperature fractions of both samples probably result from slight ^{40}Ar loss following cooling. While the K/Ca ratios for sample SBR88-27 are appropriate for a sample composed mainly of phengite, the K/Ca ratios for ABR85-17 are considerably lower, with a maximum value of about 210. These data indicate that sample ABR85-17 is a mixture of phengite with a lower K phase, possibly paragonite.

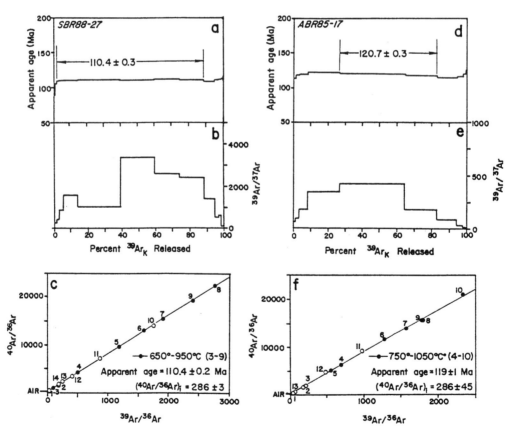

Figure 5. $^{40}Ar/^{39}Ar$ age spectra, $^{39}Ar/^{37}Ar$ spectra, and $^{40}Ar/^{36}Ar$ versus $^{39}Ar/^{36}Ar$ isochron plots for white micas from granitic orthogneiss, Schist belt. AIR = atmospheric $^{40}Ar/^{36}Ar$ ratio (295.5).

White mica and biotite from feldspathic schist. Biotite and white mica were separated from a weakly foliated feldspathic schist consisting principally of albite + white mica + biotite + chlorite + quartz, exposed on the western flank of the Middle Fork Koyukuk River. Polyphase structural fabrics also are not developed in feldspathic schist, and the relationship of white mica and biotite growth to deformational phases in the enclosing quartz-mica schists is not clear. The white mica separate (SBR88-51w) was analyzed twice because approximately 11% of the total ^{39}Ar expelled was lost in the first analysis. Both analyses produced convex-upward spectra increasing in apparent age from about 121 Ma in the low-temperature steps to a maximum apparent age of 141.9 ± 0.4 Ma on step 7 of the second analysis (Fig. 6a). A maximum ^{39}Ar$_K$/^{37}Ar$_{Ca}$ ratio of 239 obtained from the first analysis indicate that this sample probably contains a significant proportion of paragonite; because the second analysis was completed too long after irradiation for accurate measurement of small amounts of radioactive ^{37}Ar, K/Ca ratios were not calculated. A convex-upward Ar-release spectrum is common for white micas from *HP/LT* metamorphic belts, and may result from (1) the degassing of a mixture of two populations of white mica with distinct Ar retention properties (c.f. Wijbrans and McDougall, 1986; Till and Snee, 1995; Roeske et al., 1995), or (2) degassing of a white mica with variable incorporation of nonradiogenic ^{40}Ar. We believe that

^{39}Ar$_K$/^{37}Ar$_{Ca}$ and isochron data (Fig. 6b) support interpretation (1). An isochron calculated from steps 1 through 4 yields an apparent age of 122.6 ± 0.9 Ma with an atmospheric (^{40}Ar/^{36}Ar)$_i$ value (301 ± 4); thus little, if any nonradiogenic ^{40}Ar was released in the low-temperature steps. Steps 5 through 8 plot along a scattered linear trend, with an apparent age of 134 ± 7 Ma, and (^{40}Ar/^{36}Ar)$_i$ = 258 ± 141; although this isochron probably has no age significance, it suggests that the ^{40}Ar in the intermediate temperature steps is also mainly radiogenic in origin and may have been derived from an Ar reservoir that was affected by ^{40}Ar loss. In contrast, high-temperature fractions (steps 9 through 12) yield an apparent isochron age of 127 ± 4 Ma with (^{40}Ar/^{36}Ar)$_i$ = 448 ± 56, indicating that they were derived from an Ar reservoir containing significant excess ^{40}Ar. The ^{39}Ar$_K$/^{37}Ar$_{Ca}$ data indicate that the temperature step that produced the oldest apparent age also yielded the highest K/Ca ratio, with lower K/Ca ratios in the lower and higher temperature steps. We interpret this spectrum as having been derived from a mixture of white micas, probably paragonite and phengite, with the oldest white mica represented by phengite greater than 141.9 ± 0.4 Ma in age. Because little is known about the Ar retention properties of paragonite, we cannot say with confidence that the younger apparent ages have geologic meaning; however, we interpret the original age of closure of phengite in sample SBR88-51w to be older than 142 Ma. The isochron dates reported in Table 1 and

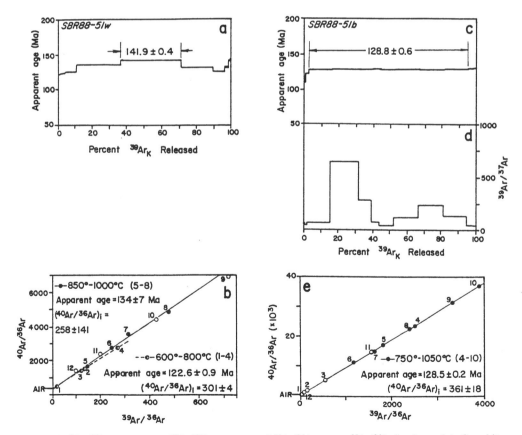

Figure 6. ^{40}Ar/^{39}Ar age spectra, ^{39}Ar/^{37}Ar spectra, and ^{40}Ar/^{36}Ar versus ^{39}Ar/^{36}Ar isochron plots for white mica and biotite from feldspathic schist, Schist belt. AIR = atmospheric ^{40}Ar/^{36}Ar ratio (295.5).

Appendix 1 were calculated from all temperature steps, and have no geologic significance.

Biotite (SBR88-51b) yielded a well-behaved plateau with an apparent age of 128.8 ± 0.6 Ma on steps 4 through 10 (Fig. 6c); a rerun of this sample (Appendix 1) produced identical results. In both cases, isochron dates are identical to plateau dates.

Amphibole. Clear to light green, unzoned actinolitic amphibole was separated from a massive metabasite composed mainly of albite + chlorite + actinolite + calcite + sphene (sample SBR88-8). Its Ar release spectrum yielded no plateau (Fig. 7a). The largest gas fraction (25.8%), the highest radiogenic yield (60.2%), and the youngest apparent age (137 ± 1 Ma) were produced from the $1,000°C$ temperature step (step 6). Apparent ages greater than 387 Ma in steps 1 through 3 result from excess ^{40}Ar; the $^{39}Ar_K/^{37}Ar_{Ca}$ plot indicates that excess ^{40}Ar in steps 1 to 3 was derived from a relatively high K/Ca phase other than actinolite (Fig. 7b). An isochron plot of steps 4 through 10 (Fig. 7c) yields an apparent age of 160 ± 13 Ma. The ($^{40}Ar/^{36}Ar$) value for this isochron is 301 ± 26, within analytical error of present-day atmospheric values, suggesting that excess ^{40}Ar is absent in steps 4 through 10. It is therefore possible that the isochron date is geologically significant. However, low $^{39}Ar_K/^{37}Ar_{Ca}$ ratios (0.01 for steps 4 through 10) indicate that the actinolite analyzed in this sample is K-poor. Furthermore the error on the ($^{40}Ar/^{36}Ar$) of the 160 ± 13-Ma isochron is sufficiently large that the presence of excess ^{40}Ar cannot be ruled out. Therefore this isochron date should be treated with appropriate caution.

Sample SBR88-56 is an amphibole separated from a massive to weakly foliated metagabbro that crops out immediately north of the Slate Creek fault, a south-dipping extensional shear zone that separates the Schist belt from the overlying Phyllite belt (Gottschalk, 1990; Little et al., 1994; Gottschalk et al., this volume, Chapter 12; Fig. 1b). Microprobe analysis indicates that this amphibole is compositionally transitional between barroisite and hornblende, and contains rare cores and patches of crossite. Barroisitic-hornblende contains between 0.20 and 0.39 wt. % K_2O; crossite is very low in K (0.03 to 0.05 wt. % K_2O) and probably contributed only small amounts of gas during isotopic analysis. Analyses of amphibole from metagabbro are presented in Gottschalk (this volume, Chapter 9, Table 2, analyses 8, 9, and 10). Sample SBR88-56 produced a surprisingly well behaved Ar release spectrum for a low-K amphibole, with a plateau date at 115 ± 1 Ma on steps 6 through 9 (Fig. 7d). Isochron analysis indicates that older apparent ages determined for the high- and low-temperature fractions result from excess ^{40}Ar; the $^{39}Ar_K/^{37}Ar_{Ca}$ plot shows that excess ^{40}Ar was released from a phase with a K/Ca ratio somewhat lower than barroisitic-hornblende, possibly crossite. An isochron on steps 6 through 9 produced a date of 114 ± 1 Ma, identical to the plateau date, with an ($^{40}Ar/^{36}Ar$)$_i$ value within analytical error of atmospheric $^{40}Ar/^{36}Ar$ (301 ± 4); these data support the validity of the plateau date.

Amphiboles from eclogite. Two amphibole samples were separated from a body of eclogite in the Schist belt exposed

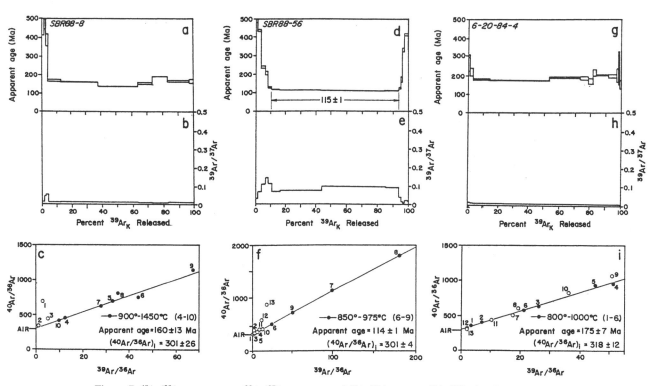

Figure 7. $^{40}Ar/^{39}Ar$ age spectra, $^{39}Ar/^{37}Ar$ spectra, and $^{40}Ar/^{36}Ar$ versus $^{39}Ar/^{36}Ar$ isochron plots for amphiboles from the Schist belt. AIR = atmospheric $^{40}Ar/^{36}Ar$ ratio (295.5).

along the eastern flank of the Middle Fork Koyukuk River. The geology of this eclogite occurrence, and microprobe analyses of amphiboles from eclogite are presented elsewhere in this volume (Gottschalk, this volume, Chapter 9). Sample 6-20-84-4 is a hornblende separated from a part of the eclogite body that was metamorphosed and recrystallized during retrograde metamorphism. Hornblende was separated from a pervasively recrystallized, weakly foliated sample consisting of hornblende + garnet + epidote + chlorite + albite + sphene; garnet-hornblende geothermometry on this sample yielded an estimated temperature of 478 ± 25°C. The amphibole separated from sample 6-20-84-4 is a blue-green ferro-hornblende; some grains contain clear actinolitic-hornblende cores that grade outward to ferro-hornblende rims. Sample 6-20-84-4 was carefully hand picked to eliminate amphiboles with obvious zoning, but minor amounts of actinolitic-hornblende were probably present in the analyzed sample. Microprobe analysis indicates that ferro-hornblende is a low-K amphibole containing between 0.02 and 0.08 wt. % K_2O.

The Ar release spectrum from sample 6-20-84-4 is slightly saddle shaped, and has a character similar to age spectra derived from samples containing excess ^{40}Ar (Fig. 7g). An isochron calculated from all temperature steps yields an apparent age of 188 ± 6 Ma, with an $(^{40}Ar/^{36}Ar)_i$ value of 304 ± 7. Because the $(^{40}Ar/^{36}Ar)_i$ of this isochron is within analytical error of atmospheric $^{40}Ar/^{36}Ar$ it may erroneously be inferred that no nonradiogenic ^{40}Ar was released. However, isochron analysis of the first six temperature steps (800 to 1,000°C, comprising 79.1% of the total released ^{39}Ar) yielded an $(^{40}Ar/^{36}Ar)_i$ value of 318 ± 12, significantly above atmospheric $^{40}Ar/^{36}Ar$ confirming the presence of excess ^{40}Ar (Fig. 7i). The 950°C gas fraction yielded the youngest apparent age (175 ± 2 Ma), the largest percentage of total gas (38.7%), and the highest radiogenic yield (68.9%).

Sample 6-20-84-7 is actinolite that was separated from late-stage monominerallic actinolite veins that crosscut both primary and retrograded parts of the eclogite body. Microprobe analysis of this actinolite indicates that it contains 0.13 to 0.20 wt. % K_2O. Sample 6-20-84-7 produced an unusual Ar release spectrum, dominated by excess ^{40}Ar in all but the 1,150°C and 1,450°C temperature steps; these fractions yielded apparent ages of 177.1 ± 0.5 Ma and 141 ± 2 Ma, respectively.

Sodic-amphibole–bearing rocks north of the MCT

Sample 8-4-85-5 is a quartz + white mica + chlorite + albite + garnet schist from immediately north of the MCT, within the belt of sodic-amphibole–bearing schists described by Till and Moore (1991). This sample was affected by both D_{1a} and D_{1b} deformation, as well as a third, semipenetrative deformational phase (D_{1c} of Gottschalk et al., this volume, Chapter 12) during which minor white mica growth took place. Our sample consisted of a mixture of syn-D_{1a} and D_{1b} white micas along with a minor amount of post-D_{1b} white micas. A white mica separate

yielded a convex-upward spectrum, with apparent ages increasing from a minimum of 117 Ma in the low-temperature steps, to a maximum of 129.1 ± 0.4 Ma in step 8 (1,000°C fraction; Fig. 8c). An isochron plot of steps 6 through 9 yielded $(^{40}Ar/^{36}Ar)_i$ values well below present-day atmospheric levels (110 ± 158), indicating significant loss of radiogenic ^{40}Ar (Fig. 8d). In contrast, an isochron plot of steps 1 through 5 yielded an apparent age of 120 ± 1 Ma, with a near-atmospheric $(^{40}Ar/^{36}Ar)_i$ value of 291 ± 8. These data suggest the degassing of two populations of white mica with differing Ar retention properties. The oldest population is greater than 129 Ma in age and has suffered severe ^{40}Ar loss. The younger generation of white mica appears to have experienced significantly less ^{40}Ar loss. The contrast in the degree of ^{40}Ar loss between the low- and high-temperature fractions suggests that ^{40}Ar loss from the population sampled in the high-temperature steps occurred during growth of the younger white micas; both appear to have suffered minor ^{40}Ar loss at 117 Ma or younger. A $^{39}Ar_K/^{37}Ar_{Ca}$ spectrum for this sample is unavailable because the time between irradiation and analysis was too long to allow for accurate measurement of radioactive ^{37}Ar. The isochron date reported in Appendix 1 was determined from all temperature steps, and has no geologic significance.

Skajit allochthon (senso stricto)

We have analyzed one white mica and two amphibole samples from rocks north of the MCT where sodic-amphiboles have not been identified. Sample MS87-14 is a white mica separate from a calcareous phyllite in the highest structural levels of the Skajit allochthon. Because sample MS87-14 was collected from a part of the Skajit allochthon affected only by syn-D_{1a} metamorphism, one generation of white mica is present, occurring as the principal foliation-forming mineral with no later overgrowths. Sample MS87-14 produced a complex Ar release spectrum, with a series of apparent ages between 103 and 127 Ma (steps 2 through 9) and no plateau (Fig. 8a). An isochron plot of steps 2 through 9 produced a scattered linear trend, with an apparent age of 125.3 ± 0.7 Ma, and $(^{40}Ar/^{36}Ar)_i$ = 260 ± 5 (Fig. 14b). Apparent ages in excess of 127 Ma, determined for the low- and high-temperature steps, are interpreted as resulting from excess ^{40}Ar; isochron analysis indicates that excess ^{40}Ar was expelled mainly in steps 1, 10, 11, and 12. We interpret this spectrum as one of ^{40}Ar loss on a white mica older than 127 Ma at 100 Ma or younger; incorporation of nonradiogenic ^{40}Ar occurred after thermal perturbation. It is possible, however, that the young apparent age in step 2 resulted from degassing of a phase other than a potassic white mica and may not represent the maximum age for a thermal perturbation resulting in Ar loss. Due to the length of time between irradiation and analysis, we were unable to accurately measure $^{39}Ar_K/^{37}Ar_{Ca}$; therefore, we cannot distinguish between these two possible explanations for the anomalously young apparent age in step 2.

Amphibole was separated from two metabasites from the southern Skajit allochthon (samples SBR88-1 and SBR88-40).

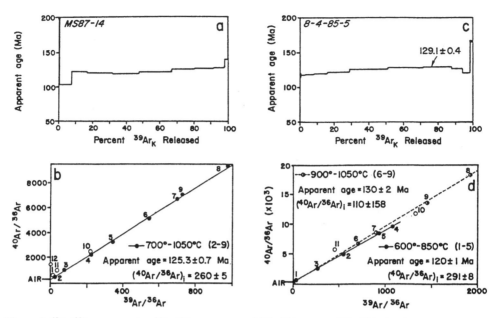

Figure 8. $^{40}Ar/^{39}Ar$ age spectra, $^{39}Ar/^{37}Ar$ spectra, and $^{40}Ar/^{36}Ar$ versus $^{39}Ar/^{36}Ar$ isochron plots for white micas from metamorphic rocks north of the Minnie Creek thrust. AIR = atmospheric $^{40}Ar/^{36}Ar$ ratio (295.5).

Sample SBR88-1 is a massive mafic schist composed of relict igneous hornblende and biotite surrounded by metamorphic albite + chlorite + quartz + epidote + sphene. A hornblende separate yielded a complex Ar release spectrum that can be interpreted either as an amphibole in excess of 420 Ma in age that experienced a severe ^{40}Ar loss during a thermal disturbance at <222 Ma, or an amphibole <222 Ma in age, whose spectrum is dominated by excess ^{40}Ar (Fig. 9a). Based upon the following relations, we favor the former interpretation: (1) An isochron plot of all data indicates that steps 5 through 17 form a scattered linear array with an apparent age of 425 ± 12 Ma, with a $(^{40}Ar/^{36}Ar)_i$ of 206 ± 16 Ma; the computed $(^{40}Ar/^{36}Ar)_i$ value lies considerably below present-day atmospheric $^{40}Ar/^{36}Ar$ levels, and confirms apparent Ar loss. (2) Steps 1 through 4 lie significantly above the linear array formed by the remainder of the temperature steps, indicating that dates greater than 600 Ma in the low-temperature fractions may result from excess ^{40}Ar. (3) An isochron plot of 12 through 17 yields an apparent age of 420 ± 3 Ma with $(^{40}Ar/^{36}Ar)_i$ at nearly present-day atmospheric values (285 ± 3; Fig. 9c).

Sample SBR88-40 is actinolite separated a weakly foliated metabasite composed of quartz + actinolite + epidote + albite + calcite + sphene + chlorite; actinolite is zoned, with light green cores grading outward to clear rims. Sample SBR88-40 yielded a saddle-shaped Ar release spectrum (Fig. 9d); an isochron plot of data shows considerable scatter, and together with the age spectrum, indicates that the age spectrum results from radiogenic and excess ^{40}Ar mixed in variable amounts. The actual age of this amphibole is probably younger than 192.5 ± 1.0 Ma, the youngest apparent age determined in step 4.

Phyllite belt

White mica was separated from a fine-grained quartz-mica schist from the Phyllite belt (sample 7-26-84-5). Although this sample displays deformation microstructures in quartz indicative of syntectonic recrystallization during D_{1a} and D_{1b} folding, white mica growth occurred almost exclusively during D_{1a}, and there is no evidence of relict detrital mica. Sample 7-26-84-5 yielded a complex Ar release spectrum (Fig. 10a) with young apparent ages in the low-temperature steps (100.0 and 113.6 Ma, steps 1 and 2), apparent ages ranging between 123 and 147 Ma on steps 3 through 9, and significantly older apparent ages (199 to 508 Ma) on steps 10 through 12. No plateau was developed. An isochron analysis of steps 3 through 9 produced an apparent age of 126 ± 2 Ma, with $(^{40}Ar/^{36}Ar)_i = 542 ± 77$, indicating a significant component of nonradiogenic ^{40}Ar. The convex-upward $^{39}Ar_K/^{37}Ar_{Ca}$ spectrum suggests that Ar is derived from a mixture of a K-bearing white mica and a low-K white mica, possibly paragonite (Fig. 10b); the low-K phase controls the low-temperature parts of the $^{39}Ar_K$ spectrum that exhibit apparent ^{40}Ar loss, and the high-temperature parts of the spectrum that contain excess ^{40}Ar. Maximum $^{39}Ar/^{37}Ar$ values occur in steps 3 to 5, indicating that gas derived from high-K white mica predominated. Steps 3 to 5 also correspond to a tight grouping of apparent ages (123.0 to 126.3 Ma), and 63.4% of the total ^{39}Ar released was produced in these temperature steps. An isochron plot on steps 3 to 5 yielded an apparent age of 124 ± 1 Ma, with $(^{40}Ar/^{36}Ar)_i = 340 ± 85$; this $(^{40}Ar/^{36}Ar)_i$ value is within analytical error of atmospheric $^{40}Ar/^{36}Ar$, indicating a possible absence of excess ^{40}Ar in steps 3 to 5. We regard the isochron date as the best date for sample 7-26-84-5.

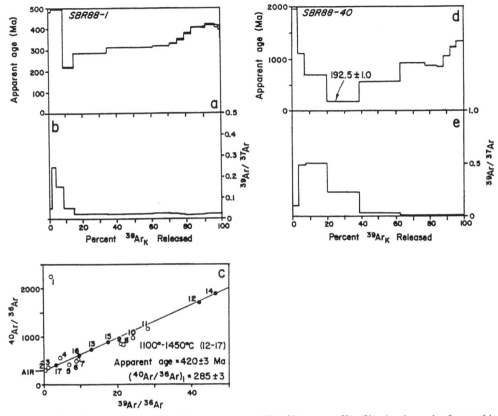

Figure 9. $^{40}Ar/^{39}Ar$ age spectra, $^{39}Ar/^{37}Ar$ spectra , and $^{40}Ar/^{36}Ar$ versus $^{39}Ar/^{36}Ar$ isochron plot for amphiboles from metamorphic rocks north of the Minnie Creek thrust. AIR = atmospheric $^{40}Ar/^{36}Ar$ ratio (295.5).

INTERPRETATION OF RESULTS

Schist belt

White mica. White mica separates from quartz-mica schist and granitic orthogneiss were a mixture of white micas that grew during *HP/LT* metamorphism and associated deformation (D_{1a}), and a second generation of white micas that grew during isothermal decompression. In the quartz-mica schists, second-generation white micas overgrew and partially replaced *HP/LT* white micas during the D_{1b} deformational event, which occurred under lower amphibolite- to greenschist-facies metamorphic conditions. In the granitic orthogneisses that we analyzed, polyphase fabrics were absent but second-generation white micas apparently grew at the expense of *HP/LT* white micas according to the reaction: high-Si phengite = low-Si phengite + biotite + K-Feldspar + quartz + H_2O (Patrick, 1995). Because quartz-mica schist and granitic orthogneiss from the Schist belt both record significant decreases in metamorphic pressure between the growth of first- and second-generation white micas, we argue that second-generation white micas in both rock types grew during the D_{1b} deformational/metamorphic event. The absence of polyphase folds in granitic orthogneisses may be explained by ductility contrast between the two rock types causing folding to be partitioned preferentially into the less rigid quartz-mica schist.

Dynamic recrystallization of granitic orthogneisses during D_{1b} appears to have been sufficiently extensive, however, to promote significant growth of white mica neoblasts.

Despite the presence *HP/LT* white mica in all samples that we analyzed, most of the white micas separated from quartz-mica schist and granitic orthogneiss yielded plateau and preferred dates ranging from 110 to 135 Ma, significantly younger than the Middle Jurassic age inferred for *HP/LT* metamorphism in the Schist belt by Armstrong et al. (1986) and Christiansen and Snee (1994). In most cases where samples yielded individual gas fractions with apparent ages older than Early Cretaceous, isochron analysis showed that these fractions contained a significant component of nonradiogenic ^{40}Ar. One possible explanation for the apparent lack of isotopic evidence of Middle Jurassic *HP/LT* metamorphism may be that blueschist-facies metamorphism took place at temperatures above the closure temperature for white mica; in this case Early Cretaceous white mica dates may record the cooling of the Schist belt through the white mica closure temperature. Alternatively, the 110- to 135-Ma clustering of white mica plateau and preferred dates may have resulted from nearly complete degassing of *HP/LT* white micas during an Early Cretaceous metamorphic/deformational event, effectively erasing much of the isotopic evidence of Middle Jurassic metamorphism. Our data indicate that later degassing of *HP/LT* white micas is the most likely scenario because one of our samples (SBR88-51w) shows evidence of par-

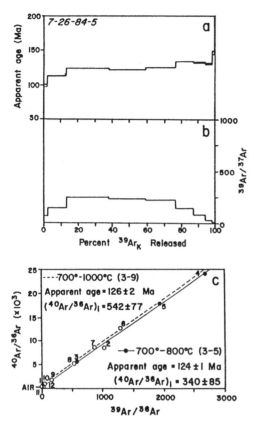

Figure 10. $^{40}Ar/^{39}Ar$ age spectrum, $^{39}Ar/^{37}Ar$ spectra, and $^{40}Ar/^{36}Ar$ versus $^{39}Ar/^{36}Ar$ isochron plot for white micas from fine-grained quartz-mica schist, Phyllite belt. AIR = atmospheric $^{40}Ar/^{36}Ar$ ratio (295.5).

tially retained Ar that may hearken back to *HP/LT* metamorphism. Sample SBR88-51w is discussed in detail below.

There is considerable controversy over the relative importance of processes that result in the complete or partial degassing of white micas in polydeformed and polymetamorphosed *HP/LT* metamorphic belts. Factors cited include (1) loss of Ar from older metamorphic phases by volume diffusion during a discrete, later thermal event (Wijbrans and McDougall, 1986; Scaillet et al., 1992); (2) recrystallization of older mineral phases during deformation; (3) increased hydrothermal activity associated with deformational processes (Chopin and Maluski, 1980; Chopin and Monie, 1984); or (4) some combination of the above processes. Although our data do not rule out the possibility that the D_{1a} and D_{1b} deformational events were each accompanied by discrete thermal pulses, there is no textural evidence to suggest a cooling phase between D_{1a} and D_{1b}; instead phase relations in the Schist belt (discussed in Gottschalk, this volume, Chapter 9) indicate that the D_{1a} and D_{1b} deformational events occurred as the Schist belt underwent nearly isothermal decompression at temperatures between 380 to 480°C. It is therefore unlikely that volume diffusion of Ar during a discrete, short-lived thermal pulse was a viable mechanism for degassing of syn-D_{1a} white micas. Instead, abundant petrologic, microstructural, and textural evidence indicates that the syn-D_{1b} recrystallization of

white mica, accompanied by growth of new white mica grains is the principal cause for the partial, and in many cases, complete degassing of older white micas.

Although most white mica samples from the Schist belt have been entirely reset during D_{1b} deformation and metamorphism, white micas separated from a sample of feldspathic schist (sample SBR88-51w) have partially retained Ar that may record the age of *HP/LT* metamorphism in the south-central Brooks Range. Samples SBR88-51w produced convex-upward Ar-release spectra with the oldest apparent ages in the most prolific gas-producing temperature steps. Spectra of this type have been produced by white micas from elsewhere in the Brooks Range (Christiansen and Snee, 1994; Till and Snee, 1995), as well as from *HP/LT* metamorphic belts elsewhere in the world (for example, Wijbrans and McDougall, 1986; Montigny et al., 1988; Hammerschmidt and Frank, 1991; Scaillet et al., 1992). Wijbrans and McDougall (1986) have demonstrated that convex-upward Ar release spectra from white micas may result from the degassing of a mixture of older, partially degassed white micas, and a younger, compositionally distinct population of white micas. We argue that Ar spectra derived from sample SBR88-51w may be similar in origin to those produced experimentally by Wijbrans and McDougall (1986) because our isochron analysis indicates that sample SBR88-51w contains a population of white mica that has experienced Ar loss, and a second population that yielded a nearly atmospheric $(^{40}Ar/^{36}Ar)_i$ value and an isochron date of 122.6 ± 0.9 Ma. We argue that the white mica population that experienced Ar loss is composed of phengite that was produced during *HP/LT* metamorphism and experienced partial Ar loss during D_{1b} deformation and metamorphism. We consider that the gas yielded in the temperature step with maximum apparent age (142 Ma) is a minimum age for *HP/LT* metamorphism in the Schist belt.

Due to uncertainties in the closure temperature of phengitic white mica the 110- to 135-Ma white mica dates cannot be uniquely interpreted. Uncertainty in the closure temperature of white mica is related, in part, to its compositional dependence that has yet to be fully quantified (Scaillet et al., 1992). (1) Early Cretaceous ages may date the cooling of D_{1b} white micas *and* D_{1a} white micas that were degassed during D_{1b} if the closure temperature(s) for Ar diffusion in D_{1a} and D_{1b} white micas (T_c) is less than metamorphic temperatures (T_c < approximately 380 to 480°C). (2) Alternatively, 110- to 135-Ma white mica dates may record the age of D_{1b} deformation and recrystallization if D_{1b} took place at temperatures below the closure temperature of phengite (i.e., T_c > approximately 380 to 480°C); closure temperatures in this range have been inferred from studies indicating that phengitic white micas in *HP/LT* metamorphic belts often retain Ar even when remetamorphosed under amphibolite-facies metamorphic conditions (Chopin and Maluski, 1980; Chopin and Monie, 1984; Wijbrans and McDougall, 1986; Hammerschmidt and Frank, 1991; Scaillet et al., 1992; Christiansen and Snee, 1994). Scaillet et al. (1992) have shown that high-Mg phengites may be particularly resistant to degassing and may have closure temperatures significantly in excess of the closure temperature of muscovite. Because white micas from quartz-mica schists and granitic orthogneisses from the Schist belt have relatively high (Mg/Mg+Fe) contents, closure temperatures in excess of 380 to 480°C are possible.

Amphibole. Of the amphiboles from the Schist belt that we analyzed, barroisitic-hornblende from metagabbro near the southern margin of the Schist belt produced the most conclusive results. Because the metagabbro from which sample SBR88-56 was separated is massive and unfoliated, the relationship of barroisitic-hornblende growth to deformation in the adjacent quartz-mica schist is not known. However, replacement of crossite by barroisitic-hornblende is indicative of metamorphic decompression and correlates well with the decrease in pressure between the D_{1a} and D_{1b} deformational events in the enclosing schistose rocks (Gottschalk, this volume, Chapter 9). Because mineral geothermometry on rocks dynamically recrystallized during D_{1b} yield estimated temperatures between 380 and 480°C (Table 2), metamorphic temperatures were probably below or near to the closure temperature for Ar diffusion in barroisitic-hornblende, here inferred to be similar to the closure temperature of hornblende (~500°C according to Harrison, 1981). These data suggest that the plateau date of 115 ± 1 Ma on sample SBR88-56 is a crystallization age that dates retrograde metamorphism of metagabbro. The date should be treated with caution, however, because barroisitic-hornblende is K-poor and radiogenic yields are low (Appendix 1).

The age of retrograde metamorphism of eclogite is constrained by our analysis of ferro-hornblende separated from retrograded eclogite. Because the Ar release spectrum of sample 6-20-84-4 is dominated by excess ^{40}Ar, we conclude that the apparent age of the youngest and most gas-prolific temperature step (175 Ma) represents a maximum age for the retrograde overprint.

The nearly atmospheric $(^{40}Ar/^{36}Ar)_i$ value yielded by isochron analysis of actinolite from massive metabasite (sample SBR88-8) suggests that the isochron date of 160 ± 13 Ma may be geologically significant, despite the relatively large errors. Although the closure temperature for Ar diffusion in actinolite is not precisely known, it may be comparable to the closure temperature of hornblende. Therefore, the isochron date of 160 ± 13 Ma may represent a crystallization age for actinolite. Because the metabasite from which actinolite was separated is massive, unfoliated, and lacks definitive evidence for retrograde metamorphism, the timing of actinolite growth with respect to deformational/metamorphic events in the enclosing schists is not known. Moreover, the assemblage albite + chlorite + actinolite + calcite + sphene is common in both blueschist- and greenschist-facies metabasites (cf. Owen, 1989; Evans, 1990). It is noteworthy, however, that the date of 160 ± 13 Ma is comparable to the Middle Jurassic age for *HP/LT* metamorphism inferred by Armstrong et al. (1986) and Christiansen and Snee (1994).

Biotite. The plateau date of 129 Ma on sample SBR88-51b is interpreted to date the cooling of biotite through its closure temperature (approximately 300°C from Harrison et al., 1985). This date is comparable to plateau dates on white micas from the nearby quartz-mica schists (120 to 125 Ma, from Blythe et al.,

this volume, Chapter 10) and younger than the 135- ± 2-Ma plateau date on paragonite from the Emma Dome area (sample SBR88-30). More significantly, the cooling age of biotite is significantly older than the 115-Ma crystallization of barroisitic-hornblende, implying that Schist belt exposed locally had cooled below 300°C while rocks along the southern flank of the Schist belt were still within the lower amphibolite to greenschist facies.

Phyllite belt

Although isochron analysis of sample 7-26-84-5 yielded an $(^{40}Ar/^{36}Ar)_i$ with fairly large error, the isochron date of 124 ± 1 Ma on white mica from the Phyllite belt may have geological significance. Based on the following relations, we suggest that the isochron of 124 ± 1 Ma may date D_{1b} deformation in the Phyllite belt: (1) The Phyllite belt was at temperatures between approximately 300 and 350°C during both the D_{1a} and D_{1b} folding events. Because white mica from the Phyllite belt is phengitic (Gottschalk, unpublished data) it is likely that its closure temperature is greater than that of muscovite (i.e., greater than approximately 325°C). Therefore white micas in sample 7-26-84-5 probably grew at temperatures near or below their closure temperature. (2) The structural style and superposition relationships of folds in the Phyllite belt and Schist belt of the south-central Brooks Range are similar (taking into account the differences in metamorphic grade between the two belts; Dillon et al., 1986; Gottschalk, 1987; Gottschalk et al., this volume, Chapter 12). If D_{1a} structures in the Schist belt and Phyllite belt formed during the same deformational event, then D_{1a} in the Phyllite belt may also be Middle Jurassic in age. Because the isochron date of 124 ± 1 Ma is significantly younger than the presumed age for D_{1a}, we suggest that the date may record the age of deformation and dynamic recrystallization associated with D_{1b} at temperatures below the closure temperature of white mica. If so, these data indicate that deformation and recrystallization during D_{1b} was sufficiently intense to cause nearly complete degassing of syn-D_{1a} white micas. Our results are similar to, but somewhat older than a plateau date of 113.3 ± 0.5 Ma on white mica from the Phyllite belt reported by Blythe et al. (this volume, Chapter 10).

Sodic-amphibole–bearing rocks north of the MCT

White mica from sodic-amphibole–bearing rocks north of the MCT (sample 8-4-85-5) yielded a convex-upward Ar-release spectrum that provides age constraints on two distinct populations of white micas. Textural and microstructural relations in sample 8-4-85-5 indicate that the presence of isotopically distinct white mica populations is probably related to growth during two folding events (D_{1a} and D_{1b}). Calcite-dolomite and garnet-biotite geothermometry carried out on samples collected from within 1.5 km of sample 8-4-85-5 yielded estimated temperatures of 486 and 438°C, respectively (Fig. 2); textural and microstructural relations in these samples suggests that estimated temperatures are representative of D_{1b} metamorphic conditions. Metamorphic temperatures during D_{1a} are not well constrained by the mineral assemblages in the schists north of the MCT, although Gottschalk (1987) reports syn-D_{1a} mineral assemblages that may have formed under either greenschist- or blueschist-facies metamorphic conditions.

The convex-upward Ar-release spectrum of sample 8-4-85-5 suggests a metamorphic history similar to sample SBR88-51w, with syn-D_{1a} white micas having formed prior to 129 Ma. Syn-D_{1a} white micas apparently suffered severe Ar loss as a result of D_{1b} folding and recrystallization. Because white micas from quartz-mica schists north of the MCT are phengitic (cf. Gottschalk, 1987, and unpublished data), their closure temperature is not known and may be in the range of temperature estimates for D_{1b}. Therefore it is not certain whether the isochron date of 120 ± 2 Ma on temperature steps 1 through 5 is a post-D_{1b} cooling age, or whether the isochron date records the age of dynamic recrystallization and white mica growth during D_{1b} folding.

Skajit allochthon (senso stricto)

White mica sample MS87-14 was collected from a portion of the Skajit allochthon affected by a single synmetamorphic phase of deformation (D_{1a}). Calcite-dolomite geothermometry on a calcareous phyllite 0.8 km to the west of the sample site yielded an estimated temperature of about $301 \pm 25°C$ (Table 4; Fig. 2). We suggest that synmetamorphic deformation occurred prior to 127 Ma because white micas in sample MS87-14 grew at temperatures at or below their closure temperature (assumed to be that of muscovite, 325°C), and therefore record a crystallization age. In addition, sample MS87-14 yielded a Ar-release spectrum indicating that white micas crystallized prior to 127 Ma suffered Ar loss at about 100 Ma. Ar loss may have occurred during one of the postmetamorphic deformational events that affected rocks in the central Skajit allochthon (cf. Oldow et al., this volume, Chapter 8).

Relict igneous amphibole separated from a metamorphosed mafic intrusive body located in the southern Skajit allochthon (sample SBR88-1) yielded a thermally disturbed spectrum with old apparent ages of >420 Ma. Thus, 420 Ma is interpreted as the minimum age of intrusion.

The saddle-shaped spectrum produced by metamorphic actinolite from sample SBR88-40 indicates that the metamorphic event during which actinolite grew occurred after 192.5 Ma.

DISCUSSION

One of the principal objectives of this study was to constrain the age of HP/LT metamorphism in the south-central Brooks Range. Although we have been unable to date the HP/LT metamorphic event directly, our analysis of white mica from a feldspathic schist in the Schist belt yielded a minimum age for HP/LT metamorphism of 142 Ma. In addition, the isochron date of 160 Ma on actinolite from the Schist belt may record the age of actinolite growth under HP/LT metamorphic conditions. These data are in good agreement with Armstrong et al. (1986), Till and Patrick (1991), and Christiansen et al. (1994) who have proposed

minimum ages for HP/LT metamorphism in the Schist belt of 160 Ma, 149 Ma, and 171 Ma, respectively.

We infer a metamorphic history similar to the Schist belt for sodic-amphibole–bearing metamorphic rocks north of the MCT. Although a direct structural link between sodic-amphibole growth and D_{1a} fabrics has not been established, we infer that HP/LT metamorphism and D_{1a} folding are concurrent, and occurred prior to 129 Ma. This age is significantly older than the 120-Ma blueschist-facies metamorphic event recorded in rocks roughly correlative to the Skajit allochthon in the western Brooks Range (Till and Snee, 1995), and implies that sodic-amphibole-bearing rocks may be part of the Schist belt in the footwall of the MCT, as argued by Till and Moore (1991). Alternatively, sodic-amphibole–bearing schists may be a portion of the Skajit allochthon that locally experienced HP/LT metamorphism prior to 129 Ma (Oldow et al., this volume, Chapter 8).

As discussed previously, our data do not allow us to uniquely determine whether Early Cretaceous white mica dates are post-D_{1b} cooling ages, or whether they record the degassing of HP/LT white micas and growth of white mica neoblasts during deformation at temperatures below the closure temperature of phengite. Based upon the following relations, we prefer the latter interpretation:

1. All of the rock types from the Schist belt that we have dated show evidence of mineral growth during isothermal decompression. In quartz-mica schists, HP/LT mineral assemblages are almost entirely obliterated by lower amphibolite- to greenschist-facies assemblages that grew during D_{1b} folding. Metamorphic decompression and associated deformation(?) of granitic orthogneisses apparently caused the growth of white micas that are significantly lower in Si than white micas produced during HP/LT metamorphism (Patrick, 1995). Massive metabasites from the Schist belt such as eclogite and crossite-bearing metagabbro have suffered relatively minor penetrative deformation, but decompression is manifested as mineral zoning and crystallization of secondary mineral assemblages in hydrothermal veins (Gottschalk, this volume, Chapter 9). In spite of textural differences in decompression reaction textures, petrologic analysis of retrograde mineral assemblages in all rock types indicates that they formed under similar pressure and temperature conditions. We infer from these relations that all rock types dated in this study experienced regional isothermal decompression that was roughly contemporaneous with D_{1b} folding, penetrative deformation, and associated dynamic recrystallization.

2. The predominance of Early Cretaceous white mica plateau, preferred, and isochron dates, rather than discordant, convex-upward Ar-release spectra, indicates that white micas that were produced during HP/LT metamorphism were, in most cases, entirely degassed during a later thermal or deformational event. Because there is no textural evidence for cooling between D_{1a} and D_{1b}, D_{1b} probably occurred as HP/LT metamorphic rocks passed through the amphibolite and greenschist facies during isothermal decompression; therefore evidence does not support Ar loss by volume diffusion of Ar during a discrete post-D_{1a}

thermal pulse. These relations, together with evidence for syn-D_{1b} dynamic recrystallization in nearly all of the rocks that we have dated, indicate that Early Cretaceous white mica dates are best explained by degassing of *HP/LT* white micas triggered by syntectonic recrystallization and growth of white mica neoblasts during D_{1b} deformation.

3. White micas and amphibole from the Schist belt yield similar dates despite differences in closure temperature. White micas from quartz-mica schists yield dates that range from 135 to 118 Ma, white micas from granitic orthogneisses yield dates of 110 and 121 Ma, and barroisitic-hornblende yields a date of 115 Ma. As discussed above, textural, microstructural, and petrologic relations in quartz-mica schists and granitic orthogneisses indicate that degassing of *HP/LT* white micas, and growth of white mica neoblasts probably occurred in response to deformation during decompression (D_{1b}). Because barroisitic-hornblende grew during decompression of crossite-bearing metagabbro and because all rock types yield similar ages, we argue that 135- to 110-Ma white mica and barroisitic-hornblende ages from the Schist belt record decompression of the Schist belt through the lower amphibolite and greenschist facies accompanied by D_{1b} penetrative deformation. A cooling age of 129 Ma on biotite may provide a local minimum age for the onset of decompression, and indicates that uplift did not occur simultaneously throughout Schist belt.

4. White mica and barroisitic-hornblende dates from the Schist belt are similar to white mica dates from the Phyllite belt and sodic-amphibole–bearing rocks north of the MCT that are inferred to record the age of D_{1b} deformation.

Altogether, our data suggest that the D_{1b} deformational event in the south-central Brooks Range took place between about 135 through 110 Ma. Our results are similar to those of Till and Patrick (1991) and Christiansen and Snee (1994) who recognize a deformational event in the Schist belt that was accompanied by greenschist- to amphibolite-facies metamorphism at 110 to 105 Ma, and 130 to 125 Ma, respectively. These data indicate that the Early Cretaceous deformational event that resulted in D_{1b} folds in the Schist belt is regional in extent, and local differences in the age of the event suggest that deformation did not affect the entire Schist belt simultaneously, but was spatially transgressive throughout the Early Cretaceous.

CONCLUSIONS

Our results indicate that the metamorphic belts that comprise the south-central Brooks Range were affected by two principal phases of synmetamorphic deformation, the oldest of which was contemporaneous with *HP/LT* metamorphism and occurred in the Late Jurassic or earlier (D_{1a}), and the youngest of which occurred during retrograde metamorphism in the Early Cretaceous (D_{1b}). White micas yielding convex-upward Ar-release spectra indicate that *HP/LT* metamorphism, isoclinal folding, and the formation of the oldest penetrative fabric (S_{1a}) in the south-central Brooks Range occurred prior to 142 Ma in the Schist belt, and prior to 129 Ma in sodic-amphibole–bearing rocks north of the MCT.

These dates place a minimum age on the timing of deformation and metamorphism of Alaskan continental margin sediments during arc-continent collision (see Gottschalk and other, this volume, Chapter 12, for complete discussion). Our minimum ages are in good agreement with the geochronologic data of Armstrong et al. (1986), Till and Patrick (1991), and Christiansen and Snee (1994), which indicates that *HP/LT* metamorphism in the southern Brooks Range probably occurred prior to the Middle to Late Jurassic. Based on the similarity in the structural style and superposition relationships of folds in the Phyllite and Schist belts, a pre-Middle to Late Jurassic age is also inferred for early isoclinal and sheath folds in the Phyllite belt.

We interpret 135- to 110-Ma plateau, preferred, and isochron dates on white mica, amphibole, and biotite to record degassing of white micas and amphibole during dynamic recrystallization related to the D_{1b} deformational event. This interpretation accords well with studies in the Schist belt, including Till and Patrick (1991), who document deformation under amphibolite-facies conditions at 110 to 105 Ma, and Christiansen and Snee (1994) who argue that the Schist belt experienced deformation and dynamic recrystallization under lower amphibolite- to greenschist-facies conditions between 125 to 130 Ma. Our analyses indicate that D_{1b} structures and fabrics in the south-central Brooks Range are significantly older than Late Cretaceous to Tertiary(?) extensional structures in the southern Brooks Range, and support the interpretation that D_{1b} structures are formed in a strain regime related to contractional deformation (Gottschalk, 1990; Till and Patrick, 1991). Early Cretaceous white mica and amphibole ages may therefore record the timing of ductile contractional deformation at deep structural levels in the Brooks Range fold and thrust belt.

Our geochronologic study of the Schist belt produced results that are remarkably similar to those of Roeske et al. (1995), who studied the metamorphic rocks of the Ruby Geanticline, and Hannula et al. (1991), who analyzed white micas from *HP/LT* rocks of the Seward Peninsula. These recent geochronologic studies, and the general similarities between the metamorphic terranes of northern and central Alaska (summarized in Dusel-Bacon et al., 1989) indicate that the Brooks Range Schist belt, the Ruby Geanticline, and the schist of the Nome Group of the Seward Peninsula may once have comprised a single, *HP/LT* metamorphic belt of Middle Jurassic age prior to tectonic dispersal along strike-slip faults in the Late Cretaceous to Tertiary.

In the Skajit allochthon (senso stricto), D_{1a} is constrained by a white mica that produced a disturbed Ar release spectrum indicating an age older than 127 Ma. A maximum age is provided by an amphibole with a saddle-shaped spectrum indicating that metamorphism occurred after 193 Ma. Relict igneous amphibole separated from a mafic intrusive body in the southern Skajit allochthon yielded a thermally disturbed spectrum with old apparent ages in the high-temperature steps of about 420 Ma. We interpreted 420 Ma as the minimum age for intrusion, with Ar-loss occurring during Mesozoic metamorphism (>222 Ma).

APPENDIX 1. ^{40}AR/^{39}AR DATA* FOR MINERAL SAMPLES FROM METAMORPHIC ROCKS IN THE SOUTH-CENTRAL BROOKS RANGE, ALASKA

Temp (°C)	Radiogenic ^{40}Ar[†]	K-derived ^{39}Ar[†]	^{40}Ar$_R$/^{39}Ar$_K$[§]	^{39}Ar$_K$/^{37}Ar$_{Ca}$[**]	Radiogenic Yield (%)	^{39}Ar (%)	Apparent Age and Error[‡] (Ma)
			7-26-84-5, WHITE MICA				
		Total-gas date 125.8 ± 0.5 Ma; no plateau; isochron date 124 ± 1 Ma					
		(^{40}Ar/^{36}Ar)$_i$ = 340 ± 85 (700–800°C); J = 0.007908 ± 0.25%; wt. 84.5 mg					
500	0.3317	0.0460	7.2	73	33.3	2.2	100 ± 2
600	1.864	0.2268	8.22	149	96.5	11.0	113.6 ± 0.5
700	4.649	0.5120	9.08	254	94.6	24.9	125.1 ± 0.3
750	3.870	0.4337	8.92	236	98.6	21.1	123.0 ± 0.3
800	3.273	0.3568	9.17	222	98.2	17.4	126.3 ± 0.3
850	2.167	0.2213	9.79	139	97.6	10.8	134.5 ± 0.4
900	1.416	0.1456	9.73	75	96.4	7.1	133.7 ± 0.9
950	0.7634	0.0798	9.57	24	94.3	3.9	131.6 ± 0.7
1000	0.2596	0.0241	10.8	3.7	84.8	1.2	147 ± 3
1050	0.0719	0.0040	17.8	0.4	73.5	0.2	237 ± 12
1100	0.0269	0.0018	14.8	0.2	41.5	0.1	199 ± 27
1200	0.0512	0.0012	41.13	0.3	60.8	0.1	508 ± 40
			SBR88-8, AMPHIBOLE				
		Total-gas date 197 ± 4 Ma; no plateau; isochron date 160 ± 13 Ma					
		(^{40}Ar/^{36}Ar)$_i$ = 301 ± 26 (900–1450°C); J = 0.00779 ± 0.25%; wt. 283 mg					
600	0.4126	0.0029	145	0.01	58.1	1.8	1363 ± 87
700	0.0756	0.0016	47.8	0.04	15.5	1.0	571 ± 55
800	0.0756	0.0025	30.7	0.05	34.6	1.5	387 ± 32
900	0.1739	0.0139	12.5	0.01	34.7	8.6	168 ± 5
950	0.4810	0.0399	12.0	0.01	57.5	24.8	162 ± 2
1000	0.4228	0.0416	10.2	0.01	60.2	25.8	137 ± 1
1050	0.1761	0.0153	11.5	0.01	52.2	9.5	155 ± 4
1100	0.2256	0.0157	14.3	0.01	63.1	9.8	190.9 ± 0.6
1200	0.2818	0.0227	12.4	0.01	73.9	14.1	166 ± 4
1450	0.1521	0.0148	13.4	0.01	29.5	3.1	166 ± 10
			SBR88-56, AMPHIBOLE				
		Total-gas date 147 ± 2 Ma; plateau date 115 ± 1 Ma; isochron date 114 ± 1 Ma					
		(^{40}Ar/^{36}Ar)$_i$ = 301 ± 4 (850–975°C); J = 0.00779 ± 0.25%; wt. 375.3 mg					
500	0.252	0.006	40.9	0.03	2.3	1.1	499 ± 20
600	0.4592	0.0131	35.0	0.06	3.4	2.4	435 ± 5
700	0.2807	0.0154	18.2	0.10	7.3	2.8	239 ± 6
750	0.1671	0.0107	15.6	0.14	13.0	1.9	207 ± 9
800	0.1230	0.0131	9.42	0.11	17.8	2.4	128 ± 4
850^P	0.7262	0.0310	8.61	0.06	40.7	5.6	117 ± 2
900^P	1.285	0.1510	8.51	0.07	74.0	27.4	115.8 ± 0.7
950^P	1.910	0.2291	8.34	0.09	83.4	41.5	113.5 ± 0.4
975^P	0.3928	0.0468	8.39	0.09	58.6	8.5	114 ± 2
1000	0.0740	0.0078	9.50	0.03	28.9	1.4	129 ± 4
1025	0.0657	0.0050	13.0	0.01	28.0	0.9	175 ± 15
1050	0.1721	0.0066	26.2	0.01	49.2	1.2	335 ± 11
1100	0.4035	0.0121	33.3	0.02	66.3	2.2	416 ± 6
1150	0.0294	0.0010	29.7	0.01	15.7	0.2	375 ± 47
1200	0.0577	0.0016	36.0	0.01	24.6	0.3	446 ± 24
1450	0.0573	0.0014	40.7	0.01	8.0	0.3	497 ± 22

APPENDIX 1. ^{40}AR/^{39}AR DATA* FOR MINERAL SAMPLES FROM METAMORPHIC ROCKS IN THE SOUTH-CENTRAL BROOKS RANGE, ALASKA (continued - page 2)

Temp (°C)	Radiogenic ^{40}Ar[†]	K-derived ^{39}Ar[†]	^{40}Ar$_R$/^{39}Ar$_K$[§]	^{39}Ar$_K$/^{37}Ar$_{Ca}$**	Radiogenic Yield (%)	^{39}Ar (%)	Apparent Age and Error[‡] (Ma)
			6-20-84-4, AMPHIBOLE				
		Total-gas date 187 ± 6 Ma; no plateau; isochron date 188 ± 6 Ma					
		(^{40}Ar/^{36}Ar)$_i$ = 304 ± 7 (all steps); J = 0.007881 ± 0.1%; wt. 313.2 mg					
800	0.0399	0.0019	20.6	0.01	18.9	1.4	271 ± 43
850	0.0535	0.0033	16.4	0.01	27.9	2.3	220 ± 22
900	0.2046	0.0154	13.3	0.01	53.8	10.8	180 ± 4
950	0.7098	0.0549	12.9	0.01	68.9	38.7	175 ± 2
975	0.4039	0.0284	14.2	0.01	68.3	20.1	191 ± 3
1000	0.1144	0.0082	14.0	0.01	49.7	5.8	189 ± 9
1025	0.0506	0.0040	12.6	0.01	42.5	2.8	171 ± 15
1050	0.0485	0.0030	16.4	0.01	51.7	2.1	219 ± 18
1100	0.1778	0.0116	15.3	0.01	72.3	8.2	206 ± 3
1150	0.1125	0.0076	14.8	0.01	64.3	5.3	200 ± 9
1200	0.0219	0.0014	15.1	0.01	35.0	1.0	203 ± 38
1250	0.0151	0.0008	18.2	0.01	12.5	0.6	242 ± 91
1450	0.0138	0.0012	11.2	0.01	7.5	0.9	153 ± 23
			6-20-84-7, AMPHIBOLE				
		Total-gas date 174.4 ± 0.6 Ma; no plateau; J = 0.007885 ± 0.25%; wt. 319.8 mg					
1150	4.147	0.3170	13.1	0.01	80.4	92.6	177.1 ± 0.5
1450	0.2613	0.0254	10.3	0.03	56.8	7.4	141 ± 2
			SBR88-57, WHITE MICA				
		Total-gas date 116.4 ± 0.3 Ma; preferred date 118.3 ± 0.3 Ma; isochron date 121.0 ± 0.7 Ma					
		(^{40}Ar/^{36}Ar)$_i$ = 176 ± 33 (750–1000°C); J = 0.007955 ± 0.25%; wt. 68.4 mg					
500	0.5470	0.0780	7.02	115	46.1	0.9	98.0 ± 0.6
600	1.697	0.2858	5.94	104	17.8	3.4	83.3 ± 0.5
700	10.03	1.212	8.28	214	78.6	14.2	115.1 ± 0.3
750^P	12.61	1.493	8.44	553	92.9	17.5	117.3 ± 0.3
800^P	11.18	1.331	8.40	607	94.1	15.6	116.7 ± 0.3
850^P	11.32	1.325	8.54	496	96.0	15.6	118.6 ± 0.3
900^P	9.201	1.062	8.67	303	96.6	12.5	120.3 ± 0.3
950^P	9.293	1.082	8.59	140	96.4	12.7	119.2 ± 0.3
1000^P	5.209	0.6094	8.55	18	92.9	7.2	118.7 ± 0.3
1050	0.1807	0.0220	8.22	0.8	58.3	0.3	114 ± 3
1100	0.0687	0.0102	6.73	0.5	39.0	0.1	94 ± 3
1250	0.0413	0.0093	4.43	1.5	20.5	0.1	62 ± 6
			SBR88-8A, WHITE MICA				
		Total-gas date 127 ± 0.5 Ma; plateau date 123.3 ± 0.7 Ma; isochron date 124.2 ± 0.3 Ma					
		(^{40}Ar/^{36}Ar)$_i$ = 225 ± 9 (750–950°C); J = 0.007925 ± 0.25%; wt. 68.4 mg					
500	0.5716	0.0616	9.28	4.9	54.5	8.1	128 ± 1
600	0.6121	0.0638	9.59	42	84.1	8.4	132.2 ± 0.8
700	1.768	0.1870	9.46	105	90.0	24.6	130.4 ± 0.6
750^P	1.046	0.1163	8.99	132	97.0	15.3	124.2 ± 0.5
800^P	1.280	0.1440	8.89	270	97.8	18.9	122.9 ± 0.4
850^P	0.7215	0.0807	8.94	104	96.2	10.6	123.5 ± 0.5
900^P	0.4238	0.0472	8.97	97	93.3	6.2	124 ± 1
950^P	0.3816	0.0435	8.77	28	85.7	5.7	121.2 ± 0.9
1050	0.1423	0.0154	9.26	10	72.0	2.0	128 ± 4
1150	0.0173	0.0007	25.7	1.6	53.4	0.1	334 ± 74
1300	0.0056	0.0005	12.1	0.6	1.8	0.1	165 ± 50

R. R. Gottschalk and L. W. Snee

APPENDIX 1. ^{40}AR/^{39}AR DATA* FOR MINERAL SAMPLES FROM METAMORPHIC ROCKS IN THE SOUTH-CENTRAL BROOKS RANGE, ALASKA (continued - page 3)

Temp (°C)	Radiogenic ^{40}Ar[†]	K-derived ^{39}Ar[†]	^{40}Ar$_R$/^{39}Ar$_K$[§]	^{39}Ar$_K$/^{37}Ar$_{Ca}$**	Radiogenic Yield (%)	^{39}Ar (%)	Apparent Age and Error[‡] (Ma)
SBR80-30, WHITE MICA							
Total-gas date 154 ± 2 Ma; plateau date 135 ± 2; isochron date 145 ± 2 Ma							
(^{40}Ar/^{36}Ar)$_i$ = 260 ± 7 (800–950°C); J = 0.00791 ± 0.25%; wt. 97.4 mg.							
500	0.1389	0.0083	16.6	2.9	14.0	2.2	223 ± 6
600	0.2725	0.0108	25.3	4.9	7.1	2.9	330 ± 6
700	0.5906	0.0434	13.62	29	17.9	11.5	154 ± 1
750	0.7611	0.0678	11.23	47	21.6	18.0	154 ± 1
800[P]	0.9238	0.0938	9.85	79	59.9	24.9	135 ± 1
850[P]	0.8117	0.0827	9.82	28	59.4	22.0	135 ± 1
900[P]	0.2726	0.0279	9.78	23	60.8	7.4	134 ± 2
950[P]	0.2305	0.0240	9.61	15	54.4	6.4	132 ± 4
1000	0.1144	0.0103	11.10	3.1	39.1	2.7	152 ± 5
1100	0.0741	0.0026	29.0	0.1	33.6	0.7	373 ± 24
1400	0.0558	0.0052	10.8	1.2	21.7	1.4	148 ± 19
SBR88-27, WHITE MICA							
Total-gas date 110.1 ± 0.3 Ma; preferred date 110.4 ± 0.3 Ma; isochron date 110.4 ± 0.2 Ma							
(^{40}Ar/^{36}Ar)$_i$ = 286 ± 3 (650–950°C); J = 0.007868 ± 0.25%; wt. 76.9 mg.							
500	0.3319	0.0516	6.43	215	44.5	0.4	89 ± 2
600	1.071	0.1406	7.62	220	83.4	1.1	105.0 ± 0.6
650[P]	1.417	0.1785	7.94	394	74.2	1.4	109.3 ± 0.4
700[P]	2.612	0.3284	7.95	842	92.9	2.5	109.5 ± 0.3
750[P]	8.655	1.082	8.00	1553	96.8	8.4	110.1 ± 0.3
800[P]	26.427	3.280	8.06	990	97.7	25.4	110.9 ± 0.3
850[P]	21.202	2.650	8.00	3337	98.0	20.6	110.1 ± 0.3
900[P]	14.751	1.829	8.07	2563	98.6	14.2	111.0 ± 0.3
950[P]	15.071	1.890	7.98	2387	98.3	14.7	109.8 ± 0.3
1000	6.210	0.7928	7.83	1386	97.8	6.1	107.9 ± 0.3
1050	2.1894	0.2740	7.99	497	95.7	2.1	110.0 ± 0.3
1100	1.5933	0.1972	8.08	597	91.7	1.5	111.2 ± 0.2
1200	1.055	0.1294	8.15	55	87.5	1.0	112.1 ± 0.4
1400	0.5676	0.0691	8.22	86	84.6	0.5	113.0 ± 0.3
ABR85-17, WHITE MICA							
Total-gas date 119.2 ± 0.3 Ma; plateau date 120.7 ± 0.3 Ma; isochron date 119 ± 1 Ma							
(^{40}Ar/^{36}Ar)$_i$ = 286 ± 45 (750–1050°C); J = 0.007850 ± 0.25%; wt. 87.0 mg							
500	0.1330	0.0274	5.85	19	45.9	0.2	81 ± 1
600	0.8239	0.1003	8.22	56	82.7	0.7	112.8 ± 0.4
700	2.782	0.3250	8.56	90	85.2	2.2	117.3 ± 0.4
750	6.697	0.7735	8.66	177	95.1	5.1	118.6 ± 0.3
800[P]	25.165	2.8320	8.89	342	94.2	18.8	121.6 ± 0.3
850[P]	49.853	5.6807	8.78	417	97.3	37.7	120.2 ± 0.3
900	24.403	2.8282	8.63	172	97.7	18.8	118.2 ± 0.3
950	14.014	1.6655	8.42	76	98.0	11.0	115.4 ± 0.3
1000	4.2347	0.4956	8.55	20	98.0	3.3	117.1 ± 0.3
1050	1.8958	0.2145	8.84	4.0	98.5	1.4	121.0 ± 0.3
1100	0.2064	0.0229	9.02	1.5	96.5	0.2	123 ± 3
1200	0.6342	0.700	9.06	0.9	93.5	0.5	123.9 ± 0.7
1450	0.3631	0.0427	8.50	2.4	64.7	0.3	117 ± 2

APPENDIX 1. ^{40}AR/^{39}AR DATA* FOR MINERAL SAMPLES FROM METAMORPHIC ROCKS IN THE SOUTH-CENTRAL BROOKS RANGE, ALASKA (continued - page 4)

Temp (°C)	Radiogenic ^{40}Ar†	K-derived ^{39}Ar†	^{40}Ar$_R$/^{39}Ar$_K$§	^{39}Ar$_K$/^{37}Ar$_{Ca}$**	Radiogenic Yield (%)	^{39}Ar (%)	Apparent Age and Error‡ (Ma)
			SBR88-51, White Mica (First Analysis)				
		Total-gas date indeterminate; no plateau; isochron date 131 ± 1 Ma					
		(^{40}Ar/^{36}Ar)$_i$ = 277 ± 9; J = 0.007965 ± 0.25%; wt. 62.1 mg					
500	0.2350	0.0252	9.31	5.1	48.2	0.4	129 ± 3
600	0.3176	0.0359	8.85	1.4	80.7	0.5	122.9 ± 1.0
700	1.6223	0.1832	8.85	2.0	83.4	2.8	122.9 ± 0.4
750	5.2779	0.5769	9.15	132	84.9	8.7	126.9 ± 0.3
800	23.197	2.3803	9.75	239	97.0	36.0	134.9 ± 0.4
850			Approximately 11% of total expected argon was lost.				
900	19.137	1.9923	9.61	35	98.3	30.1	133.0 ± 0.4
950	8.9283	0.9856	9.06	20	98.4	14.9	125.7 ± 0.3
1000	2.2874	0.2485	9.20	7.8	97.7	3.8	127.6 ± 0.4
1050	1.2402	0.1249	9.93	2.9	98.2	1.9	137.3 ± 0.4
1200	0.4859	0.0473	10.27	0.9	95.8	0.7	142 ± 2
1375	0.1605	0.0178	9.04	0.8	56.4	0.3	125 ± 12
			SBR88-51, White Mica (Second Analysis)				
		Total-gas date 135.2 ± 0.4 Ma; no plateau; isochron date 132 ± 1 Ma					
		(^{40}Ar/^{36}Ar)$_i$ = 287 ± 11; J = 0.007858 ± 0.25%; wt. 83.8 mg					
600	0.1241	0.0134	9.24	n.d.	37.0	0.1	126 ± 2
700	0.5775	0.0652	8.85	n.d.	80.3	0.6	121.3 ± 0.3
750	1.0350	0.1153	8.97	n.d.	78.7	1.1	123.0 ± 0.6
800	1.6894	0.1877	9.00	n.d.	89.0	1.9	123.3 ± 0.5
850	6.1500	0.6711	9.16	n.d.	81.9	6.7	125.4 ± 0.3
900	25.888	2.6078	9.93	n.d.	89.1	26.0	135.5 ± 0.4
950	36.450	3.5001	10.41	n.d.	91.6	34.9	141.9 ± 0.4
1000	17.701	1.8407	9.62	n.d.	93.8	18.3	131.4 ± 0.4
1050	6.2202	0.6758	9.20	n.d.	95.6	6.7	126.0 ± 0.3
1100	2.2226	0.2303	9.65	n.d.	93.2	2.3	131.8 ± 0.8
1150	0.9533	0.0919	10.37	n.d.	87.4	0.9	141.3 ± 0.6
1350	0.4214	0.0402	10.50	n.d.	78.3	0.4	143 ± 1
			SBR88-51, Biotite (First Analysis)				
		Total-gas date 128.0 ± 0.4 Ma; plateau date 128.8 ± 0.6 Ma; isochron date 128.5 ± 0.2 Ma					
		(^{40}Ar/^{36}Ar)$_i$ = 361 ± 18 (750–1050°C); J = 0.007943 ± 0.25%; wt. 72.6 mg					
500	0.3419	0.0428	8.00	56	55.6	0.4	111 ± 2
600	1.7006	0.1933	8.80	40	82.9	1.7	121.8 ± 0.3
700	14.386	1.5564	9.24	64	94.3	13.4	127.8 ± 0.3
750^P	17.798	1.9117	9.31	645	98.6	16.5	128.7 ± 0.3
800^P	7.9994	0.8570	9.33	282	98.1	7.4	129.0 ± 0.4
850^P	4.9897	0.5330	9.36	73	97.2	4.6	129.4 ± 0.4
900^P	9.1611	0.9844	9.31	38	97.9	8.5	128.6 ± 0.3
950^P	15.025	1.6124	9.32	118	98.6	13.9	128.8 ± 0.3
1000^P	16.686	1.7950	9.30	236	98.9	15.4	128.5 ± 0.3
1050^P	14.382	1.5412	9.33	129	99.1	13.3	129.0 ± 0.4
1150	3.8231	0.4036	9.47	42	97.9	3.5	130.8 ± 0.4
1300	1.7925	0.1885	9.51	39	74.8	1.6	131.4 ± 0.4

248 — R. R. Gottschalk and L. W. Snee

APPENDIX 1. ^{40}AR/^{39}AR DATA* FOR MINERAL SAMPLES FROM METAMORPHIC ROCKS IN THE SOUTH-CENTRAL BROOKS RANGE, ALASKA (continued - page 5)

Temp (°C)	Radiogenic ^{40}Ar[†]	K-derived ^{39}Ar[†]	^{40}Ar$_R$/^{39}Ar$_K$[§]	^{39}Ar$_K$/^{37}Ar$_{Ca}$**	Radiogenic Yield (%)	^{39}Ar (%)	Apparent Age and Error[‡] (Ma)
			SBR88-51, BIOTITE (SECOND ANALYSIS)				
			Total-gas date 129.0 ± 0.4 Ma; plateau date 129.3 ± 0.4 Ma; isochron date 129.7 ± 0.2 Ma				
			(^{40}Ar/^{36}Ar)$_i$ = 299 ± 2 (800–1150°C); J = 0.007807 ± 0.25%; wt. 70.6 mg				
600	0.4396	0.0507	8.68	n.d.	31.3	0.7	118.2 ± 0.7
700	3.6572	0.3950	9.26	n.d.	79.0	5.3	125.9 ± 0.3
750	7.0204	0.7447	9.43	n.d.	84.8	10.0	128.1 ± 0.3
800[P]	9.4575	0.9970	9.49	n.d.	90.8	13.4	128.9 ± 0.3
850[P]	4.8651	0.5109	9.52	n.d.	89.6	6.9	129.4 ± 0.4
900[P]	3.3087	0.3468	9.54	n.d.	81.0	4.7	129.6 ± 0.4
950[P]	4.0140	0.4218	9.52	n.d.	78.4	5.7	129.3 ± 0.4
1000[P]	7.8733	0.8243	9.55	n.d.	84.4	11.1	129.7 ± 0.4
1050[P]	9.6218	1.0104	9.52	n.d.	89.6	13.6	129.4 ± 0.4
1100[P]	10.281	1.0810	9.51	n.d.	90.5	14.6	129.2 ± 0.4
1150[P]	7.7311	0.8106	9.54	n.d.	90.2	10.9	129.6 ± 0.4
1300	2.1912	0.2268	9.66	n.d.	85.5	3.1	131.2 ± 0.4
			MS87-14, WHITE MICA				
			Total-gas date 123.3 ± 0.4 Ma; no plateau; isochron date 125.3 ± 0.7 Ma				
			(^{40}Ar/^{36}Ar)$_i$ = 260 ± 5 (700–1050°C); J = 0.007812 ± 0.25%; wt. 83.7 mg				
600	0.3470	0.0182	19.11	n.d.	49.7	0.5	251 ± 2
700	2.0462	0.2710	7.55	n.d.	36.8	7.2	103.4 ± 0.3
750	3.0070	0.2710	8.99	n.d.	70.1	8.9	122.5 ± 0.3
800	5.1186	0.5797	8.83	n.d.	86.6	15.4	120.3 ± 0.3
850	5.1564	0.5901	8.74	n.d.	90.8	15.6	119.1 ± 0.3
900	6.3904	0.7154	8.93	n.d.	94.2	19.0	121.6 ± 0.3
950	4.8037	0.5203	9.23	n.d.	95.5	13.8	125.6 ± 0.3
1000	3.7853	0.4072	9.29	n.d.	96.7	10.8	126.5 ± 0.4
1050	2.4077	0.2581	9.33	n.d.	95.7	6.8	126.9 ± 0.5
1100	0.7037	0.0665	10.27	n.d.	88.3	1.8	139.2 ± 0.5
1200	0.1458	0.0079	18.42	n.d.	70.2	0.2	242 ± 2
1350	0.2358	0.0009	259	n.d.	79.4	0.0	1996 ± 87
			8-4-85-5, WHITE MICA				
			Total-gas date 125.5 ± 0.4 Ma; no plateau; isochron date 127 ± 2 Ma				
			(^{40}Ar/^{36}Ar)$_i$ = 276 ± 23 (all T); J = 0.007856 ± 0.25%; wt. 81.0 mg				
600	0.1417	0.0165	8.57	n.d.	45.6	0.4	118 ± 3
700	1.0156	0.1190	8.54	n.d.	93.9	2.9	117.1 ± 0.4
750	1.7114	0.1990	8.60	n.d.	88.3	4.8	118.0 ± 0.3
800	2.9698	0.3413	8.70	n.d.	96.8	8.2	119.3 ± 0.4
850	4.6552	0.5238	8.89	n.d.	96.4	12.6	121.8 ± 0.3
900	3.5403	0.9303	9.18	n.d.	95.5	22.4	125.6 ± 0.3
950	7.8500	0.8390	9.36	n.d.	96.5	20.2	128.0 ± 0.3
1000	6.3879	0.6766	9.44	n.d.	98.2	16.3	129.1 ± 0.4
1050	2.4932	0.2699	9.24	n.d.	97.7	6.5	126.4 ± 0.3
1100	1.4976	0.1708	8.77	n.d.	97.3	4.1	120.2 ± 0.3
1250	0.8412	0.0687	12.25	n.d.	94.9	1.7	165.8 ± 0.8

APPENDIX 1. ^{40}AR/^{39}AR DATA* FOR MINERAL SAMPLES FROM METAMORPHIC ROCKS IN THE SOUTH-CENTRAL BROOKS RANGE, ALASKA (continued - page 6)

Temp (°C)	Radiogenic ^{40}Ar[†]	K-derived ^{39}Ar[†]	^{40}Ar$_R$/^{39}Ar$_K$[§]	^{39}Ar$_K$/^{37}Ar$_{Ca}$**	Radiogenic Yield (%)	^{39}Ar (%)	Apparent Age and Error[‡] (Ma)
			SBR88-1, AMPHIBOLE				
		Total-gas date 379 ± 2 Ma; no plateau; isochron date 425 ± 12 Ma					
		(^{40}Ar/^{36}Ar)$_i$ = 206 ± 16 (850–1450°C); J = 0.007755 ± 0.25%; wt. 330.2 mg					
500	0.9789	0.0009	1076	0.07	86.9	0.1	4031 ± 55
600	0.6355	0.0114	55.9	0.05	4.2	1.7	650 ± 16
700	1.0455	0.0166	63.01	0.25	19.3	2.5	718.9 ± 4
800	1.6532	0.0280	59.00	0.15	47.7	4.3	680 ± 4
850	0.6615	0.0391	16.93	0.05	27.7	6.0	223 ± 2
900	2.8483	0.1266	22.49	0.02	39.6	19.3	290.0 ± 0.7
950	4.3140	0.1744	24.74	0.02	43.5	26.6	316.6 ± 0.8
975	1.6264	0.0647	25.13	0.02	64.5	9.9	321 ± 2
1000	0.7912	0.0302	25.16	0.02	64.7	4.6	333 ± 2
1025	0.7330	0.0263	27.90	0.02	69.2	4.0	353 ± 2
1050	0.8230	0.0271	30.37	0.02	74.8	4.1	382 ± 4
1100	1.3459	0.0409	32.89	0.02	82.4	6.3	410 ± 2
1150	0.3336	0.0101	32.90	0.02	59.1	1.5	410 ± 4
1200	1.2570	0.0370	34.02	0.02	84.2	5.6	422 ± 2
1250	0.4035	0.0121	33.30	0.02	66.3	1.9	414 ± 6
1300	0.2083	0.0064	32.69	0.02	51.8	1.0	408 ± 10
1450	0.0800	0.0026	30.53	0.02	26.5	0.4	383 ± 33
			SBR88-40, AMPHIBOLE				
		Total-gas date 893 ± 3 Ma; no plateau; isochron date -21 ± 121 Ma					
		(^{40}Ar/^{36}Ar)$_i$ = 1060 ± 142 (all steps); J = 0.007685 ± 0.25%; wt. 316.8 mg					
600	12.696	0.0204	624	0.10	81.2	3.3	3169 ± 43
700	2.5702	0.0233	110.3	0.49	64.1	3.8	1108 ± 4
800	4.8935	0.0794	62.67	0.50	72.6	12.9	700 ± 2
850	1.7002	0.1161	14.65	0.23	64.1	18.9	192.5 ± 1.0
900	7.0888	0.1457	48.65	0.03	80.3	23.7	573 ± 1
950	8.0405	0.0914	87.92	0.01	81.2	14.9	931 ± 3
975	2.4779	0.0305	81.27	0.01	72.6	5.0	875 ± 3
1000	1.2757	0.0158	80.94	0.01	61.8	2.6	873 ± 4
1050	1.6198	0.0208	77.89	0.01	68.8	3.4	846 ± 4
1100	2.6037	0.0252	103.3	0.01	81.4	4.1	1055 ± 4
1150	2.6175	0.0207	126.4	0.01	82.8	3.4	1224 ± 9
1300	3.6559	0.0255	143.2	0.01	72.3	4.2	1339 ± 6

*Analytical data for "Radiogenic ^{40}Ar" and "K-derived ^{39}Ar" are calculated to five places right of the decimal; "^{40}Ar$_R$/^{39}Ar$_K$" is calculated to three places right of the decimal. "Radiogenic ^{40}Ar," "K-derived ^{39}Ar," and "^{40}Ar$_R$/^{39}Ar$_K$" are rounded to significant figures using analytical precisions. Apparent ages and associated errors were calculated from unrounded analytical data and then each rounded using associated errors. All analyses were done on a Mass Analyzer Products 215 rare gas mass spectrometer in the Argon Laboratory, U.S. Geological Survey, Denver, Colorado. Decay constants are those of Steiger and Jäger, 1977. The standard for this experiment is hornblende MMhb-1 with percent K = 1.555. ^{40}Ar$_R$ = 1.624 x 10^{-9} mole/gm, and K-Ar age = 520.4 Ma (Samson and Alexander, 1987).

[†]Abundances of "Radiogenic ^{40}Ar" and "K-derived ^{39}Ar" are reported in volts. Conversion to moles can be made by using 9.736 x 10^{-13} moles argon per volt of signal. Detection limit at the time of this experiment was 2 x 10^{-17} moles argon measured on a Faraday detector.

[§]"^{40}Ar$_R$/^{39}Ar$_K$" has been corrected for all interfering isotopes including atmospheric argon. Mass discrimination in our mass spectrometer was determined by determining the ^{40}Ar/^{36}Ar ratio of atmospheric argon; our measured value is 298.9 during the period of this experiment; the accepted atmospheric ^{40}Ar/^{36}Ar ratio is 295.5. Abundances of interfering isotopes of argon from K and Ca were calculated from reactor production ratios determined by irradiating and analyzing pure CaF$_2$ and K$_2$SO$_4$ simultaneously with these samples. Samples were irradiated in two separate irradiation packages. The production ratios for all but four samples are (^{40}Ar/^{39}Ar)$_K$ = 1.135 x 10^{-3}, (^{38}Ar/^{39}Ar)$_K$ = 1.3 x 10^{-2}, (^{37}Ar/^{39}Ar)$_K$ = 2.34 x 10^{-4}, (^{36}Ar/^{37}Ar)$_{Ca}$ = 2.64 x 10^{-4}, (^{39}Ar/^{37}Ar)$_{Ca}$ = 6.66 x 10^{-4}, and (^{38}Ar/^{37}Ar)$_{Ca}$ = 3.65 x 10^{-5}. The production ratios for white mica samples SBR88-51 (second analysis), 8-4-85-5, and MS87-14, and biotite sample SBR88-51 (second analysis) are (^{40}Ar/^{39}Ar)$_K$ = 1.56 x 10^{-2}, (^{38}Ar/^{39}Ar)$_K$ = 1.305 x 10^{-2}, (^{37}Ar/^{39}Ar)$_K$ = 1.46 x 10^{-4}, (^{36}Ar/^{37}Ar)$_{Ca}$ = 2.51 x 10^{-4}, (^{39}Ar/^{37}Ar)$_{Ca}$ = 7.46 x 10^{-4}, and (^{38}Ar/^{37}Ar)$_{Ca}$ = 2.96 x 10^{-5}. Corrections were also made for additional interfering isotopes of argon produced from irradiation of chlorine using the method of Roddick, 1983. Irradiation of each group of samples was for 30 hours at 1 megawatt in the TRIGA reactor at the U.S. Geological Survey, Denver, Colorado.

**"^{39}Ar$_K$/^{37}Ar$_{Ca}$" has been corrected for all interfering isotopes and for decay of radioactive ^{39}Ar and ^{37}Ar. To calculate apparent K/Ca ratios, divide the "^{39}Ar$_K$/^{37}Ar$_{Ca}$" ratios by 2.

[‡]1-sigma error. [P]Fraction included in plateau or preferred date.

ACKNOWLEDGMENTS

This study was supported by grants to J. S. Oldow, H. G. Avé Lallemant, and A. W. Bally from the Department of Energy (DE-ASO5-83ER13124), the National Science Foundation (EAR-8517384), and contributions to the Alaska Industrial Associates Program at Rice University from Arco, Amoco, Chevron, Standard Oil of Ohio, Mobil, and Gulf. Sample collection was carried out during field investigations by J. S. Oldow, H. G. Avé Lallemant, and the senior author. We are indebted to J. E. Wright and W. P. Leeman for generously providing the use of their laboratories and equipment. Ross Yeoman and Gary Davidson provided indispensable assistance during all phases of isotopic analysis. We are grateful to A. B. Till and S. M. Roeske whose thorough and thoughtful reviews significantly improved the quality of this manuscript.

REFERENCES CITED

Anovitz, L. M., and Essene, E. J., 1987, Phase equilibria in the system $CaCO_3$-$MgCO_3$.$FeCO_3$: Journal of Petrology, v. 28, p. 389–414.

Armstrong, R. L., Harakal, J. E., Forbes, R. B., Evans, B. W., and Thurston, S. P., 1986, Rb-Sr and K-Ar study of metamorphic rocks of the Seward Peninsula and southern Brooks Range, Alaska, in Evans, B. W., and Brown, E. H., eds., Blueschists and eclogites: Geological Society of America Memoir 164, p. 185–203.

Boler, K. B., 1989, Stratigraphy, structure, and tectonics of the central Brooks Range near Dietrich Camp, Alaska [M.A. thesis]: Houston, Texas, Rice University, 163 p.

Box, S. E., 1985, Early Cretaceous orogenic belt in northwestern Alaska: internal organization, lateral extent, and tectonic interpretation, in Howell, D. G., ed., Tectonostratigraphic terranes of the Circum-Pacific region: Houston, Texas, Circum-Pacific Council for Energy and Mineral Resources, p. 137–145.

Box, S. E., 1987, Late Cretaceous or younger southwest-directed extensional faulting, Cosmos Hills, Brooks Range, Alaska: Geological Society of America Abstracts with Programs, v. 19, p. 361.

Brosgé, W. P., and Reiser, H. N., 1964, Geologic map and section of the Chandalar Quadrangle, Alaska: U.S. Geological Survey Miscellaneous Geologic Investigations Map I-375, scale 1:250,000.

Brosgé, W. P., and Reiser, H. N., 1971, Preliminary bedrock geologic map: Wiseman and eastern Survey Pass Quadrangles, Alaska: U.S. Geological Survey Open File Map 479, scale 1:250,000.

Chopin, C., and Maluski, H., 1980, $^{40}Ar/^{39}Ar$ dating of high-pressure metamorphic micas from the Gran Paradiso area (Western Alps): Evidence against the blocking temperature concept: Contributions to Mineralogy and Petrology, v. 74, p. 109–122.

Chopin, C., and Monie, P., 1984, A unique magnesiochloritoid-bearing high-pressure assemblage from the Monte Rosa, Western Alps: petrologic and $^{40}Ar/^{39}Ar$ radiometric study: Contributions to Mineralogy and Petrology, v. 87, p. 388–398.

Christiansen, P. P., and Snee, L. W., 1994, Structure, metamorphism, and geochronology of the Cosmos Hills and Ruby Ridge, Brooks Range Schist belt, Alaska: Tectonics, v. 13, p. 193–213.

Dalrymple, G. B., Alexander, E. C. J., Lanphere, M. A., and Kraker, G. P., 1981, Irradiation of samples for $^{40}Ar/^{39}Ar$ dating using the Geological Survey TRIGA reactor: U.S. Geological Survey Professional Paper 1176, 55 p.

Dillon, J. T., 1989, Structure and stratigraphy of the southern Brooks Range and northern Koyukuk basin near the Dalton Highway, in Mull, C. G., and Adams, K. E., eds., Dalton Highway, Yukon to Prudhoe Bay, Alaska: Alaska Division of Geological and Geophysical Surveys Guidebook 7, v. 2, p. 157–187.

Dillon, J. T., Pessel, G. H., Chen, J. H., and Veach, N. C., 1980, Middle Paleozoic magmatism and orogenesis in the Brooks Range, Alaska: Geology, v. 8, p. 338–343.

Dillon, J. T., Brosgé, W. P., and Dutro, J. T., Jr., 1986, Generalized geologic map of the Wiseman Quadrangle: U.S. Geological Survey Open File Report OF 86-219, scale 1:250,000.

Dillon, J. T., Harris, A. G., and Dutro, J. T., Jr., 1987, Preliminary description and correlation of lower Paleozoic fossil-bearing strata in the Snowden Mountain area of the south-central Brooks Range, in Tailleur, I., and Weimer, P., eds., Alaskan North Slope geology: Society of Economic Paleontologists and Mineralogists, Pacific Section, Publication 50, p. 337–345.

Dusel-Bacon, C., Brosgé, W. P., Till, A. B., Doyle, E. O., Mayfield, C. F., Reiser, H. N., and Miller, T. P., 1989, Distribution, facies, ages, and proposed tectonic associations of regionally metamorphosed rocks in northern Alaska: U.S. Geological Survey Professional Paper 1497-A, 44 p.

Evans, B. W., 1990, Phase relations of epidote-blueschists: Lithos, v. 25, p. 3–23.

Gottschalk, R. R., 1987, Structural and petrologic evolution of the south-central Brooks Range near Wiseman, Alaska [Ph.D. thesis]: Houston, Texas, Rice University, 263 p.

Gottschalk, R. R., 1990, Structural evolution of the Schist belt, south-central Brooks Range fold and thrust belt, Alaska: Journal of Structural Geology, v. 12, p. 453–469.

Gottschalk, R. R., and Oldow, J. S., 1988, Low-angle normal faults in the south-central Brooks Range fold and thrust belt, Alaska: Geology, v. 16, p. 395–399.

Grybeck, D., Beikman, H. M., Brosgé, W. P., Tailleur, I. L., and Mull, C. G., 1977, Geologic map of the Brooks Range, Alaska: U.S. Geological Survey Open File Report Map 77-166, scale 1:2,500,000.

Hammerschmidt, K., and Frank, E., 1991, Relics of high pressure metamorphism in the Lepontine Alps (Switzerland)—$^{40}Ar/^{39}Ar$ microprobe analyses on white K-micas: Schweizerische Mineralogische Petrographische Mitteilungen, v. 71, p. 261–274.

Hannula, K. A., McWilliams, M. O., and Gans, P. B., 1991, Triassic or older blueschist facies metamorphism on the Seward Peninsula: results of $^{40}Ar/^{39}Ar$ dating of white micas from the Nome Group: Geological Society of America Abstracts with Programs, v. 23, p. 33.

Harrison, T. M., 1981, Diffusion of ^{40}Ar in hornblende: Contributions to Mineralogy and Petrology, v. 78, p. 324–331.

Harrison, T. M., Duncan, I., and McDougall, I., 1985, Diffusion of ^{40}Ar in biotite: temperature, pressure, and compositional effects: Geochimica et Cosmochimica Acta, v. 49, p. 2461–2468.

Hitzman, M. W., Proffett, J. W., Jr., Schmidt, J. M., and Smith, T. M., 1986, Geology and mineralization of the Ambler district, northwestern Alaska: Economic Geology, v. 81, p. 1592–1618.

Hodges, K. V., and Spear, F. S., 1982, Geothermometry, geobarometry, and the Al_2SiO_5 triple point at Mt. Moosilauke, New Hampshire: American Mineralogist, v. 67, p. 1118–1134.

Jones, D. L., Coney, P. J., Harms, T. A., and Dillon, J. T., 1988, Interpretive geologic map and supporting radiolarian data from the Angayucham terrane, Coldfoot area, southern Brooks Range, Alaska: U.S. Geological Survey Miscellaneous Field Studies Map MF-1993, scale 1:63,360.

Karl, S. M., and Aleinikoff, J. N., 1990, Proterozoic U-Pb zircon age of granite in the Kallarichuk Hills, western Brooks Range, Alaska: evidence for Precambrian basement in the Schist belt: U.S. Geological Survey Bulletin 1946, p. 95–100.

Law, R. D., Miller, E. L., Little, T. A., and Lee, J., 1994, Extensional origin of ductile fabrics in the Schist belt, central Brooks Range, Alaska—II. Microstructural and petrofabric evidence: Journal of Structural Geology, v. 16, p. 919–940.

Little, T. A., Miller, E. L., Lee, J., and Law, R. D., 1994, Extensional origin of ductile fabrics in the Schist belt, central Brooks Range, Alaska—I. Geologic and structural studies: Journal of Structural Geology, v. 16, p. 899–918.

Miller, E. L., and Hudson, T. L., 1991, Mid-Cretaceous extensional fragmentation of a Jurassic–Early Cretaceous compressional orogen, Alaska: Tectonics,

v. 10, p. 781–796.

Montigny, R., LeMer, O., Thuizat, R., and Whitechurch, H., 1988, K-Ar and ^{40}Ar/^{39}Ar study of metamorphic rocks associated with the Oman ophiolite: tectonic implications: Tectonophysics, v. 151, p. 345–362.

Moore, T. E., Wallace, W. K., Bird, K. J., Karl, S. M., Mull, C. G., and Dillon, J. T., 1994, Geology of northern Alaska, *in* Plafker, G., and Berg, H. C., eds., The geology of Alaska: Boulder, Colorado, Geological Society of America, The Geology of North America, v. G-1, p. 49–140.

Murphy, J. M., and Patton, W. W., Jr., 1988, Geologic setting and petrography of the phyllite and metagreywacke thrust panel, north-central Alaska: U.S. Geological Survey Circular 1016, p. 104–108.

Oldow, J. S., Seidensticker, C. M., Phelps, J. C., Julian, F. E., Gottschalk, R. R., Boler, K. W., Handschy, J. W., and Avé Lallemant, H. G., 1987, Balanced cross-sections through the central Brooks Range and North Slope, Arctic Alaska: American Association of Petroleum Geologists, Special Publication, 19 p. and 8 pl.

Owen, C., 1989, Magmatic differentiation and alteration in isofacial greenschist and blueschist, Shuksan Suite, Washington: statistical analysis of major element variation: Journal of Petrology, v. 30, p. 739–761.

Patrick, B. E., 1995, High pressure–low temperature metamorphism of granitic orthogneisses in the Brooks Range, northern Alaska: Journal of Metamorphic Geology, v. 13, p. 111–124.

Rice, J. M., 1977, Contact metamorphism of impure dolomitic limestone in the Boulder aureole, Montana: Contributions to Mineralogy and Petrology, v. 59, p. 237–259.

Roddick, J. S., 1983, High-precision intercalibration of ^{40}Ar/^{39}Ar standards: Geochimica et Cosmochimica Acta, v. 47, p. 887–898.

Roeske, S. M., Dusel-Bacon, C., Aleinikoff, J. N., and Snee, L. W., 1995, Metamorphic and structural history of continental crust at a Mesozoic collisional margin, the Ruby Terrane, Alaska: Journal of Metamorphic Geology, v. 13, p. 25–40.

Samson, S. D., and Alexander, E. C. J., 1987, Calibration of interlaboratory ^{40}Ar/^{39}Ar dating standard MMhb-1: Chemical Geology, v. 66, p. 27–43.

Scaillet, S., Ferraud, G., Ballevere, M., and Amouric, M., 1992, Mg/Fe and [(Mg,Fe)Si-Al$_2$] compositional control on argon behavior in high-pressure white micas: a ^{40}Ar/^{39}Ar continuous laser-probe study from the Dora-Maira nappe of the internal western Alps, Italy: Geochimica et Cosmochimica Acta, v. 56, p. 2851–2872.

Snee, L. W., Sutter, J. F., and Kelly, W. C., 1988, Thermochronology of economic mineral deposits: dating the stages of mineralization at Panasquiera, Portugal, by high-precision ^{40}Ar/^{39}Ar age spectrum techniques on muscovite: Economic Geology, v. 83, p. 335–354.

Steiger, R. H., and Jäger, E., 1977, Subcommision on thermochronology: convention on the use of decay constants on geo- and cosmochronology: Earth and Planetary Science Letters, v. 36, p. 259–362.

Till, A. B., 1992, Detrital blueschist-facies metamorphic mineral assemblages in Early Cretaceous sediments of the foreland basin of the Brooks Range, Alaska, and implications for orogenic evolution: Tectonics, v. 11, p. 1207–1223.

Till, A. B., and Moore, T. M., 1991, Tectonic relations of the Schist belt, southern Brooks Range, Alaska: Eos (Transactions, American Geophysical Union), v. 72, p. 295.

Till, A. B., and Patrick, B. E., 1991, ^{40}Ar/^{39}Ar evidence for a 110–105 Ma amphibolite-facies overprint on blueschist in the south-central Brooks Range, Alaska: Geological Society of America, Abstracts with Programs, v. 23, p. A436.

Till, A. B., and Snee, L. W., 1995, ^{40}Ar/^{39}Ar isotopic evidence that blueschists formed during collision, not subduction in the Nanielik antiform, western Brooks Range, Alaska: Journal of Metamorphic Geology, v. 13, p. 41–60.

Till, A. B., Schmidt, J. M., and Nelson, S. W., 1988, Thrust involvement of metamorphic rocks, southwestern Brooks Range, Alaska: Geology, v. 16, p. 930–933.

Turner, D. L., Forbes, R. B., and Dillon, J. T., 1979, K-Ar geochronology of the southwestern Brooks Range, Alaska: Canadian Journal of Earth Sciences, v. 16, p. 1789–1804.

Wijbrans, J. R., and McDougall, I., 1986, ^{40}Ar/^{39}Ar dating of white micas from an Alpine high-pressure metamorphic belt on Naxos (Greece): the resetting of the argon isotopic system: Contributions to Mineralogy and Petrology, v. 93, p. 187–194.

York, D., 1969, Least-squares fitting of a straight line with correlated errors: Earth and Planetary Science Letters, v. 5, p. 320–324.

MANUSCRIPT ACCEPTED BY THE SOCIETY SEPTEMBER 23, 1997

Geological Society of America
Special Paper 324
1998

Antithetic shear and the formation of back folds in the central Brooks Range fold and thrust belt, Alaska

Hans G. Avé Lallemant
Department of Geology and Geophysics, Rice University, Houston, Texas 77005-1892
John S. Oldow
Department of Geology and Geological Engineering, University of Idaho, Moscow, Idaho 83843

ABSTRACT

The Brooks Range of northern Alaska is a north-vergent fold and thrust belt formed in Late Jurassic to Tertiary time. Although the north-directed transport is recognized and evidenced by ubiquitous macro-, meso-, and microscopic structures, antithetic, south-vergent folds occur, attesting to relative, but not necessarily absolute, south-directed tectonic transport. These back folds are mainly the result of strain incompatibilities related to northward displacement across a ramp and the formation of an antiformal-stack duplex.

INTRODUCTION

Thrust faults in fold and thrust belts are generally synthetic; that is, the sense of displacement along these faults is the same as that along the basal decollement and, thus, of the entire belt. Similarly, the vergence of folds is generally toward the foreland and consistent with the vergence of the entire fold and thrust belt. However, many examples exist of antithetic thrust faults (e.g., Serra, 1977; Banks and Warburton, 1986; Price, 1986; Vann et al., 1986; Lawton et al., 1994) and hinterland-vergent folds (e.g., Heim, 1922; Brown et al., 1986; Macaya et al., 1991) that, apparently, have formed by displacements opposite to the general sense of tectonic transport of the fold and thrust belt. These structures, generally called "back thrusts" and "back folds," respectively, are probably the result of the same kinematic framework, but the back thrusts most likely form at shallow levels where rocks behave in a brittle fashion, whereas back folds form at greater depths where ductile flow becomes dominant.

Back thrusts and back folds are relatively less common in fold and thrust belts compared to foreland-vergent structures. However, in two separate areas in the central Brooks Range fold and thrust belt of northern Alaska (for location see Figs. 1, 2A, and 2B in Preface to this volume) such structures are very well developed. In the north (Fig. 1), near the mountain front, a major

back thrust (Atigun Gorge fault) occurs, which is part of a spectacular triangle zone (Vann et al., 1986; Oldow et al., 1987). The other area lies south of the Doonerak window where antithetic folds of micro- to megascopic scale occur. The latter structures are the focus of this chapter.

The Brooks Range fold and thrust belt of northern Alaska was formed by northward thrusting. Based on the detailed field work, as described in this volume, and on previous mapping (e.g., Brosgé and Reiser, 1964, 1971; Dillon et al., 1986, 1988), the central Brooks Range is divided into several, distinct east-west–trending belts. These are from south to north: (1) Angayucham terrane, (2) Rosie Creek allochthon, (3) Phyllite belt, (4) Schist belt, (5) Skajit allochthon, (6) Doonerak and Blarney Creek duplexes, (7) Endicott allochthon, (8) North Slope assemblage, and (9) the North Slope foredeep. The geology of these belts is briefly described in the Preface to this volume and in more detail elsewhere in this volume: belts 1 to 4 by Gottschalk (Chapter 9) and Gottschalk et al. (Chapter 12); belt 5 by Oldow et al. (Chapters 7 and 8); belt 6 by Phelps and Avé Lallemant (Chapter 4), Julian and Oldow (Chapter 5), and Seidensticker and Oldow (Chapter 6); and belt 7 by Handschy (Chapters 2 and 3) and Phelps and Avé Lallemant (Chapter 4).

The Angayucham terrane, a disrupted ophiolite complex of Devonian to Jurassic age (e.g., Patton et al., 1977; Jones et al.,

Avé Lallemant, H. G., and Oldow, J. S., 1998, Antithetic shear and the formation of back folds in the central Brooks Range fold and thrust belt, Alaska, *in* Oldow, J. S., and Avé Lallemant, H. G., eds., Architecture of the Central Brooks Range Fold and Thrust Belt, Arctic Alaska: Boulder, Colorado, Geological Society of America Special Paper 324.

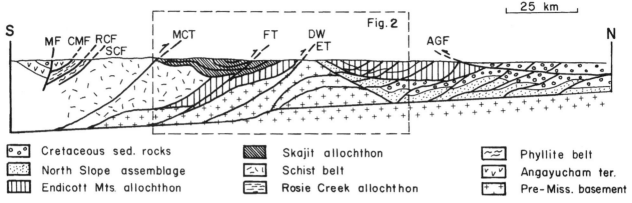

Figure 1. Restorable north-south cross section through the central Brooks Range fold and thrust belt (after Oldow et al., 1987). Location is shown in Figure 1 in Preface to this volume. Abbreviations: MF, Malamute fault; CMF, Cathedral Mountain fault; RCF, Rosie Creek fault; SCF, Slate Creek fault; MCT, Minnie Creek thrust fault; FT, Foggytop thrust fault; DW, Doonerak window; ET, Eekayruk thrust fault; AGF, Atigun Gorge thrust fault. Box represents area of Figure 2.

1988; Wirth and Bird, 1992), was the first lithotectonic unit of the Brooks Range fold and thrust belt, thrust northward during Middle Jurassic time (e.g., Roeder and Mull, 1978; Wirth and Bird, 1992; Wirth et al., 1993). Synmetamorphic deformation in the Schist belt occurred from the Middle Jurassic to the mid-Cretaceous (Gottschalk and Snee, this volume, Chapter 13). A major phase of deformation occurred in the early Tertiary (O'Sullivan et al., this volume, Chapter 11) resulting in the formation of the Doonerak and Blarney Creek duplexes (Seidensticker and Oldow, this volume, Chapter 6).

STRUCTURAL GEOLOGY

The east-northeast–trending belts in the central Brooks Range are all fault-bounded. Most bounding faults are thrust faults, but some were reactivated at a later time as normal faults (e.g., Gottschalk and Oldow, 1988; Gottschalk, 1990; Gottschalk et al., this volume, Chapter 12; Gottschalk and Snee, this volume, Chapter 13) and as strike-slip faults (Avé Lallemant et al., this volume, Chapter 15).

Megascopic structures

Based on mapping, structural analysis, lithologic correlation, and seismic reflection lines, Oldow et al. (1987) constructed balanced cross sections through the central Brooks Range. The Brooks Range fold and thrust belt is generally assumed to be a north-tapering wedge consisting of numerous thrust imbricates and duplexes (Figs. 1 and 2A in Preface to this volume; and Fig. 1). Noteworthy is the occurrence of three envelopment thrust faults (thrust faults that appear to be out-of-sequence, but have a deeper basal decollement); these are the Minnie Creek, Foggytop, and Eekayruk thrust faults. The balancing of the cross sections necessitated the presence of a major ramp and an antiformal-stack duplex underlying the Doonerak

window (Figs. 1 and 2; Seidensticker and Oldow, this volume, Chapter 6). This ramp is clearly imaged on the seismic reflection profile through the Brooks Range, as discussed by Wissinger et al. (this volume, Chapter 16).

Mesoscopic structures

In each belt, the deformation history was determined and correlated with that of the adjacent belts. Based on this correlation, the deformation history of the central Brooks Range can be divided into three phases. The first phase (D_1) is generally synmetamorphic and has been subdivided into four subphases (D_{1a} to D_{1d}), which formed sequentially during one continuous deformation event. The second is a postmetamorphic phase characterized by well-developed back folds (D_2), and the third (D_3) is a cross-folding event.

Correlation of fold generations across a fold and thrust belt based on similar folding histories and fold styles or geometries is risky. Deformation in a fold and thrust belt is very likely diachronous: that is, a particular generation of folds in the hinterland may have formed much earlier than the apparently similar generation in the foreland. Style and geometry of folds depend on several factors, such as lithology, but also finite strain (Ramsay, 1967). Handschy shows in Chapter 3 (this volume) that finite strains in the northern Endicott Mountains allochthon and the upper part of the southern Endicott Mountains are small and D_1 folds have class 1b geometries (Ramsay, 1967); toward the south and downward, the finite strains increase and fold geometries change to class 1c to 2 (Ramsay, 1967).

An additional problem for correlation of folds is that after a certain amount of finite strain, shortening does not always cause further appression of a fold but the rocks tend to be refolded by a second fold generation. These second folds may have the style and geometry of the first-generation folds in areas of high finite strain. Thus, when looking at two rock units, the lower one show-

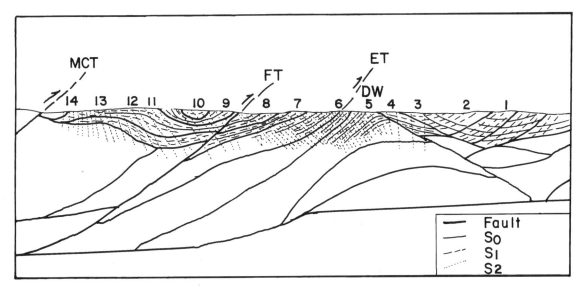

Figure 2. Detail of cross section of Figure 1. Structural data: heavy lines, faults; thin lines, bedding traces; dashes, S_1 cleavage traces; stippled, traces of S_2 fold axial planes and cleavages. Numbers 1 to 14 refer to areas from which structural data were derived: 1–3 from Handschy (this volume, Chapter 3); 4–6 from Phelps and Avé Lallemant (this volume, Chapter 4); 7–14 from Oldow et al. (this volume, Chapter 8).

ing two fold generations and the upper one only one, one might come to the erroneous conclusion that the lower unit had been folded in a much earlier orogenesis and that an unconformity must exist between the two units.

With an awareness of these problems and the limitations of structural analysis, the following deformational history is proposed for the central Brooks Range.

First-generation folds (D_1). The finite strain gradient as discussed by Handschy (this volume, Chapter 3) for the Endicott Mountains allochthon occurs throughout the central Brooks Range fold and thrust belt. In the north (the northern part of the Endicott Mountains allochthon, the Blarney Creek and Doonerak duplexes) the finite strains are relatively small and only one D_1 fold generation has been recognized. In the southern part of the Endicott Mountains and the northern part of the Skajit allochthon, two generations of D_1 folds occur (D_{1a} and D_{1b}), whereas in the southern portions of the Skajit allochthon and in the Schist belt where finite strains are very large three generations occur (D_{1a}, D_{1b}, and D_{1c}). The Phyllite and Rosie Creek allochthons have only two D_1 phases; they are structurally much higher than the Schist belt, attesting to a vertical strain gradient.

Most D_1 folds are recumbent and isoclinal and have class 1c to 2 geometry (Ramsay, 1967). Generally, they have a well developed axial-planar cleavage related to dissolution or syntectonic recrystallization. The cleavages generally dip gently to the south-southeast except for the northern Doonerak area where they dip toward the north-northwest. Fold axes and intersection lineations formed by the youngest D_1 deformation (D_{1c} in areas of three fold generations, D_{1b} in areas of two generations, and D_1 where only one generation occurs) generally trend east-northeast to west-southwest; only in high-strain areas these youngest elements may

be rotated to a north-northwest to south-southeast trend. The older D_1 fold axes and intersection lineations and the mineral and stretching lineations of all D_1 deformations trend always north-northwest to south-southeast.

In the southern part of the Schist belt, the Phyllite belt and Rosie Creek allochthon, a late synmetamorphic deformation (D_{1d}) resulted in south-dipping, south-vergent extensional kinks and shear zones (Gottschalk and Oldow, 1988; Gottschalk, 1990; Gottschalk et al., this volume, Chapter 12).

The timing of the D_1 deformations is constrained by several $^{40}Ar/^{39}Ar$ ages (Gottschalk and Snee, this volume, Chapter 13). The D_{1a} to D_{1c} deformations occurred between 175 and 135 Ma. The extensional D_{1d} deformation started at about 129 to 118 Ma and may have continued until about 115 Ma.

Second-generation folds (D_2): Back folding. The "second" generation of folds deforms all D_1 structures. The D_2 folds generally resemble kink bands and chevron folds, but folds with a larger radius of curvature occur too. The folds occur on all scales: microscopic with wavelengths of millimeters to macroscopic with wavelengths of hundreds of meters. The D_2 folds are symmetric and upright north of the Doonerak high whereas south of the high most folds are asymmetric and south-southeast–vergent (Fig. 2). Axial-planar cleavage is locally well developed. Fold axes and intersection lineations trend easterly to northeasterly. The south-southeast–vergent D_2 folds have been recognized in the entire southern portion of the region (Fig. 2), but they are best developed directly south of the Doonerak window area where they are very numerous, of large wavelength (hundreds of meters), and strongly appressed.

The timing of the D_2 deformation is conjectural. O'Sullivan et al. (this volume, Chapter 11) present zircon and apatite fission-track data for the Doonerak window area, indicating that the window rocks were uplifted from about 70 Ma to the present. As the uplift is related to deformation, and as the window is underlain by two major duplexes, the Doonerak and Blarney Creek duplexes (see Seidensticker and Oldow, this volume, Chapter 6), the formation of which postdates all D_1 structures, it seems logical to assume that the duplexes were formed during the D_2 deformation.

Third-generation folds (D_3). The D_3 deformation resulted in minor north-south–trending folds and steep north-south–striking axial planar cleavage (not shown in Fig. 2; see also in this volume: Handschy, Chapter 3; Phelps and Avé Lallemant, Chapter 4; Julian and Oldow, Chapter 5; Seidensticker and Oldow, Chapter 6; Gottschalk et al., Chapter 12). These structures are relatively rare and involve little strain. The folds may be genetically related to east-west–striking dextral strike-slip faults that crosscut all other structures (Phelps and Avé Lallemant, this volume, Chapter 4).

ORIGIN OF BACK FOLDS AND BACK THRUSTS

Back folds and back thrusts having vergences opposite to the vergence or the overall displacement direction of a fold and thrust belt have been recognized for a long time (e.g., Heim, 1922). The kinematic constraints and mechanisms of back-fold formation are probably similar to those for back thrusting: back thrusts may form at shallow levels where rocks behave in a brittle fashion and back folds at somewhat deeper levels where ductile deformation is likely to occur dominantly.

Several mechanisms for the formation of back folds and back thrusts have been proposed in the past and may be grouped into three categories: (1) displacement along a weak, low-friction basal decollement; (2) shearing along cleavage in a very anisotropic rock; and (3) displacements across a basal ramp.

Displacement along a weak, low-friction basal decollement

Potentially, two symmetric conjugate sets of faults will develop in homogeneous, isotropic rocks during coaxial contraction when the state of stress is triaxial (e.g., Anderson, 1951) and the intermediate strain (Y-axis) is zero (e.g., Reches and Dieterich, 1983). Fold and thrust belts are, however, generally characterized by only one set of faults that are foreland-vergent (e.g., Bally et al., 1966; Price, 1981, 1986; Mitra and Boyer, 1986). This asymmetry of structural development has been produced experimentally in sand box experiments (e.g., Hubbert, 1951) and in centrifuge tests (e.g., Liu and Dixon, 1991).

To evaluate the development of different types of duplexes, Mitra and Boyer (1986) made estimates about energy requirements necessary for each model duplex. Four energy terms were considered to be important: the work to form and propagate a fault (W_p), the work to overcome basal friction (W_b), the work against gravity (W_g), and the work involved in internal deformation (W_i). The same terms are considered important in evaluating whether there is a preference for the development of forethrusts or back

thrusts. As long as the bedding in a thrust sheet is horizontal, the thickness of each bed is constant, and the basal decollement is horizontal (Fig. 3A), the last three energy terms are equal for forethrusting and back thrusting. However, the first term (W_p) is different in the two cases: W_p for forethrusting is proportional to the length KPQ (Fig. 3A), and for back thrusting to KPR + PS. Thus, there should be a preference for forethrusts to develop.

In these models, the deformed material is homogeneous, at least in the horizontal plane. If horizontal heterogeneities exist, other results may be expected. If, for example, the footwall rocks underneath RP in Figure 3A are very strong (e.g., limestone, metamorphic rocks) and the hanging-wall rocks above RP are weak (e.g., foreland and forearc basin deposits), the rocks underneath RP act as a strong wedge or backstop causing delamination and back thrusting. This mechanism might be responsible for the formation of back thrusts in forearc regions (e.g., Byrne et al., 1993; Willett et al., 1993) and triangle zones (e.g., Price, 1986; Sanderson and Spratt, 1992).

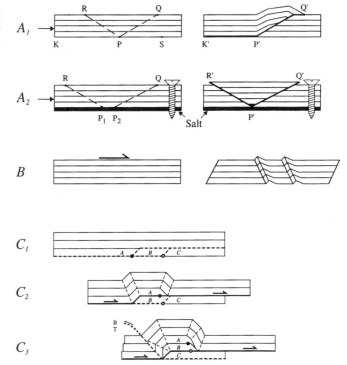

Figure 3. Origin of back thrusts. A_1, thrust fault KPQ is favored to form, because work for slip on fault KPR is larger, the difference being that additional work is needed to form fault segment PS; A_2, decollement surface is perfectly lubricated or frictionless, because of salt bed along the decollement; both foreland and hinterland thrust faults may form; B, non-coaxial or simple shear along penetrative anisotropy (cleavage, metamorphic layering, or bedding) causes antithetic kink bands to form, but layer-parallel shortening has to occur to maintain coherence; C, sequential development (1 to 3) of antiformal-stack duplex; the formation of the duplex causes back-rotation and steepening of the older horses resulting in a serious space problem that is relieved by the formation of back folds and back thrusts. If fault in foreland becomes locked, major displacement may occur on back thrust (BT).

In some fold and thrust belts, both synthetic and antithetic thrust faults formed in apparently horizontally homogeneous rock sequences such as in the Mackenzie Mountains in the Northwest Territories, Canada (Price, 1986), and on Melville Island, Canadian Arctic Archipelago (Harrison and Bally, 1987). On Melville Island, these structures are the result of southward displacement of lower and middle Paleozoic sedimentary rocks along a Silurian salt horizon. As the strength of salt is virtually zero (Carter et al., 1993), there should not be any preference for synthetic or antithetic faults (Fig. 3A$_2$) and both should form (see also Davis and Engelder, 1985). Similarly, high pore pressures along the basal decollement should result in structures of both vergence. Both antithetic and synthetic thrust faults were formed in sand box experiments (Dahlen et al., 1984; Mulugeta, 1988) where the friction along the bottom of the sand box was extremely low.

The origin of back thrusts and in particular triangle zones along mountain fronts is often related to the location of the basal decollement in a very weak layer that pinches out toward the foreland (e.g., Vann et al., 1986). Foreland-directed displacements along this basal decollement zone are inhibited where the weak layer pinches out and a back thrust or a roof thrust of a passive duplex is born by delamination of the foreland sequence (e.g., Vann et al., 1986; Banks and Warburton, 1986; Ferrill and Dunne, 1989). These roof duplexes may be deformed internally; if the strength of the roof-thrust fault is high, hinterland-vergent folds and reverse faults are expected to form, but if the roof-thrust fault is weak, approximately symmetric, upright folds and vertical cleavage may develop (Geiser, 1988; Ferrill and Dunne, 1989).

Shearing along cleavage in a very anisotropic rock

It has been shown experimentally (e.g., Donath, 1961; Paterson and Weiss, 1966) that when slates and phyllites are compressed at a moderate angle to the slaty cleavage they tend to deform by sliding along the slaty cleavage plane and by the formation of kink bands. At high strains, faults may develop along the kink-band boundaries. The rotation of the cleavage in the kink bands is opposite to the sense of shear on the slaty cleavage and, thus, the kink band is antithetic with respect to the displacement along the cleavage. The angle between the cleavage and the kink band tends to be about 60 to 70°.

Johnson (1970, p. 319) described deformation experiments of a layered sequence of cardboard and clay by layer-parallel simple shear with a small component of layer-parallel shortening. Shear along the layering was accompanied by antithetic kink bands at about 70 to 80° to the layering.

In highly deformed metamorphic rocks in the hinterland of fold and thrust belts, metamorphic cleavages and layering tend to be rotated toward parallelism with the decollement surface. Low-angle thrusting might result in simple shear along the cleavage/metamorphic layering plane and the formation of antithetic kink bands. However, a component of layering-parallel shortening (Fig. 3B) is necessary to preserve a state of constant volume.

Figure 3B also shows that the deformation is heterogeneous as small voids have to form in the fold hinges.

Displacements across a basal ramp

Ramps in the basal decollement surface of fold and thrust belts are ubiquitous structures. Suppe (1983) described how layered rocks in the hanging wall of a thrust system get folded as a result of displacement across a ramp in the footwall. In a simple situation, a kinklike, antithetic hanging-wall syncline forms at the base of the ramp. During progressive thrusting, the thrust sheet migrates up-ramp, the synform unfolds, and, at the top of the ramp, the sheet forms an antiformal structure. The fold hinge angle of the antithetic folds depends on the ramp dip angle, but generally it is large.

Serra (1977) described antithetic thrust (reverse) faults at the base of ramps in several thrust belts. He suggested that the reverse faults occurred only in areas where the mechanically strongest, load-bearing bed in the thrust sheet was as thick or thicker than the height of the ramp.

Where foreland-verging ramps occur in a fold and thrust belt, duplexes tend to form (e.g., Boyer and Elliot, 1982; Mitra and Boyer, 1986). Such ramps can be expected to form in a thrust belt, because the length of a thrust imbricate is to some extent controlled by the strength of the rocks and the friction along the decollement surface (Hubbert and Rubey, 1959). In particular, during the formation of an antiformal-stack duplex (Mitra and Boyer, 1986), the bedding planes or cleavage planes in each imbricate or horse rotate backward; the oldest imbricate rotates the most and may become vertical or even overturned to the hinterland. This rotation causes space problems, which may be relieved by back thrusting or back folding (Fig. 3C). Such a back thrust may become the main active fault plane (BT in Fig. 3C$_3$) if for some reason the basal thrust fault in the foreland area becomes locked (Ferrill and Dunne, 1989).

Several experimental studies have been carried out dealing with deformation above an active ramp. Buchanan and McClay (1991) carried out sand box tests in which they inverted down-dip displacement along a listric normal fault: during upthrusting, several low-angle back thrusts formed.

Colletta et al. (1991) also carried out sand box experiments. Ramps formed due to changes in physical properties of the material overlying the decollement. They, too, observed the formation of back thrusts.

Chester et al. (1991) presented results of room-temperature and low-pressure deformation experiments of limestone/mica and limestone/lead sandwiches overlying a sandstone layer containing a "foreland-vergent" saw-cut ramp oriented at 20° to the layering. Although some aspects of these tests were not realistic, such as separation of some of the layers, interestingly, a multitude of back thrusts formed: generally, a back thrust formed in the hanging wall at the toe of the ramp and it was transported subsequently up-ramp whereupon a new thrust fault was initiated at the toe of the ramp.

ORIGIN OF THE ANTITHETIC BACK FOLDS IN THE BROOKS RANGE

The back folds in the central Brooks Range fold and thrust belt, formed during the D_2 deformation, are only developed in the southern part of the Brooks Range, south of the Doonerak window (Fig. 2). D_2 folds north of the Doonerak window are symmetric or north-vergent.

The first model explaining the origin of back thrusts and, thus, also back folds, can be ruled out for the central Brooks Range on the basis of the fold vergences north and south of the Doonerak window. The second model is viable, but it does not explain the fact that north of the Doonerak window folds are symmetric and upright and in the south the folds are asymmetric and south-vergent. However, low-amplitude, low-wavelength back folds exist far to the south of the Doonerak window, in the Skajit allochthon (Oldow et al., this volume, Chapter 8) where no ramps or duplexes have been recognized. These kink-style folds may have formed as the result of inhomogeneous northward slip along the subhorizontal S_1 cleavages (Fig. 3B), as was suggested earlier (Avé Lallemant et al., 1983).

Detailed studies in the Doonerak window (Seidensticker and Oldow, this volume, Chapter 6) show that the Doonerak anticlinorium consists of two duplexes, the Blarney Creek and Doonerak duplexes. A cross section through the area (Figs. 1 and 2) suggests that the anticlinorium is an antiformal-stack duplex (Boyer and Elliot, 1982; Mitra and Boyer, 1986). An antiformal-stack duplex (Fig. 3C) causes large strain incompatibilities and these strains can only be released by back thrusting or back folding. The fact that the area of well-developed back folds overlaps with the area in which a major antiformal-stack duplex formed is unlikely to be a coincidence. Thus, we suggest that the back folds are mainly the result of the formation of the Doonerak and Blarney Creek duplexes.

During the formation of an antiformal-stack duplex, the amplitude of the antiform is increasing, while the radius of curvature is decreasing. This may result in bending stresses that may be released by normal faulting. As it happens, east-west–trending normal faults are ubiquitous in the Doonerak window area (e.g., Phelps and Avé Lallemant, this volume, Chapter 4).

CONCLUSIONS

Most structures in the Brooks Range fold and thrust belt are north-vergent, indicating that they were formed by northward tectonic transport. There are, however, in the central Brooks Range structures that are south-vergent, and, thus, antithetic to the general movement picture.

Three major phases of deformation have been recognized in the region. The first phase (D_1) was synmetamorphic (blueschist- to greenschist-facies conditions) in the southern hinterland to nonmetamorphic in the foreland and it was active from 175 to about 115 Ma (Gottschalk and Snee, this volume, Chapter 13). There is a strong finite strain gradient in the thrust belt, with very

high strains in the south, decreasing to the north and upward. The deformation is characterized by four generations of folds in the south, but the number of generations decreases northward and upward. Kinematic indicators of the first three synmetamorphic phases $(D_{1a}$ to $D_{1c})$ strongly suggest that the sense of tectonic transport is toward the north. The fourth synmetamorphic phase (D_{1d}) resulted in south-dipping ductile shears and extensional kink bands.

The second deformation that may have taken place since about 70 Ma (O'Sullivan et al., this volume, Chapter 11) is characterized by south-vergent folds. These folds are best developed south of the Doonerak window. The Doonerak window is underlain by an antiformal-stack duplex. The coincidence of the duplex with the area where antithetic, south-vergent folds are ubiquitous suggests that these folds were formed to release severe space problems caused by the stacking of thrust imbricates in the duplex. Minor back folds in the Skajit allochthon, south of the Doonerak window, may be conjugates to shear planes parallel to the subhorizontal S_1 cleavages.

The D_3 deformation is very minor. North-trending cross folds may be related to young (post-30 Ma) east-west–striking strike slip faults.

ACKNOWLEDGMENTS

This study was made possible by grants from the Department of Energy (DE-AS05-83ER13124), the National Science Foundation (EAR-85-17384 and EAR 87-20171), and the Rice University Alaska Industrial Associates Group (Amoco Production Company, Arco Exploration Company, Chevron U.S.A., Gulf Oil Exploration and Production Company, Mobil Exploration and Producing Services, and The Standard Oil Company). We thank our graduate students who did most of the work: Kent Boler, Rick Gottschalk, Jim Handschy, Betsy Julian, Jim Phelps, and Mike Seidensticker. Peter Geiser and Kent Nielsen reviewed the paper and made invaluable suggestions to improve it.

REFERENCES CITED

Anderson, E. M., 1951, The dynamics of faulting and dyke formation with applications to Britain (second edition): London, Oliver & Boyd, 206 p.

Avé Lallemant, H. G., Oldow, J. S., Russell, B. J., and Dillon, J. T., 1983, Structural evolution of the central Brooks Range near the Doonerak window, Alaska: Geological Society of America Abstracts with Programs, v. 15, p. 274.

Bally, A. W., Gordy, P. L., and Stewart, G. A., 1966, Structure, seismic data, and orogenic evolution of southern Canadian Rockies: Bulletin of Canadian Petroleum Geology, v. 14, p. 337–381.

Banks, C. J., and Warburton, J., 1986, 'Passive-roof' duplex geometry in the frontal structures of the Kirthar and Sulaiman Mountain belts, Pakistan: Journal of Structural Geology, v. 8, p. 229–237.

Boyer, S. E., and Elliott, D., 1982, Thrust systems: American Association of Petroleum Geologists Bulletin, v. 66, p. 1196–1230.

Brosgé, W. P., and Reiser, H. N., 1964, Geologic map and section of the Chandalar Quadrangle, Alaska: U.S. Geological Survey Miscellaneous Geologic Investigations Map I-375, scale 1:250,000.

Brosgé, W. P., and Reiser, H. N., 1971, Preliminary bedrock geologic map: Wiseman and eastern Survey Pass Quadrangles, Alaska: U.S. Geological Survey Open-File Map 479, scale 1:250,000.

Brown, R. L., Journeay, J. M., Lane, L. S., Murphy, D. C., and Rees, C. J., 1986, Obduction, backfolding and piggyback thrusting in the metamorphic hinterland of the southeastern Canadian Cordillera: Journal of Structural Geology, v. 8, p. 255–268.

Buchanan, P. G., and McClay, K. R., 1991, Sandbox experiments of inverted listric and planar fault systems: Tectonophysics, v. 188, p. 97–115.

Byrne, D. E., Wang, W. H., and Davis, D. M., 1993, Mechanical role of backstops in the growth of forearcs: Tectonics, v. 12, p. 123–144.

Carter, N. L., Horseman, S. T., Russell, J. E., and Handin, J., 1993, Rheology of rocksalt: Journal of Structural Geology, v. 15, p. 1257–1271.

Chester, J. S., Logan, J. M., and Spang, J. H., 1991, Influence of layering and boundary conditions on fault-bend and fault-propagation folding: Geological Society of America Bulletin, v. 103, p. 1059–1072.

Colletta, B., Letouzey, J., Pinedo, R., Ballard, J. F., and Balé, P., 1991, Computerized X-ray tomography analysis of sandbox models: Examples of thin-skinned thrust systems: Geology, v. 19, p. 1063–1067.

Dahlen, F. A., Suppe, J., and Davis, D., 1984, Mechanics of fold-and-thrust belts and accretionary wedges: Cohesive Coulomb theory: Journal of Geophysical Research, v. 89, p. 10087–10101.

Davis, D. M., and Engelder, T., 1985, The role of salt in fold-and-thrust belts: Tectonophysics, v. 119, p. 67–88.

Dillon, J. T., Brosgé, W. P., and Dutro, J. T., Jr., 1986, Generalized geologic map of the Wiseman Quadrangle, Alaska: U.S. Geological Survey, Open-File Report 86-219, scale 1:250,000.

Dillon, J. T., Harris, A. G., Dutro, J. T., Jr., Solie, D. N., Blum, J. D., Jones, D. L., and Howell, D. G., 1988, Preliminary geologic map and section of the Chandalar D-6 and parts of the Chandalar C-6 and Wiseman C-1 and D-1 Quadrangles, Alaska: Alaska Division of Geological and Geophysical Surveys, Report of Investigations 88-5, scale 1:63,360.

Donath, F. A., 1961, Experimental study of shear failure in anisotropic rocks: Geological Society of America Bulletin, v. 72, p. 985–989.

Ferrill, D. A., and Dunne, W. M., 1989, Cover deformation above a blind duplex: an example from West Virginia, U.S.A.: Journal of Structural Geology, v. 11, p. 421–431.

Geiser, P. A., 1988, Mechanisms of thrust propagation: some examples and implications for the analysis of overthrust terranes: Journal of Structural Geology, v. 10, p. 829–845.

Gottschalk, R. R., 1990, Structural evolution of the Schist belt, south-central Brooks Range fold and thrust belt, Alaska: Journal of Structural Geology, v. 12, p. 453–469.

Gottschalk, R. R., and Oldow, J. S., 1988, Low-angle normal faults in the south-central Brooks Range fold and thrust belt, Alaska: Geology, v. 16, p. 395–399.

Harrison, J. C., and Bally, A. W., 1987, Cross-sections of the Devonian to Mississippian fold belt on Melville Island, Canada Arctic Islands; Canadian Petroleum Geology Bulletin, v. 36, p. 331–352.

Heim, A., 1922, Geologie der Schweiz: Leipzig, Germany, C. H. Tauchnitz, 1018 p.

Hubbert, M. K., 1951, Mechanical basis for certain familiar geologic structures: Geological Society of America Bulletin, v. 62, p. 355–372.

Hubbert, M. K., and Rubey, W. W., 1959, Role of fluid pressure in mechanics of overthrust faulting, I. Mechanics of fluid-filled porous solids and its application to overthrust faulting: Geological Society of America Bulletin, v. 70, p. 115–206.

Johnson, A. M., 1970, Physical processes in geology: San Francisco, Freeman, Cooper & Company, 577 p.

Jones, D. L., Coney, P. J., Harms, T. A., and Dillon, J. T., 1988, Interpretive geologic map and supporting radiolarian data from the Angayucham terrane, Coldfoot area, southern Brooks Range, Alaska: U.S. Geological Survey Miscellaneous Field Studies Map MF-1993, scale 1:63,360.

Lawton, D. C., Spratt, D. A., and Hopkins, J. C., 1994, Tectonic wedging beneath the Rocky Mountain forcland basin, Alberta, Canada: Geology, v. 22, p. 519–522.

Liu, S., and Dixon, J. M., 1991, Centrifuge modelling of thrust faulting: structural variation along strike in fold-thrust belts: Tectonophysics, v. 188, p. 39–62.

Macaya, J., Gonzalez-Lodeiro, F., Martinez-Catalan, J. R., and Alvarez, F., 1991, Continuous deformation, ductile thrusting and backfolding of cover and basement in the Sierra de Guadarrama, Hercynian orogen of central Spain: Tectonophysics, v. 191, p. 291–309.

Mitra, G., and Boyer, S. E., 1986, Energy balance and deformation mechanisms of duplexes: Journal of Structural Geology, v. 8, p. 291–304.

Mulugeta, G., 1988, Modelling the geometry of Coulomb thrust wedges: Journal of Structural Geology, v. 10, p. 847–859.

Oldow, J. S., Seidensticker, C. M., Phelps, J. C., Julian, F. E., Gottschalk, R. R., Boler, K. W., Handschy, J. W., and Avé Lallemant, H. G., 1987, Balanced cross sections through the central Brooks Range and North Slope, Arctic Alaska: Tulsa, Oklahoma, American Association of Petroleum Geologists, Special Publication, 19 p., 8 pl.

Paterson, M. S., and Weiss, L. E., 1966, Experimental deformation and folding of phyllite: Geological Society of America Bulletin, v. 77, p. 343–374.

Patton, W. W., Jr., Tailleur, I. L., Brosgé, W. P., and Lanphere, M. A., 1977, Preliminary report on the ophiolites of northern and western Alaska, *in* Coleman, R. G., and Irwin, W. P., eds., North American ophiolites: Oregon Department of Geology and Mineral Industries Bulletin, v. 95, p. 51–57.

Price, R. A., 1981, The Cordilleran foreland thrust and fold belt in the southern Canadian Rocky Mountains, *in* Price, N. J., and McClay, K., eds., Thrust and nappe tectonics: Geological Society of London Special Publication 9, p. 427–448.

Price, R. A., 1986, The southeastern Canadian Cordillera: thrust faulting, tectonic wedging, and delamination of the lithosphere: Journal of Structural Geology, v. 8, p. 239–254.

Ramsay, J. G., 1967, Folding and fracturing of rocks: New York, McGraw-Hill, 568 p.

Reches, Z., and Dieterich, J. H., 1983, Faulting of rocks in three-dimensional strain fields. I. Failure of rocks in polyaxial, servo-controled experiments: Tectonophysics, v. 95, p. 111–132.

Roeder, D., and Mull, C. G., 1978, Tectonics of Brooks Range ophiolites (Alaska): American Association of Petroleum Geologists Bulletin, v. 62, p. 1696–1713.

Sanderson, D. A., and Spratt, D. A., 1992, Triangle zone and displacement transfer structures in the eastern Front Ranges, southern Canadian Rocky Mountains: American Association of Petroleum Geologists Bulletin, v. 76, p. 828–839.

Serra, S., 1977, Styles of deformation in the ramp regions of overthrust faults, *in* Guidebook of the Twenty-ninth Annual Field Conference—1977: Wyoming Geological Association, p. 487–498.

Suppe, J., 1983, Geometry and kinematics of fault-bend folding: American Journal of Science, v. 283, p. 684–721.

Vann, I. R., Graham, R. H., and Hayward, A. B., 1986, The structure of mountain fronts: Journal of Structural Geology, v. 8, p. 215–227.

Willett, S., Beaumont, C., and Fullsack, P., 1993, Mechanical model for the tectonics of doubly vergent compressional orogens: Geology, v. 21, p. 371–374.

Wirth, K. R., and Bird, J. M., 1992, Chronology of ophiolite crystallization, detachment, and emplacement: Evidence from the Brooks Range, Alaska: Geology, v. 20, p. 75–78.

Wirth, K. R., Bird, J. M., Blythe, A. E., and Harding, D. J., 1993, Age and evolution of western Brooks Range ophiolites Alaska: Results from ^{40}Ar/^{39}Ar thermochronometry: Tectonics, v. 12, p. 410–432.

MANUSCRIPT ACCEPTED BY THE SOCIETY SEPTEMBER 23, 1997

Geological Society of America
Special Paper 324
1998

Structural analysis of the Kobuk fault zone, north-central Alaska

H. G. Avé Lallemant
Department of Geology and Geophysics, Rice University, Houston, Texas 77005-1892
R. R. Gottschalk
Exxon Production Research Company, P.O. Box 2189, Houston, Texas 77252
V. B. Sisson
Department of Geology and Geophysics, Rice University, Houston, Texas 77005-1892
J. S. Oldow
Department of Geology and Geological Engineering, University of Idaho, Moscow, Idaho 83843

ABSTRACT

The east-west–trending Kobuk fault zone in the hinterland of the Brooks Range fold and thrust belt, north-central Alaska, consists of several poorly exposed, megascopic anastomosing faults. Rock slices between these faults are deformed by numerous meso- and microscopic faults along which only minor displacements occurred. Kinematic analysis of the small-scale faults and analysis of fluid inclusions in quartz veins emplaced synkinematically along these structures indicate that the Kobuk fault zone underwent at least four phases of deformation. The first structures (F_1) were the result of northward thrusting that started in early Middle Jurassic time. F_2 faults are related to north-south extension during the mid-Cretaceous. F_3 faults are related to right-lateral strike-slip displacements that occurred during Late Cretaceous and Tertiary time. A fourth phase (F_4; north-south extension) is indicated by the fluid inclusion study. This deformational history, however, does not apply to the entire fold and thrust belt; at the same time that lateral displacements (F_3) occurred along the Kobuk fault system, northward thrusting took place in the foreland portion of the Brooks Range fold and thrust belt, indicating that plate convergence was right-oblique resulting in displacement partitioning.

INTRODUCTION

The southern foothills of the Brooks Range are dissected by several extensive, anastomosing, east-west–striking, vertical to moderately south-dipping faults, which together form the Kobuk fault zone, one of the major fault systems in northern Alaska (Fig. 1; King, 1969). Roeder and Mull (1978) interpreted the zone as an alpine root zone. It was called the Kobuk trench by Estabrook (1985) and the Kobuk suture by Smith (1987) and Hubbard et al. (1987) who proposed it to represent the collision boundary between a "North Alaska microplate" in the north and a volcanic island arc and several igneous-metamorphic allochthonous terranes in the south (see Figs. 1 and 2 in Preface to this volume).

In the central part of the southern foothills, west of the Dalton Highway and Trans-Alaska Pipeline, the Kobuk fault zone consists of two major faults: the Angayucham fault in the north and the Malamute fault in the south (Fig. 2). The first one becomes the Slate Creek fault toward the east; the Malamute fault branches out in the east into three strands, the northern two of which are the Rosie Creek and the Cathedral Mountain faults (Figs. 2 and 3; Brosgé and Reiser, 1964, 1971; Patton and Miller, 1973; Dillon et al., 1986, 1987, 1988; Gottschalk, 1987). The Slate Creek fault separates the Schist belt in the north from the Phyllite belt in the south. The Rosie Creek fault separates the Phyllite belt from the Rosie Creek allochthon. The Cathedral Mountain fault separates the Rosie Creek allochthon from the Angayucham terrane that, in the study area, consists mainly of

Avé Lallemant, H. G., Gottschalk, R. R., Sisson, V. B., and Oldow, J. S., 1998, Structural analysis of the Kobuk fault zone, north-central Alaska, *in* Oldow, J. S., and Avé Lallemant, H. G., eds., Architecture of the Central Brooks Range Fold and Thrust Belt, Arctic Alaska: Boulder, Colorado, Geological Society of America Special Paper 324.

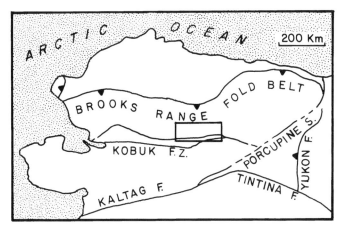

Figure 1. Major faults and shear zones in northern Alaska and northern Yukon Territory. Box is location of Figure 2.

basalt and chert (elsewhere the Angayucham terrane consists of a second tectonic slice containing peridotite and gabbro overlying the basalt-chert sheet). The Kobuk fault zone is a complex structure; it has been proposed to have acted as (1) a thrust fault zone, (2) a normal fault zone, and (3) a strike-slip fault zone.

Thrust faults. Whereas north-directed thrusting must have occurred along the zone, in view of the hundreds of kilometers of northward displacement as deduced by restoring and balancing cross sections through the Brooks Range fold and thrust belt (Oldow et al., 1987), few, if any, thrust faults can be identified because of subsequent disruption. The basal shear zone underneath the Angayucham terrane has generally been interpreted as a thrust fault (Angayucham fault; Fig. 2) along which the terrane was thrust northward (e.g., Patton and Miller, 1973; Roeder and Mull, 1978; Patton et al., 1977; Dillon et al., 1986, 1987, 1988; Jones et al., 1988). However, the rocks in this terrane are younger than the underlying metasedimentary rocks of the Rosie Creek terrane (Gottschalk, 1990; Gottschalk et al., this volume, Chapter 12); such stacking order is generally more typical of extensional tectonics than of contraction. Thus, originally, the Angayucham fault may have been a north-vergent thrust fault that was reactivated at a later time as a normal fault (e.g., Gottschalk and Oldow, 1988; Law et al., 1994; Little et al., 1994).

The time of inception of the thrusting is poorly constrained in the central Brooks Range. Turner et al. (1979) reported numerous K/Ar dates ranging from 155 to 145 Ma, which were thought to be related to the obduction. Based on $^{40}Ar/^{39}Ar$ geochronology, Gottschalk and Snee (this volume, Chapter 13) and Gottschalk et al. (this volume, Chapter 12) proposed that the obduction of the Angayucham terrane and the blueschist metamorphism occurred in early Middle Jurassic time. In the western Brooks Range, thrusting of the Angayucham terrane apparently happened somewhat later, in the Late Jurassic; this is indicated because in the foreland basin to the north sedimentary rocks of Late Jurassic age are the first to be derived from a southern source and lowermost Cretaceous clastic rocks include detritus from the Angayucham terrane (Molenaar, 1988). $^{40}Ar/^{39}Ar$ ages

of rocks in the metamorphic aureole beneath the ultramafic allochthon of the Angayucham terrane predate the obduction by a few million years (Harris, 1992; Wirth and Bird, 1992).

Normal faults. Several strands of the Kobuk fault zone (e.g., Cathedral Mountain, Rosie Creek, and Slate Creek faults) are interpreted as normal faults based on the anomalous juxtaposition of low-grade metamorphic rocks above and directly in contact with the high-grade ones (Gottschalk, 1987, 1990; Gottschalk and Oldow, 1988; Miller and Hudson, 1991; Gottschalk et al., this volume, Chapter 12; Law et al., 1994; Little et al., 1994). The age of the normal faulting has been proposed to be mid-Cretaceous (Gottschalk and Oldow, 1988; Blythe et al., this volume, Chapter 10; Gottschalk et al., this volume, Chapter 12); the age is based on anomalously young $^{40}Ar/^{39}Ar$ ages in the Schist belt near the Slate Creek fault.

Strike-slip faults. The Malamute fault (e.g., Patton and Miller, 1973; Dillon et al., 1986), one strand of the Kobuk fault system (Fig. 2), is interpreted as a dextral strike-slip fault (e.g., Jones, 1980; Dillon et al., 1986, 1987, 1988; Jones et al., 1988). The Malamute fault crosscuts and displaces several granitic plutons (Brosgé and Reiser, 1964), including the Hodzana pluton (Fig. 3), which was dated by the Rb/Sr method as 112 ± 4 Ma (Blum et al., 1987). Similar plutons in the Ruby terrane to the south were dated as Early Cretaceous (Blum et al., 1987; Miller, 1989; Arth et al., 1989a, b). This fault also cuts across and, thus, is younger than the Lower and mid-Cretaceous sedimentary and volcanic rocks of the Yukon-Koyukuk basin (Nilsen, 1989). It is not known when major movement along the fault ceased, but Patton and Hoare (1968) suggested that it was still active in early Tertiary time. Minor seismicity along the Kobuk fault zone suggests recent reactivation of the dextral strike-slip deformation (Estabrook, 1985).

The Kobuk fault zone is both geomorphically and geologically a major structure. The kinematic models of the fault zone, as discussed in previous section, are based on lithologic, geochronologic, and metamorphic data, but not on mesoscopic kinematic data. The displacement history of this fault zone is crucial for placing the entire Brooks Range fold and thrust belt in a regional framework. The goal of this study was to determine the structural evolution of the Kobuk fault zone by performing meso- and microscopic kinematic analysis. Additionally, fluid inclusion data were collected from quartz veins that were emplaced synkinematically with thrusting, extension, and strike-slip deformation in order to constrain the depth (pressure) and temperature at which the several phases of veins had formed and, thus, to establish conditions at which the different phases of faulting occurred.

STRUCTURAL GEOLOGY

The geology of the area will not be discussed here, because it is described elsewhere (e.g., in the Preface to this volume; Gottschalk et al., this volume, Chapter 12). The litho-tectonic belts in the area are, from north to south: (1) Schist belt, (2) Phyllite belt, (3) Rosie Creek allochthon, (4) Angayucham terrane, and (5) Yukon-Koyukuk basin (Fig. 2 in the Preface to this volume; Figs. 2 and 3).

Ductile deformation

Most rocks in the belts that comprise the "Kobuk fault zone" have experienced one or more phases of folding of variable intensity. Whereas rocks in the Rosie Creek allochthon and in particular in the Angayucham terrane are not very strongly folded, the rocks of the Schist and Phyllite belts were intensely deformed during three phases of folding: the first (D_1) is syn-metamorphic, whereas the second and third (D_2 and D_3) are post-metamorphic.

The D_1 phase of folding has been subdivided into four sub-phases: D_{1a} to D_{1d} (Gottschalk, 1990; Gottschalk et al., this volume, Chapter 12). The first (D_{1a}) formed during blueschist-facies metamorphism and the other three during greenschist-facies metamorphism. D_{1a}, D_{1b}, and D_{1c} folds are all interpreted as having formed during progressive deformation associated with north-directed thrusting. These contractional structures are over-printed by numerous east-west–trending extensional shears and kinks (D_{1d}), which, while ductile, postdate the peak of meta-morphism and are interpreted to have formed during north-south extension (Gottschalk and Oldow, 1988; Gottschalk, 1990).

D_2 folds generally trend east-west and are related to north-south contraction. D_3 folds are relatively rare; they trend north-south and are related to east-west contraction.

Brittle deformation

In the study area, the Kobuk fault zone consists of several major faults: the Angayucham, Slate Creek, Rosie Creek, Cathedral Mountain, and Malamute faults (Fig. 2). In the eastern parts of the area, the South Fork fault is part of the Kobuk system, but toward the west, it deflects to the southwest and becomes the southern boundary fault of the Yukon-Koyukuk basin. These faults are poorly exposed and tend to occur in long, east-west–trending valleys that are covered with alluvium and, consequently, do not allow direct determination of their movement history. However, the rocks cropping out between these pre-sumed major faults are often strongly deformed by numerous hand-sample to outcrop size mesoscopic faults, the movement history of which can be determined and may be used to speculate on the kinematics of the major faults.

To define the brittle deformation history of the Kobuk fault zone, the orientations of each mesoscopic fault (slickenside), fracture, and vein in each outcrop visited were determined. The orientations of slickenside striations (slickenlines) and quartz and calcite fibers on fault planes were measured to find the orienta-tion of the displacement direction (Gamond, 1987; Petit, 1987; Ramsay and Huber, 1987). The sense of shear was determined from vein fibers (Ramsay and Huber, 1987), from features dis-placed across the fault, and from "drag" folds. The sense of dis-placement along a fault plane was also determined from the asymmetric shape and the morphology of grooves and ridges rep-resenting the slickenside striations (Iverson, 1991). These data are plotted on equal-area nets in Figure 2. Oriented hand speci-mens were collected from several places to verify or define

microscopically the sense of slip along the mesoscopic faults. Since the outcrops in the area of study are weathered and covered with lichen, it was often impossible to determine the asym-metries, let alone the morphology of the grooves and ridges; faults with such striations are also plotted in Figure 2, without, however, the sense of displacement.

The relative ages of faults could not always be determined. In a few places, crosscutting relationships of faults were observed. More often, however, a single fault plane showed several genera-tions of slickenlines and fibrous crystals, the crosscutting relation-ships of which allowed for a relative age determination.

Domains. The study area was divided into nine domains, each of which consists of one outcrop or several adjoining ones (Fig. 2). In some domains, the fault population was divided into two (domains B, E, and I), in other into three (domains C, D, H/G, and J) separate subpopulations; each subpopulation is rep-resented by one equal-area net in Figure 2; in domains A and F, all faults were combined in one net. The criteria for assigning a fault to a particular subpopulation were (1) compatibility of the fault with an established subpopulation, and (2) relative age with respect to other faults.

A fault is considered compatible with a subpopulation of faults if the displacement along the fault is compatible with the total finite strain field of the population. This strain may be biaxial (Anderson, 1947) or triaxial (Reches and Dieterich, 1983). At this stage, no assumptions were made about partitioning.

The fault data collected in the Kobuk fault zone are not com-patible with one singular finite strain. In fact, the data require at least three separate strain fields. Of course, the data can satisfy many more strain events, but three should be the minimum. These three strain events are labeled here F_1, F_2, and F_3. No crosscutting relations were found to indicate the relative age of the F_1 thrusting event; however, the fluid inclusion study (see below) suggests that this event is the oldest one. Numerous cross-cutting relationships indicate that the F_3 strike-slip deformation postdates the F_2 normal faulting.

F_1 event. The first event is represented by a few rare thrust faults. These thrust faults are approximately east-west striking and dip north or south (Domains C, H/G, and J). One thrust fault dips moderately to the west and shows north-northeasterly tec-tonic transport (Domain D).

F_2 event. Many northeast- to east-striking faults are normal faults (Domains A, C, D, E, F, G/H, I, and J). A large number of faults of this orientation with down-dip slickenlines could be normal faults and are plotted in the F_2 nets, but, because the sense of slip could not be determined, they might be thrust faults and related to F_1. A large number of steep, northwest- to northeast-trending faults with variably oriented slickenlines (all domains) could be wrench and tear faults, and, thus, be compatible with the F_2 event as well as the F_1 event.

Although it is generally assumed that the major extensional structure is a low-angle fault (e.g., Gottschalk and Oldow, 1988), the normal faults encountered in the field are mostly high-angle faults. It may be that the low-angle faults were tilted during later

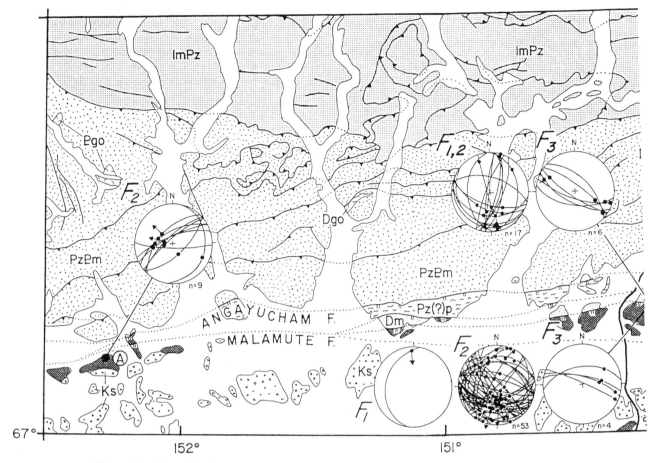

Figure 2 (on this and facing page). Simplified map of the Kobuk fault zone (location on Fig. 1; from Brosgé and Reiser, 1964; Dillon et al., 1986). Lower hemisphere, equal-area projections of mesoscopic faults are related to the north-directed thrusting event (F₁), north-south extension (F₂), and dextral strike-slip along east-west–trending faults (F₃). Symbols are: Dgo, Devonian granitic orthogneiss; Dm, Devonian metasedimentary rocks of the Rosie Creek allochthon; Kg, Cretaceous granite; Ks, Cretaceous sedimentary rocks of the Yukon-Koyukuk basin; lmPz, lower and middle Paleozoic metasedimentary rocks of the Skajit allochthon; MzPzv, Mesozoic and Paleozoic basalt, diabase, chert, and limestone of the Angayucham terrane; Pgo, Proterozoic granite; Pz(?)p, Devonian(?) rocks of the Phyllite belt; PzPm, middle and lower Paleozoic and Precambrian(?) rocks of the Schist belt.

deformation, or, more likely, that the observed normal faults are subsidiary faults and merge at depth with the low-angle detachment fault (see Gottschalk et al., this volume, Chapter 12).

F₃ event. A large number of steep, east- to northeast-striking strike-slip faults have been observed. In many cases it could be proven that motion on these faults postdates F₂. In all cases where kinematic indicators were available, these faults are shown to be right lateral.

Displacement along the mesoscopic strike-slip faults is generally very small. The total cumulative displacement, however, may be quite large: a granite body and a patch of staurolite schist, south of the Malamute fault, at about 149° 20′ west and 67° 10′ north (near A in Fig. 3), could be correlative with granite and staurolite schists, north of the fault, at about 147° 20′ west and 67° 10′ north (near A′ in Fig. 3). This would imply a right-lateral displacement of about 80 km.

PRESSURE-TEMPERATURE HISTORY

The Kobuk fault zone is strongly impregnated with quartz and calcite veins. Many of these veins occur in small, mesoscopic pull-aparts along faults; quartz and calcite grains grew as fibers parallel to the slickenlines indicating that vein material was emplaced synkinematically during the faulting events. Many veins were sampled for fluid-inclusion analysis with the goal to constrain the pressure (P)-temperature (T) history of the faulting events.

All samples were thin sectioned; many were found to be devoid of observable fluid inclusions. Five suitable samples were selected for study: two from F₁ thrust faults, one from an east-west–striking normal fault, initially thought to be related to the extension event F₂, and two samples associated with the F₃ strike-slip faults.

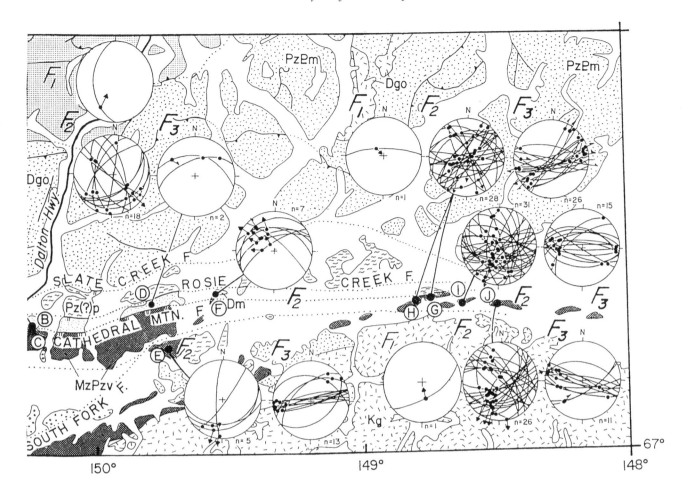

Fluid inclusion measurements were made on a USGS-type, gas-flow heating/cooling stage manufactured by Fluids Inc. Reproducibility was ±0.1°C for salinity determinations and ±1.0°C for homogenization measurements.

Fluid inclusions in all veins were found in linear trails that, in most instances, do not cross grain boundaries, indicating that they formed during crystallization of the quartz grains (Roedder, 1984). The extensional vein has fluid inclusions parallel to the long axes of the fibers also indicating that the inclusions formed during grain growth. Most fluid inclusions are subhedral and small (1 to 5 microns in length). They have uniformly low salinities ranging from 0.0 to 3.4 wt % NaCl equivalent (Fig. 4). No initial melting was observed to identify the type of salt species. One sample (AB88-41) has many fluid inclusions with metastable melting behavior. Most samples have a small range of homogenization temperatures; one vein sample (AB88-41) from a thrust fault has a second generation of inclusions with a lower homogenization temperature (Fig. 4).

The isochores for four of the five samples are shown in Figure 5. They were constructed using the MacFlinc computer program of Brown (1992). On the basis of the isochores alone, it is impossible to constrain the P-T conditions: the F_1 fluid inclusions may have formed at the same temperature as the F_3 inclusions, but at much lower pressures, or they formed at much higher tem-

perature and same pressure as the F_3 inclusions (Fig. 5). As the results do not tightly constrain the P-T history of the deformations, a cooling curve was constructed (Fig. 5) based on the assumption that the rocks of the Angayucham terrane had an overburden of only 3.5 km (about 100 MPa) during the thrusting event (Patton et al., 1977).

The maximum and minimum isochores of the earliest generation of veins (AB88-33 and AB88-41) enclose the hatched area in Figure 5, labeled F1. These veins are associated with thrusting (F_1) in the Angayucham terrane but the isochores do not coincide with the metamorphic P-T conditions (prehnite-pumpellyite facies) of the country rock. The fluids involved in the formation of the quartz veins must have been somewhat hotter (50 to 100°C) during crystallization than the country rock.

The maximum and minimum isochores of the quartz veins (AB88-37b and AB88-38) that were synkinematic with the F_3 strike-slip event (enclosing the area with crosses, labeled F3a in Fig. 5) cut the cooling curve at 160 and 110°C, respectively, and at about 50 MPa. One of the synthrusting quartz veins (AB88-41) has a second generation of fluid inclusions that formed at the same conditions as the F_3 veins (F3b in Fig. 5), suggesting reactivation of the thrust plane as a strike-slip fault. The fluid inclusions in the quartz vein from the east-west–striking normal fault (AB88-43) are one-phase inclusions, which suggests a trapping

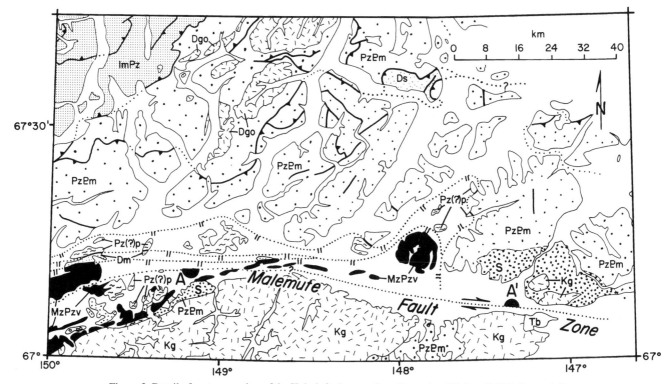

Figure 3. Detail of eastern portion of the Kobuk fault zone; from Brosgé and Reiser (1964). Dextral displacement (from A to A') along the Malamute strike-slip fault of staurolite schist (densely stippled areas labeled with "s") and Kg (Hodzana pluton) of possibly 80 km. Symbols are: Dgo, Devonian granitic orthogneiss; Dm, Devonian metasedimentary rocks of the Rosie Creek allochthon; Ds, Devonian metasedimentary rocks, undivided; Kg, Cretaceous granite; lmPz, lower and middle Paleozoic metasedimentary rocks of the Skajit allochthon; MzPzv, Mesozoic and Paleozoic basalt, diabase, chert, and limestone of the Angayucham terrane; Pz(?)p, Devonian (?) metasedimentary rocks of the Phyllite belt; PzPm, middle and lower Paleozoic and Precambrian rocks of the Schist belt; stippled lines with tick marks are south-dipping normal faults.

temperature below 75°C. It is unlikely that this vein formed during the F_2 extensional event, which preceded the F_3 strike-slip deformation. This suggests that some normal faulting (F_4) and veining may have occurred at a substantially later time.

INTERPRETATIONS AND CONCLUSIONS

The east-west–trending Kobuk fault zone in the southern Brooks Range is a complexly deformed belt, but an integral part of the Brooks Range fold and thrust belt. The fault zone consists of many east-west–striking anastomosing faults that crosscut, from south to north, the Yukon-Koyukuk basin, Angayucham terrane, Rosie Creek allochthon, Phyllite belt, and Schist belt. The style of deformation in the rocks of the southern Kobuk fault zone (the Angayucham terrane and the Rosie Creek allochthon) is very different from the style in the Phyllite and Schist belts and most other belts to the north: the deformation of the first two belts is mostly brittle, whereas the deformation toward the north is mostly ductile. Although the structural evolution of these brittle structures is not as well constrained, both kinematically and geochronologically, as that of the ductile structures in the north, sufficient evidence exists to suggest that

all structures, both brittle and ductile, formed in response to the same displacement fields.

Kinematic analysis of mesoscopic faults and fractures and fluid inclusion study of synkinematic quartz veins in the Kobuk fault zone suggest the following pressure (P)-temperature (T)–brittle deformation (F_1 to F_4) history.

The oldest brittle deformation structures (F_1) in the Kobuk fault zone are east- to northeast-striking thrust faults. Synkinematic quartz veins were emplaced at temperatures of 300 to 350°C and pressures of about 100 MPa (3.5 km depth). The F_1 thrust faults may have formed at the same time that the D_{1a} to D_{1c} ductile structures formed in the adjacent belts to the north during late Early to early Middle Jurassic time (Gottschalk et al., this volume, Chapter 12).

The second phase of brittle deformation (F_2) is well expressed by numerous mesoscopic south- and north-dipping normal faults. To the north, this phase of deformation is expressed by ductile kinks and shear zones (D_{1d}; Gottschalk, 1990). Based on $^{40}Ar/^{39}Ar$ determinations, this deformation occurred in mid-Cretaceous time (Blythe et al., this volume, Chapter 10; Gottschalk and Snee, this volume, Chapter 13).

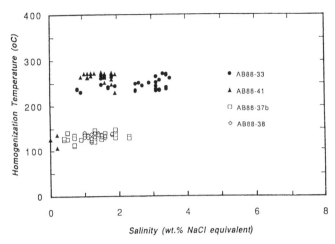

Figure 4. Homogenization temperatures versus salinity of two-phase fluid inclusions in four quartz veins from the Kobuk fault zone. Samples AB88-33 and AB88-41 are veins synkinematic with the F_1 thrusting; AB88-37b and AB88-38 are syn-F_3 (strike-slip faulting) veins.

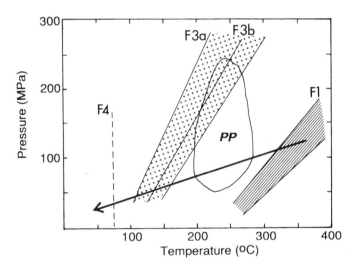

Figure 5. Maximum and minimum isochores (enclosing hatched area labeled F1) for fluid inclusions in two quartz veins (samples AB88-33 and AB88-41) related to the F_1 thrusting and for fluid inclusions in two quartz veins (AB88-37b and AB88-38) related to the F_3 strike-slip faulting (enclosing area with crosses labeled F3a); F3b is isochore for second-generation fluid inclusions in one of the synthrusting samples; F4 is 75°C line below which the vein (AB88-43) along the F_4 normal fault formed; line with arrow is estimated cooling curve; PP is prehnite-pumpellyite facies, as defined by Frey et al. (1991).

The third-phase brittle deformation structures (F_3) are related to right-lateral strike-slip displacements along the Kobuk fault zone. This deformation may have occurred at temperatures of 100 to 150°C and pressures of about 50 Ma (1.5 km depth). It must have started after the mid-Cretaceous and may have been active until early Tertiary time (Patton and Hoare, 1968). Jones (1980) suggested that the Kobuk fault was an extension of the Tintina fault (Fig. 1), which was active until about 50 Ma (Gabrielse, 1985). The Kobuk and Tintina faults were separated by right-

lateral displacement along the Kaltag fault/Porcupine shear zone (Fig. 1) during the Miocene until Recent (Fisher et al., 1982).

Although all brittle structures of the Kobuk fault zone are compatible with the three phases of faulting, as discussed above, the data do not prove that there were only three phases of faulting; more could have been possible. Low-temperature (less than 75°C) fluid inclusions in a vein along an east-west–striking normal fault suggest that normal faulting (F_4) occurred again at much later time (after the strike-slip event). The Kobuk fault zone has been shown to be seismically active; a first-motion study (Estabrook, 1985) suggests that dextral strike-slip displacement is occurring at present.

Whereas on the local scale of the southern foothills of the Brooks Range, a definite succession of thrust faulting (F_1), normal faulting (F_2), and strike-slip faulting (F_3) has been established, this succession does not apply to the entire fold and thrust belt. As shown elsewhere (O'Sullivan et al., this volume, Chapter 11), there is good evidence that north-south contraction in the Brooks Range continued from the Late Jurassic until at least the Oligocene. Thus, the right-lateral displacements along the east-west–trending Kobuk fault zone took place simultaneously with north-south shortening (northward thrusting in the Doonerak duplex and in the foreland; Figs. 2 and 3). Consequently, the Kobuk fault zone had to be displaced northward at the same time that strike-slip displacement occurred, and the fault zone had to merge at depth with the basal decollement of the Brooks Range fold and thrust belt. This can only be explained by displacement partitioning resulting from right-oblique plate convergence (Oldow et al., 1990). In this respect, the Brooks Range fold and thrust belt is very similar to the foreland fold and thrust belt in Alberta and the Tintina fault in British Columbia, which both were active simultaneously during Late Cretaceous to Eocene time (e.g., Oldow et al., 1989).

ACKNOWLEDGMENTS

This study was funded by grants from the Department of Energy (DE-AS0-83ER13124), the National Science Foundation (EAR-8517384 and EAR-8720171), and the Alaska Industrial Associates Program at Rice University and the University of Alaska, Fairbanks (Arco, Amoco, Chevron, Standard Oil of Ohio, Mobil, and Gulf). We thank Bill Patton and Ron Harris for their thoughtful comments.

REFERENCES CITED

Anderson, E. M., 1947, The dynamics of faulting and dyke formation with applications to Britain (second edition): London, Oliver & Boyd, 206 p.

Arth, J. G., Zmuda, C. C., Foley, N. K., Criss, R. E., Patton, W. W., Jr., and Miller, T. P., 1989a, Isotopic and trace element variations in the Ruby batholith, Alaska, and the nature of the deep crust beneath the Ruby and Angayucham terranes: Journal of Geophysical Research, v. 94, p. 15941–15955.

Arth, J. G., Criss, R. E., Zmuda, C. C., Foley, N. K., Patton, W. W., Jr., and Miller, T. P., 1989b, Remarkable isotopic and trace element trends in potassic through sodic Cretaceous plutons of the Yukon-Koyukuk basin, Alaska, and the nature of the lithosphere beneath the Koyukuk terrane: Journal of Geophysical Research, v. 94, p. 15957–15968.

Blum, J. D., Blum, A. E., Davis, T. E., and Dillon, J. T., 1987, Petrology of cogenetic silica-saturated and -oversaturated plutonic rocks in the Ruby geanticline of north-central Alaska: Canadian Journal of Earth Sciences, v. 24, p. 159–169.

Brosgé, W. P., and Reiser, H. N., 1964, Geologic map and section of the Chandalar Quadrangle, Alaska: U.S. Geological Survey Miscellaneous Geologic Investigations Map I-375, scale 1:250,000.

Brosgé, W. P., and Reiser, H. N., 1971, Preliminary bedrock geologic map: Wiseman and eastern Survey Pass Quadrangles, Alaska: U.S. Geological Survey Open-File Map 479, scale 1:250,000.

Brown, P. E., 1992, MacFlinc: A hypercard based, fluid inclusion tutorial: Geological Society of America Abstracts with Programs, v. 24, p. A233.

Dillon, J. T., Brosgé, W. P., and Dutro, J. T., Jr., 1986, Generalized geologic map of the Wiseman Quadrangle, Alaska: U.S. Geological Survey, Open-File Report 86-219, scale 1:250,000.

Dillon, J. T., Tilton, G. R., Decker, J., and Kelly, M. J., 1987, Resource implications of magmatic rocks and metamorphic ages for Devonian igneous rocks in the Brooks Range, *in* Tailleur, I., and Weimer, P., eds., Alaskan North Slope geology: Society of Economic Paleontologists and Mineralogists, Pacific Section, Publication 50, p. 713–723.

Dillon, J. T., Reifenstuhl, R. R., and Bakke, A. A., 1988, Geologic map of the Wiseman A-1 Quadrangle, south-central Brooks Range: Alaska Division of Geological and Geophysical Surveys Public-Data File 88.17, scale 1:63,360.

Estabrook, C. H., 1985, Seismotectonics of northern Alaska [M.Sc. thesis]: Fairbanks, University of Alaska, 139 p.

Fisher, M. A., Patton, W. W., Jr., and Holmes, M. L., 1982, Geology of Norton basin and continental shelf beneath northwestern Bering Sea, Alaska: American Association of Petroleum Geologists Bulletin, v. 66, p. 255–285.

Frey, M., De Capitani, C., and Liou, J. G., 1991, A new petrogenetic grid for low-grade metabasites: Journal of Metamorphic Geology, v. 9, p. 497–509.

Gabrielse, H., 1985, Major dextral transcurrent displacements along the northern Rocky Mountain Trench and related lineaments in north-central British Columbia: Geological Society of America Bulletin, v. 96, p. 1–14.

Gamond, J. F., 1987, Bridge structures as sense of displacement criteria in brittle fault zones: Journal of Structural Geology, v. 9, p. 609–620.

Gottschalk, R. R., 1987, Structural and petrological evolution of the southern Brooks Range near Wiseman, Alaska [Ph.D. thesis]: Houston, Texas, Rice University, 263 p.

Gottschalk, R. R., 1990, Structural evolution of the Schist belt, south-central Brooks Range fold and thrust belt, Alaska: Journal of Structural Geology, v. 12, p. 453–469.

Gottschalk, R. R., and Oldow, J. P., 1988, Low-angle normal faults in the south-central Brooks Range fold and thrust belt, Alaska: Geology, v. 16, p. 395–399.

Harris, R. A., 1992, Peri-collisional extension and the formation of Oman-type ophiolites in the Banda arc and Brooks Range, *in* Parson, L. M., Murton, B. J., and Browning, P., eds., Ophiolites and their modern oceanic analogues: Geological Society of London, Special Publication No. 60, p. 301–325.

Hubbard, R. J., Edrich, S. P., and Rattey, R. P., 1987, Geologic evolution and hydrocarbon habitat of the "Arctic Alaska microplate," *in* Tailleur, I., and Weimer, P., eds., Alaskan North Slope geology: Society of Economic Paleontologists and Mineralogists, Pacific Section, Publication 50, p. 797–830.

Iverson, N. R., 1991, Morphology of glacial striae: Implications for abrasion of glacier beds and fault surfaces: Geological Society of America Bulletin, v. 103, p. 1308–1316.

Jones, D. L., Coney, P. J., Harms, T. A., and Dillon, J. T., 1988, Interpretive geologic map and supporting radiolarian data from the Angayucham terrane, Coldfoot area, southern Brooks Range, Alaska: U.S. Geological Survey Miscellaneous Field Studies Map MF-1993, scale 1:63,360.

Jones, P. B., 1980, Evidence from Canada and Alaska on plate tectonic evolution of the Arctic Ocean basin: Nature, v. 285, p. 215–217.

King, P. B., 1969, Tectonic map of North America: Denver, Colorado, U.S. Geological Survey, scale 1:5,000,000.

Law, R. D., Miller, E. L., Little, T. A., and Lee, J., 1994, Extensional origin of ductile fabrics in the Schist belt, central Brooks Range, Alaska. II. Microstructural and petrofabric evidence: Journal of Structural Geology, v. 16, p. 919–940.

Little, T. A., Miller, E. L., Lee, J., and Law, R. D., 1994, Extensional origin of ductile fabrics in the Schist belt, central Brooks Range, Alaska. I. Geologic and structural studies: Journal of Structural Geology, v. 16, p. 899–918.

Miller, E. L., and Hudson, T. L., 1991, Mid-Cretaceous extensional fragmentation of a Jurassic–Early Cretaceous compressional orogen, Alaska: Tectonics, v. 10, p. 781–796.

Miller, T. P., 1989, Contrasting plutonic rock suites of the Yukon-Koyukuk basin and the Ruby geanticline, Alaska: Journal of Geophysical Research, v. 94, p. 15969–15987.

Molenaar, C. M., 1988, Depositional history and seismic stratigraphy of Lower Cretaceous rocks in the National Petroleum Reserve in Alaska and adjacent areas: U.S. Geological Survey Professional Paper 1399, p. 593–621.

Nilsen, T. H., 1989, Stratigraphy and sedimentology of the mid-Cretaceous deposits of the Yukon-Koyukuk basin, west-central Alaska: Journal of Geophysical Research, v. 94, p. 15925–15940.

Oldow, J. S., Seidensticker, C. M., Phelps, J. C., Julian, F. A., Gottschalk, R. R., Boler, K. W., Handschy, J. W., and Avé Lallemant, H. G., 1987, Balanced cross sections through the central Brooks Range and North Slope, Arctic Alaska: Tulsa, Oklahoma, American Association of Petroleum Geologists, Special Publication, 19 p., 8 pl.

Oldow, J. S., Bally, A. W., Avé Lallemant, H. G., and Leeman, W. P., 1989, Phanerozoic evolution of the North American Cordillera: United States and Canada, *in* Bally, A. W., and Palmer, A. R., eds., The Geology of North America—An overview: Boulder, Colorado, Geological Society of America, The Geology of North America, v. A, p. 139–232.

Oldow, J. S., Bally, A. W., and Avé Lallemant, H. G., 1990, Transpression, orogenic float, and lithospheric balance: Geology, v. 18, p. 991–994.

Patton, W. W., Jr., and Hoare, J. M., 1968, The Kaltag fault, west-central Alaska: U.S. Geological Survey Professional Paper, v. 600D, p. D147–D153.

Patton, W. W., Jr., and Miller, T. P., 1973, Bedrock geologic map of Bettles and southern part of Wiseman Quadrangles, Alaska: U.S. Geological Survey Miscellaneous Field Studies Map MF-492, scale 1:250,000.

Patton, W. W., Jr., Tailleur, I. L., Brosgé, W. P., and Lanphere, M. A., 1977, Preliminary report on the ophiolites of northern and western Alaska, *in* Coleman, R. G., and Irwin, W. P., eds., North American ophiolites: Oregon Department of Geology and Mineral Industries Bulletin, v. 95, p. 51–57.

Petit, J. P., 1987, Criteria for the sense of movement on fault surfaces in brittle rocks: Journal of Structural Geology, v. 9, p. 597–608.

Ramsay, J. G., and Huber, M. I., 1987, The techniques of modern structural geology, Volume 2: Folds and fractures: London, Academic Press, p. 307–700.

Reches, Z., and Dieterich, J. H., 1983, Faulting of rocks in three-dimensional strain fields. I. Failure of rocks in polyaxial, servo-controlled experiments: Tectonophysics, v. 95, p. 111–132.

Roedder, E., 1984, Fluid inclusions, Review of Mineralogy, Volume 12: Washington, D.C., American Mineralogical Society, 644 p.

Roeder, D., and Mull, C. G., 1978, Tectonics of Brooks Range ophiolites (Alaska): American Association of Petroleum Geologists Bulletin, v. 62, p. 1696–1713.

Smith, D. G., 1987, Late Paleozoic to Cenozoic reconstruction of the Arctic, *in* Tailleur, I., and Weimer, P., eds., Alaskan North Slope geology: Society of Economic Paleontologists and Mineralogists, Pacific Section, Publication 50, p. 785–795.

Turner, D. L., Forbes, R. B., and Dillon, J. T., 1979, K-Ar geochronology of the southwestern Brooks Range, Alaska: Canadian Journal of Earth Sciences, v. 16, p. 1789–1806.

Wirth, K. R., and Bird, J. M., 1992, Chronology of ophiolite crystallization, detachment, and emplacement: Evidence from the Brooks Range, Alaska: Geology, v. 20, p. 75–78.

MANUSCRIPT ACCEPTED BY THE SOCIETY SEPTEMBER 23, 1997

Printed in U.S.A.

Geological Society of America
Special Paper 324
1998

Seismic profiling constraints on the evolution of the central Brooks Range, Arctic Alaska

E. S. Wissinger* and A. R. Levander
Department of Geology and Geophysics, Rice University, Houston, Texas 77005-1892
J. S. Oldow
Department of Geology and Geological Engineering, University of Idaho, Moscow, Idaho 83844-3022
G. S. Fuis and W. J. Lutter
U.S. Geological Survey, 345 Middlefield Road, Menlo Park, California 94025

ABSTRACT

A seismic reflection profile from the 1990 Brooks Range seismic experiment indicates a strongly reflective upper and lower crust (0 to 50 km) throughout the northern and central parts of the range in units comprising the Brooks Range orogen. The northern range is characterized by two zones of subhorizontal reflections at 2 to 3 s and 4 to 6 s, which mark the base of the Endicott Mountains allochthon and the basal decollement of the contractional belt, respectively. The central range is characterized by a series of stacked, ~80-km-long, moderately south-dipping reflections interpreted as imbricates of the Doonerak duplex. The duplex is overlain by complex reflectivity patterns in the region of the Skajit allochthon. South of the Doonerak duplex in the metamorphic rocks of the Schist belt, the upper crust appears nonreflective. In the southern range, gently to moderately north-dipping reflectors are imaged beneath the Yukon-Koyukuk basin at 5 to 6 s and are interpreted to mark an older decollement level separating lower Paleozoic rocks of the Schist belt from Pre-Cambrian units. The boundaries of the Yukon-Koyukuk basin and the subsurface extent of the Kobuk dextral fault are not imaged.

The seismic reflection data support the existence of crustal-scale duplexing, breach thrusting, and large-scale north-directed thrusting of allochthonous crustal assemblages. Crustal thickness in the Brooks Range ranges from ~35 km in the hinterland to 50 km beneath the range front where an asymmetric crustal root is imaged. We have used the seismic reflection data and surface geologic constraints to identify the boundaries of major structural assemblages in the Brooks Range and restore three interpretations of the range to their pre-Jurassic configurations. Minimum shortening estimates along the seismic line range from 500 to 600 km of Mesozoic-Recent shortening with end-member estimates of Cenozoic shortening in the Doonerak duplex approximating 85 and 380 km.

---

*Present address: 10214 Sagebud, Houston, Texas 77089.

Wissinger, E. S., Levander, A. R., Oldow, J. S., Fuis, G. S., and Lutter, W. J., 1998, Seismic profiling constraints on the evolution of the central Brooks Range, Arctic Alaska, *in* Oldow, J. S., and Avé Lallement, H. G., eds., Architecture of the Central Brooks Range Fold and Thrust Belt, Arctic Alaska: Boulder, Colorado, Geological Society of America Special Paper 324.

INTRODUCTION

As part of the North American Cordillera, the Brooks Range is an east-west–trending fold and thrust belt that developed during the collision of the Yukon-Koyukuk island-arc complex with the southern margin of the Arctic Alaska continental microplate (cf. Mull, 1982; Mayfield et al., 1983; Box and Patton, 1985; Hubbard et al., 1987). The collision resulted in the imbrication of continental margin sediments and the north-directed obduction of oceanic crustal fragments that now form the structurally highest sheets of the range (Mull, 1982). Contraction was predominantly north directed and was initiated in the Middle to Late Jurassic (Tailleur and Brosgé, 1970; Roeder and Mull, 1978; Mayfield et al., 1983; Patton, 1984; Box and Patton, 1985; Dillon, 1989) with major shortening occurring in the Cretaceous and minor shortening continuing today in the northeastern part of the range as indicated by uplift data, Cenozoic folds, and recent earthquake activity (Stone, 1989).

Detailed mapping, stratigraphic, structural, and gravity studies have characterized the Brooks Range as a series of tectonostratigraphic terranes or lithotectonic assemblages that have been emplaced into north-central Alaska along one or more crustal detachment levels (e.g., Adkison and Brosgé, 1970; Brosgé and Tailleur, 1971; Patton, 1973; Mull, 1982; Mayfield et al., 1983; Jones et al., 1987; Oldow et al., 1987; Grantz et al., 1991; Moore et al., 1994). The large-scale distribution of allochthonous assemblages is generally agreed upon (Figs. 1 and 2A of Preface, this volume), however, the cross-sectional geometry and history of thrust sheet emplacement remain speculative. Central to this problem has been the lack of subsurface data constraining the position of the detachment levels and boundaries of the allochthons. These relationships are critical for determining tectonic emplacement histories and for reconstructing the pre-orogenic configuration of northern Alaska.

In the summer of 1990, we acquired a single deep seismic reflection/refraction profile in the north-central Brooks Range as part of the Trans-Alaska Lithosphere Investigation (TALI). The seismic line is more than 315 km long and followed the route of the Trans-Alaska pipeline (Fig. 1 of Preface, this volume). It was designed to image the internal geometry and velocity structure of the orogen, its flanking geologic provinces, and the lower crust and Moho.

The seismic reflection data obtained from this survey have been used to constrain the structure of the major tectonic assemblages comprising the range. In this chapter, we outline the major contributions provided by the seismic reflection data in the upper 30 km of crust. We combine these data with existing surface geologic and geophysical data to develop restorable cross sections describing the structural geometry and evolution of the Brooks Range and estimate magnitudes of shortening.

GEOLOGICAL AND GEOPHYSICAL SETTING

The lithotectonic assemblages comprising the Brooks Range are dominantly Paleozoic in age and have widespread lithologies. In general, the stratigraphy of the lesser disturbed northern assemblages is well known, but stratigraphic relationships are poorly understood in the south due to increasing metamorphic grade and poor age control (Mayfield et al., 1983). Continuing geologic studies in the Brooks Range are resulting in the modification of existing geologic maps. To construct our cross sections, we use a recently updated geologic map from Oldow et al. (Fig. 3 of Preface, this volume; and Fig. 1) and modify their earlier work using seismic constraint (Oldow et al., 1987). A generalized comparison of the tectonic assemblages mapped by Oldow et al. (1987) and those mapped by other researchers in the area is shown in Figure 2.

Ten major lithotectonic assemblages are crossed by the seismic line (Fig. 1). From north to south the tectonic assemblages are: (1) the Colville basin sediments—a Jurassic-Recent foreland basin succession floored by lower Mesozoic and Paleozoic sedimentary rocks and older metamorphic basement; (2) the Endicott Mountains allochthon (EMA)—an imbricate stack of Devonian-Permian marine shales, terrigenous sediments, platform carbonates, and conglomerates; (3) the Doonerak window—a northeast-trending structural high composed of lower Paleozoic to Carboniferous volcanic rocks and low-grade metamorphic rocks (Dutro et al., 1976; Mull et al., 1987) that has been deformed into a multilevel duplex structure (i.e. Doonerak duplex; Seidensticker and Oldow, this volume, Chapter 6); (4) the Skajit allochthon (Hammond subterrane of Moore and Mull, 1989)—a structurally complex lower Paleozoic and possibly Pre-Cambrian assemblage of thrusted carbonates and clastics; (5) the Schist belt (Coldfoot subterrane of Moore and Mull, 1989; and Moore et al., 1992)—a metamorphic assemblage of highly deformed, amphibolite to blueschist-facies, quartz-mica schist with a lower Paleozoic and possibly Pre-Cambrian protolith; (6) the Phyllite belt (Slate Creek subterrane of Moore and Mull, 1989; and Moore et al., 1992)—a relatively thin allochthon composed of fine-grained schists and phyllite of probable Devonian age (herein grouped with the Rosie Creek allochthon); (7) the Rosie Creek allochthon—a low-grade

Figure 1. Simplified geologic map of north-central Alaska (after Oldow et al., 1989; Beikman, 1980). A–A′ denotes the extent of the seismic reflection profiles obtained from the experiment that are used to constrain the restorable cross sections. Abbreviations: Colville basin (CB); Lisburne Group (LG) includes Kayak and Kekituk Formations; Endicott Mountains allochthon (EMA) includes the Kanayut (K), Hunt Fork (HF), and Beaucoup (B) Formations; Doonerak window (DW); Skajit allochthon (SA); Schist belt (SB); Pre-Cambrian to Devonian plutons (Dpl); Rosie Creek allochthon (RCA); Angayucham terrane (AT); Yukon-Koyukuk basin (KB); Jurassic to Cretaceous plutons (Cpl); Eekayruk thrust (ET); Foggytop thrust (FT); and Minnie Creek thrust (MCT).

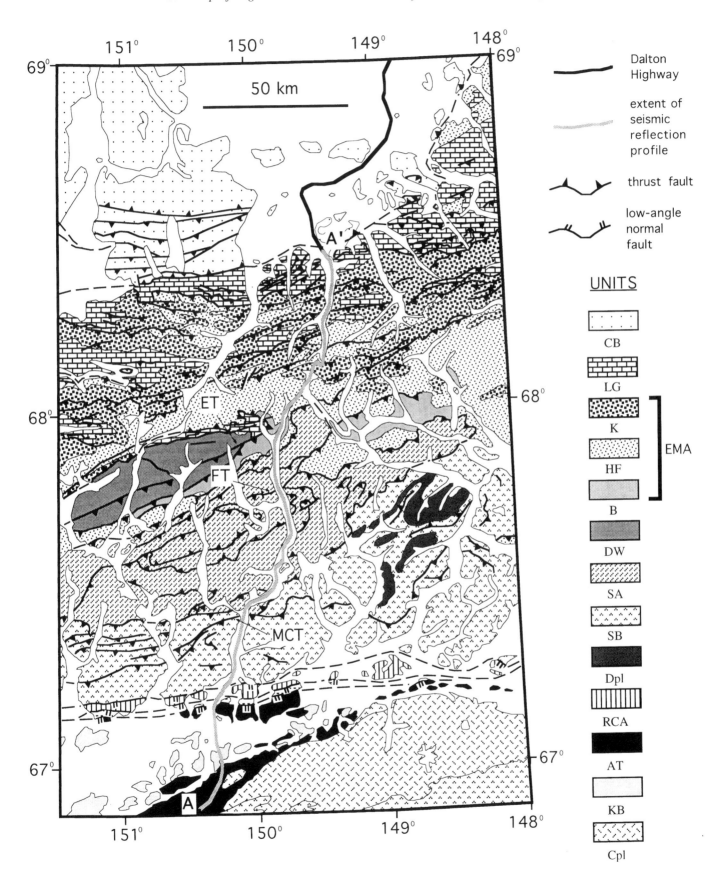

LITHOTECTONIC ASSEMBLAGES (Oldow et al., 1987)	TERRANES AND SUBTERRANES (Jones et al., 1987; Moore et al., 1994)	GEOLOGIC PROVINCES (Till et al., 1988; Moore et al., 1994)
Colville Basin	Colville Basin	Foreland belt
Endicott Mountains allochthon	Endicott Mountains subterrane	Crestal belt
Doonerak window	North Slope subterrane	Central belt
Skajit allochthon	Hammond subterrane	
Schist belt	Coldfoot subterrane	Schist belt
Phyllite belt	Slate Creek and Prospect Creek subterranes	Phyllite belt (Dillon, 1989)
Rosie Creek allochthon		
Angayucham terrane	Angayucham and Tozitna terranes	Greenstone belt (Patton, 1973)

Figure 2. Approximate tectonic equivalents in north-central Alaska and the Brooks Range from various authors. The classification of Oldow et al. (1987) is used herein for consistency with updated geologic maps.

metamorphic assemblage composed of Cambrian-Silurian clastic and carbonate rocks; (8) the Angayucham terrane—a Devonian to mid-Jurassic sequence of obducted pillow basalts, cherts, and carbonates; (9) the Yukon-Koyukuk basin—a depression comprised of sediments shed to the south from the Brooks Range that have been deposited on volcanic rocks of a Mesozoic arc complex (Patton and Box, 1989); and (10) the Ruby terrane—a metamorphic assemblage of Proterozoic (?) and Paleozoic rocks that are lithologically similar to the Schist belt and are widely intruded by Cretaceous granitic plutons (Dillon et al., 1986).

Numerous dextral fault systems that include the Kobuk, Kaltag, and Tintina faults offset rocks south of the Brooks Range. The Kobuk fault, crossed by the seismic line, extends for more than 600 km (Brosgé and Reiser, 1971) and has a reported 80 km of Cretaceous to Tertiary displacement (Avé Lallemant et al., this volume, Chapter 15). The importance of strike-slip displacement in Brookian deformation is obscure but it appears to play a significant role in the transport and assimilation of continental terranes into the Alaskan interior. It has been suggested that the Kobuk fault partially accommodates the dextral motion recorded along the much larger Tintina-Kaltag fault system (Oldow et al., 1989).

Extensional structures are recognized in the hinterland of the Brooks Range along the northern contacts of the Rosie Creek terrane and the Angayucham terrane, but the magnitude and timing of the extension are controversial. Gottschalk and

Oldow (1988) and Till et al. (1993) relate the extension to unroofing of the Schist belt ~130 to 110 Ma after or during development of the Minnie Creek breach thrust that forms the northern boundary of the Schist belt across much of the range. This scenario is similar to that proposed by other researchers to describe syncontractional extensional features seen in portions of the Alps (e.g., Ratschbacher et al., 1989; Ballévre et al., 1990). Alternatively, Miller and Hudson (1991) attribute the extensional features to regional extension (100% stretching) from 130 to 90 Ma, and relate it to the opening of the Canada Basin. In either case, the presence of ophiolitic rocks in the western range as far north as the southern Colville basin suggests that an enormous volume of oceanic crustal material and underlying passive margin sediments have overridden the southern margin of northern Alaska and have subsequently been eroded from the range and deposited in the adjacent Colville and Yukon-Koyukuk basins (Patton, 1984, 1992; Hubbard et al., 1987). High-resolution dating from fission-track analysis of apatite crystals is being analyzed to help constrain the timing of uplift episodes of Brookian units that may reflect periods during which extension was occurring (Blythe et al., this volume, Chapter 10; O'Sullivan et al., this volume, Chapter 11).

Important questions regarding the nature of Alaskan terrane interaction and the role of Brookian intracrustal detachment zones in deformation remain largely unanswered. Incorporating seismic reflection data with available geologic data provides useful constraint for interpreting the geometries of tectonic assemblages and their emplacement histories. Reflector geometries for the upper and middle crust (0 to 30 km) are discussed in the following section.

DATA ACQUISITION

The experiment was designed to provide a low-fold, near-vertical reflection image while simultaneously acquiring a crustal velocity profile from wide-angle data. A single 315-km, north-south transect was recorded with ~190 km of low-fold reflection coverage. The survey was conducted in five overlapping and abutting deployments using a 700-channel recording array (Fig. 3). Helicopters were used to reach off-road areas. Vertical-component motion from 65 explosions at 44 separate shotpoints was recorded. Receivers were spaced at 100 m over most of the survey, and the average shot interval for the entire survey is 8.3 km. The base camp for operations was the Coldfoot truck stop and deployments were located chiefly along the Dalton Highway, a gravel road used to maintain the Trans-Alaska Pipeline and transport supplies to the Prudhoe Bay oil fields.

We used 30 of the shots recorded by 180 Seismic Group Recorders (SGRs) and 210 PASSCAL Reftek channels to construct our reflection sections. The SGRs and Refteks recorded high frequencies (>20 Hz) and were more suitable for reflection profiling than other seismographs used in the experiment. The average shot interval in the reflection profiles is 6.3 km, with charge sizes ranging from 100 lbs to 1,500 lbs. The instruments were deployed in fixed arrays recording 9 to 15 shots fired through and off-end to the spread. The unique survey geometry enabled construction of low-fold common midpoint (CMP) sections while acquiring source-receiver offsets of as much as 210 km. Data quality is generally excellent, although in places near surface glacial deposits degraded signal quality.

DATA PROCESSING

Processing was heavily dependent on the use of standard reflection processing techniques although low data multiplicity necessitated the use of nonstandard processing techniques to develop stacking velocity models and to enhance signal/noise ratios. In general, time-variant filtering, automatic gaining, and coherency filtering produced the most dramatic improvements in the data while true-amplitude recovery provided a good estimation of relative reflector strength. For the purposes of this chapter we present two seismic sections. The first is an ~4-fold,

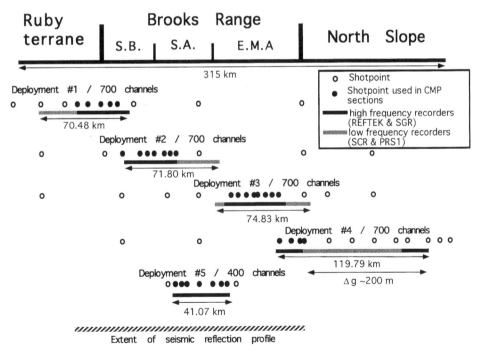

Figure 3. The Brooks Range experiment design. The experiment was conducted by deploying ~700 recording channels in five fixed arrays. The common midpoint (CMP) sections were created from data recorded by the high-frequency instruments. Average shot spacing along the high frequency spreads is 6.3 km. Reoccupation of shot holes resulted in reversed refraction coverage from the shot interval to offsets as large as 250 km. Abbreviations: Schist belt (S.B.); Skajit allochthon (S.A.); Endicott Mountains allochthon (E.M.A.); Refraction Technology seismograph (Reftek); Seismic Group Recorder (SGR); Seismic Cassette Recorder (SCR); Portable Refraction seismograph (PRSI).

Coherency Filtered Stacked Section
Max Offset = 10 km, Ave Fold = 4
Energy Display, True Amplitude

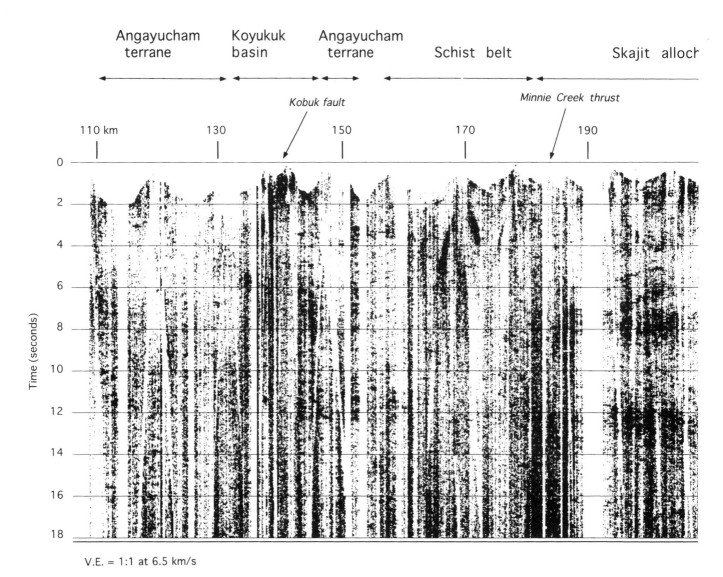

V.E. = 1:1 at 6.5 km/s

Figure 4. Coherency-filtered, true-amplitude common midpoint (CMP) section. CMP trace spacing is
100 m, average fold is 4, and offsets range from 0 to 10 km. Gaps in this section result from uneven
CMP coverage and projecting the CMP section onto a straight line. The section is shown as an energy
display (Bittner and Rabbel, 1991).

near-vertical incidence, true-amplitude section (Fig. 4) containing offsets of as much as 10 km, and the second (Fig. 5) is a
largely single-fold, automatic gain controlled (AGC) section
containing offsets of as much as 32 km. Offsets greater than 10
km were excluded in the short offset section in the interest of
improving the resolution of complex structures by reducing the
inclusion of wide-angle, high-amplitude reflections and defo-
cused energy arriving at larger offsets. The larger offset section

was produced by sorting individual shot gathers to single-fold
common midpoint gathers and examining them for signal content making no a priori assumptions about offset range. In either
case, offset/target depth ratios are <1. The procedure used in
Figure 4 compensates for amplitude losses due to spherical
spreading, attenuation, and transmission losses while equaliz-
ing amplitude differences arising from variations in charge
sizes. Amplitude decay curves for each shot were calculated

North

Doonerak window

it allochthon   Endicott Mountains allochthon

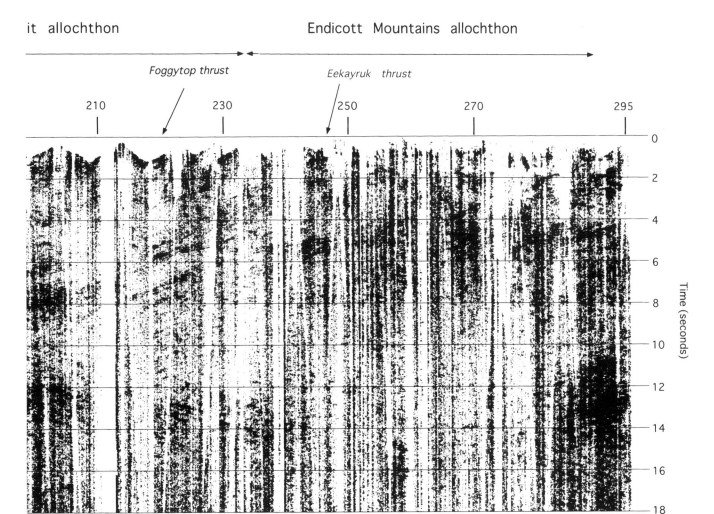

over an average offset range of 5 km and were fit with the exponential function ke^{at} (Luschen et al., 1987). Amplitude equalization was applied to the short offset section to obtain a relative measure of reflector strength and to determine areas along the profile with high reflectivity for correlation with lithologies. See Figure 6 for a summary of the processing.

VELOCITY CONTROL

Low data multiplicity precluded the determination of stacking velocities through standard techniques. Instead, interval velocities obtained through two-dimensional travel time inversion of PmP arrivals and first breaks were used to stack and migrate the data. The velocity model obtained from travel-time inversion is shown in Figure 7 (Levander et al., 1994; Fuis et al., 1995). Velocities are well constrained in the upper 6 to 8 km of crust (±0.25 km/s) but are poorly constrained with increasing depth (±0.50 km/s at 25 km depth). Velocity contours generally reflect the structural geometry of the range with high velocities present in the stacked allochthons of the range and low velocities in the sediments of the surrounding basins. The highest velocities for the upper 8 km of crust appear in the Skajit allochthon and Schist belt. Lower velocities characterize the sediments

Coherency Filtered Single Fold Section
Max Offset = 32 km, Energy Display

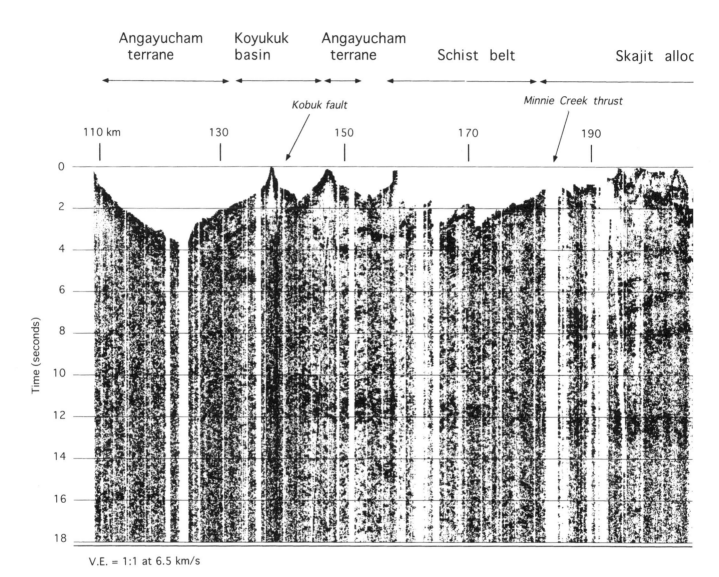

V.E. = 1:1 at 6.5 km/s

Figure 5. Coherency-filtered, single-fold 32-km offset common midpoint (CMP) section showing Moho and middle crustal reflections. CMP trace spacing is 100 m. Gaps in this section result from uneven CMP coverage and projecting the CMP section onto a straight line. The section has a 2,500-ms automatic gain control (AGC) applied and is shown as an energy display (Bittner and Rabbel, 1991).

comprising the Colville basin (3.75 km/s at the surface to 5.00 km/s at 5 km depth), and Endicott Mountains allochthon (4.25 km/s at the surface to 5.00 km/s at 5 km depth). Velocities of as much as 6.00 km/s characterize portions of the Endicott Mountains allochthon where carbonate rocks of the Lisburne Group are proximal to the surface (1 km depth) at the range front. The notable velocity peak apparent at the range front beneath the Endicott Mountains allochthon at 5 km depth is interpreted here to be resulting from an arched or upthrust body of Lisburne Group carbonate rocks (average lab velocity = 5.7 to 6.2 km/s at this depth) in contact with other Endicott units (average lab velocity = 5.4 to 6.2 km/s at this depth; Christensen, 1992). Higher velocities are recorded in the metamorphosed clastics and carbonates of the Skajit allochthon and range from 5.00 km/s at the surface to 6.00 to 6.25 km/s at 5 km depth. The average crustal velocity for the region spanned by the survey is 6.5 ± 0.25 km/s (Fuis et al., 1995).

North

Doonerak window

t allochthon   Endicott Mountains allochthon

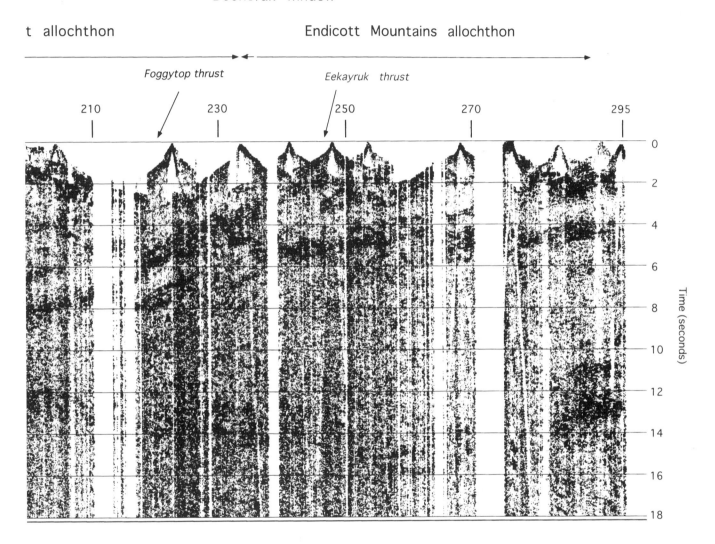

LINE DRAWING AND MIGRATION

The data were migrated in line drawing form using ray theory (Raynaud, 1988; Holliger and Kissling, 1991; Fig. 8) in order to avoid typical problems associated with the migration of deep crustal data such as low signal/noise ratios, complicated velocity structures, and apparent discontinuities in deep reflectors due to near-surface effects (Warner, 1987). Each reflector segment was migrated individually, and reflector "smiles" were reduced on the section by restricting amplitudes to those less than a uniform, empirically determined threshold value. To cre-

ate the line drawing, coincident reflections on the amplitude equalized short offset section, the 32-km-offset section, and an automatically gained short offset section (not shown; see Levander et al., 1994) were picked to reduce the inclusion of processing artifacts into the migrated section. Reflector lengths were not interpolated across areas with poor data quality. The velocities obtained by inversion were used to depth migrate the upper crust from the surface to 15 km depth (Fig. 7; Fuis et al., 1995). Average velocities of 6.5 km/s and 7.0 km/s were from 15 to 30 km depth and 30 to 50 km depth, respectively, where a detailed velocity structure was unavailable.

GENERAL PROCESSING FLOW FOR
0 - 10 KM OFFSET REFLECTION DATA

GENERAL PROCESSING FLOW FOR
0 - 32 KM OFFSET REFLECTION DATA

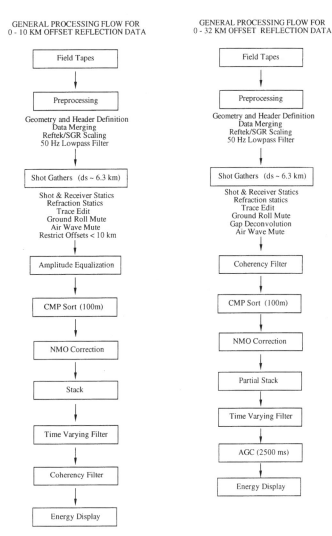

Figure 6. Processing flows used to create the 0- to 10-km offset stack and the 32-km offset stack shown in Figures 4 and 5. Abbreviations: Seismic Group Recorder (SGR); common midpoint (CMP); automatic gain control (AGC); normal moveout (NMO).

REFLECTOR GEOMETRY

The depth-migrated line drawing contains numerous events that can be correlated over tens of kilometers and can be separated into three dominant patterns of reflectivity based on reflector continuity, dip, and depth. The northern range is characterized by predominantly horizontal to subhorizontal reflections in the EMA and underlying units (model coordinates: KM 250 to 300 in Fig. 8). The central part of the range (KM 190 to 250) is dominated by a complex series of south-dipping reflections (12 to 25°) from ~10 to 25 km depth in the subsurface region of the Doonerak window. In the southern part of the range (KM 110 to 180), reflectivity patterns are more diffuse and are dominated by north-dipping reflections that extend from ~10 to ~30 km depth in the region of the Schist belt and Yukon-Koyukuk basin, and a gener-

ally synformal pattern of reflectivity throughout the upper 20 km. A notable decrease in reflectivity in the upper 30 km of crust is apparent from KM 170 to 185. Prominent reflections to the north abruptly terminate in this region, probably as a result of a combination of scattering of the seismic wavefield in the heterogeneous, metamorphic rocks of the Schist belt, and problems associated with signal generated noise (e.g., near-surface scattering).

REFLECTIVE ZONES

We identify seven major reflective zones on the migrated line drawing (Fig. 8, zones A to G) on the basis of coherent reflectivity, reflector dip, and continuity of reflective zones. In most cases, the reflective zones are interpreted to originate from the boundaries of faulted or sheared nappes as indicated by either a discordance in reflector dip or continuity with mapped faults. Such boundaries may or may not correspond to a change in lithology, however, as has been widely recognized by other researchers (Bally et al., 1966; Mooney and Meissner, 1992; Smithson et al., 1979; Cook et al., 1979, 1992; Brewer et al., 1983; Oliver, 1985; Pfiffner et al., 1990).

Zone A. A 1- to 3-km-thick zone of subhorizontal reflections at 15 km depth at the range front can be traced from the range front over 120 km to the south (KM 300 to 180) where its top deepens from 15 to 17 km to 22 to 25 km beneath the Doonerak window (zone A; Fig. 8). The zone terminates at 28 to 30 km depth beneath the Schist belt. It also appears to separate steeply dipping reflection patterns in the upper crust from gently dipping reflection patterns in the lower crust. In the northern and central parts of the range, the zone overlies generally nonreflective crust (15 to 45 km depth from KM 240 to 300).

Zone B. A shallower, 2- to 3-km-thick zone of subhorizontal reflections is visible at 5 to 6 km depth at the range front (zone B; Fig. 8). It extends more than 50 km to the south and terminates near the projected axis of the Doonerak window. The zone is highly reflective (Fig. 4); below is a 2- to 5-km-thick relatively transparent zone, and above are numerous south-dipping reflections. At the range front, south-dipping reflections (18° dips) in proximity to the surface appear to either sole into the zone or cut through it to deeper levels. The southern margin of the zone dips north at ~22°.

Zone C. Between zones A and B are sparse discontinuous reflections that dip to the south (zone C; Fig. 8). At least one coherent band of reflectivity is recognized. It is comprised of moderately south dipping reflections (28° dips) that appear to extend from the base of zone B (KM 275 to 280) to the base of zone A (KM 250 to 260).

Zone D. Numerous zones (two to four ?) of south-dipping reflectivity are present in the central part of the range at 6 to 27 km depth (zone D; Fig. 8). They are spaced 3 to 5 km apart and appear stacked atop one another. Reflector dips in the zones flatten with depth from approximately 25° at KM 230 (10 km depth) to 12° at KM 195 (25 km depth). The zones appear continuous for ~40 to 50 km, and reflectivity from at least one of the

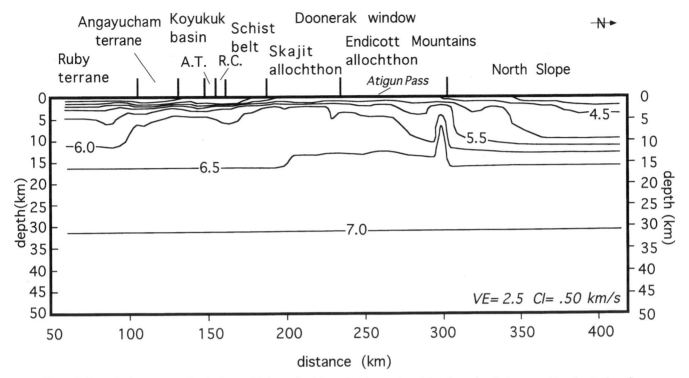

Figure 7. Smoothed upper crustal velocity model determined from direct wave travel time inversion that was used to migrate the reflection data (after Fuis et al., 1995). Lowest velocities (4.0 to 5.5 km/s) are found in the North Slope and Yukon-Koyukuk basin. Highest velocities (4.5 to 6.5 km/s) are apparent in heterogeneous crustal assemblages in the central range. The velocity peak near the northern margin of the Endicott Mountains allochthon is interpreted to represent an upthrust body of Lisburne Group carbonate rocks. Uniform migration velocities of 6.5 km/s and 7.0 km/s were used from 15 to 30 km depth and from 30 to 50 km depth, respectively. Abbreviations: A.T., Angayucham terrane; R.C., Rosie Creek allochthon.

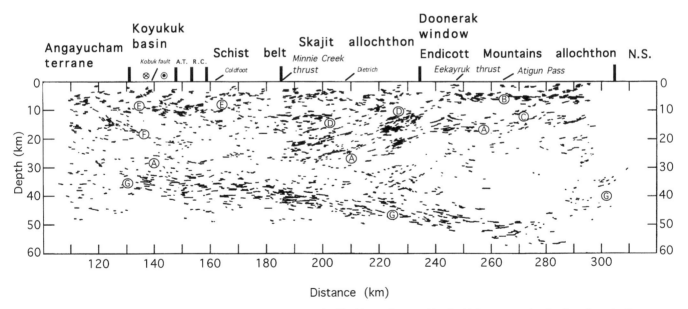

Figure 8. Migrated line drawing of the Brooks Range near-vertical incidence data combined with lower crustal reflections from the 0- to 32-km offset stack. Migration is ray based (Holliger and Kissling, 1991). Zones: A, basal decollement; B, base of Endicott Mountains allochthon (EMA); C, foreland Lisburne carbonate imbricates; D, Doonerak thrust; E, crustal duplexes in Doonerak antiform region; F, north-dipping reflection beneath Yukon-Koyukuk basin; G, lower crustal reflections that dip to the north under the southern and central range and to the south under the range front. Abbreviations: A.T., Angayucham terrane; R.C., Rosie Creek allochthon; N.S., North Slope. Locations of seismic features and surface geologic data discussed in text are referenced by kilometer distance (i.e., KM 280).

zones extends to the surface in the northern Doonerak window (KM 247). A 15- to 20-km-long subhorizontal reflector at 12 km depth (KM 180 to 205) marks the top of zone D beneath the Skajit allochthon.

Zone E. Complex discontinuous reflection patterns are present from 2 to 12 km depth in the region of the Schist belt and Yukon-Koyukuk basin (KM 110 to 170; zone E; Fig. 8). The reflection signature is weaker in this region than in the northern part of the range (Fig. 4), but reflectivity patterns suggest synformal structure beneath the basin.

Zone F. A prominent 1- to 3-km-thick band of moderately north-dipping reflections (25° dips) is observed in the southern part of the range from (KM 110 to 150) beneath the Yukon-Koyukuk basin (zone F; Fig. 8). These reflections are the strongest reflections present south of the Schist belt. The band extends from 12 km to 20 to 25 km depth and does not reach to the surface in the seismic profile.

Zone G. A prominent 3- to 5-km-thick zone of lower crustal reflections extends more than 170 km along the line from (KM 130 to 280), marking the deepest zone of reflections observed on the section (zone G; Fig. 8). The zone dips toward the north (~8°) and extends from an average depth of 35 km at the southern end of the line to an average depth of 50 km beneath Atigun Pass (KM 265). From Atigun Pass northward, the zone broadens to 10 km width and shoals to an average depth of 35 km.

SURFACE DATA

The predominance of subhorizontal and south-dipping reflections across the northern two-thirds of the range is indicative of the northward transport of crustal assemblages into north-central Alaska. We correlate the reflective zones with surface geologic and radiometric data to interpret the base of the allochthons, making the assumption that the crustal reflectivity observed on the section was generated during Brookian and younger deformation. The primary surface geologic and radiometric data used in the interpretations are outlined in the following sections. For detailed discussion of the geologic data and relationships described, the reader is referred to the references cited.

Mapped faults

1. A back-thrust triangle zone emplacing basin sediments of the Colville foredeep over carbonates of the Lisburne Group has been mapped at the range front (~KM 300; Jones, 1982; Oldow et al., 1987). Although the magnitude of displacement along the back thrust is unclear, the existence of the triangle zone is supported by a discrepancy between the large magnitudes of shortening observed in the EMA and substantially smaller magnitudes observed in the foreland basin. Previous balancing efforts have invoked the back thrust as a possible mechanism to accommodate the relatively rapid decrease in observed shortening (Oldow et al., 1987).

2. Eight south-dipping thrust faults have been mapped in the EMA along the seismic line (KM 264, 271, 277, 281, 286, 293, 296, and 300). The faults are located in outcrop exposures of the Kanayut conglomerate, Hunt Fork and Beaucoup shales, and Lisburne Group carbonate (Brosgé et al., 1979; Oldow et al., 1987; Grantz et al., 1991), which are units internal to the allochthon. The depth extent of the faults is unknown, although the fault at KM 293 appears reflective to at least 5 km depth.

3. At the northern flank of the Doonerak window is a series of high-angle, south-dipping thrust faults that form an east-northeast–striking en echelon fault system (Phelps et al., 1985). At least one of the faults (Eekayruk thrust; ~KM 247) may be traced from the surface as a continuous series of reflection events to ~15 km depth.

4. The Skajit allochthon has been mapped as a series of seven different thrust nappes with numerous lithologies that has been folded into an antiform-synform pair (Oldow et al., this volume, Chapter 7). The stratigraphic thickness of the nappes is poorly known, but aggregate thicknesses have been estimated from field observations to range from 1 to 3 km (Till, 1992). The two northernmost thrust nappes of the allochthon dip south and form the footwall of the late-stage Foggytop out-of-sequence thrust (KM 220). Along the southern margin of the allochthon, the nappes become progressively more deformed where faulting is related to earlier stage breach thrusting of the Minnie Creek thrust.

5. The Minnie Creek thrust is a mapped south-dipping breach thrust ("envelopment" thrust of Oldow et al., 1987) that emplaces rocks of the Schist belt over metasedimentary rocks of the southern margin of the Skajit allochthon along the seismic line (~KM 183) and farther west. A few kilometers to the east of the profile, the Minnie Creek thrust offsets rocks internal to the Schist belt, suggesting that the northern edge of the Schist belt was emplaced prior to breach thrusting (Oldow et al., 1987; Gottschalk, 1990).

6. South-dipping normal faults are mapped along the northern contacts of the Angayucham terrane and Rosie Creek allochthon (KM 153 and 159, respectively; Brosgé and Reiser, 1971; Oldow et al., 1987).

7. The Kobuk dextral fault zone is poorly exposed and consists of several megascopic anastomosing faults. Kinematic data indicates that the fault zone has experienced ~80 km of Cretaceous to Tertiary dextral displacement during its polyphase deformational history (Avé Lallemant et al., this volume, Chapter 15).

Radiometric dating and metamorphic conditions

1. Two deformational episodes are recognized in the Schist belt (Oldow et al., 1987). Thermobarometric constraints on the first episode indicate the Schist belt underwent recrystallization at pressures in excess of 8 kb and temperatures >450°C (Gottschalk, 1987) during subduction of the southern Alaskan continental margin beneath the Yukon-Koyukuk arc (Till et al., 1988). The second episode is related to exhumation of the Schist belt along the Minnie Creek thrust. Metamorphic decompression of the belt is

thought to be responsible for the development of low-angle normal faults along the southern border of the belt (Oldow et al., 1987). $^{40}Ar/^{39}Ar$, K-Ar, and Rb-Sr age dates indicate that the age of uplift in the Schist belt is as old as 130 to 110 Ma (Brosgé and Reiser, 1964; Turner et al., 1979; Gottschalk, 1990).

2. Peak metamorphic pressure conditions in the Skajit allochthon (central belt of Till et al., 1988) indicate that at least portions of the allochthon experienced pressures of 6 to 8 kb in ~120 Ma (Till, 1992).

3. Uplift timing estimates from apatite fission-track dating have determined that the rocks of the Doonerak antiform have undergone episodes of cooling/uplift at ~60 Ma and 25 Ma (Blythe et al., 1996). An additional episode of uplift has been recognized in the Colville basin and northeastern Brooks Range at 45 Ma (O'Sullivan et al., 1993).

Structural and stratigraphic relationships

1. Previous studies were able to constrain the position of the basal decollement at the range front using data from three industry seismic reflection lines and well data from five separate sites but all at some distance from our seismic line. One of the seismic surveys extends from the foothills into the northern range through Anaktuvuk Pass ~70 km to the west of the reflection profile shown here. The others are east-west– and north-south–trending lines recorded farther out on the North Slope (see Oldow et al., 1987). Projection of the industry seismic reflection, velocity, and well data into the region along the trend of the Brooks Range seismic line allowed the depth of the basal decollement to be constrained at ~12 km depth at the range front. A 3° southerly dip was estimated for the decollement. The seismic reflections present in zone A indicate that the decollement is actually deeper (~15 km at this locale), and has a 3 to 4° southerly dip in the northern range, provided the v(x,z) function used for migration is approximately correct.

2. Carboniferous Lisburne Group exposed along the northern flank of the EMA is a "distal" facies in contrast with the more "proximal" facies documented in the Doonerak window and northeastern Brooks Range (Churkin, 1973; Armstrong, 1974; Moore et al., 1992) suggesting that the EMA was derived from south of the Doonerak window.

3. Surface exposures along the northern margin of the EMA indicate that the structural thickness of "distal" facies Lisburne Group has been increased in this area as a result of thrust imbrication. This observation is consistent with the presence of the localized peak in the velocity data recently determined from refraction studies (Fuis et al., 1995) and the estimation of the base of the imbricate stack at 4 to 5 km depth from seismic, well data, and balancing efforts (Oldow et al., 1987).

4. The Doonerak antiform has been documented as a multi-level duplex structure based on detailed mapping studies conducted in the central part of the range (Seidensticker and Oldow, this volume, Chapter 6). The subsurface extent and number of imbricates in the duplex is unknown, although in outcrop scale,

numerous imbricates have been mapped. Formation of the duplex is thought to produce the north-dipping contacts at the southern edge of the EMA, indicating that the EMA was displaced north of the Doonerak window before or during formation of the duplex (Julian et al., 1984).

5. The Skajit allochthon contains units that are age equivalent to the Schist belt (Palmer et al., 1984), but the nature of the subsurface contact between the two is poorly known. As indicated from published cross sections, most researchers interpret uplift of the Schist belt along the southern margin of the Skajit allochthon in the central part of the range and overthrusting of at least portions of the Skajit allochthon at depth (Oldow et al., 1987; Gottschalk, 1990; Grantz et al., 1991; Till, 1992; Moore et al., 1994; Blythe et al., 1996). Similarly, we interpret the Skajit allochthon to overlie rocks of the Schist belt and be overthrust by rocks of the Schist belt in the hanging wall of the Minnie Creek thrust. This geometry is supported by the juxtaposition of penetratively foliated strata in the Schist belt with primary volcanic and sedimentary features preserved in the Skajit allochthon (Till, 1992). The contact between the assemblages in the central part of the range may have been modified by back thrusting in post-Albian times (Till, 1992; Moore et al., 1992). In two of our models, we interpret back thrusting from reactivation of an older north-vergent thrust flooring the allochthon in accordance with north-dipping fault contacts and south-vergent folds observed, although the magnitude of the back thrusting is unknown (Till, 1992).

6. South of the Minnie Creek thrust fault and ~60 km east of the seismic profile, an exposure of relatively unmetamorphosed EMA units structurally overlies high-grade metamorphic rocks of the Schist belt (Brosgé and Reiser, 1964; Beikman, 1980). This is an important structural relationship because it implies that units of the EMA may extend from the range front as far south as the Schist belt (assuming the allochthon is a coherent sheet).

7. Outcrops of Angayucham terrane and Rosie Creek allochthon along the trend of the seismic profile are relatively thin klippe of substantially thicker sequences that have been removed during uplift of the Schist belt (Oldow et al., 1987; Till et al., 1988).

INTERPRETATION OF REFLECTORS

We present three restorable cross sections (Figs. 9 to 11) that are largely consistent with seismic and surface data listed earlier. They are similar in the northern part of the range where stratigraphic relationships are well preserved but differ primarily to the south in (a) the number of imbricates interpreted in the Doonerak duplex, (b) the geometry and thickness of the nappes in the Skajit allochthon, (c) the subsurface extent of the EMA, and (d) the thickness of the Schist belt. All of the cross sections interpret the top of the subhorizontal reflections in zone B to represent the basal fault contact of the EMA and assume that the Eekayruk thrust fault, the Minnie Creek thrust fault, and the Kobuk fault sole into the basal decollement. Interpretation of the

Cross section #1

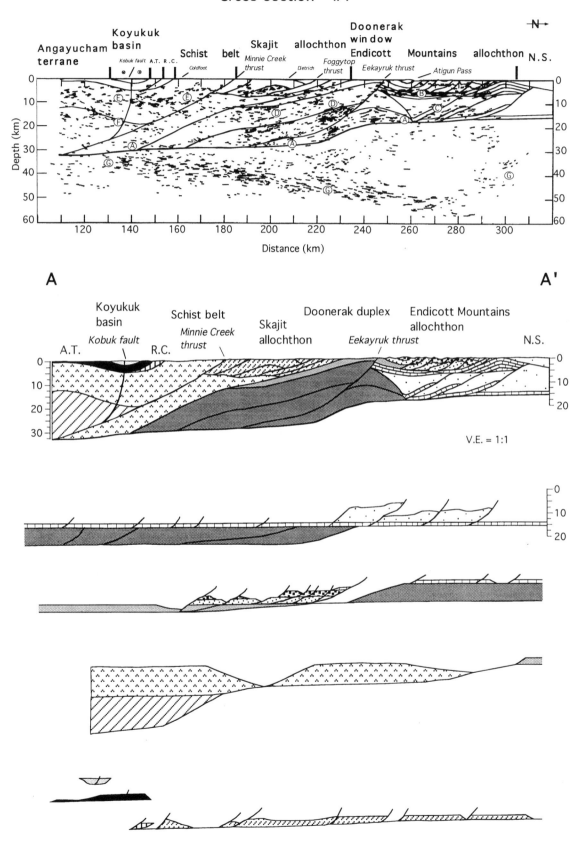

Figure 9. Cross section #1 and its restoration constructed from surface geologic data, kinematic data, and the seismic reflection data. Detail is explained in text. Major discerning features are two horses in the Doonerak duplex and the existence of foredeep sediments beneath the Endicott Mountains allochthon (EMA). Minimum shortening estimates for this model is ~600 km. See Figure 8 for definition of symbols.

reflections along the northern two-thirds of the profile have been previously published by Levander et al. (1994), and are extended to the south here for completeness (cross section #2). The interpretation of Fuis et al. (1995) in which basement horses without interleaved Lisburne Group rocks are interpreted in zones B and C is not developed here. The significance of the variations included in the interpretations is discussed in the following section.

Basal decollement

The base of the reflections in zone A is interpreted to mark the position of the basal decollement upon which the entire range formed. It can be traced with confidence for more than 60 km in the northern range where it steps from 15 km to at least 22 km depth beneath the reflective packages of the Doonerak duplex in the central part of the range (~KM 235) and appears to sole to 30 km depth in the hinterland. The subhorizontal orientation of the basal decollement in the northern range is supported by the contrast between the south-dipping reflections observed in zone C from 10 to 15 km depth and the absence of reflective structures beneath 15 km depth. In the central part of the range, the decollement bounds the base of the south-dipping reflective packages in zone D, accommodating the formation of the duplex. The presence of sparse subhorizontal reflections in the hinterland at ~32 km depth (KM 110 to 180) suggest that the decollement soles into the top of the zone G. Although the seismic evidence for the soling of the decollement in this region is poor, we extend the decollement beneath the crystalline core of the Brooks Range to accommodate the uplift and emplacement of the Schist belt.

In cross sections #1 (Fig. 9) and #2 (Fig. 10), rocks in the central part of the range are uplifted along the ramp in the decollement (~KM 235) forming the Doonerak duplex. Continued compression results in imbrication of "proximal" Lisburne Group carbonate rocks and overlying foredeep sediments beneath the EMA. An alternate interpretation of the geometry of the basal decollement is proposed in cross section #3 (Fig. 11). Here, the basal decollement is interpreted to underlie imbricates of pre-Mississippian "basement" that are involved in deformation as far north as the range front. In this model, the ramp in the decollement is less pronounced and is interpreted to result from flexural loading of the Doonerak imbricates in the region of the Doonerak window. Although this model requires the presence of large amounts of Lisburne Group carbonate rocks in the foreland, the involvement of "basement" in thrusting is compatible

with thrust imbricates of pre-Mississippian "basement" rocks that have been identified in outcrop exposures in the northeastern Brooks Range, north of the Endicott Mountains allochthon (Wallace and Hanks, 1990).

Endicott Mountains allochthon

In each of the three cross sections, the base of zone B is interpreted to represent the base of the EMA. In the northern EMA, there is a strong correlation between the bright zone of south-dipping reflectivity imaged from 1 to 6 km depth (KM 295) and the top of the imbricate stack of "distal" facies Lisburne Group carbonate rock exposed at the surface (Hawk, 1985). These reflectors may correlate with the Wolf Creek thrust fault mapped by Glenn (1991) near the top of the imbricate stack of "distal" facies Lisburne Group carbonate rocks. Farther south, north-dipping reflections imaged along the northern flank of the Doonerak window at KM 245 to 255 may correlate with the Amawk and related thrust faults at or near the top of bodies of "proximal" facies Lisburne Group These reflections can similarly can be traced to at least 5 km depth.

In the central EMA, the subsurface geometry of faults bounding surface exposures of the Kanayut, Hunt Fork, and Beaucoup Formations internal to the allochthon (Fig. 2) are seismically unconstrained as a result of coarse shot spacing. The faults are assumed to sole into the detachment at the base of the EMA to accommodate shortening occurring internal to the allochthon.

Endicott Mountains units are exposed to the north and south of the Doonerak window forming the roof of the Doonerak duplex. The top of the duplex is interpreted to be faulted along a steeply south-dipping reflector (Eekayruk thrust) and lies at a depth of 1 to 2 km along our line, consistent with the presence of subhorizontal shallow reflections seen in this region and also with structural projections of the eastward plunging (~20°) Amawk thrust fault at the top of the duplex. The exposures of Endicott units overlying rocks of the Schist belt found to the east of the profile and south of the Minnie Creek thrust (Brosgé and Reiser, 1964) suggest that at least part of the EMA was exhumed along the Minnie Creek thrust (cross sections #2 and #3).

The south-dipping reflections in the region between the basal decollement and the base of the EMA north of the Doonerak window (zone C) are interpreted as ramps of Lisburne Group carbonate rocks that have been imbricated during overthrusting of the foredeep. This interpretation is based on similar geometries previously documented in well and seismic data in the front portion of the range ~70 km to the west (see Oldow et al., 1987) and the strength of the reflections in this region in the true amplitude section. The nature of the rocks overlying the ramps is unclear and is interpreted differently in the different cross sections. Cross sections #1 and #2 model the region as consisting of foredeep sediments located in the hanging walls of thrust faults that floor imbricates of Lisburne Group rocks. This interpretation is supported by the presence of

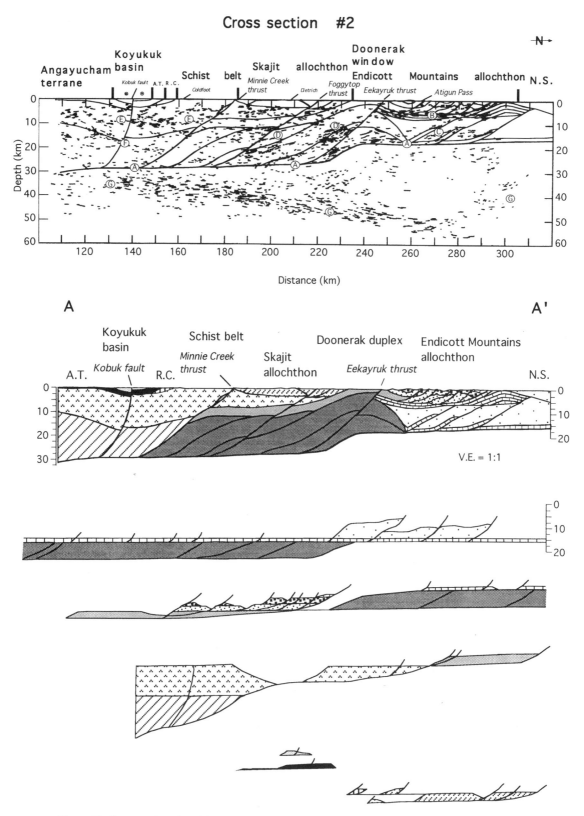

Figure 10. Cross section #2 and its restoration constructed from surface geologic data, kinematic data, and the seismic reflection data. Detail is explained in text. Major discerning features are four horses in the Doonerak duplex and the presence of foredeep sediments beneath the Endicott Mountains allochthon (EMA). Minimum shortening estimates for this model is ~500 km. See Figure 8 for symbol definition.

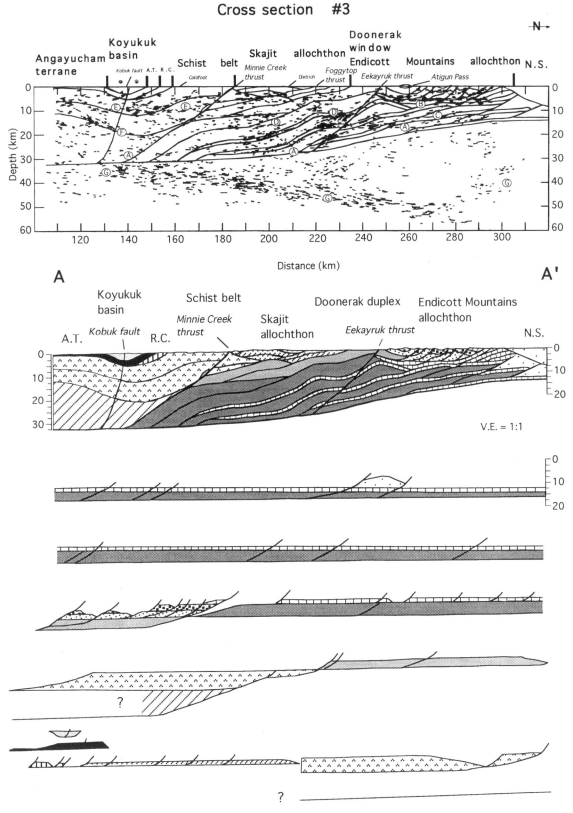

Figure 11. Cross section #3 and its restoration constructed from surface geologic data, kinematic data, and the seismic-reflection data. Detail is explained in text. Major discerning features are five horses in the Doonerak duplex and the existence of Doonerak units beneath the Endicott Mountains allochthon (EMA) near the range front. Minimum shortening estimates for this model is ~615 km. See Figure 8 for definition of symbols.

large impedance contrasts between the sediments of the Colville basin (average laboratory velocity at this depth is 5.25 to 5.50 km/s) and those of the Lisburne Group (average laboratory velocity at this depth is 5.80 to 6.20 km/s; N. Christensen, 1992). Alternatively, the detailed shallow velocity structure (0 to 15 km depth) obtained from travel time inversion shows relatively high velocities throughout this region (6.00 to 6.25 m/s) indicating possible basement imbrication beneath the EMA (cross section #3). Interpretation of basement imbrication is supported by the presence of basement cored anticlinoria in the northeastern Brooks Range (Wallace and Hanks, 1990). Accurate determination of velocities beneath the EMA remains an important problem.

Doonerak duplex

The prominent south-dipping reflections in the central range (zone D) are interpreted to represent shear zones bounding imbricates in the Doonerak duplex. The number of imbricates forming the duplex is interpreted differently in each cross section to illustrate a range of interpretations allowed by the reflectivity patterns. Cross section #1 interprets two ~7 km thick imbricates. Cross sections #2 and #3 interpret four and five imbricates that are 3 and 4 km thick, respectively. Interpreting fewer, thicker imbricates reduces the amount of shortening estimated in the duplex. Cross sections #1 and #2 approximate 90 km and 150 km of shortening respectively, while cross section #3 approximates 380 km of shortening. The comparatively large shortening estimate for cross section #3 arises from the interpretation of pre-Mississippian "basement" involvement beneath the EMA. In this interpretation, the thickness of the imbricates is inferred from prominent wide angle reflections discussed by Fuis et al. (1995). Larger magnitudes of shortening are possible for each of the cross sections if additional imbricates are interpreted to occupy the subsurface area to the south of the Minnie Creek thrust, but the numerous north-dipping reflections in the hinterland (zone F) indicate that the presence of Doonerak rocks in this area is not likely. A breach thrust that offsets rocks in the Doonerak window by ~1 km is interpreted in cross section #2 for consistency with the mapped Foggytop out-of-sequence thrust.

Schist belt

The Schist belt is characterized by a broad zone of synformal reflectivity that terminates to the north along the Minnie Creek thrust. The hinge of the synform lies in the northern Yukon-Koyukuk basin (~KM 150) in the region of zone E. Our generalized interpretations of the subsurface boundaries of the Schist belt in the cross sections do not explain the origin of the numerous cross-dipping reflectivity patterns present in zone E. We attribute these features to the polyphase deformational history of the Schist belt, and the numerous faults mapped internal to the Schist belt a few kilometers to the west of the profile (Oldow et al., 1987) and do not interpret them in any detail.

We interpret the reflections in zone F to represent the base of the Schist belt, which has been detached from basement early in the history of Brookian deformation along an older, deeper decollement level (incipient Minnie Creek thrust). Progressive displacement along this decollement is interpreted to have enveloped the Schist belt over the southern margin of the Skajit allochthon. The presence of north-dipping reflections in zone F may indicate that the Schist belt overlies a hinterland duplex, although the origin of these lower crustal rocks is unclear.

Skajit and Rosie Creek allochthons

No seismic constraint is available in the upper 3 to 4 km of the Skajit allochthon as a result of coarse shot spacing, but numerous reflectors lie between 4 and 8 km depth. Assuming the base of the allochthon is not too shallow to resolve, candidates for the base of the allochthon are the top of the reflective zone at 4 km depth (cross section #2 and #3) and the bottom of the reflective zone at 8 km depth (cross section #1). The interpretation of the base of the allochthon at 4 km depth is consistent with structural thicknesses estimated from field observations (Oldow et al., this volume, Chapter 7).

The subsurface thickness of the Rosie Creek allochthon is unknown but mapping studies indicate that it is substantially thinner than the Skajit allochthon due to the removal of section by low-angle normal faults (Gottschalk and Oldow, 1988). Since seismic constraint is poor, the depth of the allochthon is interpreted to be comparable to that of the Skajit allochthon in each of the cross sections.

Angayucham terrane and Yukon-Koyukuk basin

The dimensions of the Angayucham terrane and Yukon-Koyukuk basin are also unconstrained on the seismic section as a result of coarse shot spacing and are assumed to be <5 km thick based on surface observations (Mull, 1982; Patton and Box, 1989). We interpret the base of the Yukon-Koyukuk basin to lie at 2 to 4 km depth in each of the cross sections above the diffuse scattered reflections present in the upper 5 to 10 km of crust. Substantially larger depths are unlikely since velocity abruptly increases from 3.50 to 5.50 km/s in this region, at 4 km depth.

Lower crust

The bright, 3- to 5-km-thick zone of reflections in the lower crust (zone G) are associated with the crust-mantle boundary that is depressed in an asymmetric fashion under the northern Brooks Range. The base of the reflection band is consistent with Moho depths determined from analysis of Pmp and Pn arrivals at large offsets (Fuis et al., 1995; Levander et al., 1994). Crustal thickness ranges from 32 km in the hinterland and foreland regions with a maximum of 50 km recorded beneath the EMA.

RESTORATIONS

The models presented are a few of a number of possible geologic interpretations that are consistent with the seismic and surface geologic data. They assume that all deformation occurs in the plane of the section, line lengths representing detachments and bedding plane surfaces remain constant in both deformed and undeformed states, area is conserved, and restored stratigraphic thicknesses are approximately constant. Given a particular cross-sectional geometry these assumptions produce minimum shortening estimates. They are inadequate for completely describing the complex deformational history occurring in an orogenic belt, but they establish a framework for restoring tectonic assemblages to their pre-orogenic state, limit interpretations to those that are kinematically feasible, and allow estimation of magnitudes of orogenic shortening. The developmental effects of strike-slip motion and extension on the evolution of the Brooks Range are unclear and are ignored in the reconstructions. It should be noted, however, that the reconstructions are compatible with episodes of moderate extension predicted in the Angayucham terrane and Rosie Creek allochthon (Gottschalk and Oldow, 1988; Miller and Hudson, 1991).

The restoration of the Schist belt and Skajit allochthon (metamorphic core of the Brooks Range) is problematic due to the lack of stratigraphic control and polyphase deformational histories observed. Ductile fabrics and ductile shear zones have been documented in both of these assemblages (Till et al., 1988; Gottschalk, 1990). For simplicity, we assume that fault bounded rocks in the Schist belt (where interpreted) restore end to end above the throughgoing basal decollement interpreted from the reflections in zone A. Similarly, nappes in the Skajit allochthon are restored end to end above a shallower decollement, although their restored positions are poorly known.

The datum used in the restorations is the base of the Lisburne Group which has been imaged on the North Slope and interpreted at ~15 km depth at the range front (KM 300) based on reflective character and dip projections from industry seismic data recorded 70 km to the west of our seismic profile (Oldow et al., 1987). The pre-Jurassic positions of the allochthons are restored relative to this datum. The relative positions of the allochthons are identical in all of the restorations although shortening estimates differ in accordance with variations in the interpretation.

Northern and central part of the range

In the northern part of the range, the EMA restores south of its present position since it structurally overlies the Doonerak rocks, and contains "distal" facies Lisburne Group carbonate rocks that correlate with "proximal" facies Lisburne Group carbonate rocks identified in the Doonerak window. During contraction, the EMA is interpreted to override rocks of the Lisburne Group and underlying pre-Mississippian basement. The distal, deeper water Lisburne Group carbonate rocks

detach from the basement, imbricate at particular locations, and are incorporated into the overriding EMA where they are piled up at the northern margin of the allochthon. The carbonate pile is subsequently covered by back-thrusted foredeep sediments along its northern margin, forming a triangle zone (Jones, 1982).

North-dipping structures present in the seismic reflection data along the southern margin of the EMA suggest that the majority of the EMA units were emplaced to the north of the Doonerak rocks prior to duplexing and were subsequently folded during uplift of the duplex. Portions of the EMA are eroded above the Doonerak duplex creating the gap in the restorations that separates Endicott units exposed in the northern part of the range from those postulated in the subsurface south of the duplex. The Endicott units may restore north or south of the Schist belt. They are interpreted to restore north of the belt in each restoration to minimize shortening estimates.

As indicated by fission-track dating, uplift of the imbricates in the Doonerak duplex is bracketed by 60- and 25-Ma-age dates indicating that formation of the duplex occurred relatively late in the history of the range. The pre-Mississippian rocks in the Doonerak window form the autochthonous basement of the Colville basin below the basal decollement and restore disconformably beneath the Lisburne Group carbonate rocks (Moore et al., 1992). During deformation, the imbricates are assembled along the basal decollement from south to north during or after emplacement of the overlying EMA, terminating at the ramp in the basal decollement. They are subsequently offset by late-stage breach thrusts (Fig. 12).

Hinterland

The high-pressure, low-temperature (HP-LT) metamorphism of the Schist belt partly resulted from overthrusting of shelf margin sediments early in its deformational history (Till, 1992). Mapping studies have determined that rocks of the Skajit allochthon structurally overlie the Schist belt in the central part of the range north of the Minnie Creek thrust, suggesting that the Skajit allochthon overrode the Schist belt prior to breach thrusting along the Minnie Creek thrust (Oldow et al., 1987). We interpret the Skajit allochthon to restore to the south of the Schist belt and participate in the burial and metamorphism of the Schist belt early in its deformational history for consistency with these observations. Similarly, the Devonian Rosie Creek allochthon is interpreted to be a distal deeper water equivalent of the Skajit allochthon (Oldow et al., 1987), and therefore restores south of the Skajit allochthon, participating in the burial and metamorphism of the Schist belt.

Thermobarometric constraints in rocks of the Schist belt to the south of the Minnie Creek thrust indicate that the Schist belt was exhumed along the Minnie Creek thrust from depths as large as 35 km, eventually overriding the southern margin of the relatively unmetamorphosed EMA. Deformation was then trans-

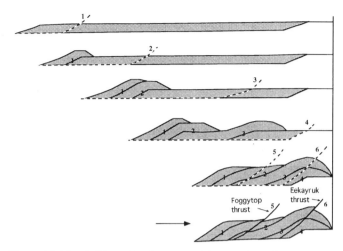

Figure 12. Sketch illustrating the sequential generation of thrust faults (1 to 4) and breach thrusts (5 to 6) in the formation of the Doonerak duplex (cross section #2).

ferred to shallower levels in the northern Brooks Range culminating in the development of the Doonerak duplex and front range imbricates.

Kinematics

The migrated section is interpreted in a style previously documented in other fold and thrust belts in which deformation proceeds from shallow to deeper detachment levels from hinterland to foreland during progressive development of a basal decollement underlying the entire system (Bally et al., 1966; Cook et al., 1979). The orogenic belt is thus interpreted to detach from the lower crust and deform as a separate but related system (Oldow et al., 1990; Cook and Varsek, 1994). The 3 to 4° southerly dip of the basal decollement beneath the northern Brooks Range is consistent with the magnitude of the dip determined previously by lithospheric loading models (Hawk, 1985; Nunn et al., 1987) and with a growing body of seismic evidence suggesting the continuation of "thin-skin" decollements into the hinterland of many orogenic belts (Cook et al., 1979; Valasek et al., 1991; Green et al., 1993; Beaumont et al., 1994; Cook and Varsek, 1994).

Our pre-Jurassic restorations indicate the following major sequence of tectonic events (some may overlap; approximate dates are proposed where possible).

1. The Angayucham terrane was obducted onto the southern margin of Arctic Alaska (pre 172 Ma; Christiansen and Snee, 1994).

2. Thrusting was initiated in the Skajit and Rosie Creek assemblages.

3. The Angayucham terrane, Skajit allochthon, and Rosie Creek allochthon overrode the Schist belt contributing to its HP-LT metamorphism (~160 Ma; Patrick et al., 1994).

4. The Schist belt detached from the underlying substrate forming the Minnie Creek thrust. Duplexing was initiated in the Schist belt (cross section #3; 130 to 110 Ma; Brosgé and Reiser, 1964; Turner et al., 1979; Gottschalk, 1990; Patrick et al., 1994).

5. The EMA was imbricated.

6. The Skajit, Rosie Creek, and Schist belt assemblages were emplaced over the southern EMA.

7. Extensional structures formed in the southern Brooks Range (130 to 90 Ma; Miller and Hudson, 1991).

8. The EMA was emplaced in the foreland.

9. Duplexing of Doonerak rocks (previously autochthonous) occurred beneath the EMA (60 and 25 Ma; O'Sullivan et al., 1993; Blythe et al., 1996; O'Sullivan et al., this volume, Chapter 11).

10. Front range carbonates were imbricated incorporating Cretaceous and younger sediments of the foreland beneath the EMA (cross sections #1 and #2; post 60 Ma).

11. The out-of-sequence Eekayruk and Foggytop thrusts formed in the Doonerak duplex (post 60 Ma).

SHORTENING

Following the obduction of Angayucham terrane, apparent minimum shortening estimates for the range along the seismic line are estimated at 600 km of Mesozoic-Recent shortening for cross section #1, 500 km for cross section #2, and 615 km for cross section #3, although significantly more shortening is suggested in the southern range from small-scale structures widely mapped at the surface in the Doonerak duplex, the Skajit allochthon, and the Schist belt. As a result of the poorly known stratigraphy in the southern range, the lack of reliable shortening estimates for deformation occurring out of the plane of the section, and the large scale of the cross sections, shortening estimates derived from the restorations are meant to serve only as first-order approximations for the geometries shown. We have selected the geometries shown to provide end-member solutions for shortening in the Doonerak duplex since the duplex has good seismic control and comprises a substantial 40 to 50 % of the total area of our seismic line. The largest estimate for Cenozoic shortening in the Doonerak units approximates 380 km (cross section #3). Our minimum estimate is ~85 km (cross section #2). In the foreland, the total amount of Cenozoic shortening of North Slope Lisburne carbonate rocks underlying the Colville basin sediments is estimated at 25 km (cross sections #1 and #2), with ~125 km estimated in the EMA for each of the three cross sections.

These shortening estimates are markedly higher than those proposed by Fuis et al. (1995), which postulate a minimum of 125 km of shortening in the Lisburne Group and a minimum of 45 km of Cretaceous-Cenozoic shortening to produce the Doonerak antiform. The estimates of Fuis et al. (1995) are lower as a result of the interpretation of minor slip along 50- to 100-km-long, south-dipping faults bounding the Doonerak imbricates and the fact that area and not line length were conserved during the restoration. Minimum shortening estimates for our models are of the same order of magnitude as those calculated by Oldow et al. (1987) and Levander et al. (1994), since we conserve both area and line length, restore Doonerak units end to end along the subhorizontal basal decollement, and assume that faults originate in subhorizontal incompetent layers in the foreland.

The presence of a basal decollement beneath the Brooks Range orogen implies a "decoupling" of upper crustal deformation from lower crustal deformation along the decollement. The large magnitudes of shortening estimated above the decollement suggest the presence of a larger crustal root to balance the distribution of shortening throughout the crust. We speculate that shortening in the rocks below the decollement is accommodated in part by the observed crustal root and also in part by movement of lower crustal rocks to other locations in or out of the plane of the cross section.

CONCLUSIONS

Seismic reflection data have been used to constrain the geometries of various terranes or lithotectonic assemblages comprising the Brooks Range orogenic belt and provide restorable kinematic models describing the evolution of the range. The primary crustal features interpreted are a basal decollement at 15 to 30 km depth, the base of the Endicott Mountains allochthon at 6 km depth, south-dipping faults bounding imbricates within the Doonerak duplex in the central part of the range, and a north-dipping reflector in the southern part of the range that is postulated to represent an older detachment level at the base of the Schist belt. Furthermore, strong Moho reflections indicate an asymmetric crustal root beneath the northern and central Brooks Range extending to 50 km depth.

The proposed cross sections are constrained by reflections in the upper 30 km of crust and provide interpretations that are kinematically feasible within the framework of the geological and geophysical data outlined above. The cross sections presented contain many similarities to the previous work of Oldow et al. (1987), Levander et al. (1994), and Fuis et al. (1995). The primary differences constrained by the seismic reflection data are the presence of a basal decollement that is deeper than was previously proposed by a factor of 1.6, the identification of the base of the Endicott Mountains allochthon at 6 km depth, seismic constraint of the Doonerak duplex, and seismic constraint to 15 km depth for a thrust fault mapped in rocks of the Doonerak window. The addition of new geologic data has also helped to constrain geometries in the Skajit allochthon and Endicott Mountains allochthon, and new geochronological data has placed constraints on the relative timing of tectonic events.

Jurassic-Recent shortening estimates for our cross sections approximate 500 to 600 km with end-member shortening estimates for the Doonerak duplex of ~85 km and 380 km.

The restorations presented have the least amount of kinematic constraint in the restored position of the Endicott Moun-

tains allochthon that may lie north or south of the Schist belt. Interpreting the allochthon to lie to the south of the Schist belt would result in an even larger estimate of shortening.

ACKNOWLEDGMENTS

We would like to thank the U.S.G.S. Deep Crustal Studies Program, the PASSCAL Program of Incorporated Research Institutes for Seismology (subaward agreements 0127 and 0140), and National Science Foundation (grant numbers EAR8905222 and EAR9105002) for funding the 1990 Brooks Range seismic experiment. The U.S. Air Force Geophysics lab also provided funds for the field work. The Geophysical Institute of Alaska at Fairbanks provided logistical support and personnel for seismic fieldwork in 1988 and 1990. Special thanks to Ed Criley for managing explosives, Isa Asudeh at the Geological Survey of Canada for managing Canadian recorders, Jim Fowler for supervising operation of the Reftek instruments, and to all others who participated in the field work. We also thank Bill Lutter and Stuart Henrys for painstakingly merging the seismic data and Klaus Holliger for developing the code to perform the migrations. Many thanks to Bob Page, Lorraine Wolf, Pat Hart, Will Kohler, Chris Humphreys, Bradd Carr, and David Stone for their contributions to the experiment. Special thanks to Wes Wallace and Ron Harris for constructive reviews of this manuscript.

REFERENCES CITED

Adkison, W. L., and Brosgé, M. M., eds., 1970, Proceedings, Geological Seminar on the North Slope of Alaska: Los Angeles, Pacific Section of the American Association of Petroleum Geologists, 203 p.

Armstrong, A. K., 1974, Carboniferous carbonate depositional models, preliminary lithofacies and paleotectonic maps, Arctic Alaska: American Association of Petroleum Geologists Bulletin, v. 58, no. 4, p. 621–645.

Ballévre, M., Lagabrielle, Y., and Merle, O., 1990, Tertiary ductile normal faulting as a consequence of lithospheric stacking in the western Alps: Memoîre de la Société Géologique de France, v. 156, p. 227–236.

Bally, A. W., Gordy, P. L., and Stewart, G. A., 1966, Structure, seismic data, and orogenic evolution of the southern Canadian Rocky Mountains: Bulletin of Canadian Petroleum Geology, v. 14, p. 337–381.

Beaumont, C., Fullsack, P., and Hamilton, J., 1994, Styles of crustal deformation in compressional orogens caused by subduction of the underlying lithosphere: Tectonophysics, v. 232, p. 119–132.

Beikman, H. M., 1980, Geologic map of Alaska: U.S. Geological Survey, scale 1:2,500,000.

Bittner, R., and Rabbel, W., 1991, Energy and power sections in seismic interpretation, *in* Meissner, R., Brown, L., Durbaum, H. J., Franke, W., Fuchs, K., and Seifert, F., eds., Continental lithosphere: Deep seismic reflections: Geodynamics, v. 22, p. 409–415.

Blythe, A. E., Bird, J. M., and Omar, G. I., 1996, Deformational history of the central Brooks Range, Alaska: Results from fission-track and $^{40}Ar/^{39}Ar$ analyses: Tectonics, v. 15, p. 440–455.

Box, S. E., and Patton, W. W., Jr., 1985, Early Cretaceous evolution of the Yukon Koyukuk basin and its bearing on the development of the Brookian orogenic belt, Alaska: American Association of Petroleum Geologists–Society of Economic Paleontologists and Mineralogists–Society of Exploration Geophysicists Pacific Section, 60th Annual Meeting, Abstracts with Programs, p. 40–41.

Brewer, J. A., Matthews, D. H., Warner, M. R., Hall, J. R., Smythe, D. K., and Whittington, R. J., 1983, BIRPS deep seismic reflection studies of the British Caledonides: Nature, v. 305, p. 206–210.

Brosgé, W. P., and Reiser, H. N., 1964, Geologic map and section of the Chandalar Quadrangle, Alaska: U.S. Geological Survey Miscellaneous Geological Investigations Map I-375, scale 1:250,000.

Brosgé, W. P., and Reiser, H. N., 1971, Preliminary bedrock geologic map: Wiseman and eastern Survey Pass Quadrangles, Alaska: U.S. Geological Survey Open-File Map 479, 2 sheets, scale 1:250,000.

Brosgé, W. P., and Tailleur, I. L., 1971, Northern Alaska petroleum province, *in* Cram, I. H., ed., Future petroleum provinces of the United States—Their geology and potential: American Association of Petroleum Geologists Memoir 15, p. 68–99.

Brosgé, W. P., Reiser, H. N., Dutro, J. T., Jr., and Detterman, R. L., 1979, Bedrock geological map of the Philip Smith Mountains Quadrangle, Alaska: U.S. Geological Survey Miscellaneous Field Studies Map MF-879B, scale 1:250,000.

Christensen, N., 1992, Trans-Alaska compilation of seismic velocities and densities: U.S. Geological Survey Report, 24 p. (unpublished).

Christiansen, P. P., and Snee, L. W., 1994, Structure, metamorphism, and geochronology of the Cosmos Hills and Ruby Ridge, Brooks Range Schist belt, Alaska: Tectonics, v. 13, p. 193–213.

Churkin, M., Jr., 1973, Paleozoic and Precambrian rocks of Alaska and their role in its structural evolution: U.S. Geological Survey Professional Paper 740, 62 p.

Cook, F. A., and Varsek, J. L., 1994, Orogen-scale decollements: Reviews of Geophysics, v. 32, p. 37–60.

Cook, F., Albaugh, D., Brown, L., Kaufman, S., Oliver, J., and Hatcher, R., Jr., 1979, Thin-skinned tectonics in the crystalline southern Appalachians: COCORP seismic reflection profiling of the Blue Ridge and Piedmont: Geology, v. 7, p. 563–567.

Cook, F. A., Varsek, J. L., Clowes, R. M., Kanasewich, E. R., Spencer, C. S., Parrish, R., Brown, R. L., Carr, S. D., Johnson, B. J., and Price, R. A., 1992, Lithoprobe crustal reflection cross-section of the southern Canadian Cordillera I: Foreland thrust and fold belt to Fraser River fault: Tectonics, v. 11, p. 12–35.

Dillon, J. T., 1989, Structure and stratigraphy of the southern Brooks Range and northern Koyukuk basin near the Dalton Highway, *in* Mull, C. G., and Adams, K. E., eds., Dalton Highway, Yukon River to Prudhoe Bay, Alaska: Alaska Division of Geological and Geophysical Surveys Guidebook 7, v. 2, p. 157–187.

Dillon, J. T., Brosgé, W. P., and Dutro, J. T., Jr., 1986, Generalized geologic map of the Wiseman Quadrangle, Alaska: U.S. Geological Survey Open-File Report 86-219, scale 1:250,000.

Dutro, J. T., Jr., Brosgé, W. P., Lanphere, M. A., and Reiser, H. N., 1976, Geological significance of the Doonerak structural high, Central Brooks Range, Alaska: American Association of Petroleum Geologists Bulletin, v. 60, p. 952–961.

Fuis, G. S., Levander, A., Lutter, W. J., Wissinger, E. S., Moore, T. E., and Christensen, N. I., 1995, Seismic images of the Brooks Range, Arctic Alaska, reveal crustal scale duplexing: Geology, v. 23, p. 65–68.

Glenn, R. K., 1991, Range front structure and lithostratigraphy of the Atigun Gorge area, north-central Brooks Range, Alaska [M.S. thesis]: Fairbanks, Alaska, University of Alaska, 86 p., 2 pl., scales 1:24847, 1:24000.

Gottschalk, R. R., 1987, Tectonics of the Schist belt metamorphic terrane near Wiseman, Alaska: Geological Society of America Abstracts with Programs, v. 19, p. 383.

Gottschalk, R. R., 1990, Structural evolution of the Schist belt, south-central Brooks Range fold and thrust belt, Alaska: Journal of Structural Geology, v. 12, p. 453–469.

Gottschalk, R. R., and Oldow, J. S., 1988, Low-angle normal faults in the south-central Brooks Range fold and thrust belt, Alaska: Geology, v. 16, p. 395–399.

Grantz, A., Moore, T. E., and Roeske, S. M., 1991, North American Continent-Ocean transect A-3: Gulf of Alaska to Arctic Ocean, Geological Society of America Continental/Ocean Transect A-3: Boulder, Colorado, Geological Society of America, 2 p., 3 sheets, scale 1:500,000.

Green, A. G., Levato, L., Valasek, P., Olivier, R., Mueller, St., Milkereit, B., and Wagner, J. J., 1993, Characteristic reflection patterns in the Southeast Canadian Cordillera, Northern Appalachians, and Swiss Alps: Tectonophysics, v. 219, p. 71–92.

Hawk, J. M., 1985, Lithospheric flexure, overthrust timing, and stratigraphic modeling of the central Brooks Range and Colville Foredeep in Alaska [M.S. thesis]: Houston, Texas, Rice University, 179 p.

Holliger, K., and Kissling, E., 1991, Ray theoretical depth migration: methodology and application to deep seismic reflection data across the eastern and southern Swiss Alps: Eclogae Geologicae Helveticae, v. 84, p. 369–402.

Hubbard, R. J., Edrich, S. P., and Rattey, R. P., 1987, Geologic evolution and hydrocarbon habitat of the Arctic Alaska microplate, *in* Tailleur, I., and Weimer, P., eds., Alaskan North Slope geology: Society of Economic Paleontologists and Mineralogists, Pacific Section, Publication 50, p. 61–74.

Jones, P. B., 1982, Oil and gas beneath east-dipping underthrust faults in the Alberta foothills, *in* Powers, R. B., ed., Geologic studies of the Cordilleran thrust belt: Denver, Colorado, Rocky Mountain Association of Geologists, v. 1, p. 61–74.

Jones, D. L., Silberling, N. J., Coney, P. J., and Plafker, G., 1987, Lithotectonic terrane map of Alaska: U.S. Geological Survey Miscellaneous Field Studies Map MF-1874A, scale 1:2,500,000.

Julian, F. E., Phelps, J. C., Seidensticker, C. M., Oldow, J. S., and Avé Lallemant, H. G., 1984, Structural history of the Doonerak window: Geological Society of America Abstracts with Programs, v. 16, p. 326.

Levander, A., Fuis, G. S., Wissinger, E. S., Lutter, W. J., Oldow, J. S., and Moore, T. E., 1994, Seismic images of the Brooks Range fold and thrust belt, Arctic Alaska, from an integrated seismic reflection/refraction experiment: Tectonophysics, v. 232, p. 13–31.

Luschen, E., Wenzel, F., Sandmeier, K. J., Menges, D., Ruhl, T., Stiller, M., Janoth, W., Keller, F., Sollner, W., Thomas, R., Krohe, A., Stenger, R., Fuchs, K., Wilhelm, H., and Eisbacher, G., 1987, Near-vertical and wide-angle seismic surveys in the Black Forest, SW Germany: Journal of Geophysics, v. 62, p. 1–30.

Mayfield, C. F., Tailleur, I. L., and Ellersiek, I., 1983, Stratigraphy, structure, and palinspastic synthesis of the western Brooks Range, northwestern Alaska: U.S. Geological Survey Open-File Report 83-779, 58 p.

Miller, E. L., and Hudson, T. L., 1991, Mid-Cretaceous extensional fragmentation of a Jurassic–Early Cretaceous compressional orogen, Alaska: Tectonics, v. 10, p. 781–796.

Mooney, W. D., and Meissner, R., 1992, Multi-genetic origin of crustal reflectivity: a review of seismic reflection profiling of the continental lower crust and Moho, *in* Fountain, D. M., Arculus, R., and Kay, R. W., eds., Continental lower crust: Amsterdam, Elsevier, p. 45–79.

Moore, T. E., and Mull, C. G., 1989, Geology of the Brooks Range and North Slope: Alaskan Geological and Geophysical Transect Field Trip Guidebook T104, 28th International Geologic Congress, p. 107–131.

Moore, T. E., Wallace, W. K., Bird, K. J., Karl, S. M., Mull, C. G., and Dillon, J. T., 1992, Geology of northern Alaska: U.S. Geological Survey Open-File Report 92-220, 191 p.

Moore, T. E., Wallace, W. K., Bird, K. J., Karl, S. M., Mull, C. G., and Dillon, J. T., 1994, Geology of northern Alaska, *in* Plafker, G., and Berg, H. C., eds., The Geology of Alaska: Boulder, Colorado, Geological Society of America, The Geology of North America, v. G-1, p. 49–140.

Mull, C. G., 1982, The tectonic evolution and structural style of the Brooks Range, Alaska: an illustrated summary, *in* Powers, R. B., ed., Geological Studies of the Cordilleran thrust belt: Denver, Colorado, Rocky Mountain Association of Geologists, v. 1, p. 1–45.

Mull, C. G., Adams, K. E., and Dillon, J. T., 1987, Stratigraphy and structure of the Doonerak fenster and Endicott Mountains allochthon, central Brooks Range, Dalton Highway, *in* Tailleur, I., and Weimer, P., eds., Alaskan North Slope geology: Society of Economic Paleontologists and Mineralogists, Pacific Section, Publication 50, p. 663–679.

Nunn, J. A., Czerniak, M., and Pilger, R. H., 1987, Constraints on the structure of the Brooks Range/Colville basin, northern Alaska: Eos (Transactions, American Geophysical Union), v. 67, p. 1196.

Oldow, J. S., Bally, A. W., and Avé Lallemant, H. G., 1990, Transpression, orogenic float, and lithospheric balance: Geology, v. 18, p. 991–994.

Oldow, J. S., Bally, A. W., Avé Lallemant, H. G., and Leeman, W. P., 1989, Phanerozoic evolution of the North American Cordillera: United States and Canada, *in* Bally, A. W., and Palmer, A. R., eds., The geology of North America—An overview: Boulder, Colorado, Geological Society of America, Geology of North America, v. A, p. 139–232.

Oldow, J. S., Seidensticker, C. M., Phelps, J. C., Julian, F. E., Gottschalk, R. R., Boler, K. W., Handschy, J. W., and Avé Lallement, H. G., 1987, Balanced cross sections through the central Brooks Range and North Slope, Arctic Alaska: American Association of Petroleum Geologists, Special Publication, 19 p., 8 pl.

Oliver, J., 1985, Tracing surface features to great depths: a powerful means for exploring the deep crust: Oil and Gas Journal, v. 83, p. 132–137.

O'Sullivan, P. B., Green, P. F., Bergman, J., Decker, I. R., Duddy, A. J., Geadow, W., and Turner, D. L., 1993, Multiple phases of Tertiary uplift and erosion in the Arctic National Wildlife Refuge, Alaska, revealed by apatite fission track analysis: American Association of Petroleum Geologists Bulletin, v. 77, p. 359–385.

Palmer, A. R., Dillon, J., and Dutro, J. T., Jr., 1984, Middle Cambrian trilobites with Siberian affinities from the central Brooks Range, northern Alaska: Geology of Alaska and Northwestern Canada Newsletter No. 1, p. 29–30.

Patrick, B. E., Till, A. B., and Dinklage, W. S., 1994, An inverted metamorphic field gradient in the central Brooks Range, Alaska, and implications for exhumation of high-pressure/low temperature metamorphic rocks: Lithos, v. 33, p. 67–83.

Patton, W. W., Jr., 1973, Reconnaissance geology of the northern Yukon-Koyukuk province, Alaska, *in* Shorter contributions to general geology: U.S. Geological Survey Professional Paper 774-A, p. A1–A17.

Patton, W. W., Jr., 1984, Timing of arc collision and emplacement of oceanic crustal rocks on the margins of the Yukon-Koyukuk basin, western Alaska: Geological Society of America Abstracts with Programs, v. 16, p. 555–566.

Patton, W. W., Jr., 1992, Ophiolitic terrane bordering the Yukon-Koyukuk basin, Alaska: U.S. Geological Survey Open-File Report 92-90D, 8 p.

Patton, W. W., Jr., and Box, S. E., 1989, Tectonic setting of the Yukon-Koyukuk basin and its borderlands, western Alaska: Journal of Geophysical Research, v. 94, p. 15807–15820.

Phelps, J. C., Oldow, J. S., Avé Lallemant, H. G., Julian, F. E., Seidensticker, C. M., Boler, K. W., and Gottschalk, R. R., 1985, Late-stage high-angle faulting, eastern Doonerak window, central Brooks Range, Alaska: American Association of Petroleum Geologists–Society of Economic Paleontologists and Mineralogists–Society of Exploration Geophysicists Pacific Section, 60th Annual Meeting, Programs and Abstracts, p. 33–34.

Pfiffner, O. A., Frei, W., Valasek, P., Stauble, M., Levato, L., Dubois, L., Schmid,

S., and Smithson, S., 1990, Crustal shortening in the alpine orogen: Results from deep seismic reflection profiling in the eastern Swiss Alps, line NFP20 EAST: Tectonics, v. 9, p. 1327–1355.

Ratschbacher, L., Frisch, W., Neubauer, F., Schmid, S. M., and Neugebauer, J., 1989, Extension in compressional orogenic belts: The eastern Alps: Geology, v. 17, p. 404–407.

Raynaud, B., 1988, A 2-D, ray-based, depth migration method for deep seismic reflections: Geophysical Journal, v. 93, p. 163–171.

Roeder, D. H., and Mull, C. G., 1978, Tectonics of the Brooks Range ophiolites, Alaska: American Association of Petroleum Geologists Bulletin, v. 62, p. 1696–1702.

Smithson, S. B., Brewer, J. A., Kaufman, S., Oliver, J. E., and Hurich, C., 1979, Structure of the Laramide Wind River uplift, Wyoming, from COCORP deep reflection profiling in the Wind River Range, Wyoming: Earth and Planetary Science Letters, v. 46, p. 295–305.

Stone, D. B., 1989, Geology of the Brooks Range and North Slope, *in* Nokleberg, W. J., and Fisher, M. A., eds., Alaskan Geological and Geophysical Transect Field Trip Guidebook T104, 28th International Geologic Congress: American Geophysical Union, p. 78–83.

Tailleur, I. L., and Brosgé, W. P., 1970, Tectonic history of northern Alaska, *in* Adkinson, W. L., and Brosgé, M. M., eds., Proceedings, Geological Seminar on the North Slope of Alaska: American Association of Petroleum Geologists Pacific Section, p. E1–E19.

Till, A. B., 1992, Detrital blueschist-facies metamorphic mineral assemblages in Early Cretaceous sediments of the foreland basin of the Brooks Range, Alaska, and implications for orogenic evolution: Tectonics, v. 11, p. 1207–1223.

Till, A. B., Schmidt, J. M., and Nelson, S. W., 1988, Thrust involvement of metamorphic rocks, southwestern Brooks Range, Alaska: Geology, v. 16, p. 930–933.

Till, A. B., Box, S. E., Roeske, S. M., Patton, W. W., Jr., Miller, E. L., Hudson, T. L., 1993, Mid-Cretaceous extensional fragmentation of a Jurassic–Early Cretaceous compressional orogen, Alaska: Comment and Reply: Tectonics, v. 12, p. 1076–1086.

Turner, D. L., Forbes, R. B., and Dillon, J. T., 1979, K-Ar geochronology of the southwestern Brooks Range, Alaska: Canadian Journal of Earth Sciences, v. 16, p. 1789–1804.

Valasek, P., Mueller, S., Frei, W., and Holliger, K., 1991, Results of NFP20 seismic reflection profiling along the alpine section of the European Geotraverse: Geophysical Journal International, v. 105, p. 85–102.

Wallace, W. K., and Hanks, C. L., 1990, Structural provinces of the northeastern Brooks Range, Arctic National Wildlife Refuge, Alaska: American Association of Petroleum Geologists Bulletin, v. 74, p. 1100–1118.

Warner, M. R., 1987, Migration—Why doesn't it work for deep seismic data?: Geophysical Journal of the Royal Astronomical Society, v. 89, p. 21–26.

MANUSCRIPT ACCEPTED BY THE SOCIETY SEPTEMBER 23, 1997

Geological Society of America
Special Paper 324
1998

Origin and tectonic evolution of the metamorphic sole beneath the Brooks Range ophiolite, Alaska

R. A. Harris*

Department of Geology, West Virginia University, Morgantown, West Virginia 26506

ABSTRACT

Structural field relations, petrologic and geochemical studies, and radiometric age analysis of metamorphic rocks at the base of the Brooks Range ophiolite provide a record of the time, tectonic setting, and initial conditions of ophiolite emplacement during the Middle Jurassic western Brookian orogeny. The metamorphic sole of the Misheguk, Siniktanneyak, and Avan Hills ophiolite bodies is transitional with and geochemically similar to basalt, chert, limestone, and psammitic material mostly of the Copter Peak allochthon. Dynamothermal metamorphism of these rocks produced garnet-bearing amphibolite and quartz-mica schists that decrease in metamorphic grade and ductile strain intensity downward. These high-grade rocks are separated by thin, fault-bounded intervals from underlying greenschist facies and unmetamorphosed basalt and sedimentary rocks. The maximum total thickness of both parts of the metamorphic sole is <500 m.

Minimum temperature estimates, from garnet-biotite and garnet-amphibole geothermometric studies of rocks currently 5–50m below the ophiolite, range from 500 to 560°C at around 5 Kb. These conditions are similar to closing temperatures estimated for Ar retention in hornblende near the base of the ophiolite, which yields $^{40}Ar/^{39}Ar$ ages of 164–169 Ma. This age is concordant with, but slightly younger than the mean age of ophiolite crystallization. The limited time between crystallization and tectonic emplacement of the Brooks Range ophiolite, and compositional similarities between the metamorphic sole and continental margin material imply spacial and temporal proximity of the ophiolite with the Brookian orogen prior to emplacement.

Localization of strain at the base of the ophiolite during emplacement produced a ductile shear zone in the metamorphic sole that is characterized by mylonitic textures overprinted by postmetamorphic deformation. Microstructures document synkinematic crystal growth at high flow stress and strain rates. Kinematic indicators show mostly top-to-northwest sense of shear at the base of the ophiolite. Postmetamorphic deformation associated with the Brookian orogen shuffled various parts of the metamorphic sole and thrust it over shelf facies sedimentary material.

INTRODUCTION

Nappes of mafic and ultramafic igneous sequences commonly form the uppermost thrust sheets of imbricated continental margin successions in many collisional mountain systems. These allochthonous igneous sequences are loosely termed "ophiolites" or "Alpine-type" ultramafics (Coleman, 1977). Other classification schemes such as "Tethyan- or Cordilleran-type" ophiolites (Moores, 1982) are also used to distinguish between ophiolites thrust over continental margins (Tethyan-type) and those that form terrane collages (Cordilleran-type). Although ophiolites

*Present address: Department of Geology, Brigham Young University, Provo, Utah 84602.

Harris, R. A., 1998, Origin and tectonic evolution of the metamorphic sole beneath the Brooks Range ophiolite, Alaska, *in* Oldow, J. S., and Avé Lallemant, H. G., eds., Architecture of the Central Brooks Range Fold and Thrust Belt, Arctic Alaska: Boulder, Colorado, Geological Society of America Special Paper 324.

represent a diverse assortment of tectonomagmatic settings (Coleman, 1984), most Tethyan-type ophiolites are chemically similar to oceanic island arc-complexes (Pearce et al., 1984). The similar structural position and tectonic associations of these ophiolites in different continental fold-thrust belts suggest common mechanisms for emplacement.

A very important feature of many Tethyan-type ophiolite nappes are the thin sheets of metamorphic rock at their structural bases (i.e., Williams and Smyth, 1973; Searle and Malpas, 1980; Ghent and Stout, 1981; Moores, 1982; Jamieson, 1986). These metamorphic rocks (herein referred to as metamorphic soles) form part of the basal shear zone of ophiolite nappes and are characterized by an inverted sequence of amphibolite- and greenschist-facies metabasalt (mostly E- and T-type mid-oceanic ridge basalt [MORB]), metachert, and marble. The shear zone commonly juxtaposes the metamorphic sole with structurally underlying distal facies continental margin deposits. Locally these deposits are structurally omitted placing the metamorphic sole directly on proximal facies shelf units. This structural contact represents an early collisional plate boundary that was uplifted by subsequent accretion of lower plate material in the collision zone.

Rocks incorporated into the metamorphic sole represent some of the first material to underthrust the "hot" ophiolite after detachment from its roots and before cooling. Determining the protolith, structural evolution, and timing and conditions of metamorphism of the metamorphic sole may constrain the tectonic setting, temperature, transport direction, and even the minimum age of initial ophiolite emplacement. These data are critical for resolving various models for the tectonic affinity of ophiolites and determining the temporal and spatial relations between ophiolites and the orogenic belts where they reside.

The Brooks Range ophiolite has a metamorphic sole of garnet-bearing amphibolite and quartz-mica schist, greenschists, and metabasalt, chert, marble, and psammitic material. Since the first general field descriptions of the Brooks Range ophiolite were published by Snelson and Tailleur (1968) and Martin (1970), interpretations of its ophiolitic associations and tectonic significance have relied on sparse data from reconnaissance-style studies (i.e., Patton et al., 1977; Roeder and Mull, 1978). The purpose of this paper is to present data collected along several structural transects through the metamorphic sole of the Brooks Range ophiolite, which is exposed mostly at the base of the Misheguk Mountain and Avan Hills ophiolite bodies (Fig. 1).

REGIONAL OVERVIEW

The Brooks Range allochthon belt (Mayfield et al., 1983) comprises the western 200–300 km of the Brooks Range fold-thrust mountain system of Arctic Alaska (Fig. 1). The western Brooks Range exposes the highest structural levels that are eroded from the central and eastern part of the orogen. The stratigraphic record of the western Brooks Range represents a long interval of passive continental margin development. These conditions persisted from early Paleozoic to the Middle Jurassic onset of orogenesis. The passive margin was shortened by collision with arc

terranes of the Yukon-Koyukuk province (Roeder and Mull, 1978; Box, 1985; Harris, R. A., et al., 1987), which is presently south of the Brooks Range (Fig. 1). The initial phases of this collision involved detachment and tectonic emplacement of large ophiolitic thrust sheets, which are partially preserved in the western Brooks Range. The final pulses of the orogen are associated with basement involved thrusting and rapid hinterland unroofing (Gottschalk and Oldow, 1988; Till et al., 1988). An extensive blueschist belt is exposed in the hinterland (Patton et al., 1977).

The shear zone at the structural base of the Brooks Range ophiolite served as the initial plate boundary between underthrust continental material and impinging arc terranes of the upper plate. Metamorphic rocks in this shear zone provide an important record of these events. Relocation of the plate boundary during and after accretion determines to a large extent how well the earlier tectonic features associated with ophiolite emplacement are preserved. In the western Brooks Range, as in Oman (Searle and Malpas, 1980; Harris, 1992), postcollision deformation was minimal and subjected the fold-thrust mountains to only minor subsequent structural and thermal overprints. This differs from most other parts of the North American Cordillera that were intensely modified by Laramide and subsequent phases of deformation (e.g., Moores, 1970).

The structural relations preserved in the western Brooks Range indicate hundreds of kilometers of passive margin contraction associated with emplacement of extensive ophiolite sequences (Martin, 1970; Tailleur, 1970; Mull, 1982; Mayfield et al., 1983). Arctic Alaska is now bounded by two principal geologic features: a passive continental margin to the north and a collisional margin known as the Kobuk suture to the south. The Kobuk suture separates Arctic Alaska from arc terranes of the Yukon-Koyukuk province (Fig. 1) in the hinterland of the Brooks Range. The metamorphic sole at the base of the Brooks Range ophiolite is also considered part of this suture zone.

Brooks Range ophiolite

The Brooks Range ophiolite consists of mafic and ultramafic thrust sheets that form five different klippen-like massifs of consanguineous composition, internal organization, structure, and age (Harris, 1995). The massifs are known from west to east as Iyokrok, Asik, Avan Hills, Misheguk, and Siniktanneyak (Fig. 1). The ophiolite bodies most likely represent fragmented and eroded remnants of what was initially an extensive roof thrust perhaps as large as 350 km in strike length, 50 km in width, and as much as 3–4 km thick.

The composition of the Brooks Range ophiolite consists of the following igneous sequence in ascending order: (1) tectonized peridotites, (2) ultramafic cumulates, (3) layered gabbro, (4) massive gabbro and high-level intrusives, (5) out-of-sequence intrusives, and (6) rare sheeted dikes, lavas, and sediments (Harris, 1995). The various units are most commonly transitional except for local intrusive relations. Chemically, the Brooks Range ophiolite is typical of ophiolites that are transitional between MORB- and arc-types (Harris, 1992, 1995). Transitional-type ophiolites are considered by Pearce et al. (1984) to represent tholeiitic magmas derived from

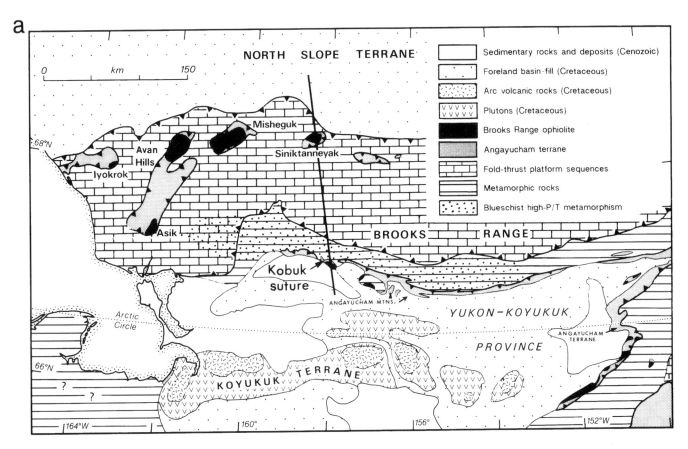

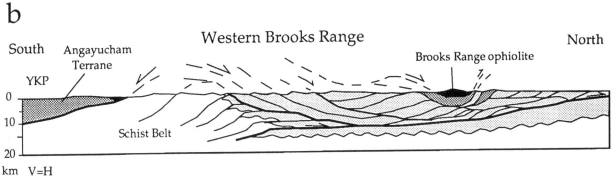

Figure 1. Geology of the western Brooks Range and northern Yukon-Koyukuk province (modified from Patton, 1973). a, Simplified geologic map. b, Schematic cross section showing the Brooks Range ophiolite nappe (black), Copter Peak allochthon/Angayucham terrane (dark gray), pre-Mississippian basement (white), and sedimentary cover sequences (light gray). The roots of the ophiolite nappe are interpreted as the mafic and ultramafic plutonic rocks that structurally overlie basalts and chert of the Angayucham Mountains region. YKP = Yukon-Koyukuk province.

partial melts of a moderately depleted mantle. Modern examples of thin oceanic crust with transitional-type geochemistry are documented from small ocean basins in the western Pacific, Scotia, and Andaman Seas. Most of these basins formed by processes of intra-arc spreading (Moores et al., 1984; Pearce et al., 1984).

Angayucham terrane

Along the Kobuk suture zone of the southern Brooks Range the Angayucham Mountains (Fig. 1) expose a narrow, curvilinear belt of Devonian to Early Jurassic pillow basalt, diabase, basaltic tuff, and radiolarian chert (Patton et al., 1977; Pallister et al., 1989). Structurally overlying the basalt-chert terrane are fragments of mafic and ultramafic thrust sheets of the Brooks Range ophiolite (Patton and Box, 1989; Harris, 1992). The gross similarity of the two separate allochthonous masses initially led to incorrect correlations that lump the Brooks Range ophiolite with the Angayucham terrane (Jones et al., 1981; Box, 1985; Moore and Grantz, 1987; Plafker, 1988).

Copter Peak allochthon

In the western Brooks Range fold-thrust zone, Angayucham-like volcanic and sedimentary rocks are known as the Copter Peak allochthon (Ellersieck et al., 1982). Thrust sheets of the Copter Peak allochthon commonly form the uppermost structural elements of the western Brooks Range fold-thrust belt, except where they are structurally overlain by Brooks Range ophiolite massifs. The Copter Peak allochthon has remnants of thick mafic igneous crust that consist of amygdaloidal and vesicular pillow and massive basalt and, to a lesser extent, felsic lavas and diabase. Chert, tuff, carbonate, and some clastic detritus are tectonically interleaved with the igneous rocks. Chert layers yield Pennsylvanian–Upper Triassic radiolaria (Nelson and Nelson, 1982; Ellersieck et al., 1982) and may be correlative with distal rise deposits of the Arctic Alaskan passive continental margin (Etivluk Group). Most of the carbonate is similar in facies and fauna to Devonian sections of para-autochthonous shallow-water limestones (Baird Group limestone, see Dumoulin and Harris, 1987). The structural relations between limestone and volcanic rocks of the Copter Peak allochthon are ambiguous. Mayfield et al. (1983) considered the carbonate as part of continental margin material that was tectonically slivered with volcanic rocks. Nelson and Nelson (1982) claim stratigraphic continuity from pillow lavas into carbonate with Devonian fauna and rounded pebbles of basalt.

Lithologic, biostratigraphic, and structural similarities justify including the Copter Peak allochthon as part of the Angayucham terrane. The basalt, chert, and other rocks associated with this terrane were accreted to the base of the Brooks Range ophiolite at the onset of the Brookian orogeny.

Geochemistry of the Angayucham/Copter Peak igneous rocks

Geochemical studies of Copter Peak allochthon igneous rocks yield similar results to basalts of the Angayucham terrane in the Kobuk suture (Harris, 1987; Moore, 1987; Wirth et al., 1987). Both complexes consist of tholeiites and some compositions transitional to alkali-basalt. Trace-element discriminant analysis of the lavas yields N- and T-type MORB trends with mostly "within-plate basalt" associations. These chemical affinities are found in many intraplate tectonic settings. Geochemical data from volcanic material in the Kobuk suture zone is interpreted by Barker et al. (1988) and Pallister et al. (1989) as ocean-plateau and ocean-island basalts. These rocks are also similar to basalts from distal continental margins. Examples of thick (10 km) deposits of N- and T-type MORB lavas exist along the Norwegian-Rockall and conjugate east Greenland banks; outer Scott, Wallaby, and Naturaliste plateaus of Western Australia; and the southeast Weddell Sea of Antarctica (Mutter et al., 1988). A distal continental margin origin for the basalt-chert terrane is also supported by stratigraphic relations between the sedimentary sequence interbedded with lavas and continental margin sequences of Arctic Alaska (Dumoulin and Harris, 1987; Alexander, 1990).

Although igneous geochemistry is limited in distinguishing between possible tectonic settings of transitional-type ophiolites, it is useful for discriminations between igneous rocks of significantly different lineage. Geochemical differences between the igneous rocks from the Brooks Range ophiolite and the Angayucham terrane preclude any genetic relation between them (Harris, 1992, 1995). The melts that formed the Brooks Range ophiolite had a moderately depleted source compared with the normal and enriched sources for Angayucham terrane lavas. Although a range of possibilities exists for the tectonic setting of both igneous complexes, there is little potential for overlap.

The Angayucham igneous complex documents a long interval of intraplate-type magmatism associated, at least in part, with the development of the Arctic Alaskan passive continental margin. The formation of the Brooks Range ophiolite immediately proceeds the contraction of this margin and juxtaposition of the two different igneous terranes. The timing and conditions of accretion of the two terranes are constrained to some extent by the thin metamorphic complex that locally occurs between them.

SUBOPHIOLITE METAMORPHIC COMPLEX

Glacial incisions through klippen of the Brooks Range ophiolite expose its structural base and metamorphic sole at several locations over a distance of 130 km. Reconnaissance descriptions of these metamorphic rocks are available for the Avan Hills (Zimmerman and Frank, 1982; Harding et al., 1985; Wirth et al., 1986), Misheguk (Harris, 1987), and Iyokrok (Boak et al., 1987) ophiolite bodies. The most extensive and complete occurrence is at Misheguk, which together with the Avan Hills metamorphic sole are the focus of this investigation.

At Misheguk the metamorphic sole forms a series of northeast-southwest–trending dark-colored hills along the northwest flank of the ophiolite body (Fig. 2). The metamorphic sole is characterized by intercalations of black amphibolite, maroon and black metasedimentary schists, dark green metavolcanic rocks, and cream-colored metachert. The rocks are generally fine grained with a strong, commonly mylonitic, layer-parallel foliation.

Field relations

Table 1 provides a description of mineral parageneses from twelve traverses through the metamorphic sole at Misheguk shown in Figure 2. Mineral parageneses are listed relative to distances from the structural base of the Brooks Range ophiolite, which is signified by "gz" (gouge zone) on Figure 2.

Traverses 1–4. Northwest of Misheguk Mountain metamorphic sole rocks underlie tectonized peridotite and dunite (Fig. 3a). The complex consists of 0–40 m of garnet amphibolite, which overlies a low-grade (greenschist facies) assemblage with mafic volcanic and pelagic sedimentary protoliths. The low-grade rocks are less than 200 m thick and decrease in grade downward toward unmetamorphosed distal facies continental margin deposits and melange. Faults structurally thin and eventually eliminate the metamorphic sole along strike, sometimes juxtaposing the ophiolite complex with Devonian and younger Baird Group carbonate, Etivluk Group pelagic material, and Cretaceous Okpikruak Formation.

Traverses 5–9. On the northeast flank of Misheguk Mountain metamorphic sole rocks underlie cumulate ultramafic rocks and layered gabbro (Fig. 3b). The mantle sections of the Brooks Range ophiolite are structurally omitted. Immediately below the cumulate sequence is a thick (as much as 50 m) fault gouge zone composed of phacoidal blocks (as much as 5 m in diameter) of mylonite, serpentinite, tectonized peridotite, gabbro, and rodingite floating in a dominantly serpentinite matrix. Veins of calcite and prehnite intrude the gouge zone. The fault gouge is underlain by a thin (0–30 m) band of garnet-bearing amphibolite and quartz-mica schist. The schists grade in broken succession downward into greenschists followed by altered melange, volcanics, pelagics, and carbonates of the Copter Peak allochthon. The metamorphic complex strikes east-northeast toward the north flank of peak 1217, maintaining a structural thickness of about 1.5 km.

Traverses 10–12. Dark green amphibolitic metabasalt with locally preserved igneous textures is exposed for about 1 km along a continuous ridge on the north flank of peak 1217 (Fig. 2 and 3c). The emplacement fault and inverted metamorphic complex dip steeply and are locally overturned by as much as 30° (dip 60° north). Faulted slivers of porphyroclastic harzburgite and lherzolite, and 2.5 km of massive dunite, which originally overlay the metamorphic complex, now underlie it along an overturned thrust.

The overturned metamorphic complex consists of an imbricate sequence of fault slices that range in thickness from 50 to 120 m (Fig. 3c). Metamorphic grade and deformation intensity decrease upward away from the Brooks Range ophiolite in each overturned slice and throughout the imbricate stack toward altered mafic lavas of the Copter Peak allochthon. Copter Peak volcanic rocks are juxtaposed with greenschists of the metamorphic complex along a north-dipping fault, which is also overturned. Scaly clay melange with blocks of Copter Peak volcanic rocks, Etivluk Group chert, and some metamorphic clasts surround the overturned sequences (Fig. 3c).

Traverse of the Avan Hills. The Avan River drains a glacial valley cut through the Avan Hills ophiolite body into the structurally underlying Copter Peak allochthon (Fig. 3d). Subophiolite metamorphic rocks, serpentine, and tectonized harzburgites are found at several locations along the flat-lying structural base of the ophiolite. Thin layers (few decimeters) of lower-greenschist-facies metabasalt, garnet amphibolites, and granitic mylonites were found during this study. The granitic material intrudes the base of the ophiolite nappe only for a distance of a few meters and is subsequently deformed. Mylonitic fabrics in these rocks are most intense at the structural base of the nappe.

Structure

The metamorphic sole documents a complex multiphase history of ductile and brittle deformation associated with emplacement and fragmentation of the Brooks Range ophiolite. The largest remnants of the metamorphic sole are found on the north flank of Misheguk Mountain, where most of the measurements for this study were obtained. In this region metamorphic rocks form a thrust and folded stack as much as 1,200 m thick that trends east-northeast to west-southwest. High-angle oblique slip faults offset the structural stack of metamorphic rocks locally (Fig. 2). Basalt, chert, and limestone of the Copter Peak allochthon form thrust sheets and broken formation within the stack. Structures were also analyzed from poorly exposed outcrops of the metamorphic sole at Avan Hills and Siniktanneyak.

The internal structure of the metamorphic sole is dominated by a layer-parallel mylonitic schistosity (S1), which is associated with D1 and was subsequently folded and fractured by post-peak metamorphism deformation, D2. Poles to S1 foliation plot along a north-northwest to south-southeast girdle with a pole near to the mean of F2 axes (Fig. 4a). Layer parallelism is demonstrated by the overlap between the poles of foliations and folded chert layers near the base of the metamorphic sole (Fig. 4b).

Three phases of folding are recognized in the metamorphic sole. The first phase (F1) is found in helicitic garnet inclusions (Fig. 5), rotated porphyroclasts, and northwest-vergent isoclinal microfolds associated with S1. F2 folds are more obvious in the field because they modify the entire thrust stack. These folds have southeast-dipping axial planes (Fig. 4c), which indicate vergence to the northwest. Locally these folds overturn sections of the ophiolite and the metamorphic sole (Fig. 3c). F3 folds are poorly developed, have a low amplitude, long wavelength, and plunge steeply to the southeast (Fig. 4c).

Two major sets of faults are found (Fig. 4d). North-south–striking faults are commonly associated with S1 foliations and are interpreted as forming during D1. East-west–striking faults include those forming the base of the Brooks Range ophiolite at Misheguk, Avan Hills, and Siniktanneyak. Major discontinuities in metamorphic grade occur across these faults, which are associated with top-to-the-northwest, F2 folds. Grooves on fault surfaces associated with the basal thrust of the Brooks Range ophiolite trend north-south. A variety of groove orientations were found on other faults, which produce an ambiguous kinematic pattern. Grooved fault surfaces indicate that latest motion along many faults was near horizontal. Crosscutting and abutting mode 1 fractures mostly trend north-south, which may indicate the orientation of stresses after faulting.

Microstructures show various effects of inhomogeneous rotational shear overprinted locally by static crystal growth (Fig. 5). Most early, foliation-forming mineral grains show strong dimensional preferred alignment, streaky mylonitic foliation, and lenticular porphyroclasts of strained plagioclase, amphibole, and snowball and helicitic garnet that are enclosed by sheety and fibrous silicates with rotational asymmetry.

Shear-sense indicators such as asymmetric augen structures and pressure shadows, composite or s-c planar fabrics, mica "fish," asymmetric microfolds, shape-preferred orientations, mineral stretching lineations, and displaced broken grains were inspected in thin sections oriented parallel to D1 lineations. These structures show a top-to-the-northwest displacement during the earliest recorded and most penetrative phase of deformation.

Deformation mechanisms

Dynamic recrystallization of quartz formed subgrains and neoblasts by mechanisms of grain boundary migration and subgrain rotation distortions. In amphibolite-grade sections of the metamorphic sole, quartz shows evidence of near-complete recovery. Many layers have equant, strain-free subgrains with 120 triple junctions. Other layers have more irregular, bulging grain boundaries, and preserve deformation bands. Polycrystalline quartz ribbons are common (Fig. 5c and d). Quartz grains in lower amphibolite and greenschist facies parts of the metamorphic sole are larger, with more common deformation bands, and serrated grain boundaries.

The softening effect of dynamic recrystallization caused localization of strain that is documented in the metamorphic sole by changes in quartz subgrain diameter. Under steady-state conditions grain size reduction is a function of flow stress and is essentially temperature independent (Twiss, 1986). Quartz subgrain diameters were measured in oriented thin sections from throughout the metamorphic sole (Table 1). Most samples yield a unimodal distribution of grain size where 60–70% of the grains vary by only 3 μm. Quartz subgrain sizes range from 4.13 to 8.14 μm within 20 m of the ophiolite, and from 6.00 to 10.70 μm between 20 and 70 m from the ophiolite (Table 1). Traverse 11 shows the most systematic trend of reduced grain size toward the basal contact of the ophiolite.

Applying the results of laboratory piezometric studies (Mercier et al., 1977; Schmid, 1982; Twiss, 1986) to the size of quartz subgrains in the metamorphic sole predicts flow stresses of 150–200 MPa within 20 m of the ophiolite, and 80–160 MPa 20–70 m below the contact. The deformation mechanism map of Rutter (1976) predicts that at temperatures of 500–600°C and flow stresses greater than 100 MPa that quartz will deform by dislocation glide at rapid strain rates of 10^{-10} s^{-1} to 10^{-9} s^{-1}.

Crystal plastic deformation of other minerals indicate similar dynamothermal metamorphic conditions to those of quartz. Calcite grains between 40–100m below the ophiolite range from 0.5–3 μm in diameter, which is consistent with yield stresses 40–70 MPa at room temperature (Olsson, 1974). K feldspar is slightly recrystallized within 20 m of the base of the ophiolite, which initiates in laboratory tests at temperatures of more than 500°C (Kronenberg and Shelton, 1980).

Crystal plastic deformation was followed locally by thermal annealing producing blastomylonitic and hartschier textures in some amphibolitic sections of the metamorphic sole. Crossmicas, helicitic and idioblastic garnet and amphibole, polygonization, and chlorite and epidote pseudomorphism locally crosscut foliations. Some clasts have much coarser, randomly oriented grains than the matrix fabric, and preserve various relict phases of the metamorphic history. The transition from D1 layer-parallel flattening to D2 folds accompanies the annealing and subsequent passage of the rocks through the brittle-ductile transition.

Brittle disruption of ductile fabrics by faults, fractures, cataclasis, and veining indicate low temperatures or high strain rates during the final phases of ophiolite emplacement. Low-grade mineral-filled veins are associated with metamorphic retrogression and high fluid pressures. Early crosscutting and some layer-parallel fractures are sealed by secondary mineral deposits (albite, prehnite, calcite, chlorite, and sometimes epidote and pumpellyite) possibly associated with serpentinization of the overlying Brooks Range ophiolite.

Petrology and mineral paragenesis

Lithologic descriptions of samples from traverses made through the metamorphic sole and into immediate underlying rocks are provided in Table 1. Lateral discontinuities and postmetamorphic faulting make it difficult to correlate individual lithologic units between traverses. However, where thick sections of the metamorphic sole are exposed, a similar lithologic succession is found. This succession consists, in descending order, of: (1) a zone of cataclasis and partial melt intrusion at the structural base of the ophiolite, (2) amphibolite and quartz-mica schists, (3) greenschist facies metavolcanic and metasedimentary rocks, and (4) volcanic and sedimentary rocks of the Copter Peak allochthon and underthrust continental margin sequences (described above). Metamorphic rocks form two distinct units that are referred to as the upper or amphibolite unit and lower or greenschist units of the metamorphic sole.

The initial thickness and metamorphic gradient of the succession are difficult to estimate due to postmetamorphic deformation. North of Misheguk Mountain the metamorphic sole is commonly structurally thickened, whereas west of Misheguk Mountain, and at Avan Hills and Siniktanneyak, most of the

Figure 2. Geologic map of the western Misheguk Mountain ophiolite body, its metamorphic sole, and underlying thrust sheets. Positions of traverses through the metamorphic sole are shown in stippled areas and numbered for reference to text. Lines of section in figure 3 are shown. Abbreviations are for figures 2 and 3: *Ophiolite units* (in descending stratigraphic order): mii - mafic and intermediate (high level) intrusives; um - ultramafic out-of-sequence intrusives; lgb - layered gabbro; PGM - Platinum group metal; tz - transition zone; wh - wehrlite; cum - cumulate ultramafics; du - dunite: pd - peridotite; tpd - tectonized peridotite; serp - serpentinite; gz - gouge zone. *Metamorphic Sole units (MS):* ma - amphibolite-facies rocks; mgs - greenschist-facies rocks. *Copter Peak complex (CP):* CPv - volcanics; CPmm - metasediments and metavolcanics; CPc - chert. *Pre-rift continental margin sequences:* Db - Devonian Baird Group carbonates; MDn - Mississippian and Devonian Noatak sandstone and conglomerates; Mko - Mississippian Kogruk Formation deep-water limestones and chert; Mls - Mississippian limestone; Mt - Mississippian Tupik Formation limestone and chert; Mu - Mississippian Utukok formation limestone, sandstone, and shale. *Post-rift passive continental margin sequence:* PMc - Pennsylvanian and Mississippian radiolarian chert; JPe - Jurassic to Pennsylvanian Etivluk Group chert and shale; JPm - mafic dikes and sills. *Syn- and post-orogenic sedimentary rocks:* KJo - Cretaceous (Ko) and Jurassic Okpikruak Formation (melange and broken formation with interbedded conglomerate, sandstone, and remobilized shale); Q - Quaternary alluvium. *Structural symbols:* Faults - shown in heavy lines. Faults with filled square teeth have omitted lithotectonic units. Faults with open, nested teeth are overturned thrusts.

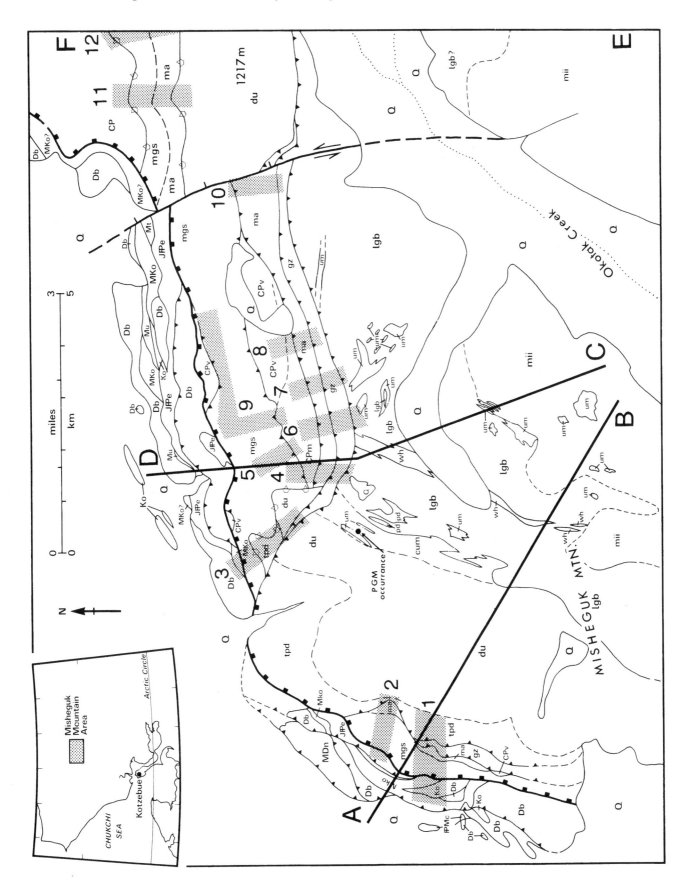

TABLE 1. MINERAL PARAGENESES AND QUARTZ SUBGRAIN SIZE

Distance (m)	Mineral Parageneses*	Qtz SGS	N
Traverse 1			
30 up	Dunite		
30 up	Lherzolite		
30 up	Lherzolite + harzburgite		
20 up	Harzburgite		
3 up	Grt+pl+chl+ep+op		
0	Harzburgite		
1 dn	Qtz+chl+wm+grt	4.55	200
1 dn	Qtz+fld+am		
1 dn	Grt+am+cal+qtz+chl+pl+spn	4.72	200
1 dn	Am+ep		
1 dn	Grt+bt+qtz+chl+am+ab+ep+/-wm	4.84	225
10-20 dn	Act+ab+chl+ep (relict px+ab)		
20-30 dn	Cherty argillite		
30 dn	Fossiliferous marble		
Traverse 2			
10 up	Harzburgite		
3 dn	Qtz+chl+act		
20 dn	Ab-or+act+chl+ep+/-spn		
40 dn	Marble		
40 dn	Purple and green cherty argillite		
Traverse 3			
30-40 up	Lherzolite		
1 up	Lherzolite		
1 up	Srp+ol+chr		
10 dn	Relict ol?+cpx+pl+chl+ep+am		
40 dn	Relict ol+cpx+pl/ab+chl+ep		
Traverse 4			
50 up	Srp		
40 up	Harzburgite		
40-30 up	Olivine gabbro		
10 dn	Am+pl(An40)+spn+chl+ep+cal+hrn +/-wm+ap		
10 dn	Qtz+chl+ep+grt+/-wm	5.23	125
10 dn	Wm+qtz+op+ep+grt	8.14	50
10-15 dn	Basalt		
15 dn	Qtz+chl+wm+op+ep	6.00	44
Traverse 5†			
30-50	Grt+qtz+bt/chl+ab+ep	4.13	164
200-300	(Rd+bk) qtz-mica schist		
300-350	Act+chl schist		
350-400	Act+chl+ep+fld		
500-600	(Rd+bk) qtz-mica schist	5.55	38
700-800	Act+chl schist		
1100-1200	Chl+qtz schist		
Traverse 6			
20-30 up	Gabbro		
20 up	Gabbro+clinopyroxenite layers		
0	Altered gabbro+wehrlite cataclasis		
0	Opx+srp		
10-20 dn	Hrn+opx+ep+ab		
30-50 dn	Grt+hrn+wm+pl(An23)+ep+qtz	5.20	100
150-200	Meta pillow basalt		
200-250	Meta basalt		
Traverse 7			
20 up	Gabbronorite		
0	Dunite (+cpx+sp)		
0	Cpx+pl (Gabbro)		
0	Tectonized harzburgite		
3	Cataclastite		
4-5 dn§	Qtz+ab+grt+wm+chl schist hrn+grt+pl (An29-34)+opx+chl	4.70	200
20 dn			
100-150 dn	Marble		
100-150 dn	Cpx+am+srp		
150-200 dn	Chl+ab+qtz		
200 dn	Spilitic basalt		
200 dn	Radiolarian chert		
Traverse 8			
0-10 up	Websterite + lherzolite		
10-20 dn	Am+chl+pl(An30)+ep?		
40-50 dn	Chl+act+ab+ep+qtz	8.24	425
Fault			
50-60 dn	Volcaniclastic w/ chert frag		
60-100 dn	Spilitic basalt		
Traverse 9†			
20-50**	Qtz+bt+grt+pl(An29-36)+chl+py	6.00	200
20-50	Act+chl+ab+ep+cal+op	8.10	365
800	Meta cpx basalt		
800	Act+pl+chl+qtz+ep+op		
800	Qtz-chl schist		
1200	Am+pl+ep+chl+op		
1400	Am+pl+ep+chl+op		
1400	Am+pl+ep+chl+op		
1600	(Rd+bk) knarly chert		
1600	Fossilliferous limestone		
800-1200	Bt+cal+qtz+op		
800-1200	(Grn)cherty argillite		
800-1200	(Bk+rd)chert+argillite		
800-1200	Cal+chl+qtz+ep+grt		
800-1200	Am+ab+ep+chl		
800-1200	Interlayered chert + mafic rock		
Traverse 10			
10 up	Srp harzburgite		
10 dn	Am+qtz+ep		
10-20 dn	Am+pl+qtz		
50-60 dn	Chl+act+ep+qtz		
Traverse 11‡			
5-10 dn	Srp harzburgite		
Fault gorge	Am+ep+chl+qtz+cal	3.80	90
Block	Meta chert	5.32	330
Block	Pl+am+/-ep+/-qtz	5.90	81
10 dn	Qtz+am+chl+ep	6.92	25
20 dn	Am+pl+qtz	7.08	36
27 dn	Ep+qtz+am+chl	8.50	120
32 dn	Am+ep+qtz+cal	8.40	25
40 dn	Grt+qtz+wm+py+chl	9.50	249
65 dn	Am+ep+qtz		
70 dn	Am+ep+qtz (relict grains)	10.70	184
90 dn	Am+pl+ep+chl		
90-100 dn	Qtz+am+chl+ep		
105-125 dn	Meta-chert+meta-marls		
135 dn	Meta-basalt		
147 dn	Meta-chert		
165 dn	Meta-basalt		
250 dn	Spilitic basalt + ls blocks		
Traverse 12			
0-5 dn	Serpentinite		
0-5 dn	Am+ep+qtz		
15 dn	Am+pl (meta-basalt)		
17 dn	Act+chl+pl		

*Abbreviations after Kretz, 1983: ab-albite; act-actinolite; am-amphibole; ap-apatite; bt-biotite; cal-calcite; chl-chlorite; chr-chromite; cpx-clinopyroxene; ep-epidote; fld-feldspar; grt-garnet; hrn-hornblende; ol-olivine; op-opaque; opx-orthopyroxene; or-orthoclast; pl-plagioclase; px-pyroxene; py-pyrite; qtz-quartz; sp-spinel; spn-sphene; srp-serpentine; wm-white mica. SGS-subgrain size; N-number of grains counted; rd-red; bk-black; grn-green; up-above contact; dn-below contact; ls-limestone. †Horizontal distance; §Sample 132b; **Sample 150; ‡Overturned.

metamorphic sole is structurally omitted. The most complete section, representing what may be the original thickness of the metamorphic sole, is traverse 11 where a systematic reduction of grain size and increase in metamorphic grade is found toward the structural base of the ophiolite. At this location the base of the ophiolite is separated from the Copter Peak allochthon by 250 m of metamorphic rock.

1. Structural base of the Brooks Range ophiolite. The structural base of the Brooks Range ophiolite, immediately above the metamorphic sole, is characterized by foliated and fractured serpentinite with phacoidal blocks of porphyroclastic gabbro and peridotite that are encased in an orange, serpentine-rich fault-gouge. Where relict grains are preserved, they are recrystallized, undulose, bent, and fractured from the combined effects of crystal plastic and cataclastic flow-type deformational mechanisms. Intense veining and rodingization in this zone indicates the presence of high fluid pressures and temperatures.

At Siniktanneyak and Avan Hills minor granitic intrusions are found at the base of the Brooks Range ophiolite mantle sequence. Nelson and Nelson (1982) report small (1–5 m width), two-mica alaskite dikes (S-type granites) that intrude mylonitic dunite and peridotite immediately above the structural base of the Brooks Range ophiolite at Siniktanneyak. Further study of these dikes as part of this investigation found granodiorite, quartz-diorite, and tonalite compositions. Glassy quartz and albite(?) veins are also abundant. Most of the dikes are near-vertical, wedge-shaped prongs that narrow upward from the base of the ophiolite and pinch out only a few decimeters into the mantle sequence hanging wall. Some dikes have mylonitic textures, indicating that intrusion coincided with various stages of tectonic emplacement.

The concentration of granitic dikes along the structural base of the Brooks Range ophiolite mantle sequence, and the geometry and composition of the dikes, show that some partial melting was associated with thrusting of the hot ophiolite over felsic footwall rocks, as suggested in other settings by Clemens and Wall (1981). Harding et al. (1985) interpreted granitic dikes at the base of the Avan Hills peridotite as segregations of anatectic granitic melt. However, it is also possible that the dikes and particularly the veins are a product of solid-state diffusional processes related to solution mass transfer during metamorphism.

2. Amphibolite facies schists. Garnet-bearing amphibolite-grade schists and mylonites are widely distributed immediately below the structural base of mantle peridotites of the Brooks Range ophiolite. Locally these rocks are "welded" to the structural base of the ophiolite and occur as part of the hanging wall above fault-gouge zones. Compositional bands in the schists and mylonites are common and consist of alternating bands of mafic- and felsic-rich material. The most common rock types are garnet-bearing amphibolites and quartz-mica schists. A discontinuous zone of epidote-amphibolite occurs in traverse 9 (Table 1).

The most common mineral parageneses in amphibolite grade schists are hornblende-plagioclase ± garnet ± epidote ± quartz ± chlorite ± sphene ± orthoclase ± ilmenite. Accessory minerals include zircon, tourmaline, apatite, and iron oxides. The

relative proportions of hornblende to plagioclase, low abundance of quartz, relict igneous textures, and bulk chemistry suggest that the amphibolites have basaltic and gabbroic protoliths. At Avan Hills, Zimmerman and Frank (1982) reported the assemblage amphibole-plagioclase-quartz-orthopyroxene ± garnet. However, orthopyroxene has not been found in any samples collected from the base of the Avan Hills body during this investigation. Along Kisimilot Creek, at the structural base of Iyokrok, Boak et al. (1987) report metabasite with hornblende-plagioclase ± sphene ± ilmenite mineral assemblages. The metabasites are associated with more dominant para-amphibolites with layered mineral assemblages of hornblende-plagioclase-quartz ± biotite ± garnet ± ilmenite ± sphene. Distinctions of para-amphibolite are based on the occurrence in some nonmylonitized layers of rounded plagioclase porphyroclasts interpreted as relict clastic feldspars.

Banded quartz-mica schists are commonly interlayered with amphibole-schists. At Misheguk, quartz-mica schists are fine grained and consist of folded, alternating layers rich in quartz, plagioclase, mica, and graphite with garnet porphyroblasts. Other minerals of minor abundance include calcite, zircon, iron oxides, and pseudomorphs of chlorite and epidote after amphibole and plagioclase porphyroclasts. At Iyokrok, schists with plagioclase-quartz-biotite-garnet ± muscovite ± staurolite(?) mineral assemblages were reported (Boak et al., 1987). These assemblages most likely have sedimentary protoliths with low abundances of Al and K (little to no muscovite). Similar clay-rich sandstones, siltstones, and silica-rich claystones are documented in the Etivluk Group of the distal Arctic Alaska passive margin (Mayfield et al., 1983).

3. Greenschists. Greenschist-facies metamorphic rocks structurally underlie and are tectonically intermixed with amphibolites of the upper metamorphic sole. The basal contact of the metamorphic sole is transitional in places where greenschists grade in broken succession into underlying unmetamorphosed sequences. Banded schists and phyllites, similar to those interlayered with amphibolite, constitute the upper part of the greenschists. The intermixing is interpreted as a function of postmetamorphism faulting. The bulk of the schists have mineral assemblages of quartz-plagioclase-biotite ± garnet-chlorite-pyrite (metasediments) or plagioclase-quartz-actinolite-chlorite ± garnet ± calcite ± magnetite (metabasites). Metamorphic grade decreases structurally downward where banded schists grade into, and are locally mixed with, chlorite-epidote-actinolite-albite metavolcanics with decreasing degrees of recrystallization and increasing prominence of relict igneous mineral phases (clinopyroxene and plagioclase) and textures. Metasedimentary rocks include recrystallized radiolarian chert, fossiliferous marble, argillite, shale, and tuff.

Metamorphic textures in greenschists are dramatically influenced by original textures and grain size. Mylonitic fabrics, like those described above for amphibolite, are most common in rocks near the top of the greenschists. In underlying lower grade rocks, effects of dynamic metamorphism are less common and more localized. Original textures are generally preserved in units with limited dynamic recrystallization. Thermal metamorphism of these units produces blasto-intersertal, blasto-ophitic, and blasto-porphyritic recrystallized primary igneous textures. Relict pyroxenes commonly have actinolitic overgrowths and plagioclase is albitized and clouded by clay inclusions. Rocks with finer grained original textures are more recrystallized than adjacent coarser grained ones.

Unmetamorphosed Copter Peak volcanic and sedimentary rocks most commonly underlie the greenschist unit. Etivluk Group pelagic material, melange, and broken formation are also found at Misheguk and Siniktanneyak. Locally these units form gradational contacts with the overlying metamorphic sole. Although disrupted by faults, some contacts are gradual enough to trace a single unit into the metamorphic sole. The best example is the transition from greenschist to albite-actinolite hornfels to unmetamorphosed Copter Peak tholeiite exposed along traverses 7, 8, 9, and 11 (Fig. 2). These relations indicate that the protolith for the greenschist-facies metavolcanic rocks and intercalated sedimentary material is locally stratigraphically linked to the Copter Peak allochthon and Etivluk Group pelagics, which presently structurally underlie the Brooks Range ophiolite and its metamorphic sole.

Whole-rock geochemistry

Whole-rock geochemical analyses were conducted to test compositional similarities between rocks of the metamorphic sole and possible protolith units and to provide some constraints for the tectonic affinity of these units (Table 2). Major and rare earth elements were analyzed by inductively coupled plasma emission spectrometry (ICP) by methods of Walsh (1980) and Walsh et al. (1981). SiO_2 was determined by the methods of Shapiro and Brannock (1962). Other trace elements and some rare earth elements (REE) were analyzed with x-ray fluorescence (XRF) using methods similar to those of Thirwall and Burnard (1990).

The abundance of "immobile" trace elements and REE in amphibolite schists is very similar to that in mafic igneous material of the Copter Peak allochthon (Fig. 6). MORB-normalized immobile trace element abundances of mafic schists mimic the enriched N-type MORB trends of Copter Peak allochthon transitional tholeiites, which are shaded in Figure 6A. Mafic schist Y/Nb ratios (3.6–2.1, 2.55 ave.) and TiO_2 wt% (1.46 ave.) are also similar to the transitional tholeiite group of Copter Peak allochthon lava compositions. REE abundances in the mafic schists are also akin to more transitional-type lavas (Ce/Yb of 1.8–7.2) with slightly enriched LREE patterns (shaded in Fig. 6B). In general, the most immobile element abundances of mafic schists in the metamorphic sole overlap with those of Copter Peak allochthon lavas; both have transitional tholeiite compositions.

Mineral chemistry

Routine microprobe analyses of metamorphic minerals were conducted to provide temperature/pressure and protolith constraints. Minerals analyzed include amphibole, feldspar, garnet, mica, epidote, chlorite, sphene, and relict grains. The minerals were analyzed at University College of London using a

Cambridge Instruments Microscan V electron microprobe with a Link System energy dispersive system. Natural silicates and pure metals were used for standards.

Garnet. Almandine garnet is present but not abundant in most medium-grade metamorphic rocks within 50 m of the Brooks Range ophiolite contact. The crystals commonly occur as less than 0.8 mm idioblastic grains with inclusion-rich cores. Some inclusion trails are S-shaped and, along with snowball garnets, indicate synkinematic growth (Fig. 5 and 6). The geometry of inclusions within garnet cores are commonly discordant to fabrics encompassing the grains. This may be a function of static fabric development after garnet rotation or rotation not accompanied by garnet growth.

Garnet composition varies at a crystalline scale (core to rim of zoned crystals) and with distance from the Brooks Range ophiolite contact (Table 3). Almandine-rich garnet of Alm 60.5–62.3, Sp 5.5–6.7 and Pyr 27.2–30.0 is found in metagabbro 3 m above the basal fault contact of the ophiolite along traverse 1. At the emplacement fault contact garnets are typically reversely zoned with an increase in spessartine toward crystal rims. Below the fault, garnet has prograde zoning patterns (spessartine-rich cores) and exhibit an increase in Mn from Sp 17.8 to Sp 24.3 and decrease in Mg from Pyr 12.9 to Pyr 4.9 with increasing distance from the contact. The presence of garnet throughout the metamorphic sole is strongly controlled by rock chemistry, particularly layers rich in Mn (traverse 11, Table 1).

Amphibole. Amphibole compositions vary systematically throughout the metamorphic sole and within individual samples. In the upper section of the metamorphic sole poikiloblastic porphyroblasts 0.5–2.2 mm in diameter of magnesio- and tschermakitic-hornblende are mantled by smaller neoblasts of actinolitic- and magnesio-hornblende. Si and Ti cation abundances (23 oxygen basis), which reflect grade of metamorphism (Raase, 1974; Miyashiro, 1975), indicate porphyroblasts formed at lower amphibolite facies (6.40–6.72 Si and 0.09–0.13 Ti) and neoblasts at greenschist-amphibolite transition facies (6.93–7.30 Si and 0.03–0.07 Ti). The reverse of this trend is found in the underlying greenschists and epidote-amphibolites: fine-grained, foliation-forming amphiboles are usually a higher temperature variety (6.67–6.90 Si and 0.05–0.08 Ti) than amphibole porphyroblasts that are actinolitic.

Amphibole compositions provide only general constraints for interpreting the pressure and temperature of metamorphism. Pressure-sensitive relations of Al^{vi} to Si and Al^{iv} in metamorphic

Figure 3. Structural cross sections through the Brooks Range ophiolite at Misheguk Mountain and Avan Hills. Positions of lines A–B and C–D are shown in Figure 2. Line through Peak 1217m is at east edge of Figure 2 near traverse 12. East-west line through Avan Hills is field modified from Curtis et al. (1984). Surface dip is shown by lines intersecting the topographic surface: these lines indicate dip directions of various planar structural markers (magmatic flow fabric in ophiolite, foliation planes in metamorphic sole, and bedding planes in subophiolite thrust sheets). Metamorphic sole is black with arrows indicating direction of increasing metamorphic grade. Ultramafic portions of ophiolite are shaded. Abbreviations are as in Figure 2.

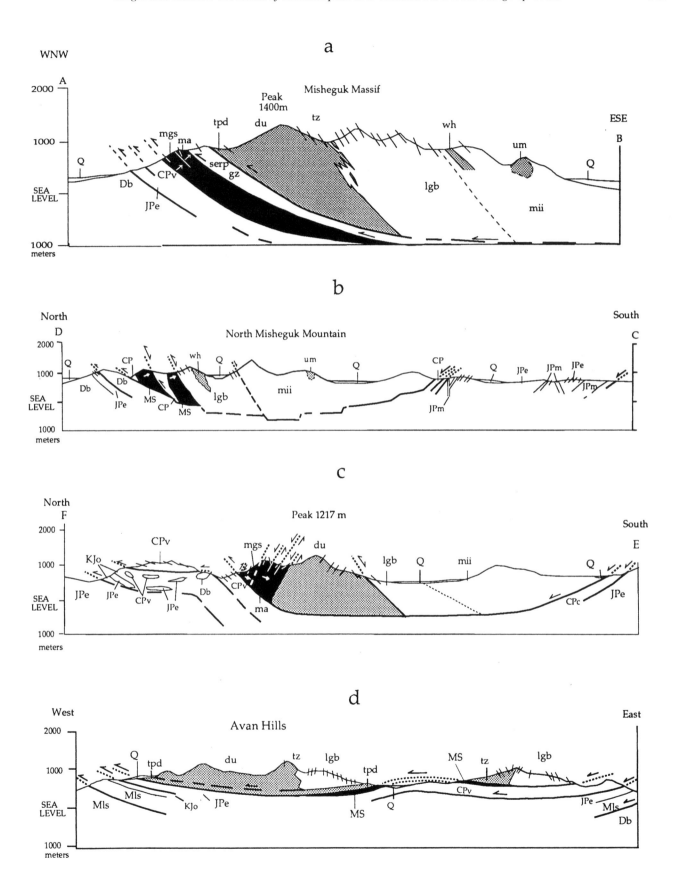

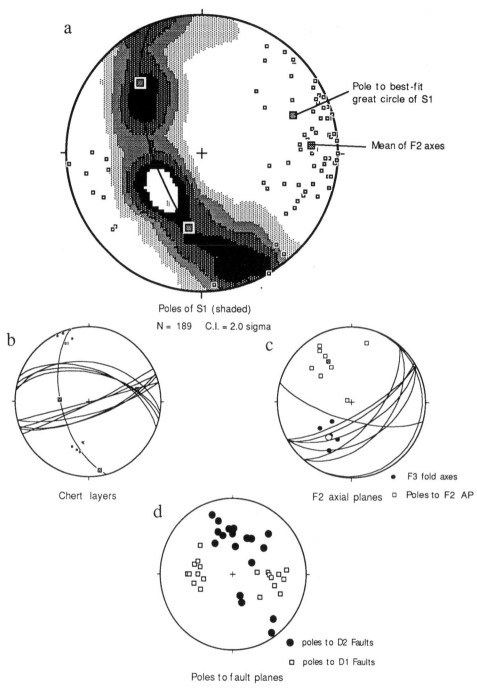

Figure 4. Lower hemisphere equal-area stereographs of structural features in the metamorphic sole. a, shaded contours of poles to S1 foliation planes (Kamb contour method) and F2 fold axes (small open squares). Pole to best-fit great circle of S1 foliation (large open circle) is near mean direction of F2 axes (large shaded square). b, Folded chert layers in transition between lower greenschist-facies metamorphic sole and the Copter Peak complex. Bedding planes are shown as great circles and poles to planes. Fold orientation is similar to foliation planes of Figure 4A. c, F2 axial planes plotted as great circles and poles to planes with F3 fold axes. Mean directions of poles to F2 axial planes are shown in shaded square, and open circle for F3 fold axes. d, Poles of fault planes in the metamorphic sole. Faults associated primarily with D1 were most likely near horizontal before folding along a north-south axis. Faults associated with D2 dominantly dip south.

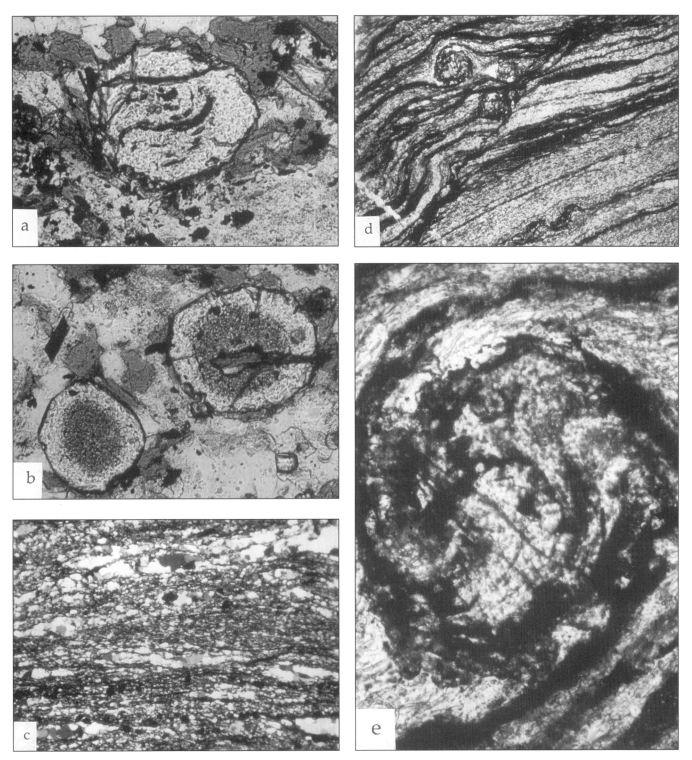

Figure 5. Photomicrographs of microstructures of the metamorphic sole related to synkinematic growth indicators in quartz-mica schist. a, Helecitic garnet with curved inclusion of previous foliation (0.5 mm field of view). b, Idioblastic garnets indicative of late static crystal growth (0.5 mm field of view). c, Polycrystalline quartz ribbons with polygonized subgrains in polarized light (3 mm field of view). Quartz ribbon characteristics mostly fit the type 2A classification of Boullier and Bouchez (1978). Locally a grain shape preferred orientation of type 2B occurs. d, Snowball garnet with graphitic bands. Layer-parallel S1 foliation is disrupted by F2 folds and postmetamorphic shear and tension fractures (3 mm field of view). e, Detail of snowball garnet in A″ (0.5 mm field of view).

TABLE 2. METAMORPHIC SOLE WHOLE-ROCK CHEMISTRY

Sample	73	79	80	115	124a	124b	132b	134a	150	154	170b
Major Elements											
SiO_2	47.48	47.81	48.36	46.73	46.21	46.39	47.62	50.62	45.26	42.64	44.99
TiO_2	1.80	1.84	1.87	2.56	1.26	1.23	1.26	1.01	0.65	0.92	0.11
Al_2O_3	14.32	14.61	13.65	13.41	13.06	13.95	15.54	17.16	12.37	15.53	2.74
FeO	8.79	7.91	0.82	7.04	6.37	5.90	8.84	6.19	3.70	3.37	2.61
Fe_2O_3	2.55	4.64	4.57	7.80	5.44	5.43	2.85	2.57	1.18	4.98	3.19
MnO	0.17	0.16	0.17	0.21	0.17	0.24	0.40	0.15	0.11	0.13	0.11
MgO	8.57	7.17	7.09	6.30	7.10	7.21	8.61	5.96	3.29	5.70	25.91
CaO	10.42	11.41	10.09	8.31	16.84	14.81	7.51	6.61	14.85	15.01	12.64
Na_2O	2.19	2.34	2.87	4.01	0.80	1.48	2.62	5.31	0.42	2.08	0.07
K_2O	0.41	0.41	0.67	0.18	0.41	0.48	1.13	0.11	2.37	0.70	0.00
P_2O_5	0.17	0.16	0.15	0.23	0.14	0.18	0.23	0.12	0.12	0.13	0.03
LOI	3.09	2.06	1.47	3.31	1.64	2.11	2.94	4.47	15.67	7.99	6.98
Total	99.96	100.52	99.78	100.09	99.48	99.41	99.55	100.28	99.99	99.18	99.38
Trace Elements											
Li	17	8	9	14	0	44	28	39	60	17	2
Sc	41	44	43	43	41	40	31	32	14	35	55
Ti	10791	11031	11211	15347	7554	7374	7554	6055	3897	5515	659
V	315	337	353	434	233	300	260	247	115	216	139
Cr	293	313	120	77	234	245	313	59	70	164	1298
Co	45	55	56	52	59	65	60	41	24	46	57
Ni	104	130	79	76	86	86	230	66	62	87	513
Cu	145	118	56	238	10	92	16	30	24	41	6
Zn	93	94	97	123	96	89	129	80	88	73	34
Rn	12	12	14	6	10	12	24	5	112	15	0
Sr	194	194	202	105	330	172	317	410	253	217	8
Y	23	23	26	33	21	21	34	18	23	15	2
Zr	58	25	16	36	32	24	49	25	113	24	5
Nb	10	10	11	13	10	10	11	5	13	9	6
Ba	169	224	608	75	121	158	903	202	446	142	7
REE											
La	7.60	12.20	13.50	9.60	13.40	13.00	34.10	11.20	27.70	10.90	1.30
Ce	17.70	15.70	17.80	25.20	13.00	11.80	49.30	11.60	56.20	8.60	0.20
Pr	2.60	2.60	2.80	3.80	2.00	1.90	6.40	2.10	7.00	1.50	0.10
Nd	12.80	12.00	12.90	17.50	9.20	9.10	26.10	9.90	26.20	6.70	0.50
Sm	3.60	3.40	3.80	5.00	2.80	2.70	6.10	2.90	5.50	2.00	0.20
Eu	1.30	1.30	1.40	1.80	1.10	1.10	1.60	0.90	1.20	0.80	0.10
Gd	4.50	4.70	4.80	6.40	3.60	3.80	6.60	3.20	5.00	2.60	0.30
Dy	4.60	4.60	5.00	6.60	4.00	4.20	6.40	3.50	4.30	2.80	0.40
Ho	0.90	0.90	1.10	1.40	0.80	0.90	1.30	0.70	0.90	0.60	0.10
Er	2.60	3.00	3.00	3.80	2.50	2.30	3.90	2.30	2.50	1.70	0.30
Yb	2.10	2.20	2.60	3.00	2.20	1.00	3.20	2.00	2.20	1.50	0.20
Lu	0.30	0.30	0.40	0.40	0.40	0.10	0.50	0.40	0.30	0.20	0.00

amphiboles (Raase, 1974; Fleet and Barnett, 1978) indicate pressures for the entire metamorphic sole of around or slightly above 5 Kb. Ti-rich amphiboles increase in abundance upward toward the Brooks Range ophiolite contact. This trend is most likely a function of increased temperature rather than compositional variation, due to the presence of other coexisting Ti-bearing phases like sphene and ilmenite.

Plagioclase. Plagioclase occurs throughout the metamorphic sole as 0.5–2.0 mm porphyroblasts and less than 0.5 mm xenoblastic grains. Original plagioclase compositions, particularly of porphyroblasts, are difficult to determine due to secondary effects of albitization and alteration to cloudy aggregates of white mica, epidote and clays. Least altered parts of plagioclase grains show some compositional variation with distance from the Brooks Range ophiolite contact. More calcic plagioclase of An (46–24) is found within 20 m of the Brooks Range ophiolite contact, although the range overlaps with that of plagioclase found 20–50 m from the contact (An 36–25). Late albite is present in most samples and is commonly associated with prehnite in veins that truncate all other deformational fabrics.

Geothermometry

Geothermometric estimates were obtained using coexisting phases of garnet, amphibole, and biotite. High Mn contents in

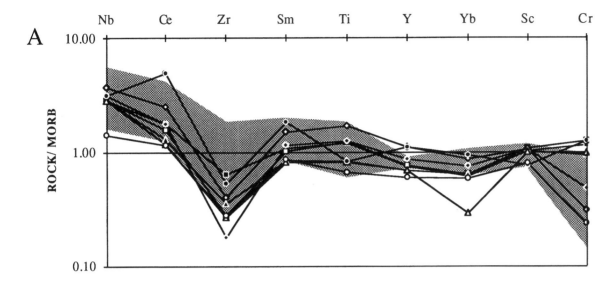

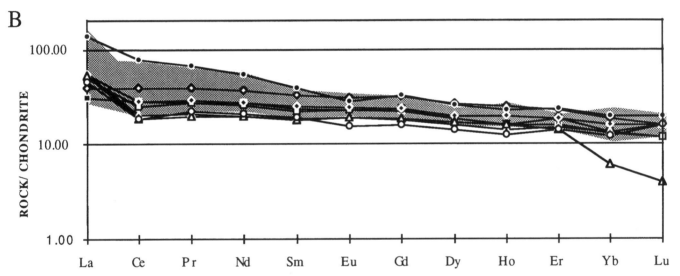

Figure 6. Geochemical comparison between metamorphic sole mafic schist (metabasalt) and Copter Peak basalt (shaded). A, MORB-normalized trace element abundances. B, chondrite-normalized REE abundances. Shaded area is trend of eight Copter Peak basalt analyses. Trends are consistent with T-Type MORB lavas.

most garnets (12.0 mole % average with values as much as 24.3 mole %) precluded obtaining estimates from most of the metamorphic units. The 5-Kb pressure estimate obtained from relations of Alvi to Si and Aliv in metamorphic amphiboles (Raase, 1974; Fleet and Barnett, 1978) was assumed for the geothermometric model. Chemical analysis of homogenous pairs of low-Mn garnet and biotite from a metapelite 40–50 m below the ophiolite were used with TWEEQU thermobarometry software (Berman, 1991) to yield temperature estimates of 550°C. The mixing models of Berman (1990) for garnet and McMullin et al. (1991) for biotite account for nonideal behavior in the solid solutions. Coexisting phases of low-Mn garnet and amphibole were also used from mylonites 4–5 m below the structural base of the

ophiolite. These rocks yield temperature estimates of 500–560°C using the garnet-hornblende geothermometer of Graham and Powell (1984).

Boak et al. (1987) estimated temperatures of 560°C for a garnet rim–biotite pair in amphibolite below the Brooks Range ophiolite at Iyokrok. This estimate applies only to the medium-grade metamorphic rocks "a few tens of meters" from the ophiolite contact. Partial melts of material thrust beneath the Brooks Range ophiolite at Avan Hills and Siniktanneyak indicate that local temperatures along the base of the ophiolite must have been greater than 650°C, which is the intersection between the 5-Kb line and the melting curve for water-saturated peraluminous granite (Clemens and Wall, 1981).

Compositional variation in mineral chemistry between the low- to medium-grade and higher grade metamorphic mineral assemblages (systematic changes in plagioclase, hornblende, and garnet composition upward) indicate a general increase in metamorphic grade upward toward the base of the Brooks Range ophiolite. However, the detailed thermal evolution of the "inverted" metamorphic sequence is unclear. The occurrence of retrograde metamorphism and metasomatism only in the highest

grade rocks nearest the ophiolite contact, and the presence of blocks of amphibolite within greenschists, suggest the amphibolite unit may have formed before the lower grade rocks it now overlies. It is also possible that mixing may have resulted entirely from post-peak metamorphism faulting.

Age of metamorphism

K-Ar and $^{40}Ar/^{39}Ar$ radiometric age analyses were conducted on hornblende from garnet amphibolite collected 5 m below the emplacement thrust of the Brooks Range ophiolite at Mishcguk. Hornblende and plagioclase layers comprise most of the rock in equal amounts with minor garnet and sphene. Compositional layers are disrupted by veins associated with retrograde mineral assemblages (epidote + chlorite + white mica + albite). Porphyroblasts (as large as 1.2 mm) of magnesio- and tschermakitic-hornblende are locally mantled with smaller neoblasts of magnesio-hornblende, which are more enriched in Si and depleted in Ti than the porphyroblasts. Plagioclase porphyroblasts range in composition from An 28 to An 38 and are commonly saussuritized. Almandine garnet with Mn-rich rims is distributed unevenly in the sample and is most closely associated with chlorite- and white mica–rich layers. The low Fe content of the hornblende renders it most suitable for $^{40}Ar/^{39}Ar$ dating (O'Nions, 1969).

Hornblende from sample 132b (Table 2) yields an $^{40}Ar/^{39}Ar$ plateau age of 165 ± 3 Ma (Fig. 7). This age is similar to other plateau ages of 164–169 Ma of hornblende from the structural base of Avan Hills and 163–171 Ma at the base of the Asik (Wirth et al., 1987). These ages are significantly older than hornblende K/Ar ages from the sample 132 (157 ± 5 Ma) at Misheguk and those reported from Iyokrok of 153–154 ± 5 Ma by Boak et al. (1987). Age discrepancies between hornblende separates, some from the same sample, may be due to difficulties of the K-Ar method in (1) estimating the small amounts of K present in the amphiboles used and (2) detecting Ar loss.

Cooling ages from the metamorphic sole are concordant with hornblende $^{40}Ar/^{39}Ar$ plateau ages from high-level gabbro and diorite at Misheguk (Harris, 1998), Asik, and Avan Hills (Wirth and Bird, 1992) of 163–171 Ma. Plagiogranite from Siniktanneyak yielded a U-Pb age of 170 ± 3 Ma (Moore et al., 1993). These ages are also concordant with a white mica $^{40}Ar/^{39}Ar$ Ar plateau age of 171 ± 4 Ma from Ruby Ridge in the metamorphic core of the Brooks Range (Christiansen and Snee, 1994). Age concordance may be associated with a common tectonic uplift event that cooled both the igneous and metamorphic rocks of the Brooks Range ophiolite and the continental material beneath them. An alternative explanation may be that conductive cooling of the ophiolite slab coincided with high pressure metamorphism in the Schist belt.

The cooling history of subophiolite metamorphic rocks has implications for the tectonic processes and uplift rates associated with ophiolite emplacement. $^{40}Ar/^{39}Ar$ age data from metamorphic minerals with different closure temperatures provide independent time-temperature control points that help constrain the timing of

TABLE 3. GARNET COMPOSITIONS*

Number	Distance	Alm	Gro	Py	Sp	C/R
78.1	1m	0.619	0.034	0.280	0.067	-
78.2		0.614	0.048	0.281	0.057	
78.3		0.623	0.049	0.272	0.055	-
78.6		0.605	0.033	0.300	0.062	-
78.7		0.613	0.051	0.280	0.057	-
78.8		0.618	0.034	0.289	0.060	-
83A.3	1m	0.357	0.214	0.028	0.401	C
83A.5		0.340	0.275	0.000	0.385	Micro
83A.7		0.539	0.204	0.086	0.170	-
83A.8		0.402	0.336	0.016	0.247	C
83A.9		0.547	0.246	0.038	0.168	R
132SP.1	4-5m	0.619	0.075	0.129	0.178	C
132SP.2		0.395	0.099	0.222	0.284	Int
132SP.3		0.626	0.061	0.125	0.187	R
132SP.9		0.649	0.044	0.136	0.649	Micro
132B.1		0.537	0.126	0.199	0.138	C
132B.1		0.555	0.121	0.164	0.159	R
132B.2		0.488	0.210	0.159	0.143	C
132B.2		0.535	0.167	0.168	0.130	R
132B.3		0.560	0.153	0.195	0.091	C
132B.3		0.547	0.160	0.194	0.098	R
9C.3	10m	0.535	0.172	0.028	0.265	-
9C.8		0.491	0.156	0.023	0.329	C
9C.9		0.531	0.157	0.035	0.277	R
9C.10		0.498	0.161	0.000	0.341	C
9C.11		0.515	0.166	0.032	0.288	R
9C.15		0.529	0.170	0.029	0.271	C
9C.17		0.540	0.175	0.034	0.251	R
150DP.1	20-50m	0.663	0.086	0.051	0.201	C
150DP.5		0.508	0.215	0.034	0.243	C
150DP.13		0.591	0.164	0.049	0.196	Inc. in 14
150DP.14		0.616	0.120	0.045	0.218	C
150DP.14		0.707	0.070	0.073	0.150	R
123B.1a	30-50m	0.589	0.073	0.085	0.253	Int
123B.1b		0.652	0.063	0.100	0.184	C
123B.1c		0.593	0.072	0.081	0.255	R
123B.2a		0.511	0.087	0.065	0.336	C
123B.2b		0.577	0.083	0.070	0.270	Int
123B.2c		0.645	0.063	0.093	0.199	R
123B.3a		0.548	0.084	0.065	0.304	C
123B.3b		0.602	0.067	0.081	0.249	Int
123B.3c		0.649	0.070	0.092	0.190	R
123B.15a		0.554	0.175	0.076	0.195	R

*Alm = Almandine; Gro = grossularite; Py = pyrope; Sp = spessartine; C/R = core/rim; Micro = small grain; Int = intermediate between C/R; Inc. = inclusion.

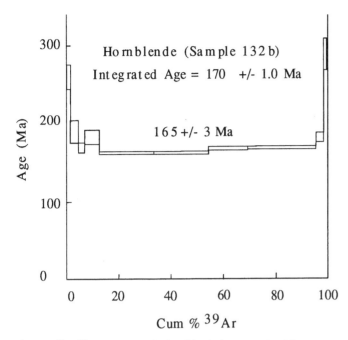

Figure 7. $^{40}Ar/^{39}Ar$ age spectra for hornblende from sample 132b (traverse 7 at 4–5 m below the structural base of Brooks Range ophiolite). Plateau age has slight saddle-shaped spectrum, which may be a function of excess argon in unstable sites within the mineral grains.

various pressure-temperature metamorphic conditions. Age spectra from $^{40}Ar/^{39}Ar$ analyses of hornblende, biotite, and K-feldspar from the metamorphic sole of the Brooks Range ophiolite provide constraints for effective closure temperatures of these minerals. Plotting these temperatures against time yields an empirical cooling curve for subophiolite metamorphism (Harris, 1992; Wirth and Bird, 1992). Maximum temperatures for the cooling curve are provided by Ar closure temperatures for hornblende, which are mostly controlled by cooling rate, composition, and grain size. An intermediate-high cooling rate of 10–15°C/Ma is assumed for the metamorphic sole based on differences in age between subophiolite metamorphism, basement uplift, and deposition of detritus from high-pressure metamorphic rocks (Till et al., 1988). At these rates, the argon closure for hornblende is about 530 ± 40°C (Harrison and McDougall, 1980). The overlap of closure temperatures and geothermometric estimates from garnet-amphibole pairs within the dated sample and from garnet-biotite pairs several decimeters below the base of the ophiolite, indicate that the $^{40}Ar/^{39}Ar$ age is close to the age of metamorphism.

ORIGIN AND TECTONIC EVOLUTION

The metamorphic sole of the Brooks Range ophiolite represents the initial plate boundary suture between an arc-related upper plate and the underthrust continental margin of Arctic Alaska. The plate boundary moved to lower structural levels as the Brookian orogen progressed, which caused uplift and cooling of the metamorphic sole suture. In this process the metamorphic sole was structurally modified by postmetamorphic deformation.

However, the metamorphic sole preserves a general record of its origin, conditions of dynamothermal metamorphism, and tectonic evolution.

Exposures of the metamorphic sole have a consistent succession of garnet-bearing amphibolite and quartz-mica mylonitic schists that decrease in metamorphic grade downward into greenschist facies and unmetamorphosed protoliths. The protolith of amphibolite grade mylonite that is "welded" to the structural base of the Brooks Range ophiolite document the crustal composition of footwall rocks that first underthrust the hot ophiolite as it was detached and moved upward. Geochemical similarity between mafic amphibolites and Copter Peak allochthon basalt is interpreted as evidence that the basalt was one of the first crustal units to underthrust the ophiolite after detachment. Mn-rich garnet in pelitic material are consistent with a chert and psammitic sedimentary protolith that is abundant in the Copter Peak allochthon and Etivluk Group. Chemical, age, and stratigraphic links between the metamorphic sole and distal parts of the Arctic Alaska continental margin indicate that the Brooks Range ophiolite was formed and emplaced very near to the continental margin.

Boudier and Nicolas (1988) propose a "compressed" spreading ridge tectonic model for the detachment of Oman-type ophiolites. If the Brooks Range ophiolite and other ophiolites formed at a spreading ridge that overthrust itself, metamorphic soles would be composed of rocks similar in composition and age to the ophiolites. However, differences in geochemistry and age between ophiolite and metamorphic sole rule out this tectonic scenario for the Brooks Range.

A more likely scenario that is consistent with geochemical, structural, and petrologic data is that the Brooks Range ophiolite formed above a young subduction zone that developed near the Arctic Alaska continental margin (Fig. 8). As the continental margin was pulled into the subduction zone it underthrust the hot ophiolite (young arc). Thrust stacking beneath the leading edge of the ophiolitic upper plate may have been what caused the ophiolite to detach from its roots, forming a passive roof thrust above the developing orogen.

Strain associated with the structural emplacement of the Brooks Range ophiolite was localized in the metamorphic sole, where strain rates may have been as much as 10^{-10} s^{-1} to 10^{-9} s^{-1}. Ductile strain produced blastomylonitic fabrics where crystal plastic flow was locally overprinted by thermal annealing. Flow stress estimates, based on progressive quartz subgrain size reduction, yield a stress gradient of 2–3 MPa/m in the metamorphic sole.

Temperature sensitive chemical changes in plagioclase, amphibole, and garnet overgrowths indicate that metamorphic grade increases upward toward the ophiolite. Variations in prograde and retrograde patterns exist. Most crystal growth was synkinematic at temperatures near that predicted for hornblende Ar retentivity. Geothermometric estimates from garnet-bearing amphibolite near the ophiolite yield minimum temperatures of 500–560°C. These temperatures are consistent with the type of crystal plastic deformational mechanisms (dislocation glide and creep) observed in microstructures. Assuming an initial thickness of around 500–600m, from amphibolite-grade schist to unmetamorphosed basalts, a minimum thermal gradient of 500–600°C/km is predicted. This estimate does not take into account the higher temperatures necessary (650–700°C) to produce partial melts along the base of the ophiolite, which approach the maximum static temperature for contact metamorphism.

An estimate of the maximum temperature attainable by linear flow from a hot hanging wall of peridotite (T^h=1,100–1,200°C) into a cool footwall (T^f=0–100°C) is given by 0.5(T), where T = T^h – T^f (Jaeger, 1961; Spray, 1984). Applying this constraint to probable conditions at the base of the ophiolite yields maximum temperatures of 500–600°C, which are not high enough to produce partial melts. Frictional heating may account for some excess heat, but its input is limited by the low strength of the hot hanging wall and weak material in the footwall (Spray, 1984). Another possible explanation are the "iron board" models proposed by Pavlis (1986) and Smith (1988). These models account for the interplay of thermal and mechanical processes associated with a hot, moving ophiolite slab over wet upper crustal material. The models predict that preheating, either of the footwall or hanging wall by the leading edge of the slab can raise temperatures an additional 200°C and cause local melting. Each viable model is constrained by the temperatures recorded in the metamorphic sole, which limits the time available for the Brooks Range ophiolite to cool before coming into contact with the Copter Peak allochthon to a few million years.

The inverted metamorphic gradient, overlap between ages of ophiolite cooling and subophiolite metamorphism, and structural and petrologic associations indicate that the metamorphic sole formed during tectonic emplacement of the Brooks Range ophiolite while it was still hot over mafic igneous and sedimentary rocks of the Copter Peak allochthon and perhaps other continental margin units. The timing of this event (164–169 Ma) provides a minimum age for ophiolite genesis and maximum age for the initiation of the western Brookian orogen. The development of the metamorphic sole during ophiolite emplacement documents the thermal and mechanical conditions of the event. The narrow time and temperature windows between ophiolite genesis and emplacement, implies limited travel; that the Brooks Range ophiolite was near its birth place at the time of subophiolite metamorphism and tectonic emplacement. The general synchroneity between cooling ages of the ophiolite and its metamorphic sole is a common feature of many Oman-type ophiolites (Spray, 1984). This feature is a characteristic of ophiolite genesis by pericollisional extension (Harris, 1992).

Incorporation of the Brooks Range ophiolite and Copter Peak allochthon into the Brookian fold-thrust zone resulted in cooling of the metamorphic sole. This event coincided with the first evidence of north-directed orogenic (Brookian) sedimentation (Mayfield et al., 1983) and of metamorphism associated with the orogen (Christiansen and Snee, 1994; Gottschalk and Snee, this volume, Chapter 13). The transition from ductile to more brittle deformation mechanisms follow within 5–10 m.y. of peak-metamorphism as evidenced by the cooling ages of mica schists in the metamorphic sole (Wirth and Bird, 1992). Continental underthrusting beneath the ophiolite produced a hinterland belt of high-pressure blueschists that was subsequently overprinted by a greenschist-grade event. Metamorphic cooling occurred from 130 to 100 Ma (Armstong et al., 1986; Gottschalk, 1990; Christiansen and Snee, 1994; Gottschalk and Snee, this volume, Chapter 13).

Uplift and unroofing of the hinterland Schist belt may be responsible for postemplacement fragmentation and remobilization of the Brooks Range ophiolite nappe and its metamorphic sole. This event is documented by the fault juxtaposition of amphibolite with greenschist-facies rocks within the metamorphic sole, and the stacking of these rocks above continental margin sedimentary rocks with conodont alteration values of only 1–2 (Harris, A. G., et al., 1987). The final phase of brittle deformation recorded by the metamorphic sole may be associated with strike-slip faulting.

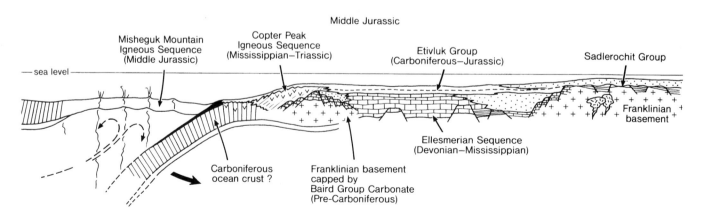

Figure 8. Middle Jurassic schematic reconstruction of the Arctic Alaska passive continental margin. The Misheguk Mountain Igneous Sequence becomes the Brooks Range ophiolite as the Brookian continental margin is thrust beneath it. The Copter Peak Igneous Sequence is interpreted in part to represent a rift-related volcanic pile similar to many other volcanic margins (see Mutter et al., 1988). The width of the intracratonic basin sequence is underestimated for illustration purposes. The dashed line marks the breakup unconformity.

ACKNOWLEDGMENTS

Support for this research was provided by grants from the American Chemical Society–Petroleum Research Fund, U.S. Geological Survey, U.S. Bureau of Mines, and the Rice University/University of Alaska Industrial Research Program. I wish to thank Mike Audley-Charles, Robert Hall, Helen Lang, Roger Mason, Mike Searle, and David Stone for direction and encouragement. Cal Wescott, Jeff Foley, and Tom Light provided field support. Paul Ballantyne and J. Walsh assisted with whole-rock geochemical analyses. Dengliang Gao assisted with piezometric measurements. Helen Lang, Rick Gottschalk, Jay Zimmerman, Aley El-Shazly, and Brian Patrick provided careful reviews of various drafts of the manuscript. Access to the Noatak Wilderness was granted by the U.S. Park Service.

REFERENCES CITED

Alexander, R. A., 1990, Structure and lithostratigraphy of the Kikiktat Mountain area, central Killik River Quadrangle, north-central Brooks Range, Alaska [M.A. thesis]: Fairbanks, Alaska, University of Alaska, 248 p.

Armstrong, R. L., Harakal, J. E., Forbes, R. B., Evans, B. W., and Thurston, S. P., 1986, Rb-St and K-Ar study of the metamorphic rocks of the Seward Peninsula and southern Brooks Range, Alaska, *in* Evans, B. W., and Brown, E. H., eds., Blueschists and eclogites: Geological Society of America Memoir 164, p. 185–203.

Barker, F., Jones, D. L., Budahn, J. R., and Coney, P. J., 1988, Ocean plateau-seamount origin of basaltic rocks, Angayucham Terrane, central Alaska: Journal of Geology, v. 96, p. 368–374.

Berman, R. G., 1990, Mixing properties of Ca-Mg-Fe-Mn garnets: American Mineralogist, v. 75, p. 328–344.

Berman, R. G., 1991, Thermobarometry using multiequilibrium calculations: a new technique, with petrologic applications: Canadian Mineralogist, v. 29, p. 833–855.

Boak, J. L., Turner, D. L., Henry, D., Moore, T. E., and Wallace, W. K., 1987, Petrology and K-Ar ages of the Misheguk igneous sequence: An allochthonous mafic and ultramafic complex and its metamorphic aureole, western Brooks Range, Alaska, *in* Tailleur, I., and Weimer, P., eds., Alaskan North Slope geology: Society of Economic Paleontologists and Mineralogists, Pacific Section, Publication 50, p. 737–745.

Boudier, F., and Nicolas, A., 1988, The ophiolites of Oman: Tectonophysics, v. 151, 401 p.

Boullier, A.-M., and Bouchez, J.-L., 1978, Le quartz en rubans dans les mylonites: Bulletin de la Société géologique de France, v. 7, p. 253–262.

Box, W., 1985, Early Cretaceous orogenic belt in northwestern Alaska: internal organization, lateral extent, and tectonic interpretation, *in* Howell, D. G., ed., Tectonostratigraphic terranes of the Circum-Pacific region: Houston, Texas, Circum-Pacific Council for Energy and Mineral Resources Earth Science Series, v. 1, p. 137–147.

Christiansen, P. P., and Snee, L. W., 1994, Structure, metamorphism, and geochronology of the Cosmos Hills and Ruby Ridge, Brooks Range Schist belt, Alaska: Tectonics, v. 13, p. 193–213.

Clemens, J. D., and Wall, V. J., 1981, Origin and crystallization of some peraluminous (S-type) granitic magmas: Canadian Mineralogist, v. 19, p. 111–131.

Coleman, R. G., 1977, Ophiolites—Ancient oceanic lithosphere?: New York, Springer-Verlag, 299 p.

Coleman, R. G., 1984, The diversity of ophiolites: Geologie en Mijnbouw, v. 63, p. 141–150.

Curtis, S. M., Ellersieck, I., Mayfield, C. F., and Tailleur, I. L., 1984, Reconnaissance geologic map of the southwestern Misheguk Mountain Quadrangle, Alaska: U.S. Geological Survey Open-File Report OF 82-612, scale 1:63,360.

Dumoulin, J. A., and Harris, A. G., 1987, Cambrian through Devonian carbonate rocks of the Baird Mountains, western Brooks Range, Alaska: Geological Society of America Abstracts with Programs, v. 19, p. 373–374.

Ellersieck, I., Curtis, S. M., Mayfield, C. F., and Tailleur, I. L., 1982, Reconnaissance geologic map of south-central Misheguk Mountain Quadrangle, Alaska: U.S. Geological Survey Open-File Report OF 82-612, scale 1:63,360.

Fleet, M. E., and Barnett, F. L., 1978, $Al^{IV}Al^{VI}$ partitioning in calciferous amphiboles from the Frood Mine, Sudbury, Ontario: Contributions to Mineralogy and Petrology, v. 16, p. 527–532.

Ghent, E. D., and Stout, N. Z., 1981, Metamorphism at the base of the Samail ophiolite, SE Oman Mountains: Journal of Geophysical Research, v. 86, p. 2557–2572.

Gottschalk, R. R., 1990, Structural evolution of the Schist belt, south-central Brooks Range, Alaska: Journal of Structural Geology, v. 12, no. 4, p. 453–469.

Gottschalk, R. R., and Oldow, J. S., 1988, Low-angle normal faults in the south-central Brooks Range fold and thrust belt, Alaska: Geology, v. 16, p. 395–399.

Graham, C. M., and Powell, R., 1984, A garnet-hornblende geothermometer: calibration, testing, and application to the Pelona schist, southern California: Journal of Metamorphic Geology, v. 2, p. 13–31.

Harding, D. J., Wirth, J. M., Bird, J. M., and Shelton, D. H., 1985, Ophiolite emplacement, western Brooks Range, northern Alaska: Eos (Transactions, American Geophysical Union), v. 46, p. 1129.

Harris, A. G., Lane, H. R., Tailleur, I. L., and Ellersieck, I. F., 1987, Conodont thermal maturation patterns in Paleozoic and Triassic rocks, northern Alaska, *in* Tailleur, I., and Weimer, P., eds., Alaskan North Slope geology: Society of Economic Paleontologists and Mineralogists, Pacific Section, Publication 50, p. 181–194.

Harris, R. A., 1987, Structure and composition of sub-ophiolite metamorphic rocks, western Brooks Range ophiolite belt, Alaska: Geological Society of America Abstracts with Programs, v. 19, p. 38.

Harris, R. A., 1992, Peri-collisional extension and the formation of Oman-type ophiolites in the Brooks Range and Banda arc, *in* Parsons, L. M., Murton, B. J., and Browning, P., eds., Ophiolites and their modern oceanic analogues: Geological Society of London, Special Publication No. 60, p. 301–325.

Harris, R. A., 1995, Geochemistry and tectonomagmatic origin of the Misheguk Massif, Brooks Range ophiolite belt, Alaska: Lithos, v. 35, p. 1–25.

Harris, R. A., 1998, The Brooks Range ophiolite, NW Alaska: U.S. Geological Survey Professional Paper (in press).

Harris, R. A., Stone, D. B., and Turner, D. L., 1987, Tectonic implications of paleomagnetic and geochronologic data from the Yukon-Koyukuk basin, Alaska: Geological Society of America Bulletin, v. 99, p. 362–375.

Harrison, T. M., and McDougall, I., 1980, Investigations of an intrusive contact, northwest Nelson, New Zealand—1. Thermal, chronological and isotopic constraints: Geochimica et Cosmochimica Acta, v. 44, p. 1985–2003.

Jaegar, J. C., 1961, The coding of irregularly shaped igneous bodies: American Journal of Science, v. 259, no. 10, p. 721–734.

Jamieson, R. A., 1986, P-T paths from high temperature shear zones beneath ophiolites: Journal of Metamorphic Geology, v. 4, p. 3–22.

Jones, D. L., Silberling, N. J., Berg, J. C., and Plafker, G., 1981, Tectonostratigraphic terrane map of Alaska: U.S. Geological Survey Open-File Report 81-792, scale 1:2,500,000.

Kretz, R., 1983, Symbols for rock-forming minerals: American Mineralogist, v. 68, p. 277–279.

Kronenberg, A. K., and Shelton, G. L., 1980, Deformation microstructures in experimentally deformed Maryland diabase: Journal of Structural Geology, v. 2, p. 341–353.

Martin, A. J., 1970, Structure and tectonic history of the western Brooks Range De Long Mountains and Lisburne Hills, northern Alaska: Geological Society of America Bulletin, v. 81, p. 3605–3622.

Mayfield, C. F., Tailleur, I. L., and Ellersieck, I., 1983, Stratigraphy, structure and palinspastic synthesis of the western Brooks Range, northwestern Alaska: U.S. Geological Survey Open-File Report OF 83-779, 58 p., 5 pl.

McMullin, D. W. A., Berman, R. G., and Greenwood, H. J., 1991, Calibration of

SGAM thermobarometer for pelitic rocks using data from phase-equilibrium experiments and natural assemblages: Canadian Mineralogist, v. 29, p. 889–908.

Mercier, J. C., Anderson, D. A., and Carter, N. L., 1977, Stress in the lithosphere; inferences from steady state flow of rocks: Pure and Applied Geophysics, v.115, no. 1–2, Stress in the Earth, p. 199–226.

Miyashiro, A., 1975, Classification, characteristics and origin of ophiolites: Journal of Geology, v. 83, p. 249–281.

Moore, T. E., 1987, Geochemical and tectonic affinity of basalts from the Copter Peak and Ipnavik River allochthons, Brooks Range, Alaska: Geological Society of America Abstracts with Programs, v. 19, p. 434.

Moore, T. E., and Grantz, A., 1987, Crustal structure of central Alaska along Geodynamics continent/ocean transect (A-3): Geological Society of America Abstracts with Programs, v. 19, p. 434.

Moore, T. E., Aleinikoff, J. N., and Walter, M., 1993, Middle Jurassic U-Pb crystallization age for Siniktanneyak Mountain ophiolite, Brooks Range, Alaska: Geological Society of America Abstracts with Programs, v. 25, p. 124.

Moores, E. M., 1970, Ultramafics and orogeny, with models for the US Cordillera and the Tethys: Nature, v. 228, p. 837–842.

Moores, E. M., 1982, Origin and emplacement of ophiolites: Reviews of Geophysics and Space Physics, v. 20, p. 735–760.

Moores, E. M., Robinson, P. T., Malpas, J., and Xenophontos, C., 1984, Model for the origin of the Troodos massif, Cyprus, and other mideast ophiolites: Geology, v. 12, p. 500–503.

Mull, C. G., 1982, The tectonic evolution and structural style of the Brooks Range, Alaska: An illustrated summary, in Powers, R. B., ed., Geological studies of the Cordilleran Thrust Belt: Denver, Colorado, Rocky Mountain Association of Geologists, v. 1, p. 1–45.

Mutter, J. C., Buck, R. W., and Zehnder, C. M., 1988, Convective partial melting. 1. A model for the formation of thick basaltic sequences during the initiation of spreading: Journal of Geophysical Research, v. 93, p. 1031–1048.

Nelson, S. W., and Nelson, W. H., 1982, Geology of Siniktanneyak Mountain ophiolite, Howard Pass Quadrangle, Alaska: U.S. Geological Survey Miscellaneous Field Studies Map MF-1441, scale 1:63,360.

O'Nions, R. K., Smith, D. G., Baadsgaard, H., and Moreton, R. D., 1969, Influence of chemical composition on argon retentivity in metamorphic calcic amphiboles from south Norway: Earth and Planetary Science Letters, v. 5, p. 339–345.

Pallister, J. S., Budahn, J. R., and Murchey, B. L., 1989, Pillow basalts of the Angayucham terrane: Oceanic plateau and island crust accreted to the Brooks Range: Journal of Geophysical Research, v. 94, p. 15901–15923.

Patton, W. W., Jr., 1973, Reconnaissance geology of the northern Yukon-Koyukuk province, Alaska: U.S. Geological Survey Professional Paper 774-A, p. A1–A17.

Patton, W. W., Jr., and Box, S. E., 1989, Tectonic setting of the Yukon Koyukuk basin and its borderlands, western Alaska: Journal of Geophysical Research, v. 94, no. 11, p. 15807–15820.

Patton, W. W., Jr., Tailleur, I. L., Brosgé, W. P., and Lanphere, M. A., 1977, Preliminary report on the ophiolites of northern and western Alaska, in Coleman, R. G., and Irwin, W. P., eds., North American ophiolites: Oregon Department of Geology and Mineral Industries Bulletin, v. 95, p. 51–57.

Pavlis, T. L., 1986, The role of strain heating in the evolution of megathrusts: Journal of Geophysical Research, v. 91, p. 12407–12422.

Pearce, J. A., Lippard, S. J., and Roberts, S., 1984, Characteristics and tectonic significance of supra-subduction zone ophiolites, in Kokelaar, B. P., and Howells, M. F., eds., Marginal basin geology: Oxford, United Kingdom, Blackwell Scientific Publications, Geological Society of London, Special Publication, v. 16, p. 77–94.

Plafker, G., 1988, Synopsis of the phanerozoic tectonic evolution of Alaska: Geological Society of America Abstracts with Programs, v. 20, p. A133.

Raase, P., 1974, Al and Ti contents of hornblendes, indicators of temperature and pressure of regional metamorphism: Contributions of Mineralogy and Petrology, v. 45, p. 231–236.

Roeder, D. H., and Mull, C. G., 1978, Tectonics of Brooks Range ophiolites, Alaska: American Association of Petroleum Geologists Bulletin, v. 62, p. 1696–1702.

Rutter, E. H., 1976, The kinetics of rock deformation by pressure solution: Philosophical Transactions of the Royal Society of London, v. 283, p. 203–219.

Schmid, S. M., 1992, Microfabric studies as indicators of deformation mechanisms and flow laws operative in mountain building, in Hsue, K. J., ed., Mountain building processes: London, Academic Press, p. 95–110.

Searle, M. P., and Malpas, J., 1980, Structure and metamorphism of rocks beneath the Semail ophiolite of Oman and their significance in ophiolite obduction: Transactions of the Royal Society of Edinburgh, Earth Sciences, v. 71, p. 247–262.

Shapiro, L., and Brannock, W. W., 1962, Rapid analysis of silicate, carbonate and phosphate: U.S. Geological Survey Bulletin, v. 1144-A, 56 p.

Smith, A. G., 1988, Temperatures at the base of a moving ophiolite slab: The geology and tectonics of the Oman region: International Discussion Meeting of the Geological Society of London, Edinburgh, United Kingdom: Abstracts, p. 59.

Snelson, S., and Tailleur, I. L., 1968, Large-scale thrusting and migrating Cretaceous foredeeps in the western Brooks Range and adjacent region of northwestern Alaska [abs.]: American Association of Petroleum Geologists Bulletin, v. 52, p. 567.

Spray, J. G., 1984, Possible causes and consequences of upper mantle decoupling and ophiolite displacement, in Gass, I. G., Lippard, S. J., and Shelton, A. W., eds., Ophiolites and oceanic lithosphere: Oxford, United Kingdom, Blackwell Scientific Publications, Geological Society of London, Special Publication, v. 13, p. 225–268.

Tailleur, I. L., 1970, Structure and stratigraphy of western Alaska [abs.]: American Association of Petroleum Geologists Bulletin, v. 54, p. 2508.

Till, A. B., Schmidt, J. M., and Nelson, S. W., 1988, Thrust involvement of metamorphic rocks, southwest Brooks Range, Alaska: Geology, v. 16, p. 930–933.

Thirwall, M. F., and Burnard, P., 1990, Pb-Sr-Nd isotope and chemical study of the origin of undersaturated and oversaturated shoshonitic magmas from the Borralan pluson, Assynt, NW Scotland: Journal of the Geological Society of London, v. 147, p. 259–269.

Twiss, R. J., 1986, Variable sensitivity piezometric equations for dislocation density and subgrain diameter and their relevance to olivine and quartz, in Hobbs, B. E., and Heard, H. C., eds., Mineral and rock deformation: Laboratory studies: American Geophysical Union Geophysical Monograph 36, p. 247–262.

Walsh, J. N., 1980, The simultaneous determination of the major, minor and trace constituents of silicate rocks using inductively coupled plasma spectrometry: Spectrochemica Acta, v. 35B, p. 107–111.

Walsh, J. N., Buckley, F., and Barker, J., 1981, The simultaneous determination of the rare earth elements in rocks using inductively coupled plasma source spectrometry: Chemical Geology, v. 33, p. 141–153.

Williams, H., and Smyth, W. R., 1973, Metamorphic aureoles beneath ophiolite suites and alpine peridotites: Tectonic implications with western Newfoundland examples: American Journal of Science, v. 273, p. 594–621.

Wirth, K. R., and Bird, J. M., 1992, Chronology of ophiolite crystalization, detachment, and emplacement: evidence from the Brooks Range, Alaska: Geology, v. 20, p. 75–78.

Wirth, K. R., Harding, D. J., Blythe, A. K., and Bird, J. M., 1986, Brooks Range ophiolite crystallization and emplacement ages from $^{40}Ar/^{39}Ar$ data: Geological Society of America Abstracts with Programs, v. 18, p. 792.

Wirth, K. R., Harding, D. J., and Bird, J. M., 1987, Basalt geochemistry, Brooks Range, Alaska: Geological Society of America Abstracts with Programs, v. 19, p. 454.

Zimmerman, J., and Frank, C. O., 1982, Possible ophiolite obduction-related metamorphic rocks at the base of the ultramafic zone, Avan Hills complex, De Long Mountains, in Coonrad, W. L., ed., The United States Geological Survey in Alaska—Accomplishments during 1980: U.S. Geological Survey Circular, v. 844, p. 27–28.

MANUSCRIPT ACCEPTED BY THE SOCIETY SEPTEMBER 23, 1997

Index

[Italic page numbers indicate major references]